UNDERSTANDING MICROBES

A LABORATORY TEXTBOOK FOR MICROBIOLOGY

G. WILLIAM CLAUS

With contributions by
DAVID BALKWILL

W. H. FREEMAN AND COMPANY · NEW YORK

Cover: Yeasts are microorganisms that form visible and often colorful structures when they alight or are deposited on a suitable medium. Pure cultures of several yeasts are shown 8 to 10 days after they were seeded on a nutrient agar in a glass dish. The map identifies the 15 yeasts. Five of the yeasts *(1, 2, 6, 9, 10)* yield commercially useful products. One of the yeasts, *Phaffia rhodozyma (12)*, is being tested as a food supplement for hatchery-raised fish, the flesh of which tends to be white. The yeast synthesizes a carotenoid, astaxanthin, which turns the flesh to the normal orange pink. (Courtesy Herman J. Phaff)

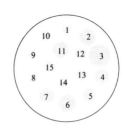

Yeasts

1 *Saccharomyces cerevisiae*
2 *Candida utilis*
3 *Aureobasidium pullulans*
4 *Trichosporon cutaneum*
5 *Saccharomycopsis capsularis*
6 *Saccharomycopsis lipolytica*
7 *Hanseniaspora guilliermondii*
8 *Hansenula capsulata*
9 *Saccharomyces carlsbergensis*
10 *Saccharomyces rouxii*
11 *Rhodotorula rubra*
12 *Phaffia rhodozyma*
13 *Cryptococcus laurentii*
14 *Metschnikowia pulcherrima*
15 *Rhodotorula pallida*

Illustration Credits: Pages 1, 117, 211, 347, 395, 513: Z. Yoshii, J. Tokunaga, and J. Tawara. 1976. *Atlas of Scanning Electron Microscopy in Microbiology.* Tokyo, Japan: Igaku-Shoin Ltd.; page 11: From H. J. Phaff, *Scientific American,* September 1981. Photograph by Erika A. Hartwieg; page 49: W. L. Dentler, University of Kansas/BPS; page 97: Courtesy of Nucleopore Corporation, Pleasanton, California; page 141: Courtesy of James A. Shapiro, University of Chicago; page 233: Courtesy of Carl E. Shively; page 313: From N. K. Robakis, N. J. Palleroni, C. W. Despreaux, M. Boublick, C. A. Baker, P. J. Charn, and G. W. Claus, *J. Gen. Microbiol.* 131: 2467–2473 (1985). Page 453: Courtesy of M. A. Daeschel and H. P. Fleming, USDA; Figure 3 in *Appl. Environ. Microbiol.* 42: 1111–1118 (1981); page 477: Courtesy C. A. Baker, G. W. Claus, and P. A. Taylor, *Appl. Environ. Microbiol.* 46: 1214–1223 (1983).

Library of Congress Cataloging-in-Publication Data

Claus, G. William.
    Understanding microbes: A laboratory textbook for microbiology

    Includes bibliography and index.
    1. Microbiology—Laboratory manuals. I. Title.
QR63.C55 1988   576'.07'8   87-33152
ISBN 0-7167-1809-X

Copyright © 1989 by W. H. Freeman and Company

No part of this book may be reproduced by any mechanical, photographic, or electronic process, or in the form of a photographic recording, nor may it be stored in a retrieval system, transmitted, or otherwise copied for public or private use, without written permission from the publisher.

Printed in the United States of America

1 2 3 4 5 6 7 8 9 VB 6 5 4 3 2 1 0 8 9 8

This book is dedicated to W. R. Lockhart and C. L. Baugh, who first introduced me to the microbes; to Marcella Claus, who has willingly shared me with the microbes ever since; and to the many past and present students who always want to know why.

Contents

Preface — xi
A Note to Students — xv

I HOW TO HANDLE MICROORGANISMS — 1
1 Principles of Aseptic Technique — 3
2 Aseptic Method for Entering Culture Tubes and Transferring Cells — 5

II INTRODUCTION TO MICROSCOPY AND TO OBSERVING MICROORGANISMS — 11
3 Principles and Care of the Light Microscope — 13
4 Introduction to Staining Microorganisms — 21
5 Smear Preparation, Fixation, and Simple Staining with Basic Dyes — 27
6 Observing Live Microorganisms: The Wet-Mount Method and the Phase-Contrast Microscope — 35

III MICROSCOPIC TECHNIQUES FOR DETERMINING MICROBIAL TYPE AND STRUCTURE — 49
7 The Gram Stain: A Differential Stain — 51
8 The Acid-Fast Stain: A Differential Stain — 59
9 Bacterial Endospores — 67
  A Observing endospores with a phase-contrast microscope — 69
  B Observing endospores with simple stains and the gram stain — 69
  C Structural staining of endospores with hot malachite green — 69
10 Bacterial Flagella and Motility — 79
  A Structural staining of flagella — 80
  B Observing motility with the phase-contrast or brightfield microscope — 82
  C Observing the consequence of motility on plates and in soft-agar deeps — 82
11 Bacterial Capsules: Indirect Observation with India Ink — 89
  A Wet-mount method — 92
  B Dried-smear method with counterstain — 93

IV STERILIZATION PRINCIPLES AND CULTURE MEDIA PREPARATION — 97
12 Sterilization Principles and Methods — 99
13 Preparing Culture Media — 107
  A Preparing liquid media (broths) — 108
  B Preparing solid media: Deeps, slants, and plates — 110

V TECHNIQUES FOR DETERMINING CULTURE PURITY — 117

14 Separating Microbes on Streak Plates — 119
15 Determining Culture Purity — 131
    A Testing the purity of a colony — 131
    B Separating cultures from an unknown mixture — 133

VI CULTURING AND QUANTITATING MICROORGANISMS — 141

16 Agar-Slant and Agar-Deep Cultures — 143
17 Broth Cultures — 17
18 Culturing Anaerobes: Thioglycolate Use and the Anaerobe Jar — 159
19 Measuring Turbidity Changes during Growth of Broth Cultures — 167
20 Counting Viable Cells: Serial Dilution and Spread Plates or Pour Plates — 175
21 Selective and Differential Media — 195
22 Enrichment Techniques — 203

VII THE EFFECT OF ENVIRONMENT ON MICROBIAL GROWTH AND VIABILITY — 211

23 The Effects of Temperature on Growth — 213
24 The Effects of Elevated Sugar and Sodium Chloride Concentrations On Growth — 219
25 The Effects of Free-Oxygen Concentrations on Growth: Agar-Deep Cultures — 225

VIII PHYSIOLOGICAL CHARACTERISTICS OF MICROORGANISMS — 233

Hydrolysis and Use of Large Extracellular Materials

26 Introduction to Extracellular Degradation — 235
27 Microbial Degradation of Polysaccharides (Starch) — 241
28 Microbial Degradation of Proteins (Casein and Gelatin) — 247
    A Degradation of milk protein (casein) — 248
    B Degradation of animal protein (gelatin) — 250
29 Microbial Degradation of Lipids — 257

Transport and Use of Small Molecules

30 Differential Utilization of Citrate by Enteric Bacteria — 263

Measuring Products and Effects of Metabolism

31 Acid and Gas Production from Sugar Fermentation — 269
32 Acid and Neutral Products from Sugar Fermentation: Methyl-Red and Voges-Proskauer Tests — 277
33 Microbial Hydrogen Sulfide (H_2S) Production from Thiosulfate and Sulfur-Containing Amino Acids — 285
34 Indole Production from the Amino Acid Tryptophan and Catabolite Repression — 291
35 Products Formed in Milk: The Litmus Milk Test — 297
36 Test for Cytochrome *c* (Oxidase) and Catalase Activities — 303

IX BACTERIAL VIRUSES AND MICROBIAL MUTATIONS — 313

37 Enumeration of Lytic Viruses: The Plague Assay — 315
38 Using Mutants to Detect Carcinogens: The Ames Test — 323
39 Effects of Ultraviolet Light on DNA, Cell Viability, and Mutation Frequency — 333

X	**THE FUNGI**	**347**
40	Introduction to the Fungi	349
41	Morphology and Reproduction of the Molds	363
42	Morphology and Reproduction of the Yeasts	379

XI	**MEDICAL MICROBIOLOGY AND IMMUNOLOGY**	**395**
43	Evaluating the Effectiveness of Common Antiseptics and Disinfectants	397
44	Antibiotic Evaluation by the Kirby-Bauer Method	405
45	Coagulase Production by Pathogenic Staphylococci	413
46	Hemolysis of Red Blood Cells	421
47	Bacteria on Human Skin	431
48	Bacteria in the Human Throat	441
49	Slide-Agglutination Test for Serotyping Pathogens	447

XII	**FOOD MICROBIOLOGY**	**453**
50	Milk Preservation—Yogurt	455
51	Solid Food Preservation—Sauerkraut	463
52	Numbers of Bacteria on Solid Foods	469

XIII	**ENVIRONMENTAL MICROBIOLOGY**	**477**
53	Detecting Coliforms Bacteria in Water	479
54	Introduction to the Nitrogen Cycle	487
55	Nitrogen Fixation—Reduction of Dinitrogen to Ammonia by Procaryotes	493
56	Ammonification—Microbial Deamination of Nitrogenous Organic Compounds	499
57	Denitrification—Complete Reduction of Nitrate and Nitrite to Dinitrogen	507

XIV	**MICROBIAL IDENTIFICATION**	**513**
58	Identification of Unknown Microorganisms	515

XV	**APPENDIXES**	**521**
A	Media	523
B	Stains and Staining Reagents	531
C	Other Reagents	533
D	Cultures	537
Index		543

Preface

Understanding Microbes is designed for an introductory course in microbiology in which students are expected to learn more than just techniques and what constitutes a positive or negative reaction. This text shows students *why* procedures are done a certain way and the *basic principles* behind morphological and physiological tests. In many cases these basic principles cannot be found in today's lecture texts or laboratory manuals.

This text will be most useful for colleges and universities that offer a well-rounded quarter- or semester-length course to students who are majoring in the life sciences and who have received a previous introduction to the principles of biology.

FEATURES

Background Information

One purpose of this text is to provide students with sufficient information to understand why microorganisms behave as they do and the significance of microbial behavior to diagnostic, industrial, and environmental microbiology.

I have found that if the necessary background information is presented in the laboratory text and if assignments are made and incentives given for studying those assignments, students come to the laboratory prepared. The instructor can then devote more time to class demonstrations, individual assistance, and challenges for students, and the students accomplish and understand more. See *A Note to Students,* p. xv

Learning Objectives

Specific learning objectives are provided for each procedural exercise so that students can see immediately what is to be gained from reading that unit and performing that technique or experiment.

Illustrated Procedures

Much of a student's time in the average laboratory may be devoted to listening to the instructor explain how to perform new procedures and how to proceed through the assigned exercise. I have found that students can and will prepare themselves for a laboratory period when they are provided with adequately written procedures, and if they are given incentive to do so. Students who study these clearly written exercises before coming to class accomplish more in the assigned time.

In studying laboratory exercises prior to class, students find it helpful to have line drawings accompanying written instructions. This is especially true in the beginning of the academic term when students are handling microorganisms for the first time. These sequential illustrations help the student visualize the procedure, before class and serve as a visual summary of the procedure while performing that technique or experiment in class.

Boxed Safety and Technique Notes

Notes dealing with student safety are boxed and set within the Procedures; these notes deal with things that will directly affect the safety of each student and those around that student. Their importance should be emphasized by each instructor.

Technique notes either deal with why we do things a certain way to elaborate on what to look for when performing a laboratory manipulation. I feel strongly that the principles (reasons behind) a technique are more important than how to do a procedure. If a student learns why a technique is performed, then that student will also learn that there are several adequate ways to achieve a desired goal and will always be receptive for a better way to achieve that objective.

Interpretation Notes

These boxed notes are usually inserted into the procedure for laboratory periods that follow the one initiating an exercise. They are usually attempts to explain why an observed reaction occurs in a certain way. Often they reemphasize points made in the introduction, but they are placed where students can easily refer to them as they see the results of an experiment.

Test Summaries

Certain procedures taught in all introductory microbiology laboratory courses, such as the Gram stain and tests for certain physiological characteristics, are used several times during an academic term. For future reference, the student will find boxed test summaries near the ends of many procedure sections; Each summarizes the conditions needed to perform a specific test, the reagents used, and the appearance of positive and negative reactions. These summaries enable students to quickly review the critical details of a test before repeating that procedure. For example, test summaries will be helpful when students need to use these tests for identifying unknown microorganisms.

Results Sections

Over the years, I have found that students do a better job of recording their observations if the results section is located near the procedure and test summary. This placement has another advantage. As students progress through an academic term, they occasionally need to review their previously obtained results. This is especially true when they are asked to identify an unknown microorganism. For example, what did a previously observed positive test for H_2S production look like, and which microbes gave that positive reaction? With the results section near the procedure and test summary, a student need only look in the table of contents or index to find a reference to all needed information.

Questions

Each exercise includes questions for which answers may be found in that exercise or a previous exercise. These questions provide students with a systematic review of the more important aspects of background information, experimental technique, procedure, and interpretation of results.

Instructor's Manual

If you adopt this text, I encourage you to closely examine the instructor's manual that accompanies *Understanding Microbes*. In addition, I would strongly recommend that you request an instructor's manual for each person who teaches a laboratory section (graduate-teaching assistants included).

The instructor's manual has a separate listing for each laboratory exercise. Each listing contains a detailed preparation list (cultures, media, and supplies) for each laboratory exercise. Each listing also contains suggestions for what to emphasize, common mistakes and how to avoid them, what results to expect, suggested organizational methods, possible demonstrations, and other helpful hints.

The instructor's manual also contains lists of cultures, media, stains, and reagents with recommendations on how to obtain these items. The culture list contains details on the growth characteristics, gram-staining properties, cell structure, and the exercise(s) in which each culture is used. The lists of media, stains, and reagents include details on the preparation and commercial availability of each item.

The instructor's manual contains the collective thoughts and suggestions of our preparation-room staff, curators of our culture collection, faculty colleagues, and countless graduate-teaching assistants covering a period of 21 years.

Suggestions

I welcome all comments (positive or negative) and all suggestions for improvements and additions in the future. Please do not hesitate to send your written thoughts (or even call) directly to the author.

ACKNOWLEDGMENTS

A project of this magnitude cannot be the work of only one person. Countless friends, colleagues, graduate students, secretaries, and undergraduates have assisted in various ways. It is impossible to acknowledge all who have contributed, but I have not forgotten their efforts. Credit for that which is good I gladly share with others, but I alone accept responsibility for the final content.

I especially thank Alice Kelling who contributed advice and council based on her many years in clinical and teaching laboratories. In addition to her excellent technical suggestions, Alice also spent countless hours editing early forms of manuscript for clarity and correctness. Many thanks also go to Allen Yousten who offered suggestions, helped to edit early drafts, and was the first additional faculty member willing to class-test most of these exercises.

I appreciate the efforts of David Balkwill (Florida State U.) who wrote exercises 11, 12, 21, 22, 29, and 46 contributed part of Exercise 39, critically reviewed with enthusiasm the entire manuscript, and was the first to class test all exercises outside of Virginia Tech.

My thanks also go to the many graduate teaching assistants who were fellow pioneers with early mimeographed exercises. Their thoughtful criticism and constructive suggestions helped greatly in developing the text. I wish especially to acknowledge Carol Baker for the many hours she devoted to duplication and assembly of early forms of these exercises, for help in editing, and for suggesting several important pedagogic features. My thanks also to Bruce Micales for his assistance and to Barbara Harris-Feshami who suggested adding the test summary sections. Special appreciation is due to Lynn Lewis for exceptional and voluminous editorial assistance and to Patrice Boerman for her many suggestions.

I am also grateful to Mary Catherine Thomas who, as an undergraduate at Virginia Tech, began with the primary literature on the Ames test and, with the advice of Roger VanTassal, tested the media and strains now used by students in Exercise 38.

The students who enrolled in my introductory microbiology courses have helped in many ways. I acknowledge each of them for their thoughtful constructive criticisms and corrections as well as for their words of encouragement. One student *anonymously* submitted many pages of detailed editorial corrections and suggestions, but her graduate-teaching assistants could not allow Josephine Reed's efforts go unrecognized.

Many were involved in the construction of earlier forms of this text. Margie Lee and Linda White enthusiastically helped with typing and figure construction. I am also indebted to our departmental secretaries Carolyn Furrow, Teresa Jones, and to Cheryl Mallan, who contributed countless hours to typing and duplicating many early drafts of the manual; thanks also to Shirley Hale, Connie Melton, Pam Pettry, Carolyn Poyer, and Geri Stock for their efforts.

To the kind people at W. H. Freeman and Company, I am very grateful to Kathleen Dolan, associate editor, for her continuous and enthusiastic encouragement given during the preparation and reviewing of the manuscript and to Georgia Lee Hadler, project editor, for her firm yet kind guidance and capable leadership during production of the book. Thanks are also given to sponsoring editors John H. Staples, James A. Dodd, and Gary Carlson; to copy editors Yvonne Howell and Glenn Cochran; to Elizabeth Marrafino, proofreader; to Lynn Pieroni, designer; Patricia Holtz, illustration coordinator; and to all other staff members at Freeman who worked on this book.

My gratitude is also extended to David Balkwill (Florida State University), Mary Jane Tershack (Pennsylvania State University), and Patricia Edelmann (California State University, Chico) for constructively reviewing the entire manuscript, and to Thoyd Melton (North Carolina State University) and J. J. Gauthier (University of Alabama) for reviewing more than half of the exercises.

Most importantly, I wish to thank Marcella Claus, without whose patience, encouragement, love, and understanding this manual would not have been written.

G. William Claus

A Note to Students

Please take a few minutes to read the following suggestions! If you follow them, I believe they will help you get the most from this text and from your introductory laboratory experience in microbiology.

Thoroughly read the assigned material before coming to class. This text was designed to be a self-guided approach to the introductory microbiology laboratory and to contain enough background information so that you can understand why things are done a certain way and why microbes behave as they do.

If you read the assigned material *before* coming to class, your instructor will not need to use laboratory time to give you another lecture and can instead demonstrate critical techniques and give you individual attention. Moreover, you will have more time to work independently and gain first-hand experience in the microbiology laboratory.

One effective way to encourage you to prepare for class is for the instructor to give simple quizzes at the beginning of most class periods that cover the assigned material. Instructors who use this method find that their laboratory classes run more smoothly and that students accomplish more. This method also rewards those who are prepared for class.

Use the index. You may periodically need to review terms, techniques, procedures, and test results. The index at the back of this text contains a list of these items and the page numbers where each can be found. Make use of this index for frequent review.

Use the interpretation notes and test summaries. The interpretation notes are occasionally included to help you understand concepts or test results that are often difficult for beginning students to understand. Test summaries are included for each test you may later need to identify microorganisms. This should save you time when, later, you must repeat these tests. You should then only need to examine the summary to remind yourself how the test is performed and interpretated.

Use the technique illustrations. These accompany many steps in the procedures. When you read over the exercise before class, the illustrations allow you to visualize the steps while you read the assigned material; they also provide you with a quick, visual summary of the procedure while you are performing that procedure in class.

Answer the study questions. These are written to give you a review of the more important concepts for each portion of the text. In most cases you will find answers to each question in the corresponding text. Occasionally questions will deal with introductory material given in a previous exercise. Use the questions to see if you have learned the important concepts. Some students suggest that it is helpful to read the study questions *before* reading the exercise because it helps you focus on the key points of the exercise.

Thoroughly record your results. Accurate and careful observations and the thorough recording of those observations are *essential* for all who work in the laboratory. Do not rely on your memory. At the minimum, you should clearly record (1) the appearance of each positive and negative test and (2) how each tested microbe behaved in that test. *Remember that you will rely on your notes when asked to perform these tests to identify unknown microorganisms.*

UNDERSTANDING MICROBES

A LABORATORY TEXTBOOK FOR MICROBIOLOGY

I

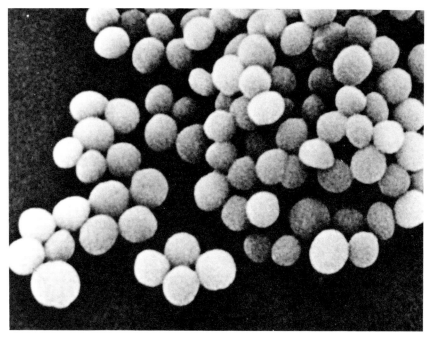

Typical morphology of *Staphylococcus aureus*.

Do there exist many worlds, or is there but a single world? This is one of the most noble and exalted questions in the study of nature.
—*St. Albertus Magnus*

HOW TO HANDLE MICROORGANISMS

The purpose of Part I is to explain the importance of using pure cultures in the study of microorganisms and to explain why it is necessary for you to learn aseptic technique: how to enter culture vessels aseptically and transfer cells to a microscope slide or sterile medium for further study.

Principles of Aseptic Technique

A **pure culture** contains only one species of microorganism. During this laboratory course you will work with many pure cultures. If another species of microorganism is accidentally introduced into a pure culture, the culture is said to be *contaminated* and is called a **mixed culture**. The possibility of contamination is a constant concern to microbiologists who study pure cultures because results from contaminated cultures are not usually the same as those from pure cultures. The threat of contamination is always present because microorganisms are found everywhere: on our skin, on dust particles in the air, in tap water, and on everything we touch. Therefore you cannot obtain pure cultures unless you begin with a sterile growth medium and keep all unwanted microorganisms out of that medium.

By definition, **sterilization** is the *destruction* or *removal* of all forms of microorganisms (eucaryotes, procaryotes, and viruses). Once you have a pure culture, it is essential (1) that all materials that will touch the culture are first sterilized, (2) that all media to which the cells are transferred are sterile, and (3) that no contaminants are allowed to enter the growing culture. The methods of obtaining and maintaining pure cultures are collectively called **aseptic technique.**

Skill with aseptic technique comes only with practice, but understanding *why* things are done in certain ways helps. The exercises in this text emphasize the reasons why you should do things in certain ways. There often are several ways to reach a goal; and if you understand why one way may be better than another you will be much better able to make the best choices among alternatives.

The following general principles of aseptic technique must *never* be violated:

The growth medium and its container must be sterilized as soon as the medium is prepared.

The container must be covered to prevent entrance of microorganisms on dust particles and in aerosols. If you must remove the cover, do not leave it off any longer than necessary. You may use the cover to shelter the container opening during the aseptic transfer. *Never* set the cover down on a contaminated surface.

Instruments (such as loops or pipettes) **and fluids** (such as saline or diluents) **that touch the inside of the container, the sterile medium, or the culture must first be sterilized.** If you touch your sterilized loop or pipette tip to a contaminated surface, the loop or pipette will transfer contaminants to your pure culture.

You must not contaminate your work area with cultures. Your inoculating loop must be sterilized before it leaves your hand. The pipettes you use to transfer cultures must be placed in disinfectant solution immediately after making the transfer. Avoid creating aerosols.

All these principles must be followed anytime you are working with pure cultures. Think about what you are doing, and question why you are doing it that way. Practice categorizing *everything* as either sterile or contaminated and develop your technique accordingly.

2

Aseptic Method for Entering Culture Tubes and Transferring Cells

Before you begin to study microorganisms, you need to know how to transfer a microbial culture properly. Be guided by the following simple rules:

Never enter a culture with an inoculating loop that has not been sterilized. This prevents the loop from introducing contaminants into your culture.

Always flame the lip of the tube both before you insert the loop and after you withdraw the loop. The first flaming heats the lip of the tube so that convection currents carry airborne dust particles and aerosols away from the tube opening, thus preventing the entrance of bacteria that are attached to these particles. The second flaming incinerates cells that may have been deposited on the lip of the tube when the loop was inserted or withdrawn.

Never leave the tube open any longer than the amount of time needed to transfer the culture. This lessens the possibility of contamination by dust particles.

Never place the tube cap (or plug) on the work surface or touch it to anything except the flamed lip of the culture tube. Doing so contaminates the cap (plug).

Always sterilize the inoculating loop after you complete the transfer. This prevents microorganisms on your loop from contaminating your work area.

LEARNING OBJECTIVES

- Develop skills in transferring cultures aseptically.
- Develop an understanding of aseptic technique.

MATERIALS

Cultures   Any available tubes that contain microbial cultures

Media   None

Supplies   Inoculating loop
Bunsen burner

PROCEDURE

Note that the number of a procedural step corresponds to the number of the adjacent illustration.

1. **Grasp the inoculating loop** as if it were a pencil.

    If you are making smears from **broth** (liquid) cultures, *go to step 2*.

    If you are making smears from an **agar slant,** first transfer a loopful of water to the center of the microscope slide or coverslip, and *then go to step 2*.

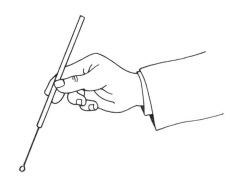

2. **Hold the inoculating loop** at a 60° angle in the hottest part of a bunsen burner flame. Heat the entire wire to redness.

    Alternatively, you may be asked to hold the loop so that the handle points downward and the wire points upward in the hottest part of the flame. This method is used in many hospital laboratories, because some people feel that it prevents spattering of microbes.

    Allow the loop to cool before it touches a culture.

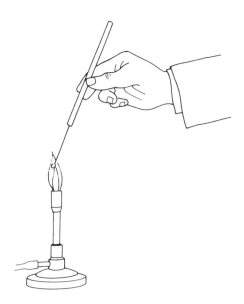

3. **Pick up a tube of culture** with your free hand. With broth cultures, hold the tube near the top and swirl the bottom of the tube to suspend the microbes. No mixing is necessary with slant cultures. If you are working with screw-capped tubes, loosen the cap now.

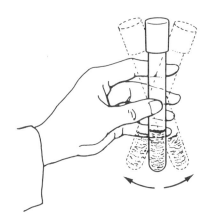

4. **Remove the plug or cap of the tube** with the free fingers of the hand holding the inoculating loop.

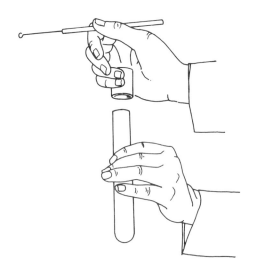

5. **Flame the lip of the tube** by momentarily rotating it within the upper cone of the flame.

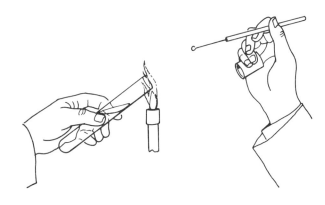

6. **Remove a small amount of culture from the tube** with the sterilized inoculating loop, either by taking a loopful of broth or by gently touching the loop to the solid growth on a slant. **Make sure the loop is cool before you touch the culture.** Wait about 15 seconds after removing the loop from the flame. If the loop is too hot, it spatters the culture and creates an aerosol. Microorganisms are easily spread by these fine droplets of liquid; so avoid creating them.

9. **Flame your inoculating loop** and set it down in a safe place after completing each aseptic transfer.

7. a. **Flame the lip of the tube again.**
   b. **Replace the plug or cap,** and return the tube to the test-tube rack.

8. **Transfer the culture** to the desired location. Take care not to touch the culture to anything else during this transfer.

   Techniques for transferring cells to various types of microscope slides or culture vessels are given in the following exercises:

Microscope slides	Exercises 5 and 6
Streak plates	Exercise 14
Agar slants and deeps	Exercise 16
Broth cultures	Exercise 17

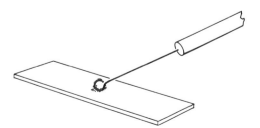

2 ASEPTIC METHOD FOR TRANSFERRING CELLS

NAME _____

DATE _____

SECTION _____

QUESTIONS

Completion

1. Never enter a culture tube without first _____ your inoculating loop.

2. The lip of the culture tube is flamed prior to entering it because _____ _____ .

3. The lip of the tube is flamed after the sample has been removed because _____ _____ .

4. After the transfer has been made and you have capped the culture vessels, you should always _____ _____ .

True – False (correct all false statements)

1. The term *aseptic technique* refers to handling sterile materials or known cultures without contaminating them. _____

2. Many of the procedures of aseptic technique are designed to prevent dust from entering the culture vessel or from contacting sterile materials because microorganisms are almost always associated with dust particles. _____

3. The heated surface of a test tube causes convection currents that carry dust particles into the tube. _____

4. The term *sterilization,* as used by microbiologists, means the complete destruction or removal of all forms of life. _____

5. The term *contamination* may refer to either the introduction of microorganisms into sterile materials or to the introduction of unwanted microorganisms into a pure culture. _____

II

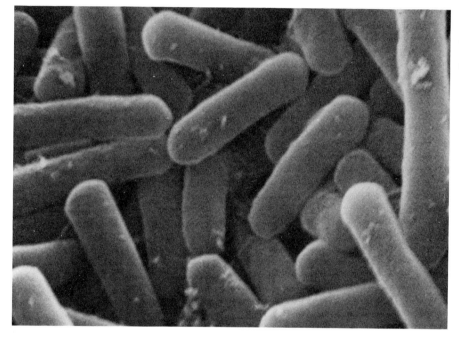

Bacteria of the species *Bacillus brevis,* which manufactures the antibiotic gramicidin *S*.

... see what everybody else has seen, and think what nobody else has thought.

—*A. Szent-Gyorgyi*

INTRODUCTION TO MICROSCOPY AND TO OBSERVING MICROORGANISMS

The purposes of Part II are (1) to teach you how to use the light microscope with an oil-immersion objective and (2) to show you how to increase the contrast between microbial cells and the material around them so that you can study the shapes and sizes of cells. In this series of exercises, you will be using only common bacteria; however, these techniques can also be used when you study other types of microorganisms.

3

Principles and Care of the Light Microscope

Microbiologists use two basic types of microscopes: light microscopes and electron microscopes. All **light microscopes** use photons of visible light to form images, but the type of light used and how that light is manipulated can vary. The most common types of light microscopes are the **brightfield, darkfield, phase-contrast,** and **fluorescent** microscopes. As you progress through these exercises, you will be introduced to the theory, mechanics, operation, and limitations of two of these types of microscope.

The operating and handling instructions suggested here are derived from years of experience with college students. Please follow these instructions—for your own benefit and for the sake of other students who will be using the same microscope. Failure to do so will result in damaged microscopes and costly repairs. You may then be without a microscope until the repairs are completed. Please use care and good common sense.

Parts of the brightfield microscope

Examine Figure 3-1 and your microscope. You should know each part of the microscope, its function, and how it is used.

The light source **bulb** is in the base of the microscope. The **on-off switch and brightness control** turns on the current to the bulb and controls brightness. *Never* adjust brightness in any other way!

The lenses of the **lamp condenser** collect the light into a column and direct it upward to the **substage condenser,** which focuses it on the specimen. Your microscope is adjusted so that the condenser automatically stops when you raise it to its topmost position by turning the **condenser focusing knob.** This is the condenser's correct position. The **condenser diaphragm** works like the iris of your eye. The size of the hole in the center of this mechanical iris is controlled by a lever on the side of the condenser. This diaphragm confines the column of light entering the condenser so that the correct amount enters the objective lens.

The platform on which you place the microscope slide is called the **stage.** A **mechanical stage** has **coaxial knobs** that slowly move the slide horizontally or vertically across the field so that you can center your slide preparation in the light path before you rotate an objective into position.

Both the **coarse- and fine-focus adjustment knobs** raise and lower the microscope stage, but they move it at different rates. *The coarse-focus knob should **never** be used while you are looking into the microscope unless you are using it to lower the microscope stage.* Learn which direction to turn these knobs to lower the stage. The function of both knobs is to position the stage so that the objective lens is focused on the specimen.

The **objectives,** are compound lenses that determine the magnification of the specimen. Several objectives are positioned around the **revolving nosepiece,** which rotates to bring the desired objective into the light path and allows you to easily change the microscope magnification.

The **eyepiece tube** is a lighttight compartment that receives the light coming through the objective and bends it with a prism before sending it to the eyepiece. The **eyepiece** contains several lens elements that collect the light, focus it, and then send it to your eye.

Lenses, magnification, and working distance

The magnification of a microscope depends on the **objective-** and **eyepiece-lens systems,** which create a

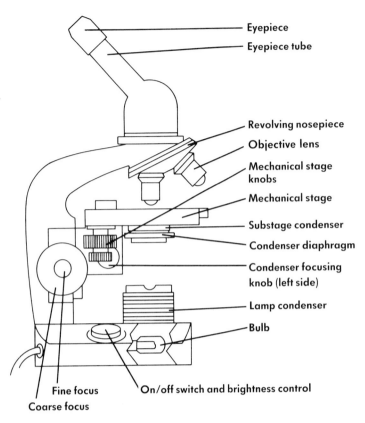

FIGURE 3-1 Basic components of a brightfield microscope.

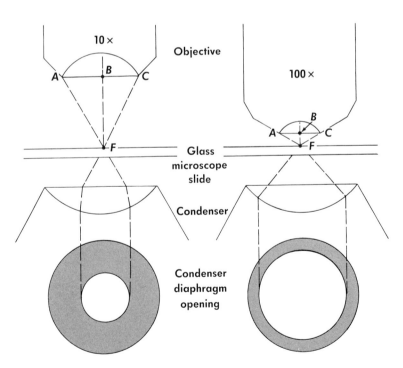

FIGURE 3-2 Characteristics of objective lenses when they are in focus on the specimen. For illustration, the condenser diaphragm is shown tilted 90° forward from its normal position. Also shown are the angular aperture (*AFC*), working distance (*BF*), and the necessary condenser-diaphragm opening for both a low- and high-magnification objective.

pattern on your retina that is called the *virtual* or *retinal image*. The **total magnification** of this image equals the product of the magnifications of the objective and the eyepiece lenses, that is,

total magnification = objective × eyepiece (3-1)

Examine your microscope and calculate the magnifications that are possible.

Figure 3-2 illustrates a microscope objective and its relation to the specimen when the microscope is in focus. **Focus**, a fixed property of any lens, depends on how the lens maker grinds the simple lenses and assembles them into a compound objective. When an objective is in focus, a specimen at point *F* is observed and the objective lens and specimen are always the same distance apart: a distance called the **working distance** (distance *BF* in Figure 3-2). Objective lenses are constructed so that the working distance decreases as the magnification of the objective lens increases (Figure 3-3). This relationship is important if you want to examine thick specimens or if your specimen is covered with a thick coverslip. The fixed working distance of the objective may limit the types of equipment you can use between the objective and the stage.

Resolution

Microbiologists frequently refer to the **resolving power** or **resolution** of a microscope. This is the ability to distinguish two very small and closely spaced objects: The better the resolving power of the microscope, the smaller and closer together these objects can be and still be seen as separate objects. The **limit of resolution** is defined as the minimum distance by which two small objects can be separated and still be perceived as separate objects. Therefore, it can be said that the *smaller* this minimum distance, the *better* the resolving power of the lens system. Measurement of the limit of resolution depends on the length of the electromagnetic waves used for illumination (for instance, light or electrons) and the **numerical aperture** of the lens. This relationship is expressed as

$$\text{limit of resolution} = \frac{\text{wavelength}}{2 \times \text{numerical aperture}} \quad (3\text{-}2)$$

The equation shows that you can decrease the limit of resolution (that is, increase the resolving power) by either *shortening* the wavelength of the illumination or *increasing* the numerical aperture.

The wavelength of visible light ranges from about 400 to 750 nanometers (nm). Since visible light contains all wavelengths in this range, an average value of 600 nm is often used to calculate the resolution limit. If you place a filter in the light path so that only light having a shorter wavelength is transmitted, the limit of resolution *decreases*. Your microscope probably contains a blue filter that passes 450-nm light. This shorter wavelength improves resolution by decreasing the limit of resolution.

The value of the numerical aperture (NA) depends on the **refractive index** (n) of the medium through which the light passes after it leaves the specimen and before it enters the lens, and on the **angular aperture** (θ) of the lens. Equation 3-2 shows that the limit of resolution becomes smaller as the numerical aperture is increased. Thus the larger the NA (the higher its value) the better the resolving power of the microscope. The NA is expressed by the formula

$$\text{NA} = n \times \frac{\text{sine}[\theta]}{2} \quad (3\text{-}3)$$

The angular aperture (θ), shown as the angle *AFB* in Figure 3-2, is a measure of the cone of light delivered to the specimen by the condenser and gathered by the objective when it is in focus. Figure 3-2 shows that higher-magnification lenses are designed to accept larger cones of light (have larger angular apertures) and thus have a higher NA. The angular aperture is a fixed property of the lens system that is determined during construction of the lens; it cannot be altered by the person using the microscope.

Light passes through the glass slide and through air before it enters the objective lens. The refractive index of glass (about 1.5) is much higher than that of air (about 1.0); thus, when light passes from glass to air, much of the light is lost due to refraction (Figure 3-4). Equation 3-3 shows that the refractive index (n) of the material that lies between the glass slide and the objective lens affects the NA. If this refractive index has a value greater than that of air, there is less refraction and thus an increase in the NA value. You can increase this refractive index in the laboratory by placing immersion oil between the microscope slide and an **oil-immersion objective**. In fact, the immersion oil

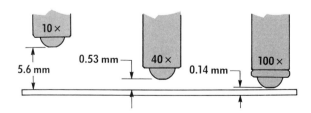

FIGURE 3-3 Relationship between the working distance and the objective lens magnification when the lens is in focus.

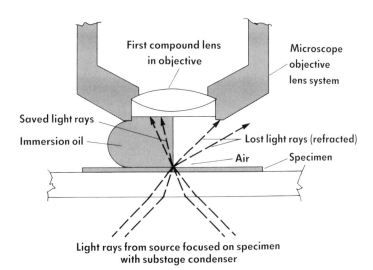

FIGURE 3-4 Light lost due to refraction and the reduction of refraction with the use of immersion oil. To be most effective, the immersion oil should have the same refractive index as the glass slide.

should have a refractive index identical to that of the glass slide so that all loss of light due to refraction is eliminated.

If you use a blue filter to shorten the wavelength of the illumination and immersion oil to increase the NA, the resolving power of a good brightfield microscope is about 0.2 micrometers (μm). If you consider that a typical bacterial cell is about 0.5 μm wide, it is obvious that a brightfield microscope cannot resolve structures in the cell. The best one can hope for is to be able to distinguish the sizes and shapes of cells and how they are associated with one another.

Magnification and resolution are equally important to the microbiologist. You can greatly magnify an image; but unless the resolution is also good, you will have only a very large fuzzy image. Likewise, you may obtain fantastic resolution; but unless you magnify the image, you cannot see the resolved detail.

Condenser, condenser diaphragm, and illumination

The **condenser lens** system collects the light from the source and focuses it on the specimen (see Figure 3-4). Like any lens system the condenser has a focal point, which is a point at which the light coming through the condenser is focused. Therefore the position of the condenser lens is important to correct image formation. Many microscopes have the condenser preset to focus on the specimen when the condenser is moved to its uppermost position. If the condenser is in a lower position, the resolving power of the microscope is reduced.

The **condenser diaphragm** controls the diameter of the light column that enters the condenser. Moving the lever increases or decreases the size of this light column (see Figure 3-2). The effects of opening and closing the condenser diaphragm are most noticeable when the lower-magnification objectives are used. *Closing the diaphragm increases image contrast but decreases resolution. Opening the diaphragm decreases contrast but increases resolution.* If the condenser diaphragm is open too wide when you are using a low-magnification objective, the light will be too intense (it will hurt your eyes). With the low-magnification objective, the correct condenser diaphragm position is intermediate between best resolution (open) and best contrast (closed). With the high-magnification (oil-immersion) objective, it is best to open the condenser diaphragm completely, adjust the light intensity with the brightness control, and then slightly close the condenser to improve the image contrast.

Always control the illumination with the brightness control, never with the condenser diaphragm. Controlling brightness with the condenser diaphragm diminishes the resolving power of the microscope.

Required handling procedures

Before you use your microscope, do the following:

1. Make sure the last person who used it put it away correctly.

2. Use clean lens paper to check all objectives and make sure they are clean.

3. Visually inspect the stage and the condenser for oil.

4. Promptly report any discrepancies to your laboratory instructor.

Follow these commandments while using your microscope:

1. Never touch the lenses with anything but lens paper. If you suspect that any of the lenses are dirty, gently wipe them with a clean, dry piece of lens paper. If they still seem dirty, ask your instructor for assistance.

2. Never remove any parts of the microscope.

3. Never use oil with any but the oil-immersion objective. Use immersion oil *only* between the objective lens and the microscope slide.

4. Never remove a slide while the oil-immersion objective lens is in the light path.

5. Never *decrease* the distance between the slide and the oil-immersion objective while you are looking through the eyepiece.

6. Use clean lens paper to remove oil that spills on the stage and seeps under the mechanical-stage parts.

7. Never force any part of the microscope. All adjustments should work easily. If this is not the case, call your instructor.

8. Never allow an objective lens to touch a cover glass or slide.

9. Always watch carefully from the side of the microscope when you change from one objective lens to another.

Before you put your microscope away, always do the following:

1. Turn off the bulb and unplug the microscope; then wrap the cord around the microscope base.

2. Lower the stage or offset the objective to remove the slide.

3. Place the low-power objective in the light path.

4. Remove excess oil from the lens with lens paper; then gently wipe the lens with another piece of clean lens paper. Check all lenses this way.

5. Check to see that there is plenty of lens paper for the next class. If not, get more from your instructor.

6. Remove spilled oil from the other parts of the microscope.

7. Raise the stage to its highest position. This places the microscope in its most stable configuration.

8. Center the movable part of the mechanical stage on its platform.

9. Grip the microscope with both hands when you return it to the storage cabinet.

10. Replace the dust cover on the microscope.

LEARNING OBJECTIVES

• Learn the names of the major parts of the light microscopes and how they function.

• Understand the importance of resolution and learn what affects the resolving power of a microscope.

• Learn how to use the light microscope properly, especially at high magnification with the oil-immersion lens.

• Learn how to check your microscope before you use it and before you put it away.

MATERIALS

Cultures None

Media None

Supplies Brightfield microscope
Immersion oil
Lens tissue
Microscope slide containing stained microorganisms

PROCEDURE

1. **Place the microscope in front of you,** and check to see that the condenser is in the uppermost position. Place the low-power objective (_____×) in the light path directly above the condenser.

2. **Lower the stage of the microscope** using the coarse-focusing knob. Place a stained slide on the stage, right side up, and use the spring clip to hold it in place. Turn on the light source, and adjust it to a level that is comfortable when you look through the eyepiece.

3. **Position the slide** so that one of its long edges is directly over the center of the condenser lens. Now, looking at the objective and stage from the side of the microscope, use the coarse-focusing knob to raise the stage toward its uppermost position. *Do not allow*

the lens to touch the slide. (There may be a built-in mechanical stop to keep this from happening.)

4. **With the condenser diaphragm almost closed, and while looking at the side of the microscope,** move the mechanical stage so that the long edge of the slide moves back and forth across the condenser lens. Now look into the eyepiece while continuing to move the slide back and forth. You should see a dark (but fuzzy) line moving back and forth across the microscope field.

5. **Lower the stage** with the coarse-focusing knob until the edge of the slide is in focus. You are now focused on the *bottom edge* of the slide. Continue lowering the stage until you again see a sharp edge. Your microscope is now focused on the *top edge* of the slide, which is where the microorganisms should be located. Open the condenser diaphragm *halfway,* and move the mechanical stage so that the microbes are in the light path. You should be able to see them. *If you are having trouble, please ask your instructor for help.*

6. You are now using the lowest-power objective lens. **Watching from the side of the microscope,** change to the next higher-power objective. Adjust the focus. The large globs you see are probably clumps of stain or massive clumps of stained cells. Focus on this material.

7. **As soon as your instructor says to continue,** *carefully* position the oil-immersion objective (_____×) in the light path directly over the specimen. *Do this only while watching the objective and slide from the side* of the microscope (to make sure that the oil-immersion objective does not touch the slide). If the objective lens touches the slide, either you have not focused on the microorganisms or your slide is upside down (with the smear on the bottom side). *Do not be afraid to ask for help.*

8. **After you have completed step 7,** turn the nosepiece so that the objective lens is to one side of the light path, and *carefully* place 1 or 2 drops *(only)* of oil on top of the slide. Rotate the nosepiece so that the oil-immersion objective lens is again in the light path. *Before you touch the focusing knobs,* move the mechanical stage back and forth slowly, while looking through the eyepiece, until you see something (out-of-focus) move across the field. Place this out-of-focus material in the viewing field, and *use only the fine-focus knob to focus the image. When you are looking into the eyepiece, never use the coarse-focus knob with the oil-immersion objective in the light path.* When the image is in focus, *open the condenser diaphragm just until the field no longer gets any brighter.* This is the proper opening for this objective; you are now ready to study these microorganisms.

Note that all modern light microscopes are **parfocal,** which means that you can switch from the low-power objective to the oil-immersion objective without having to change the focus very much. If you do not find this to be the case, please ask for assistance from your instructor.

3 PRINCIPLES AND CARE OF THE LIGHT MICROSCOPE

NAME _____

DATE _____

SECTION _____

QUESTIONS

Completion

1. There are several different types of light microscopes. The type that you will most often use in this laboratory is called a _____ microscope.

2. Blue substage filters are often used because they emit a _____ wavelength, which effectively _____ the microscope's resolving power.

3. Use the words *increase* or *decrease* to show how each of the following are affected by opening the condenser diaphragm when using the oil-immersion objective:

   a. Image brightness _____
   b. Image contrast _____
   c. Resolving power _____
   d. Limit of resolution _____

4. The limit of resolution of a good brightfield microscope is about _____ μm.

5. When you are using an 8× eyepiece and a 97× objective, the apparent magnifications of your specimen are _____ × and _____ ×.

6. The substage condenser should always be placed in the _____ position before you use the microscope.

7. The refractive index of immersion oil should be about the same as _____, which is numerically expressed as about _____.

8. Before you put your microscope away, you should turn the _____ lens so that it is in the light path; you should check all objectives with _____ to make sure they are clean, turn the condenser to its _____ position, and turn the mechanical stage to its _____ position.

9. When you begin to observe a specimen, you should first place the slide on the stage with the specimen-side up; then position the _____ objective in the light path, move the stage to its _____ position, and focus on the _____ edge of the slide with the _____ focus knob.

10. After the specimen is in focus with a low-power objective, it is possible to turn the nosepiece to position another objective in the light path; but you should do this only while watching the objectives from _____.

11. The _____× objective is most commonly used to examine the size and shape of bacteria and the way in which these cells cling together.

12. A microscope that allows you to switch from one objective to another without drastically changing focus is called _____.

13. When you want to observe a specimen on a slide, it is always best to focus first on the edge of the slide using the _____ objective.

True–False (correct all false statements)

1. You can improve the resolving power of your light microscope by changing from a 10× to a 20× objective lens. _____

2. The working distance decreases as the objective lens magnification increases. _____

3. All objective lenses have variable angular apertures. _____

4. The limit of resolution can be mathematically described as a distance, and the greater this distance the better the resolution. _____

5. When using an oil-immersion objective, you should always use immersion oil with a refractive index of about 2.0. _____

6. You will rarely find a laboratory microscope that is parfocal. _____

7. You should always change illumination intensity by changing the height of the condenser. _____

8. If you suspect that your substage condenser is dirty, you should take it apart and clean it with lens paper. _____

9. Immersion oil improves the resolving power of all of your objective lenses, and it should be used for all objectives. _____

10. Always inspect your microscope before using it, and report any discrepancies to your instructor. _____

11. Always place the stage in its lowest position before you put the microscope away. _____

12. The specimen should always be on top of the slide when you examine it with the microscope. _____

13. While changing from the low-power to the oil-immersion objective, you should be looking through the eyepiece. _____

14. One part of your microscope that you should never touch while you are looking through the eyepiece with the oil-immersion objective in place is the coarse-focus knob. _____

4

Introduction to Staining Microorganisms

Most microbial cells are either *colorless* or they have so little pigment that they cannot easily be seen with the brightfield microscope. One way to increase contrast between microbes and the surrounding fluid is to stain the cells, that is, to fill them with a dye. However, dyes do not readily penetrate living cells; so microorganisms are usually killed before dye is applied. Dead cells soak up dye and retain it in much the same manner as a sponge soaks up water. Once the dead cells absorb a dye, a chemical association is usually formed between the dye and materials inside the cell; so the dye remains when the cells are rinsed with water. Therefore the cells appear colored against a colorless background.

A common method for killing cells prior to staining is **heat fixation.** Cells are first smeared onto a microscope slide, and the resulting thin film is allowed to air dry. Then the slide is gently heated, killing the cells so that the dye can penetrate and fixing the cells to the glass slide so they are not washed off by the staining process.

Microbiological staining procedures are categorized according to the job that they do. **Simple stains** color the cell or the background in a way that enables you to observe whether the cell is a straight or curved rod or a sphere and whether the cells are arranged in pairs or clusters. But simple stains do not enable you to differentiate among various types of rod-shaped or sphere-shaped cells. **Differential stains** use a combination of dyes that take advantage of chemical differences among cells. A differential stain dyes the entire cell of only certain types of bacteria. **Structural stains** preferentially color only one part of the cell, so that it can be distinguished from the rest of the microbe, making it possible to test cells for the presence or absence of that structure. Examples of these staining procedures follow.

Simple stain (stains whole cells or background, regardless of cell type)
    Basic dyes
    Acidic dyes

Differential stain (stains whole cells but only those of a certain type)
    Gram stain
    Acid-fast stain

Structural stain (stains only one part of the cell)
    Endospore stain
    Flagella stain
    Capsule stain
    Inclusion stains (chemically specific for glycogen, polymetaphosphate, lipids, starch, and so on)

To understand how a dye colors a bacterial cell, you must first know what a dye is. Dyes are salts containing an organic ion and an inorganic ion. From inorganic chemistry, you know that salts are composed of positively charged ions (cations) and negatively charged ions (anions); for example, the powdered dyes **methylene blue** and **crystal violet** are salts called methylene-blue chloride and crystal-violet chloride. Like other salts, dyes dissociate when you place them into solution; the dissociation products are positively charged organic ions and negatively charged inorganic ions. It is the positively charged organic cation that gives these dyes their color. Note in Figure 4-1 that these cations are organic compounds that contain more than one closed ring. This is characteristic of dyes.

The colored part of a dye (the organic ion) is called the **dye base.** Depending on its charge, the dye base is associated with either an anion (a negatively charged ion) or a cation (a positively charged ion) to form the

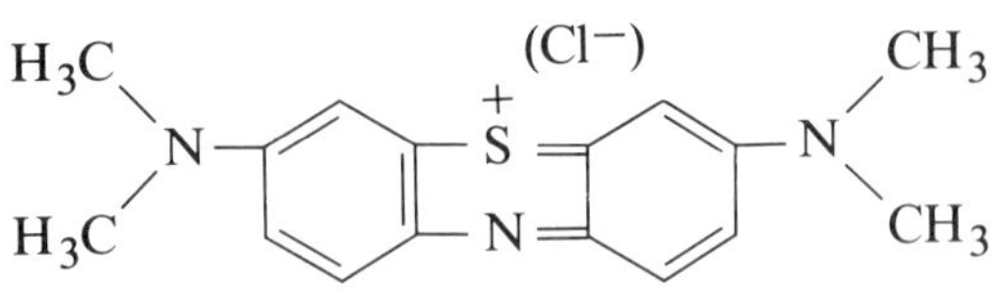

FIGURE 4-1 Chemical structure of two basic dyes used in the microbiological laboratory.

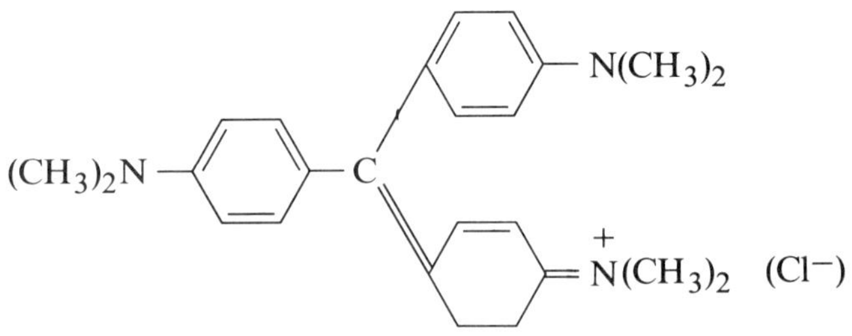

dye salt. If the dye base is positively charged, it is called a **basic dye** (both methylene blue and crystal violet are basic dyes). If the dye base is negatively charged, the dye is called an **acidic dye.** The relationships between the charge on the organic dye base and the type of dye are shown in Table 4-1.

A dye base that has a *positive* charge (such as methylene blue or crystal violet) is attracted to the negatively charged surfaces of the cells, and some of it actually binds to components on their surfaces. If the cells are dead, the dye base penetrates the cell envelope and binds with negatively charged interior parts. Therefore, dead cells that are stained have dye bound to their surfaces and also to the material inside them; so they appear entirely stained. The ionic bonds between the dye base and various parts of a cell are rather strong; so the dye is not easily washed away. Basic dyes are commonly used for simple stains and for more complex differential and structural stains.

All microbial cells have a negative charge on their surface—called the **surface charge**—when the pH of their surroundings is either near neutral (as in most environments) or alkaline. You may remember that *pH is defined as the negative log of the hydrogen-ion concentration.* This is expressed mathematically as

$$pH = \log \frac{1}{[H^+]} = -\log[H^+]$$

This equation shows that the numerical value for pH decreases as you *increase* the hydrogen-ion concentration ($[H^+]$). That is, as you increase $[H^+]$, you increase the acidity of the environment (lower the pH). Alternatively, when you *decrease* $[H^+]$, you raise the numerical value of the pH and make the environment more alkaline. The concept of pH and its effect on microbial growth are discussed in Exercise 21.

Many bacterial cultures excrete acids, thereby adding hydrogen ions to a culture medium and decreasing its pH. These hydrogen ions (H^+) interact with the surface of the negatively charged cell, resulting in fewer free negative charges on the surface. When this happens, the cell surface no longer strongly attracts positively charged dye ions (basic dyes). Thus the microbes from acidic environments stain poorly with basic dyes. For this reason, basic dyes are often made up as alkaline solutions. For example, potassium hydroxide (KOH) is added to solutions of methylene blue to form the stain called **Loeffler's methylene blue.**

Conversely, a decrease in $[H^+]$ increases the alkalinity and the numerical value of the culture pH. Some bacteria excrete alkaline materials during growth, and this decreases the number of available hydrogen ions in the culture medium. Under these conditions, the cell surface has a greater negative charge, which is more attractive to basic dyes and, therefore, allows greater binding, penetration, and internal staining of the microbe. Basic dyes stain microorganisms better under neutral or alkaline conditions.

If the dye base molecule has a *negative* charge, it is repelled by the cell's negatively charged surface. Thus, negatively charged dyes neither bind to the cell's surface nor are they able to penetrate into the cell. These are called *acidic dyes.*

Acidic dyes are used differently from basic dyes. Usually, an acidic dye is mixed with a drop of culture, smeared on a microscope slide, and allowed to dry. This preparation is not washed before it is observed with the microscope. Since the negatively charged cells are not stained by the negatively charged dye, they appear as clear areas surrounded by a colored

TABLE 4-1 Relationship between the type of dye and its charge when dissociated

Dye salt		
Organic ion (dye base)	Inorganic ion	Dye type
Positively charged (cation)	Negatively charged (anion)	Basic
Negatively charged (anion)	Positively charged (cation)	Acidic

background. Negatively charged dyes used in this way are called **negative stains.** Negative stains (acidic dyes) also work best under neutral or alkaline conditions because these conditions allow the surface charge to be most negative. Negative stains are of limited usefulness for those using light microscopes, but they can be used to avoid some of the disadvantages of staining with basic dyes.

Remember that the use of an acidic dye or basic dye *alone* is called a simple stain because either procedure enables you to study the relative sizes and shapes of microorganisms regardless of type. Both of these simple stains increase the contrast between the cell and its environment, as viewed with the brightfield microscope.

4 INTRODUCTION TO STAINING MICROORGANISMS

NAME _____

DATE _____

SECTION _____

QUESTIONS

Completion

1. Microorganisms must be _____ before a basic-type dye will penetrate and stain them.

2. The main purpose of staining microorganisms is to _____
   _____.

3. Both simple and differential stains dye the entire cell, but only a _____ stain enables you to see what *type* of rod or sphere you are examining.

4. If the organic part of a dye salt is positively charged, this dye should be called a(n) _____ dye, and if it is negatively charged, it should be called a(n) _____ dye.

5. The negative charge found on the surface of all particles, including microbial cells, is called the _____. This should have an attraction to _____-type dyes.

6. After a microorganism has been stained with a basic dye, where might the organic part of this dye salt be found associated with the cell?

7. Basic dyes stain _____ (better or worse) as the pH of the culture medium is lowered (becomes more acidic).

8. The organic ion portion of an *acidic* dye has a _____ charge.

9. Complete the following sentence by adding the words *increases* or *decreases*: When a microbe releases acids into the growth medium, this _____ the pH, _____ the hydrogen-ion concentration, _____ the acidity, and _____ the numerical value for the negative log of the hydrogen-ion concentration.

True-False (correct all false statements)

1. Simple stains enable you to determine the sizes and shapes of microbes and the type of rod or sphere you are examining. _____

2. Both methylene blue and crystal violet are examples of acidic dyes. _____

3. Simple stains color whole cells but only certain types of cells. _____

4. Structural stains are so named because they preferentially stain only one part of a cell. _____

5. The term *pH* represents the negative log of the hydrogen ion concentration. _____

6. Adding alkaline substances to a stain or culture medium should increase the surface charges of the cells and allow acidic dyes to stain better. _____

7. One difference between procedures used in staining with acidic or with basic dyes is that no washing is needed when using acidic dyes. _____

5

Smear Preparation, Fixation, and Simple Staining with Basic Dyes

Morphology is the study of the size, shape, and arrangement of cells. The purpose of this exercise is to show you how to prepare, fix, and stain a microbial culture on a glass microscope slide so you can determine the morphology of cells in a culture. Generally speaking, the shapes and sizes of microorganisms are genetically determined, and most bacteria have very rigid cell walls. Although changes in environment may have some small effect on their sizes and shapes, such as shortening or elongating a rod-shaped cell, spheres almost never change into rods or visa versa. Therefore, the shapes and sizes of microbes are important characteristics that you can use for their identification.

Most **bacteria** are either cylinders or spheres (Figure 5-1). Cylindrical bacteria can be either straight or curved. Straight cylinders are called **rods**, which vary from long and thin to short and plump. Curved cylinders vary from short and slightly curved (like a comma) to long spirals. Rod-shaped bacteria also vary in how tightly they stick together after cell division, which determines whether they are found as single cells, pairs, or long chains.

A bacterium called *Escherichia coli* normally appears as a short plump rod, found separate or in pairs. When grown in chemically complex media, each rod is about $1.0\,\mu m$ long. Cultures of *E. coli* are frequently used in this manual, so you will become accustomed to seeing this cell in your microscope. Memorize its size, as observed with the oil-immersion objective, then you can use its 1.0-μm length as a *ruler* to estimate the sizes of other microorganisms.

Spherical bacteria are frequently referred to as **cocci**. Different types of cocci do not vary much in diameter, but these bacteria often have characteristic planes of cell division and characteristic adhesion patterns after division. One division plane will produce a pair of cells (diplococci) or long chains (streptococci). Three or more division planes produce cubelike packets of cells (sarcinae) or irregularly shaped grapelike clusters (staphylococci). Cell arrangement is not as stable a characteristic as cell shape because you often can disrupt arrangements during the handling of cultures or during the preparation of samples for microscopic observation. However, the arrangement of cocci in a growing culture can help to identify the organism.

A few types of bacteria are long filaments. They have about the same diameter as average bacteria but are much longer. These filamentous bacteria sometimes show evidence of branching. Except for their diameters and internal structures, filamentous bacteria appear similar to eucaryotic fungi.

Fungi can also be dyed with simple stains. Most microscopic fungi have a spherical or a filamentous shape. Single-celled fungi are commonly called *yeasts* and are usually spherical or elliptical. Fungi that are characterized by very long and often-branching filaments are usually referred to as *molds*. In general, the diameters of these eucaryotic spheres or filaments are about 10-times greater than the diameters of procaryotic cells. Here is an easy way to remember these size relationships: the *diameter* of a yeast or the *width* of a mold filament is about 5 to 10 times larger than the *length* of an *E. coli* cell. This method for estimating cell size will be useful in the future when you must determine whether you are examining eucaryotic or procaryotic microorganisms.

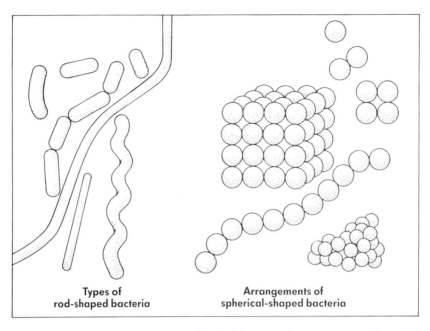

FIGURE 5-1 Representative types of typical shapes and arrangements of bacterial cells. No attempt has been made to draw these representative cell types to scale.

LEARNING OBJECTIVES

• Develop an understanding of how basic dyes stain cells.

• Understand the purposes and limitations of simple stains.

• Learn the sizes and shapes of typical procaryotic and eucaryotic microorganisms.

• Begin the development of good staining technique.

• Learn to recognize a well-stained and properly prepared smear of microbial cells.

MATERIALS

Cultures   *Staphylococcus epidermidis* (a bacterium)
           Escherichia coli (a bacterium)
           Saccharomyces cerevisiae (a yeast)

Media   None

Supplies   Crystal violet
           Microscope slides
           Staining racks
           Wash bottles
           Methanol in dropper bottles
           Nonabrasive soap

PROCEDURE

Review Exercise 2 before you begin.

1. **Clean all microscope slides** if so directed by your instructor.

CLEANING USED SLIDES

Most teaching laboratories recycle used microscope slides, but these slides are seldom clean enough for student use. Even new slides may not be clean enough for proper use. If your instruction directs you, clean slides in the following way. Make a paste with water and a nonabrasive soap (such as Bon Ami). Thoroughly scrub both sides of the slide with this wet paste. *Do not rinse the paste off the slide.* Lean the slide against a vertical object so that the bottom edge of the slide is resting on a paper towel. Allow the paste to dry. Use a clean paper towel to wipe off the dry soap powder and to polish the slide. The dried Bon Ami will not scratch the glass, but it will remove the previously stained microbes from the glass of used slides, and the soap will remove residual immersion oil. *Helpful hint:* If you know that you will be using several microscope slides during a class, scrub a sufficient quantity of slides with Bon Ami paste early in the class period so that the paste will be dry when you are ready to use the slides.

5 SMEAR PREPARATION AND SIMPLE STAINING

Use your marking pen to place a *B* near the end of the slide (on the side that you will smear later). This will let you know which side contains the smear. Try not to get stains and reagents on this mark because many of these liquids contain solvents that will remove the mark.

2. **Prepare a smear** of microorganisms *on the marked side* of the microscope slide as follows:

   a. *Liquid cultures* may be smeared directly on a clean microscope slide. Aseptically enter the tube containing the turbid broth, remove one loopful of culture, and smear it over the middle of the slide. Sterilize your loop before setting it down.

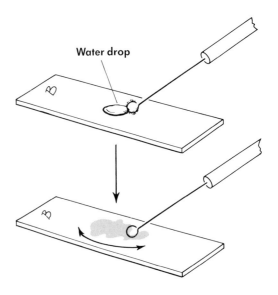

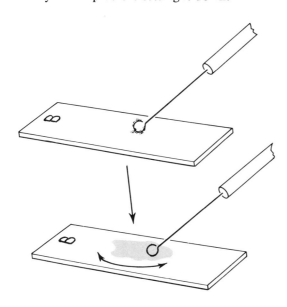

SPECIAL NOTES ON SMEAR PREPARATION

If the slide is clean, you should have no problems smearing the cells evenly over the surface. The more turbid the suspension, the greater the area over which it should be spread. The most common problem is making the smear too thick. *The dried unstained smear should appear as a barely visible film on the slide.* If your smear is quite cloudy, it contains too many cells, and they are so close together that you will not be able to distinguish their individual sizes and shapes using the microscope; moreover, it will be hard to wash excess dye out of the smear. This extra dye often crystalizes on the slide, making cell observations more difficult. Your instructor should show you how to judge the proper smear thickness.

---

   b. *Agar cultures* require that you place one loopful of tap water on the center of your clean slide. Aseptically enter the agar culture, and just barely touch the sterile loop to the cells on the agar surface. Mix the cells thoroughly with the water drop, and smear this mixture over the middle of the slide.

3. **Air dry the smear.** Blowing on the smear or waving the slide in the air hastens drying. Do not apply heat while the slide is wet because it may distort the shapes of the cells. Make sure the smear is dry before fixing it. This is especially important with smears from broth cultures. Materials in the broth take awhile to dry; so you must be patient.

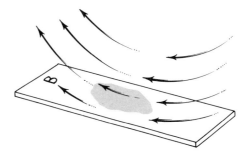

4. **Fix the cells** using one of the following procedures.

> **SPECIAL NOTES ON FIXATION**
>
> Fixation accomplishes two things: It kills the cells so that the dyes can penetrate easily, and it makes the cells stick tightly ("fixes" them) to the microscope slide. Cells removed from a slant culture are mixed with water and then smeared on the slide surface, from which the water evaporates leaving the cells in contact with the glass. *Heat fixation works well with smears prepared from slant cultures.* However, the beginning student often has difficulty with heat-fixed smears prepared from broth cultures. When a broth culture is smeared directly on a slide, the water evaporates but leaves the cells embedded in a film of organic goo. Heat often does not fix many of these cells to the glass; consequently, many are washed off during staining. *Methanol fixation works well with smears prepared from broth cultures.* In comparison with heat fixation, more cells remain fixed to the slide after staining. Thus, the student is advised to methanol-fix broth cultures and heat-fix slant cultures. Note that this is meant as a general guide and not as an absolute rule.

> **SAFETY NOTE**
>
> You can injure yourself during the heat fixation procedure. If you do not hold the slide at 45° angle, as illustrated, the flame can travel along the slide toward your fingers. If you keep the slide in the flame too long, it will become so hot that it will burn the back of your hand. When you first try this procedure, pass the slide across the flame three times, using one second for each exposure. Then adjust your technique depending upon whether the slide is too hot or not hot enough.

a. *Heat fix cells* by holding the slide between your fingers at about a 45° angle to the flame and then passing the slide rapidly across the flame about three times. Always test the slide with the back of your hand after heat-fixing to make sure that you have not overheated them (see Safety Note). The slide should feel warm, but not hot enough to burn the back of your hand. Overheating shrinks cells and often distorts their shapes. If you find that you have overheated the slide, do not stain that slide; prepare a new smear and try again.

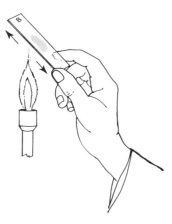

b. *Methanol fix cells* by placing the completely dry slide on the staining rack, smear side up, and flood the smear with methanol. Immediately drain the methanol, and gently shake the excess from the slide. Allow the methanol to air dry before staining. Methanol will remove the pen mark; so you will need to replace it after fixing the cells.

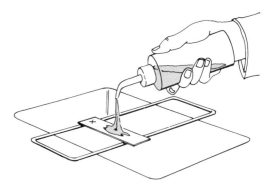

5. **Stain the cells** using the following procedure.
    a. *Flood the smear with crystal violet* after it has been placed on the staining rack. Stain for 30 seconds.

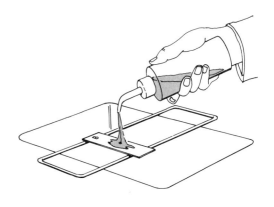

b. *Wash the slide.* Pick it up with a slide holder, and while holding it at about a 60° angle, rinse it thoroughly with water from the rinse bottle or tap. But *be gentle.* Too much rinsing or too strong a stream of water may dislodge the cells from the slide.

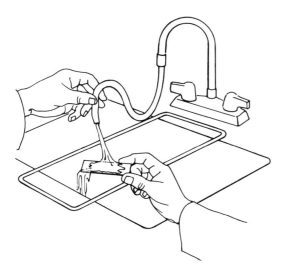

c. *Blot the excess water* from the slide with bibulous paper, and allow it to air dry. If you rub the slide with the bibulous paper, you may rub off most of the cells. Do not examine the slide until it is completely dry.

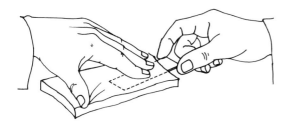

SPECIAL NOTES ON STAINING TIMES

The time of dye application is important. Various dyes differ in the rate and degree to which they stain a cell. Methylene blue is one of the slowest. Crystal violet is more reactive and will stain satisfactorily in 15 to 30 seconds. Carbol fuchsin is even more powerful and requires only a few seconds; but it is easy to overstain with this dye, which makes cell sizes and shapes difficult to determine.

---

d. *Always write the name of the culture on the top of your slide* (the same side on which you placed the mark) after the slide has been stained and dried. This good habit will enable you to easily identify the side of the slide on which the cells are located, and it will help you avoid misidentification when you are working with more than one culture.

6. **Place the microscope slide on the stage** so that the marked side is up.

7. **Focus the low-power objective on the pen mark.** This assures you that you are focused on the side of the slide that contains your smear.

8. **Move the slide so that the smear is under the low-power objective,** and you should be almost exactly in focus on the cells. A slight movement of the fine-focus adjustment knob should bring them into exact focus. Make sure that you have focused on the cells and that these are on the top side of the slide. If you experience difficulty at this point, please ask your instructor for assistance.

9. **Carefully position the oil-immersion objective over the smear** after adding immersion oil to the smear. Slight movement of the mechanical stage should reveal an out-of-focus image moving across the field of view. You can sharpen that image by slightly moving *only* the *fine-focus* adjustment knob. If you have difficulty, please seek assistance from your instructor.

10. **Observe all three microbes** under the oil-immersion objective as described in steps 6–9 above. Look for significant differences among these cultures in cell size, shape, and arrangement.

    If you have *too many cells* on your slide, you will see masses of cells and excess stain and few if any individual cells.

    If the slide is not *perfectly clean* or if the smear is not *completely dry* before staining, *too few cells* may remain on the slide.

11. **Record your observations, and write your conclusions.** Criticize and analyze each stained smear and the technique used to prepare it.

5 SMEAR PREPARATION AND SIMPLE STAINING

NAME _____

DATE _____

SECTION _____

RESULTS

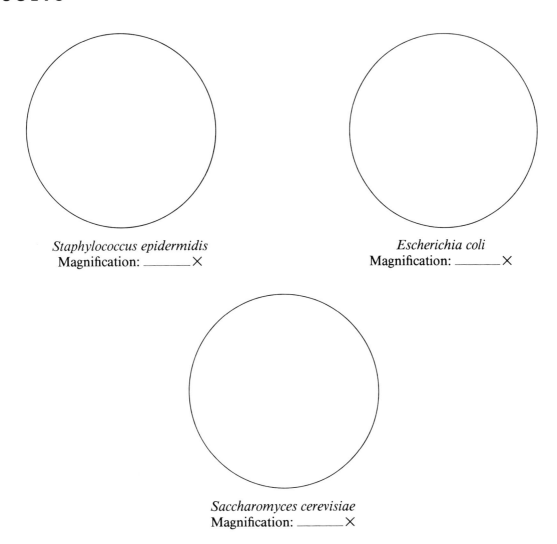

Staphylococcus epidermidis
Magnification: _____ ×

Escherichia coli
Magnification: _____ ×

Saccharomyces cerevisiae
Magnification: _____ ×

CONCLUSIONS

QUESTIONS

Completion

1. The main reason that microorganisms are stained with basic dyes is _____.

2. Heat fixation accomplishes two things necessary for staining: _____ and _____.

3. Most bacteria are of two basic cell shapes: cylinders and _____. Cylindrical cells may be straight rods or some variation such as _____ or _____.

4. _____ is the *length* of an average *Escherichia coli* cell. In contrast, most eucaryotic cells, such as yeast and molds, have a *diameter* that is about _____ times larger than this.

5. Spherical fungi are called _____, and filamentous fungi are called _____ _____. In contrast, spherical bacteria are called _____.

6. Some procaryotic bacteria and eucaryotic fungi grow in the shape of filaments. However, using the brightfield microscope, filamentous bacteria can be distinguished from filamentous fungi by _____.

7. The procedures given in this exercise make it easy to tell which side of the slide contains your smear because _____.

8. A stained smear that is properly made should show about _____ cells per oil-immersion field and _____ color when examined with the unaided eye. It should also contain _____ written at the end of the slide on the same side as the stained smear.

9. *Staphylococcus epidermidis* has the shape of a _____ and has a _____ (procaryotic or eucaryotic) cell type.

True–False (correct all false statements)

1. Most bacteria have flexible cell walls. _____

2. Spherical bacteria are called *cocci* and are typically five to ten times larger than the typical *Escherichia coli* cell. _____

3. Procaryotic cells are usually much smaller than eucaryotic cells. _____

4. A properly made smear should look very cloudy after air-drying. _____

5. It is generally advisable to heat-fix smears prepared from slants and methanol-fix smears from broth cultures. _____

6. All basic dyes require 30 seconds staining time. _____

7. *Escherichia coli* is a type of fungus that has the shape of a long cylinder. _____

8. *Saccharomyces cerevisiae* is a type of eucaryote known as a filamentous fungus (a mold), and it is much larger than most bacteria. _____

6

Observing Live Microorganisms: The Wet-Mount Method and the Phase-Contrast Microscope

Brightfield versus phase-contrast microscopy

Before stains were developed and before specialized light microscopes were invented to increase the contrast between microorganisms and the surrounding fluid, microbiologists used the brightfield microscope and unstained preparations to examine microorganisms. They found it very difficult to study transparent cells in a transparent medium with a brightfield microscope. Nevertheless, you should see how living microorganisms appear when examined with a brightfield microscope so that you will appreciate the value of staining and the phase-contrast microscope.

Living cells can be easily studied with the **phase-contrast microscope**. Not all microbiology laboratories have phase-contrast microscopes; but, hopefully, yours will have at least one that can be used for demonstration.

Observing microorganisms live is the best way to determine their true size and shape. Size and shape cannot be accurately determined with stained microbes because the staining procedures invariably reduce the size of cells and may also slightly distort their shape.

Principles of the phase-contrast microscope

Our ability to see things depends on the existence of sufficient contrast between the object and its surroundings. Our eyes perceive contrast as differences in light intensity or color. When we observe a *stained* cell with the brightfield microscope, the light passing through the cell is **diffracted** (dashed line, Figure 6-1), and the **energy** of that diffracted light ray is partially **absorbed**. When this energy is absorbed by the stain, the **amplitude** (intensity) of the light is decreased (Inset A). The denser and thicker the microbe, the more energy it absorbs. The greater the absorption, the smaller the amplitude of the light ray, and the darker the cell appears (Inset B). Stained cells absorb some wavelengths more than others, and the result is perceived as color.

Light that passes through an *unstained* cell is also diffracted (dashed line, Figure 6-2), but energy is not absorbed; thus there is no change in amplitude. However, unstained cells are more **refractile** (have a higher refractive index) than the surrounding medium. This *slows* the diffracted light (dashed line, Inset A) so that it is a quarter wavelength (λ) out-of-step, or **out-of-phase,** with the direct (incident) light passing through the surrounding medium (solid line, Inset A). This **phase shift** is used to advantage in the design of the phase-contrast microscope.

The **phase-contrast microscope** was invented by the Dutch physicist Frederick Zernike in the early 1930s. The main differences between this microscope and the brightfield microscope are (1) the placement of an **annular stop** below the condenser lens system and (2) a **phase plate** inside the objective lens (Figure 6-2). Light enters the condenser only through the open ring in the annular stop and approaches the top surface of the microscope slide as a hollow cone of light.

As shown in Inset B of Figure 6-2, direct light rays (those that do not go through a cell) pass through the phase ring (a component of the phase plate). In the phase-dark type of phase-contrast microscope, the phase ring is coated with a substance that advances

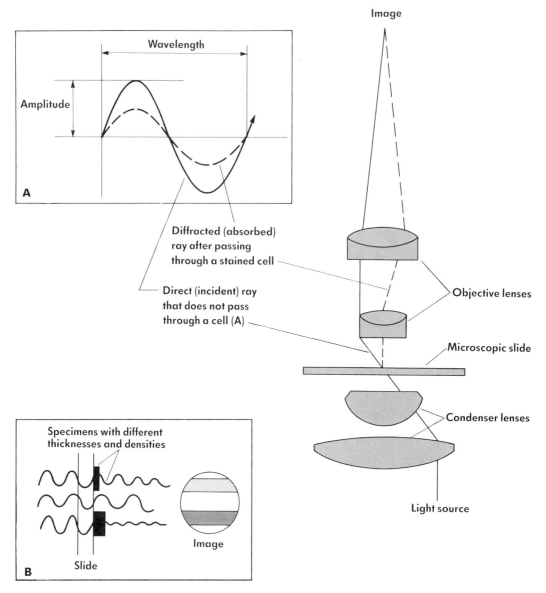

FIGURE 6-1 Paths taken by a direct light ray (solid line) and a ray diffracted by a cell (dashed line) through a brightfield microscope. A diffracted ray (dashed line) has less energy and a shortened amplitude (Inset A). The decrease in amplitude is proportional to the density and/or thickness of the specimen (Inset B). The greater the energy loss, the shorter the amplitude, and the darker the specimen appears.

the direct ray by a quarter wavelength. Diffracted light rays (those that are diffracted by the cell and retarded by a quarter wavelength) completely miss the phase ring and pass unaffected through the clear-glass portion of the phase plate.

When the diffracted ray is retarded by a quarter wavelength and the direct ray is advanced by a quarter wavelength, and both rays converge at the image plane, one obtains a half-wave shift and an **interference** between the two rays (Inset C of Figure 6-2). Therefore, a cell appears dark against a lighter background. This type of microscope has so-called **phase-dark** optics, which seem to be the most commonly used phase optics.

For phase optics to demonstrate the proper contrast between direct and diffracted light, the annular stop and phase plate must be properly **aligned.** The way you do this depends on the microscope's design, but all microscopes have some way to focus on the back (uppermost) lens of the objective. When this is done, you will simultaneously see the coated phase ring on the phase plate and the ring of light coming from the

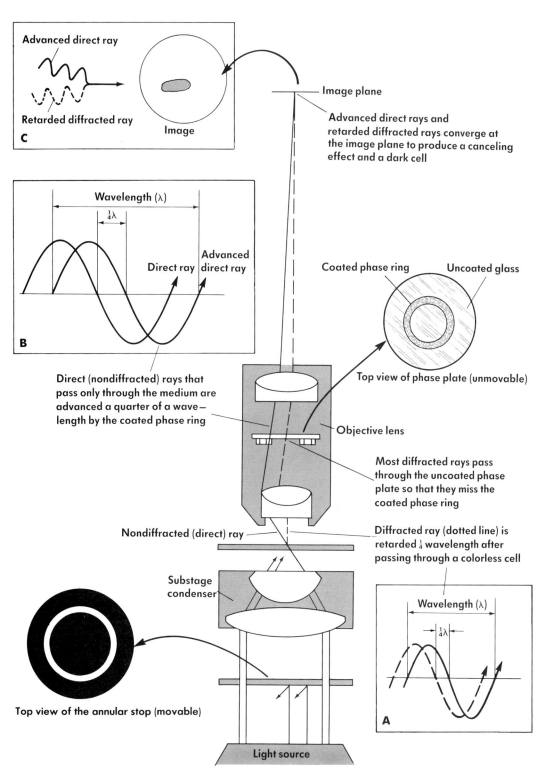

FIGURE 6-2 Paths taken by a direct light ray (solid line) and a ray diffracted by a cell (dashed line) through a phase-contrast microscope. A diffracted ray is retarded (shifted out of phase) by a quarter wavelength, but its amplitude is not changed (Inset A). As the direct ray passes through the phase plate, its wavelength is advanced by a quarter wavelength (Inset B); this means that the advanced direct wave is a half wavelength ahead of the diffracted ray (Inset B). This *phase shift* causes an interference that creates the appearance of a darker cell against a lighter background (Inset C).

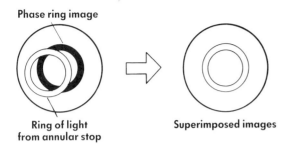

FIGURE 6-3 Appearance of the superimposed annular stop and the phase-plate images as observed through a special focusing eyepiece or lens. The left illustration shows the annular stop and phase plate to be out of alignment; you will not achieve phase contrast under these conditions. When you adjust the annular stop so that it aligns with the phase plate, you will see the aligned image shown on the right and achieve a phase contrast image.

annular stop (Figure 6-3). Proper phase contrast is achieved only when the images of these two rings are superimposed, which is done by adjusting the annular stop until the ring of light is directly under the phase ring.

If you have access to a phase-contrast microscope, ask your instructor to demonstrate this alignment procedure for you. Hopefully, you will be able to compare the appearance of live microbes as seen by brightfield and phase-contrast microscopes.

Observing and inhibiting cell movement

Live microbes in liquid always appear to be moving. The amount of movement is determined by whether they are motile or nonmotile. A **motile** organism moves rapidly in one direction or appears to be tumbling rapidly. This type of motility is caused by the cell's **flagella** or **cilia,** which propel the cell through the fluid (see Exercise 10). Such rapid movement, however, can be a disadvantage if you want to study the exact movements of a cell. One way to overcome this problem is to increase the **viscosity** of the fluid, which causes the cell to slow down so that you can study its contortions. It is like making a human swim in molasses.

It is difficult to study the sizes and shapes of motile cells while they are rapidly swimming through a fluid. However, you can stop motility by killing cells in a nondamaging way, such as by adding **formaldehyde** to the culture fluid. Formaldehyde not only kills cells but also acts as a biological **fixative,** which makes the cell rigid and also preserves it from physical and chemical damage (somewhat like embalming fluid).

Even if cells are killed or nonmotile, you will observe some movement when they are suspended in a fluid. This movement appears as random jiggling, in which the cells never move very far in one direction (usually less than one-eighth of a microscope field as seen with the $100\times$ objective). All **colloidal particles,** whether they are live or dead cells or very small nonliving particles, exhibit this type of motion, called **brownian movement.** It occurs as the soluble molecules in the surrounding fluid constantly bombard the colloidal particles. Although the cell does move, this is *not* motility because the cell is not propeling itself. You should be able to determine whether or not the cell is motile by using the **wet-mount** method.

Measuring live microorganisms

Brownian movement and motility both interfere with measurement of cell size because you cannot measure moving cells. One way to prevent movement of cells in a liquid environment is to **sandwich** them between a coverslip and an agar-covered microscope slide (Figure 6-4). If you prepare the agar properly, it will absorb the liquid until the cells are immobilized, and all cells will be in the same focal plane.

Trapped cells can be measured with an **ocular micrometer,** a device that is inserted into the microscope eyepiece (Figure 6-5). The ocular micrometer is a flat piece of glass with a scale etched on its surface (Inset B). This scale is not calibrated because the apparent calibration would change as magnification changes. However, at any one magnification, the ocular micrometer can be calibrated by comparing its scale with that of a **stage micrometer** (Inset C). The stage micrometer is a glass microscope slide that contains a metric rule with accurate subdivisions etched on its surface. For example, Inset C shows that the distance

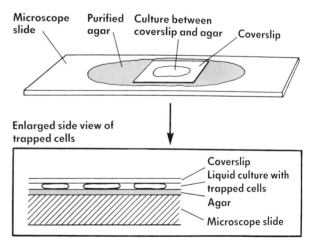

FIGURE 6-4 Technique for trapping cells between a coverslip and an agar-covered microscope slide (the agar sandwich). This stops motility and brownian motion and keeps cells in the same focal plane so that all cells are in focus.

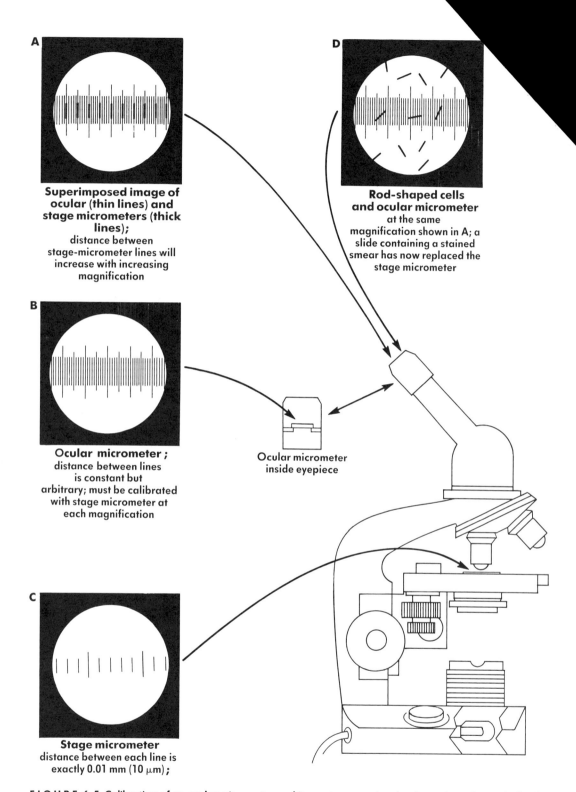

FIGURE 6-5 Calibration of an ocular micrometer and its use in measuring the dimensions of microbial cells. When you view the ocular micrometer (Inset B) and the stage micrometer (Inset C) simultaneously, they appear as in Inset A. The appearance of the ocular micrometer will not change with changes in magnification (objective lenses), but the distance between the lines on the stage micrometer will increase with each change in magnification. For this reason, the ocular micrometer should be separately calibrated with each objective lens. Considering the distance between stage-micrometer lines (10 μm), how much distance is there between ocular-micrometer lines at the magnification shown in Inset A?

is 0.01 mm (10
eters are both in
th any objective
of Figure 6-5).
neter scale can
. For example,
n the stage-mi-
is equal to the
rometer lines.
wo ocular-mi-
... mm / (0.0014 mm, or 1.4 μm).

After you calibrate the ocular micrometer at that magnification, you remove the stage micrometer and place a microscope slide with microorganisms on the microscope stage. The ocular micrometer is left in the eyepiece so that you can see both the micrometer and the microbes at the same time. Using the information in the preceding paragraph, find the length of the rod-shaped procaryotic cell shown in Inset D of Figure 6-5 and write that length on the following line: _____

Relative sizes of eucaryotic and procaryotic microbes

One purpose of this exercise is to help you gain an appreciation of differences in **cell sizes** among various microorganisms: **eucaryotic** microorganisms such as yeasts (single-celled fungi), molds (filamentous fungi), algae (photosynthetic plants), and protozoa (single-celled animals) and **procaryotic** organisms such as bacteria. Procaryotes average about five to ten times smaller than eucaryotes, and their internal structure appears less complex. You should review the way in which *E. coli* cells can be used to estimate the size of other microorganisms (see the introduction to Exercise 5).

In this exercise, you should carefully observe the relative sizes of eucaryotic yeasts and procaryotic bacteria with the oil-immersion objective, and draw your observations to scale in the Results section. In later exercises, you will be expected to differentiate eucaryotic and procaryotic cells by size alone.

LEARNING OBJECTIVES

• Learn the appearance of living microbes as observed with the brightfield microscope.

• Gain an appreciation of differences in sizes between eucaryotic and procaryotic microorganisms.

• Learn the advantages and disadvantages of the wet-mount method for observing microorganisms.

• Learn to differentiate between brownian movement and motility.

• Gain an appreciation for the phase-contrast microscope for observing live cells.

MATERIALS

Cultures   Hay infusion
           The yeast *Saccharomyces* species (a eucaryotic cell type)
           The bacterium *Pseudomonas* species (a procaryotic cell type)

Media   None

Supplies   Inoculating loop
           Bunsen burner
           Microscope with immersion oil and lens paper
           Glass coverslips
           Microscope slides
           Vaseline and toothpicks
           Phase-contrast microscope (if available)

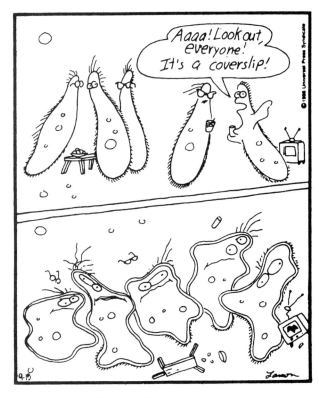

Life on a Microscope Slide

PROCEDURES

Preparing the wet mount

The wet-mount procedure enables you to study the sizes and shapes of live or preserved microorganisms whose shapes have not been distorted by staining or drying. It also enables you to determine if live cells are motile. The main advantages of the wet-mount method are that it is quick and easy to prepare and that you do not need special microscope slides or other special equipment. The main disadvantage is that you may disrupt the arrangements of cells, such as chains or clusters, when you place the coverslip on the slide. It is assumed that you know how to remove cells aseptically from a culture tube (see Exercise 2).

If you are making a wet mount of an **agar culture** *begin with steps 1, 2, and 3, and then do steps 6 through 8.*

If you are making a wet mount of a **broth** (liquid) **culture,** *do steps 4 through 8.*

1. **Place a drop of tap water** on the center of your slide with your inoculating loop.

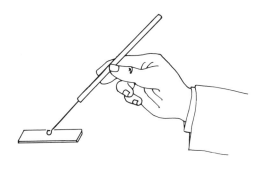

2. **Aseptically enter the culture tube,** *gently touch* the growth on the agar surface, and remove the cells.

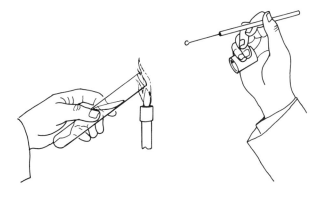

3. **Mix the cells with the water drop** on the slide. Do not smear the drop; keep it as a drop.

4. **Aseptically remove one drop** of culture from the tube with a loop.

5. **Place this drop on the center of the slide** without smearing it.

6. **Flame your loop before setting it down.**

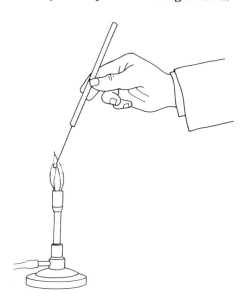

7. **Mark a coverslip** about one-fourth of the way in from one edge. This will be your focusing aid.

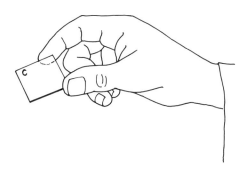

8. **Apply the coverslip** as follows. Pick up the coverslip and turn it so that the *marked side faces downward;* then place one edge of the coverslip on the microscope slide, and gently lower it so that it contacts the culture. This wll spread the drop of culture between the coverslip and slide.

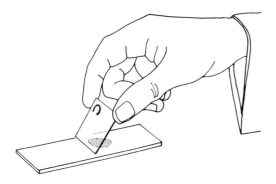

SPECIAL NOTES ON PREVENTING EVAPORATION

If you want to observe a culture for a long time by using the wet-mount method, you must do something to retard evaporation of the culture fluid. Otherwise, the fluid will decrease in volume, and the suspended cells will flow across the field of view so fast that you will not be able to study them. One way to limit evaporation is to rim the coverslip edge with Vaseline before placing the coverslip on the slide. However, Vaseline is difficult to clean from glassware, and many laboratories avoid its use when possible. Another way to retard evaporation after preparing a wet mount is to place a small bead of immersion oil along the edge of the coverslip. This oil forms a seal between the edge of the coverslip and the microscope slide and prevents evaporation of the culture fluid. Immersion oil is easier to clean off glassware than Vaseline. Neither method is necessary if you want to observe the preparation for a short time. Ask your instructor about the policy in your laboratory. If oil or Vaseline is recommended, ask your instructor for a demonstration of the proper way to apply these materials.

Observing live cells with the brightfield microscope

1. **Place the slide on the microscope stage** so that the coverglass is on top; then focus on the *top* edge of the coverslip with the *low-power* objective.

2. **Move the stage** so that the low-power objective is directly over mark on the coverslip; then focus on this mark. Once you have the mark in focus, move the slide to an edge of the culture fluid and refocus if necessary. When the droplet edge is in focus, you may see some of the microorganisms held to the edge by surface tension.

3. **Close the condenser diaphragm** almost all the way, and place the image slightly out of focus. This increases the contrast of transparent objects, although some resolving power is lost. This is a *time-honored technique.* Until development of the phase-contrast microscope, nearly all work on living cells was done with this procedure.

4. **Observe** this culture with the **low-power objective,** and record your observations.

5. **When you are ready to move to the oil-immersion objective,** please proceed as follows:
    a. Go back to the mark on the coverslip, and make sure it is focused.
    b. Turn the revolving nosepiece so that no objective is directly over the specimen.
    c. Place one drop of immersion oil on top of the coverslip so that it is in the light path.
    d. Make sure you know which is the oil-immersion objective; then place this objective directly over the specimen. Watch from the side so that you do not jam the objective lens into the coverslip.

6. You should now be able to see the pen mark through the eyepiece. **Focus on the pen mark with the fine-focus adjustment. Move the slide so that the edge of the droplet comes into view;** then make your observations.

7. **Repeat this procedure with all cultures.**

HELPFUL HINTS

Hay infusion Use the low-power objective to observe the hay infusion. This infusion may contain bacteria, protozoa, and algae. With the low-power objective, the larger eucaryotic yeast, protozoa, and algae are the most noticeable. When you switch to the oil-immersion objective, you should be able to see the much smaller bacteria.

Pseudomonas **sp.** These are bacteria. Some of the cells should be very motile. Some nonmotile cells may be seen at the edge of the droplet. Remember that all bacteria are procaryotic cells. Pick a typical field of view, and draw some of these cells to scale in the Results section of this exercise.

Saccharomyces **sp.** This culture is a yeast which is a type of single-celled fungus. All fungi are eucaryotic cell types. These cells are not motile, but you should be able to detect brownian movement. Compare the size of these cells with that of the bacteria, and record your observations in the Results section by drawing the cells to scale.

6   OBSERVING LIVE MICROORGANISMS   45

NAME _____

DATE _____

SECTION _____

RESULTS

Review the box entitled *Helpful Hints* in this exercise.

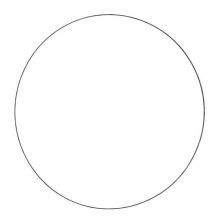

Hay infusion
(low-power objective)
Magnification: _____ ×

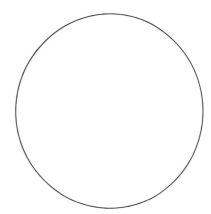

Hay infusion
(oil-immersion objective)
Magnification: _____ ×

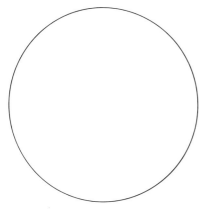

Saccharomyces sp.
(oil-immersion objective)
Magnification: _____ ×

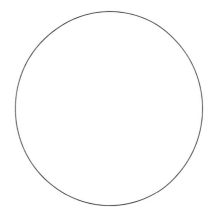

Pseudomonas sp.
(oil-immersion objective)
Magnification: _____ ×

CONCLUSIONS

QUESTIONS

Completion

1. One problem with trying to study live microbes with a brightfield microscope is _____ _____.

2. Rapid tumbling or movement in a single direction by microbes is called _____ _____.

3. _____ is the term given to slight movement of live or dead particles caused by bombardment by molecules in the suspending fluid.

4. One way to increase the contrast of live cells when studying them with a brightfield microscope is to _____.

5. If you assume the *Pseudomonas* sp. is the size of a typical procaryotic rod, then about how much larger (in cell widths) are the eucaryotic yeast cells used in this study? _____ ×

6. How might you accurately describe the appearance of microbial cells when you observe them at high magnification using the wet-mount method and the brightfield microscope?

7. What is the purpose of using Vaseline to rim the edges of the coverslip?

8. There are two types of microbial movement when cells are observed with the wet-mount method. These are _____ and _____.

9. The major function of the phase-contrast microscope is _____ _____.

True – False (correct all false statements)

1. Some but not all bacteria have a procaryotic cell structure. _____

2. It is a good practice to always focus first on the upper edge of the coverslip when setting up the microscope to observe cells with a wet-mount preparation. _____

3. *Saccharomyces* is a procaryotic bacterium; therefore, it is smaller than cells in the genus *Pseudomonas*. _____

4. Cells in the genus *Pseudomonas* are spherical. _____

6 OBSERVING LIVE MICROORGANISMS

NAME _____

DATE _____

SECTION _____

5. The mark placed on the specimen side of the coverslip is meant to serve as a focusing aid to help you be sure that you are focused on the top side of the coverslip in a properly prepared wet mount. _____

6. Microbiologists can measure the size and determine the correct shape of a microorganism once it has been stained. _____

7. Bacteria are about five to ten times larger than yeasts and protozoa. _____

8. If dead microbes have flagella, or cilia, the cells will be motile. _____

9. A special type of microscope slide, called a depression slide, is required for making a wet mount. _____

10. *Saccharomyces* appears generally spherical in shape. _____

11. Vaseline or immersion oil should always be used to seal the coverslip edge when preparing wet-mount slides. _____

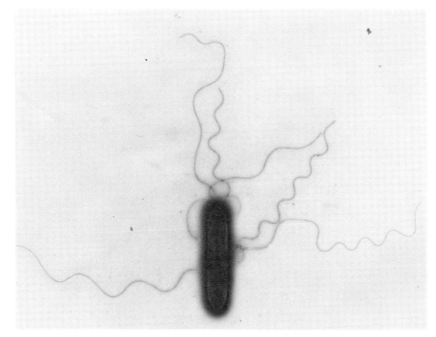

Salmonella species showing peritrichous flagellation.

Dear God, what marvels there are in so small a creature.
— *Leeuwenhoek's draughtsman*
(Letter 76, 15 October 1693)

MICROSCOPIC TECHNIQUES FOR DETERMINING MICROBIAL TYPE AND STRUCTURE

The primary purpose of Part III is to demonstrate how you can use the light microscope to examine structural features of the procaryotic cell and thereby differentiate various cell types. The techniques involve two types of microbiological stains: differential and structural.

Differential stains selectively dye certain types of cells; that is, these dyes or their solvents react differently with various cells. This is often due to differences in structure and/or chemical composition of the bacterial cell wall. Perhaps the two most important differential stains are the Gram stain (named after Christian Gram, a Danish physician, who developed this stain in 1864 to detect bacteria in diseased tissues) and the acid-fast stain (an important tool for diagnosing tuberculosis).

Structural stains selectively dye only one part of the cell (for example, the procaryotic endospore). Many staining procedures exist for dyeing particular structures of the microbial cell.

The exercises that follow explain how to use differential stains and how to stain endospores and flagella.

Microbiologists also use nonstaining methods to examine cells for the presence of certain structures or the accumulation of cell-produced materials. For example, a phase-contrast microscope can help determine motility (indirect evidence of active flagella) without using stains; the phase-contrast microscope can also help detect the presence of endospores. India ink particles, along with the brightfield microscope, can be used to indirectly detect the presence of capsules. These nonstaining techniques also are presented in this part of your manual.

7

The Gram Stain: A Differential Stain

It is difficult to overemphasize the importance of the Gram stain to microbiology. The Gram stain is one of the first steps in identifying procaryotic cells, which is why you will devote so much time to becoming proficient in using it. The first part of this exercise will help you Gram stain *known* gram-positive and gram-negative cultures, and the second part will test your skill in determining the Gram-staining characteristics of *unknown* cultures.

The following information should be committed to memory:

Staining reagents The **primary stain** must be crystal violet or another type of methyl pararosaniline dye. Other basic dyes do not work. The **mordant,** Gram's iodine (Lugol's solution), increases the affinity between the primary stain and the reactive substances in the cell. It is believed that a crystal-violet–iodine complex is formed inside the cell. The **decolorizing agent** can be any chemical that is a good lipid solvent and a good dehydrating agent. The **counterstain** can be any dye as long as it is a color that contrasts with the deep blue-violet of the stained gram-positive cells. Basic *fuchsin* or *safranin* are red dyes that are commonly used as counterstains.

Common staining problems An incorrect Gram reaction is usually caused by *improper* decolorization. Gram-positive cells can be decolorized too long and appear gram-negative. It is also possible to decolorize gram-negative cells for too short a time so that they appear gram positive. Make sure that you understand why this is so.

One way to check for improper decolorization is to prepare three smears on the same slide: one smear a known gram-positive culture, another a known gram-negative culture, and another the bacterium that you are studying. Such knowns are called **standards.** If you use them routinely, you can always determine if you have properly decolorized the smear.

A smear of a gram-negative culture that is *too thick* will not decolorize in a reasonable length of time. Such a smear contains masses of cells very close together, which the decolorization agent cannot penetrate in the normal time for decolorizing. As a consequence, this thick smear containing gram-negative cells will appear to contain masses of gram-positive cells.

Generalizations If the smear of your unknown culture is thin and yet appears to contain both gram-negative and gram-positive cells, and if your standard cultures are correctly stained, then you should suspect that your unknown culture is either an old gram-positive culture or it is contaminated. If the morphology of the cells is also different, then it is almost certain that your unknown is not a pure culture.

Bacteria are usually either gram positive or gram negative regardless of culture age. Occasionally, however, cultures show a tendency to go from gram positive to gram negative as the culture ages. The reverse is almost never true. *It is always best to work with young (exponentially growing) cultures.*

Some bacteria do not stain by Gram's method. This is particularly true of some genera in the order *Spirochaetales,* such as the bacterium that causes syphilis. These cells need only simple staining by Giemsa's method, the technique frequently used to stain protozoa. Simple stains work well for identifying bacteria that cause syphilis because of the bacteria's unusual spiral shape and very small diameter.

Microbiologists often have to identify unknown bacterial cultures that are isolated from infected plant or animal tissue or from environments such as soil and water. Determining whether bacteria are gram positive or gram negative is one of the first steps in identifying an unknown bacterium. If this determination is *not* made correctly, all hope for accurate identification is lost. Therefore, it is essential that you learn to accurately perform the Gram stain and correctly interpret the results.

In past laboratory periods, you have probably used both the brightfield and the phase-contrast microscopes. The phase-contrast microscope should *not* be used to examine stained specimens because phase-contrast optics do not enable you to see the true color of the dyed cells. Always use a brightfield microscope for examining and interpreting Gram-stained smears.

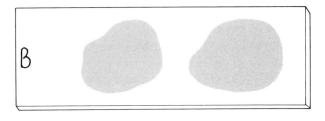

FIGURE 7-1 Placement of the focusing-aid mark (B) and two thin bacterial smears on the same side of a single glass microscope slide for determining the Gram morphology of *Bacillus cereus* and *Pseudomonas fluorescens*.

LEARNING OBJECTIVES

- Become proficient with the Gram-stain technique.
- Become skilled in the interpretation of results of the Gram stain, especially in the ability to recognize the difference between gram-positive and gram-negative bacteria, and to recognize improperly decolorized slides.

MATERIALS

Cultures   *Pseudomonas fluorescens* (gram-negative standard)
Bacillus cereus (gram-positive standard)
Unknown culture. Your instructor may ask you to examine cultures for which you do not know the gram-staining characteristics.

Media   None

Supplies   Gram-stain reagents
Possible demonstrations: Gram-stained slides of the *Pseudomonas* and *Bacillus* standard cultures.

PROCEDURES

The Gram stain

Review Exercise 5 for procedures on how to prepare and fix culture smears on glass slides. Use only thoroughly cleaned slides for all cultures.

1. **Place a pen mark *(B)* near the edge of one slide.** This will serve as a focusing aid. Heat this mark with the bunsen burner to fix it to the slide.

2. **Prepare two smears on the same side of that slide,** as shown in Figure 7-1. Smear the known gram-positive culture *(B. cereus)* on the left side of the slide and the known gram-negative culture *(P. fluorescens)* on the right side. Thoroughly air dry these smears. Remember that *properly prepared smears are so thin that they are almost invisible.*

3. **Heat-fix this slide** if you are working with slant cultures; **methanol-fix the slide** if you are working with broth cultures or a combination of smears from both slants and broth cultures on the same slide (see Exercise 5 for a review). Note that these are suggestions rather than absolute rules. *Use whatever fixation procedure works best for you.*

During the following steps, avoid placing stains or alcohol on the pen mark because they may dissolve the ink.

4. **Stain and decolorize,** as follows:
   a. **Cover smears with crystal violet** for 30 seconds. (The primary stain.)
   b. **Wash off** liquid stain with water (no more than 2 to 3 seconds).
   c. **Cover smears with iodine solution** for 60 seconds. (The mordant.)
   d. **Wash off iodine solution** with water as before.
   e. **Blot dry** (without rubbing) to remove all excess water.
   f. **Decolorize smears with 95 percent ethyl alcohol** (EtOH) using one of the following techniques: (1) *Flood* the smear with EtOH and let stand for no more than 15 to

25 seconds; (2) *Drip* EtOH onto the smear with the slide held at a slight angle, and stop application when dye stops coming off the smear. Both of these techniques are acceptable. Personal preference varies, but neither works well if the smear is too thick.

g. **Flood the slide with water immediately after decolorizing** with EtOH. This dilutes the EtOH and stops the decolorization process. Blot dry without rubbing.

h. **Flood the smears with basic fuchsin** for 20 seconds (the counterstain.)

i. **Rinse briefly** with water to remove all excess stain. Blot without rubbing, and air dry. If you have properly prepared the smears, they should be so thin that little or no color will appear on the slide surface.

5. **Place the slide right-side up on the microscope stage;** then use the low-power objective to focus on the pen mark. This should assure you that you are focused on the side of the slide that contains your smear.

6. **Move the slide so that the smear is under the low-power objective;** you should be almost exactly in focus on the cells. Slight movement of the fine-focus adjustment knob should bring the cells into exact focus. Make sure that you have focused on the cells and that these are on the top side of the slide. If you experience difficulty, please ask your instructor for assistance.

7. **Carefully add oil and position the oil-immersion objective over the smear.** Slight movement of the mechanical stage should reveal an out-of-focus image moving across the field of view. Now sharpen the image by slightly moving *only the fine-focus adjustment knob.* If you have difficulty, please seek assistance from your instructor.

8. **Examine many fields of each smear** with the oil-immersion objective. Gram-positive cells should appear dark blue (blue violet); gram-negative cells should appear light red.

9. **Examine demonstration slides** of *Pseudomonas* and *Bacillus* (if provided), and compare them with your own preparation.

10. **Repeat this procedure** and, if necessary, alter your technique until your standard cultures show the correct Gram reaction.

SPECIAL NOTES ON INTERPRETING THE GRAM STAIN

It is important to adjust your microscope correctly before you examine these smears. Beginning students often incorrectly label cells gram negative because they have too much light blazing through their microscopes. Make sure your condenser diaphragm is correctly adjusted (see Exercise 1). Then adjust the illumination with the brightness control so that it is comfortable for your eyes and you can easily see the difference in color between your gram-positive and gram-negative standard cells. Look back and forth between these two standards until you have this difference firmly fixed in your mind, and then examine the stained smear of your unknown culture.

The gram-negative standard will help you determine whether you have kept the 95 percent ethanol on the slide for too short a time *(underdecolorization).* If some or all of the cells in your thinly spread gram-negative standard appear dark blue-purple like gram-positive cells, then you have underdecolorized your slide.

The gram-positive standard will help you determine whether you have kept the 95 percent ethanol on the slide too long *(overdecolorization).* If some or all of the cells in your thinly spread gram-positive standard appear distinctly red like gram-negative cells, then you have overdecolorized your slide.

If your technique correctly stains both of these known bacteria, you can be quite sure that your unknown culture has also been stained correctly. This will be a good technique for you to follow in the future.

If your slide has been stained correctly yet your unknown contains a mixture of gram-positive and gram-negative cells, it may be either that your culture is *contaminated* or that it is an *old* (stationary-phase) culture. Only *young* (exponentially growing) cultures should be used for this stain. Gram-variable pure cultures do occur, but they are rare; you will not study this type of culture during this academic term.

Remember, only very thin smears can be relied upon to show correct Gram-staining characteristics.

---

11. **Stain and microscopically examine all unknown cultures** in this same manner.

12. **Discuss** your results with your laboratory instructor.

13. **Record your results** at the end of this exercise.

14. **Keep a list of all microorganisms that you Gram stain in the laboratory.** You will be expected to know the morphologies and Gram reactions of the cultures that you study in the laboratory.

> **GRAM-STAIN SUMMARY**
>
> | | |
> |---|---|
> | Test name | Gram stain |
> | Cultures used | Must be very young |
> | Reagents and times | Crystal violet, 30 seconds
Gram's iodine, 60 seconds
95 percent ethanol, 15–25 seconds
Basic fuchsin, 20 seconds |
> | Positive reaction | Dark blue (blue-violet) cells |
> | Negative reaction | Light red cells |

Determining the Gram-staining characteristics of an unknown microorganism

Review the Gram-stain procedure and summary.

1. **Randomly select one numbered tube** from the rack of unknown cultures.

2. **Record that code number** where indicated on your Results form for this procedure.

3. **Obtain a known gram-positive and a known gram-negative culture.**

4. **Place a pen mark *(B)* at one end of the slide** near the edge. This will serve as a focusing aid. Heat-fix the mark to the slide. If possible, avoid touching it with stains or decolorizing agents.

5. **Prepare three smears on one slide,** as shown in Figure 7-2.

6. **Gram stain all three smears simultaneously.** Make sure that the dyes, mordant, and decolorizing agent cover all three smears for the same length of time at each step in the staining procedure.

7. **Examine each smear carefully** with the oil-immersion lens. Assess the gram reaction of the standard cultures and your unknown. Make other slides if the standard cultures do not appear as they should.

8. **Record your final results and analysis** in the Results section for this part of Exercise 7.

9. **Write your name on your best slide above** the gram-positive smear. Carefully blot off the excess oil with a paper towel.

10. **Tape this slide** to the bottom of the Result section. One small piece of masking tape on each end of the slide should be sufficient. This is done so that your instructor can microscopically analyze your results if necessary.

11. **Turn in the completed report form** to your laboratory instructor.

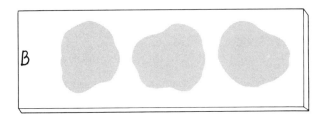

| Gram (+) standard | Unknown | Gram (−) standard |

FIGURE 7-2 Placement of the focusing-aid mark *(B)* and three bacterial smears on the same side of a single glass microscope slide for determining the Gram morphology of an unknown bacterium.

7 THE GRAM STAIN: A DIFFERENTIAL STAIN

NAME _____

DATE _____

SECTION _____

RESULTS: THE GRAM STAIN

Pseudomonas fluorescens
Magnification: _____ ×
Gram reaction: _____
Cell morphology: _____

Bacillus cereus
Magnification: _____ ×
Gram reaction: _____
Cell morphology: _____

* Unknown Culture A
Magnification: _____ ×
Gram reaction: _____
Cell morphology: _____

* Unknown Culture B
Magnification: _____ ×
Gram reaction: _____
Cell morphology: _____

* If you used cultures isolated from Exercise 14 as unknowns, also record these observations in the Results section of Exercise 14!

CONCLUSIONS

7 THE GRAM STAIN: A DIFFERENTIAL STAIN

NAME _____

DATE _____

SECTION _____

RESULTS: GRAM STAINING AN UNKNOWN MICROORGANISM

Code number of your unknown: _____

Appearance of unknown culture
(exaggerate size for detail)

Gram-stain reaction of your unknown:

Cellular morphology (shape and arrangement):

Notes on appearance of Gram-stained standard cultures:

Blot off all excess oil with a paper towel; then tape your best slide below.

QUESTIONS

Completion

1. As far as decolorization is concerned, the purpose of the gram-positive known culture is to _____, and the purpose of the gram-negative known culture is to _____.

2. If gram-positive cells are decolorized too long, they appear to be colored _____ after completing the Gram stain. If gram-negative cells are decolorized too long, they are colored _____.

3. If gram-negative cells are not decolorized long enough during the Gram stain, they are the color _____. If gram-positive cells are not decolorized long enough, they are the color _____.

True – False (correct all false statements)

1. *Pseudomonas fluorescens* is a gram-positive rod. _____

2. *Bacillus cereus* is a gram-negative coccus. _____

3. The Gram stain is one of the last tests performed in identifying a pure culture. _____

4. Smears should always be heat fixed before they are Gram stained. _____

8

The Acid-Fast Stain: A Differential Stain

Acid-fastness is characteristic of only a few types of bacteria. These bacteria have **cell walls** containing unusually large amounts of **lipid** (up to 60 percent of the wall's dry weight). These walls (often referred to as **waxy**), are **hydrophobic** (water repelling) and form relatively impermeable barriers to stains and other chemicals in aqueous solution. For example, acid-fast bacteria are unusually resistant to acid and alkaline solutions. Unusual methods must be used to force the stain through these lipid-rich walls, and, once inside, it is more difficult to remove than it is from other types of bacteria.

The acid-fast staining procedure begins with a strong, red, basic dye called **basic fuchsin.** This is the same red dye we used as a counterstain for the Gram stain (Exercise 7). This dye is purchased as a powder and dissolved in an aqueous solution containing phenol. The resulting staining solution is called **carbol fuchsin,** which you apply to heat-killed bacteria on the microscope slide. Then you heat the slide so that the stain steams for 5 to 10 minutes. If you do not heat the stain, it will not easily penetrate the waxy wall of the acid-fast bacteria.

After heating the slide, you cool it, wash it with water, and decolorize the cells with **acid alcohol.** Acid alcohol is 95 percent ethyl alcohol (in water) to which you add 2.5 percent (weight per volume) of either nitric or sulfuric acid. All but acid-fast bacteria are decolorized by acid alcohol. Although the biochemical mechanism is not precisely known, it is obvious that these waxy-walled bacteria are more slowly decolorized by acid alcohol than other bacteria. Thus such cells are called *acid-fast.*

Cells that are not acid fast must be counterstained if they are to be seen with the brightfield microscope, because they are decolorized with acid alcohol. Any dye works as a counterstain as long as it contrasts with the red fuchsin in the stained acid-fast cells. Most microbiologists use **methylene blue** for this purpose (see Exercise 4).

Most acid-fast bacteria are placed in the genus *Mycobacterium.* The acid-fast stain is important because it is used for diagnosing two diseases caused by species of mycobacteria: (1) **Tuberculosis,** caused by the bacterium *Mycobacterium tuberculosis,* still affects a significant number of people in the United States, especially now that acquired immune deficiency syndrome (AIDS) is causing infected people to be more susceptible to this disease; and (2) **Leprosy,** caused by *Mycobacterium leprae,* which is not common in the United States but very common in other parts of the world.

Cells of the genus *Mycobacterium* are rod-shaped, from 0.2 to 0.6 μm in diameter, and from 1.0 to 10 μm long. *Mycobacterium* rods are sometimes branched, especially when first isolated from a diseased animal. Long filamentlike chains of cells may also be formed in liquid media, but these are easily separated by shaking. Growth of *Mycobacterium* species is often slow, but species vary widely in their rates of growth.

Mycobacterial cultures do not stain well with the Gram stain because of their resistance to penetration by the Gram-staining reagents. Therefore, microbiologists do not refer to them as either gram-positive or gram-negative. However, when these cultures are sectioned and examined with the electron microscope, their cell walls look more like a gram-positive type than a gram-negative type.

LEARNING OBJECTIVES

• Learn how to perform the acid-fast stain.

• Understand the theory of why *Mycobacterium* species are acid fast.

• Learn to differentiate acid-fast bacteria and those that are not acid fast.

MATERIALS

Cultures *Mycobacterium smegmatis* (*not* a pathogen)
Bacillus subtilis (a young nonsporulating culture)

Media None

Supplies Acid-fast staining reagents
Crystal violet
Paper towels, staining racks, forceps, and bunsen burners
Demonstration slides (if available)

PROCEDURE

Review how to prepare culture smears and how to fix smears on microscope slides in Exercise 5.

1. **Prepare two slides** so that each has one smear of *M. smegmatis,* one smear of *B. subtilis,* and one smear that contains a mixture of *M. smegmatis* and *B. subtilis,* as shown in Figure 8-1.

   If you prepare the mixed smear from broth cultures, transfer a drop (loopful) of one culture to the slide with your loop, but do not smear it. Flame the loop, and then place a drop of the second culture on top of the first drop. Mix these two drops with your loop, and then smear them on the slide. Flame your loop before setting it down.

   If you prepare the mixed smear from slant cultures, transfer a loopful of tap water to the slide. Flame the loop, and then transfer a very small amount of one culture to the water drop. Mix these cells in the water, but do not smear this drop. Flame the loop again, and then transfer a very small amount of the second culture to the drop containing the first culture. Mix this second culture in the drop, and then spread the drop over a larger area to make the smear. Flame your loop before setting it down.

   Remember to make these smears *very thin.*

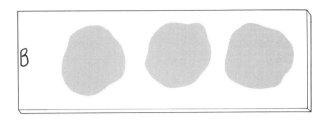

B. subtilis   M. smegmatis   B. subtilis and M. smegmatis

FIGURE 8-1 Placement of the focusing-aid mark (B) and three bacterial smears on the same side of glass microscope slides to determine either the cellular morphology (as defined by crystal violet) or the acid-fast characteristics of *Bacillus subtilis* and *Mycobacterium smegmatis.*

2. **Air-dry and fix these smears with heat or methanol.**

3. **Acid-fast stain one slide** as follows:
   a. **Place this slide on a tall staining rack** that has a large tray underneath (if available). This rack should be high enough to hold slides while they are being heated and also catch any spilled stain in the tray below.
   b. **Flood the entire slide with carbol fuchsin, and place a cut piece of paper towel over the smear.** The paper towel should be saturated with stain because its purpose is to keep the smear thoroughly coated with liquid stain during the heating process. The paper should be cut smaller than the slide so that the stain does not spill over the edge.

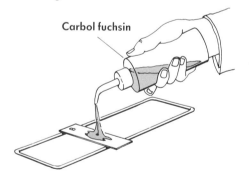

CONSERVATION NOTE

Carbol fuchsin is very hard to remove from fingers and can be removed from clothing only with scissors. You may want to wear old clothing or bring a laboratory coat.

8 THE ACID-FAST STAIN: A DIFFERENTIAL STAIN

c. **Begin to gently heat the underside of the slide with a bunsen burner.** Use a small flame, and pass it back and forth under the slide until the stain begins to steam.

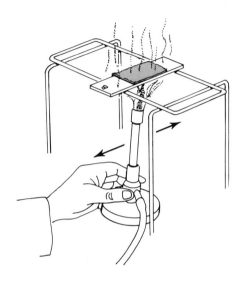

d. **Keep the stain steaming for 5 to 10 minutes.** To do this, you must periodically apply heat while keeping the towel saturated with stain. *Do not let the toweling dry during this staining period.*

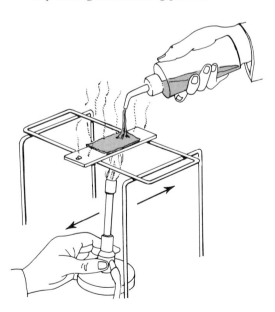

e. **Carefully remove the paper towel with a forceps.** Hold another paper towel under the stain-soaked paper, and discard both in a waste basket.

f. **Wash the slide gently with water** until all excess stain is removed. Blot off excess water.

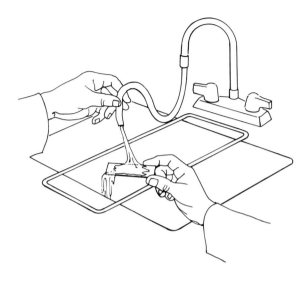

g. **Decolorize in acid alcohol for 20 to 30 seconds.** To accomplish this, first flood the slide with acid alcohol. Then raise one side of the slide with a forceps and drip acid alcohol over the slide for the remaining time or until no more stain comes off.

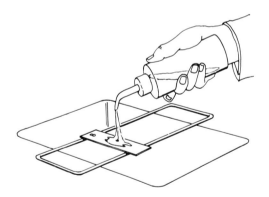

h. **Wash with water** to stop the decolorization. Blot off excess water.

i. **Counterstain with methylene blue for 60 seconds.**

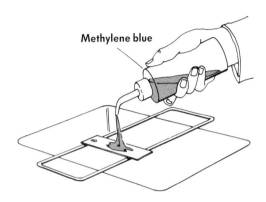

j. **Wash with water, blot, and air-dry.**

4. Simple stain the second smear with crystal violet only for 30 seconds.

5. Examine both slides with the oil-immersion objective.

6. Examine the demonstration slides.

7. Record your observations.

ACID-FAST STAIN SUMMARY	
Test name	Acid-fast stain
Cultures used	*Mycobacterium* species are acid fast.
Reagents and times	Carbol fuchsin, 5–10 minutes Acid alcohol, 20–30 seconds Methylene blue, 60 seconds
Positive reaction	Red cells
Negative reaction	Light blue cells

8 THE ACID-FAST STAIN: A DIFFERENTIAL STAIN

NAME _____

DATE _____

SECTION _____

RESULTS: ACID-FAST STAINS

Use colored pencils to illustrate the appearance of these cells. Verbal descriptions may also be appropriate.

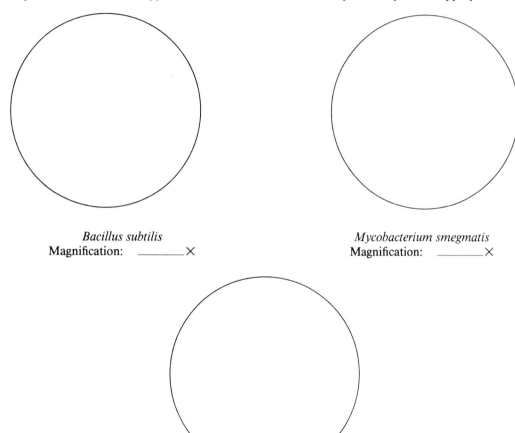

Bacillus subtilis
Magnification: _____ ×

Mycobacterium smegmatis
Magnification: _____ ×

Bacillus subtilis and
Mycobacterium smegmatis
(exaggerate size for detail)

CONCLUSIONS

III TECHNIQUES FOR DETERMINING MICROBIAL TYPE

RESULTS: SIMPLE STAINS

Use colored pencils to illustrate the appearance of these cells. Verbal descriptions may also be appropriate.

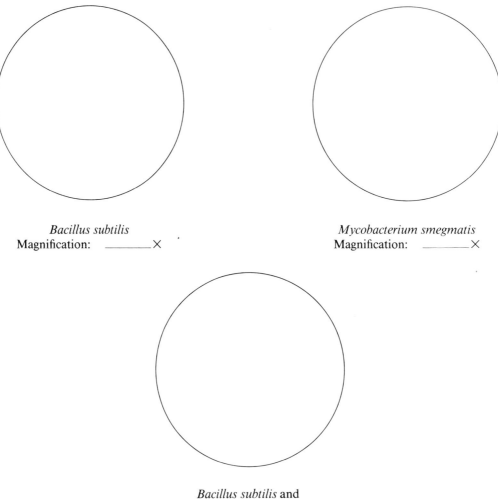

Bacillus subtilis
Magnification: _____ ×

Mycobacterium smegmatis
Magnification: _____ ×

Bacillus subtilis and
Mycobacterium smegmatis
(exaggerate size for detail)

CONCLUSIONS

8 THE ACID-FAST STAIN: A DIFFERENTIAL STAIN

NAME _____

DATE _____

SECTION _____

QUESTIONS

Completion

1. The primary dye used for the acid-fast stain is a solution containing phenol and _____. For decolorization, this staining method uses a mixture of _____ and _____.

2. _____ is an acid-fast genus of bacteria that includes a species that causes tuberculosis.

3. Two morphological features that are characteristic of this genus are _____ and _____.

4. After proper staining using the acid-fast method, *M. tuberculosis* should exhibit a _____ color and non-acid-fast bacteria should be stained a _____ color.

True – False (correct all false statements)

1. Acid-fast bacteria have cell walls that structurally look like those in gram-positive cells. _____

2. The decolorization step in the acid-fast stain removes the dye forced into non-acid-fast cells by heating. _____

3. The acid-fast stain is important in diagnosing tuberculosis and pleurisy. _____

Bacterial Endospores

Terms

An **endospore** is a structure that is formed inside *(endo-)* certain types of bacteria. Figure 9-1 illustrates some of the terms used to describe the location of procaryotic endospores in the cells. A cell that lacks an endospore is called a **vegetative cell.** Once a mature endospore is formed, the rest of the cell surrounding it is called the **sporangium** (singular; plural, sporangia). Various types of endospore-forming bacteria, usually rod-shaped, characteristically form endospores at particular locations in the rod. Only the **central** and **terminal** locations are shown in the figure. After forming an endospore, the sporangium may die and eventually disintegrate (cell **lysis**). If the resulting spore retains no visible evidence of the sporangium, it is called a **free spore.** Endospores and free spores are large enough to be seen (resolved) with a good light microscope.

Taxonomy and identification

There are five currently known genera of bacteria that form endospores. Most are rod-shaped and gram-positive. A key to the endospore-forming bacteria is in Figure 9-2. Of the five genera shown, *Bacillus* and *Clostridium* are the most commonly encountered by the beginning microbiology student. Microbiologists frequently refer to all spores formed inside these bacteria as endospores, regardless of whether they are observed as free spores or as surrounded by a sporangium.

It is possible to use endospore formation as a characteristic to help identify an unknown bacterium. Obviously, if you examine an unknown bacterium and find endospores, this tells you that the bacterium is probably one of the few spore-forming bacteria shown in Figure 9-2. However, the *absence* of endospores does not necessarily mean that the unknown bacterium is not one of these genera. It may simply mean that the culture is too young and has not yet formed endospores, or it may mean that you have not provided the correct nutrition for endospore formation. For example, cultures do not form endospores during rapid growth, which is another way of saying that bacteria produce spores only at the end of exponential growth. Also, cells need the correct nutrients before they form endospores. If critical nutrients are lacking, endospores are not formed. Therefore, cells may have the genes necessary for endospore formation yet not express them. Many investigators find it difficult to produce sporulating bacterial cultures consistently.

Physical properties of endospores

From your lectures and textbook, you will learn that endospores are very resistant to environmental stresses that kill bacterial cells. Bacterial endospores are resistant to extreme desiccation, high temperatures, ionizing radiation, and penetration by chemicals.

As an endospore matures inside the sporangium, it becomes **desiccated** (loses water); so mature endospores contain much less water than vegetative cells. It is thought that the components of a mature endospore are in a very dry, almost crystallinelike state, even when suspended in a fluid. Since spores in this state totally lack metabolic activity, it is not surprising that they resist drying (desiccation) by the environment. These properties are useful in preparing mature endospores for long-term storage in the microbiology laboratory. You can quick-freeze a preparation by immersing a glass vial containing a spore suspension in liquid nitrogen and then freeze-dry **(lyphophilize)**

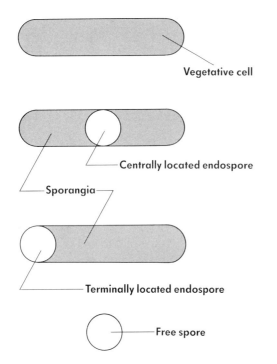

FIGURE 9-1 Terms associated with endospore-forming bacteria.

the frozen block. Properly prepared and stored, endospores maintain their viability indefinitely.

Perhaps the most significant endospore characteristic, from a microbiologist's standpoint, is its resistance to **high temperatures.** Scientists believe that this resistance is related to the dry, crystallinelike state of the mature endospore. Endospore suspensions can be heated to 63°C (145°F) for 35 minutes (one type of pasteurization condition) without killing the endospores. Some types of endospores even withstand boiling water (100°C, or 212°F) for extended periods of time. On the other hand, these amounts of heat kill vegetative cells and non-spore-forming bacteria. The ability of spores to resist high temperatures is of particular concern to those preparing and storing food and to those who need sterile materials and environments to protect public health.

Rod-shaped cells:
    Genus I   *Bacillus*
    Genus II *Sporolactobacillus*
    Genus III *Clostridium*
    Genus IV *Desulfotomaculum*

Spherical-shaped cells:
    Genus V *Sporosarcina*

FIGURE 9-2 Morphological key to common endospore-forming bacteria.

The vegetative cells of *Bacillus* and *Clostridium* are only moderately resistant to **ionizing radiation,** but their spores are extremely resistant. This resistance is probably due to the spore's low water content, which appears to reduce the efficiency of ionization events (see Exercise 39).

Mature endospores also resist **penetration of chemicals** that inhibit growth or kill vegetative cells and non-spore-forming bacteria. For example, endospores are quite resistant to many antiseptics and disinfectants (see Exercise 43), a characteristic that concerns those involved in public health. Scientists believe that this resistance to chemical penetration is due to the presence of several impenetrable layers **(coats)** of protein that surround the mature spore.

Significance of endospore-forming bacteria

Bacteria capable of endospore formation are important to us in many ways. For example, some are helpful because they produce toxins that kill insects; endospores from these bacteria are currently being mass-produced and used as **bioinsecticides** that are harmless to humans. The spores germinate inside the insect, and the resulting vegetative cell produces a toxin that kills the insect. Other types of spore-forming bacteria produce **toxins** that are poisonous to humans. Still other spore-forming bacteria may infect wounds and cause diseases such as **tetanus** and **gangrene.**

Endospore-forming bacteria are common contaminants on Petri dishes or in other types of laboratory cultures. This is because spores survive most natural stresses and are commonly found on airborne dust particles.

Because endospores are so common and because they have such notable effects, microbiologists and other health professionals go to extremes to render solutions and other materials sterile (see Exercise 12). Moreover, microbiologists must be able to recognize endospore formation in microbial cultures whenever it occurs.

LEARNING OBJECTIVES

• Learn the appearance of mature endospores, surrounding sporangia, and vegetative cells when observed with the phase-contrast microscope.

• Demonstrate the resistance of endospores to the penetration of basic dyes in the absence of elevated temperatures, and learn to recognize spores in simple-stained or Gram-stained preparations.

- Demonstrate the stained appearance of endospores after applying the spore structural stain that uses heated malachite green.

A · Observing endospores with a phase-contrast microscope

The mature endospore has a **greater density** (mass per unit volume) than the surrounding sporangium or vegetative cells. This property makes the endospores look very bright **(refractile)** when cultures containing endospores are observed using a wet-mount preparation and a brightfield light microscope. The refractility of the spores makes them easier to see, but it is very difficult to see the sporangium surrounding the endospore with a brightfield microscope because the sporangium density is very similar to the that of the suspending fluid.

That problem is overcome by a phase-contrast microscope, which accentuates the differences in density between the sporangium and the suspending fluid and increases the contrast between them. In addition, you can easily see the endospore in the sporangium. The phase-contrast microscope is especially helpful because staining is not necessary; one need only prepare a wet mount and observe the cells directly in their growth medium. This avoids shrinking and distorting the cells during the staining process.

MATERIALS

Cultures *Bacillus sphaericus* grown on nutrient sporulation medium (NSM)
Lactobacillus plantarum

Media None

Supplies Phase-contrast microscope
Possible demonstrations: wet mounts of both sporulating and nonsporulating culture, separately observed with phase-contrast microscopes

PROCEDURE

1. **Obtain or prepare two clean microscope slides** and **two cover slips.**
2. **Prepare a wet mount of *B. sphaericus*** (see Exercise 6) **and observe** this preparation with a phase-contrast microscope.
3. **Sketch your observations** in the Results section (A) following this exercise.
4. **Prepare a wet mount of *L. plantarum* and observe** this preparation with a phase-contrast microscope.
5. **Sketch your observations** in the Results section.
6. **Discard these slides and coverslips and clean the phase-contrast microscope** as directed.

B · Observing endospores with simple stains and the Gram stain

The great resistance of the mature procaryotic endospore to the penetration of chemicals (including dyes) is a property that can be used for detecting them. Dyes, such as those used for simple stains (see Exercise 5), do not penetrate endospores under normal staining conditions. After you perform a simple stain, the sporangium is colored and the endospore appears because of the absence of dye. Stained cells look very much like those shown in Figure 9-1; that is, they look like a dyed cell surrounding a colorless endospore. If your laboratory lacks a phase-contrast microscope, you can use this principle to observe endospores in a culture.

After you perform a Gram stain, cells that contain endospores appear similar to those that have been stained with simple basic dyes because all dyes used for the Gram stain are basic dyes, and these do not penetrate the endospores. The only advantage of a Gram stain is that you can check for two things at once: the presence of endospores (noticeable because of dye exclusion) and the gram characteristics of the sporangium or vegetative cell.

Although both the simple stain and the Gram stain enable you to visualize endospores, neither is considered to be a differential spore stain because neither stains the endospore.

MATERIALS

Cultures *Bacillus sphaericus* grown on nutrient sporulation medium (NSM)
Lactobacillus plantarum
Culture for gram-negative standard

Media None

Supplies Basic fuchsin (or other basic dye)
Gram-staining reagents
Possible demonstrations: sporulating and nonsporulating cultures stained with a single basic dye

PROCEDURE

1. **Obtain or prepare two clean microscope slides.**
2. **Prepare two smears on one slide,** so that one smear contains the *B. sphaericus* culture, and the other smear contains the *L. planatarum* culture.

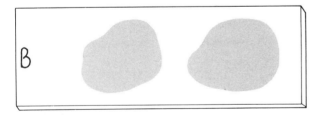

B. sphaericus   L. plantarum

3. **Simple stain the slide that contains two smears** according to the method described in Exercise 5.
4. **Prepare three smears on another slide,** so that the middle of the slide contains a smear of the *B. sphaericus* culture. To the left of the center smear, make a smear of the *L. plantarum* culture. To the right of the center smear, make a smear of the gram-negative control culture.

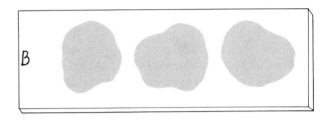

L. plantarum   B. sphaericus   Gram-negative control culture

5. **Gram stain the slide that contains three smears** according to the method described in Exercise 7.
6. **Observe both slides** with a brightfield microscope.
7. **Record your observations** in the Results section (B) following this exercise.
8. **Discard these slides** and **clean the microscope** as directed.

C · Structural staining of endospores with hot malachite green

No dye will penetrate a bacterial endospore under normal environmental temperatures. There are dyes, however, that you can *force* into the spore by raising the dye temperature to about 100°C (212°F). One such dye is **malachite green,** whose chemical structure is given in Figure 9-3. Malachite green is a weak basic dye; that is, it has a weak positive charge and will not bind strongly to the surface or interior of the cell. The

FIGURE 9-3 Structure of malachite-green chloride, a weak basic dye commonly used for the endospore staining procedure. See Exercise 5 for a review of basic dyes and types of stains.

dye easily penetrates bacterial cells but is also easily washed out of cells with water.

The spore, however, is another matter. If you apply malachite green to the smear and boil it for a period of time, the heat forces the dye into the endospores. Once the dye is inside the endospore, you cannot easily remove it, even though you can easily wash it out of sporangia and vegetative cells. Thus, after you stain cells with hot malachite green and then cool and rinse them, the spore remains green and the sporangium becomes colorless.

To be able to see a stained endospore clearly inside a sporangium with a brightfield microscope, the sporangium must be *counterstained* with a dye of a contrasting color. Red dyes such as **safranin** or **basic fuchsin** are commonly used. The result of counterstaining is a green endospore inside a red sporangium.

This staining procedure was first developed by **Schaeffer and Fulton** and modified in 1950 by Bartholomew and Mittwer. It produces a colorful result and is an excellent example of structural staining.

In laboratory practice, however, the malachite-green staining procedure is seldom used. Other procedures for observing spores are less time-consuming and less messy. For example, the use of a wet mount and the phase-contrast microscope is a quick and easy way to detect endospores and free spores. If only a brightfield microscope is available, most experienced microbiologists rely on the Gram stain or a simple stain to help them detect endospores.

MATERIALS

Cultures *Bacillus sphaericus* grown on nutrient sporulation medium (NSM)
Lactobacillus plantarum

Media None

Supplies Staining solutions: malachite green and basic fuchsin
Bunsen burner and staining rack
Cut paper toweling and forceps
Possible demonstrations: sporulating and nonsporulating cultures stained for spores with malachite green and counterstained with basic fuchsin

PROCEDURE

1. **Obtain or prepare one clean microscope slide.**
2. **Prepare two smears on this slide** so that one smear contains the *B. sphaericus* culture and the other contains the *L. plantarum* culture.
3. **Air dry and heat fix** this slide.
4. **Prepare a malachite-green spore stain** using the following procedure:
   a. **Place the slide on the staining rack.**
   b. **Flood the smear with malachite green, and place a cut piece of paper toweling over the smear** so that it completely covers the smear and soaks up most of the stain. This piece of cut toweling should be slightly smaller than the slide. The toweling should now appear saturated with stain, and it should remain that way throughout the staining process. The procedure from here on is similar to that used in preparing an acid-fast stain (see Exercise 8).

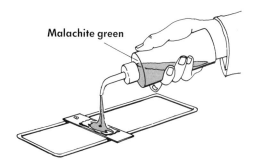

CONSERVATION NOTE

Malachite green is very difficult to remove from hands and fingers. It can be removed from clothing only with a scissors. You may want to wear old clothes or bring a laboratory coat to the laboratory period when this exercise is performed.

c. **Begin to gently heat the underside of the slide with a bunsen burner.** Use a small flame, and pass it back and forth under the slide until the stain begins to steam.

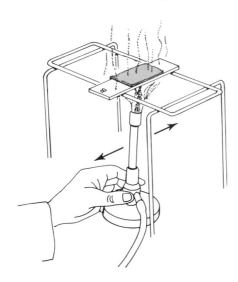

d. **Keep the stain steaming for a full 5 minutes.** To do this, you must periodically apply heat and add stain to keep the toweling saturated. Do not allow the stain to boil or the paper toweling to dry. *Keep the towel saturated with stain.*

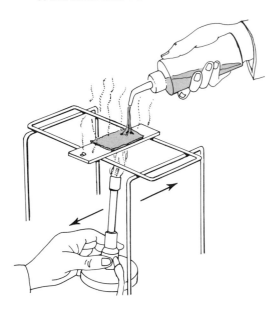

e. **Carefully remove the cut piece of paper towel with a forceps.** Hold a clean paper towel underneath the stain-soaked piece, and transfer both to the waste basket.

f. **Wash the slide gently with water until all excess stain is removed.**

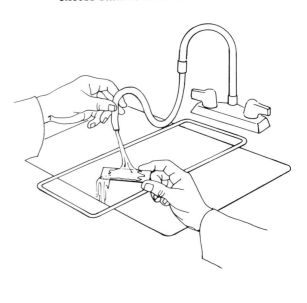

g. **Counterstain with basic fuchsin for 30 seconds.**

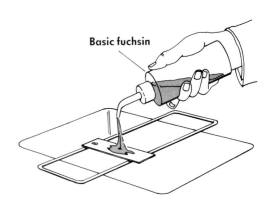

h. **Wash gently with water, blot, and air dry.**

5. **Examine this slide using the oil-immersion objective.**

6. **Record your results** in the Results section (C) following this exercise.

7. **Compare your results with a demonstration slide.**

8. **Discard your slides and clean your microscope** as directed.

9 BACTERIAL ENDOSPORES

SPORE STAIN SUMMARY	
Test name	Spore stain
Cultures used	*Bacillus* or *Clostridium* species
Reagents and times	Malachite green, >5 minutes Basic fuchsin, 30 seconds
Positive reaction	Green-colored spores alone or inside red-colored cells
Negative reaction	No green structures present

9 BACTERIAL ENDOSPORES

NAME _____

DATE _____

SECTION _____

RESULTS A: OBSERVING ENDOSPORES WITH PHASE-CONTRAST MICROSCOPE

Draw the appearance of the cells that you observed with the phase-contrast microscope so that these cells appear larger than they actually appeared with the oil-immersion objective. Label the drawing of the sporulating culture to show examples of endospores, free spores, vegetative cells, and sporangia.

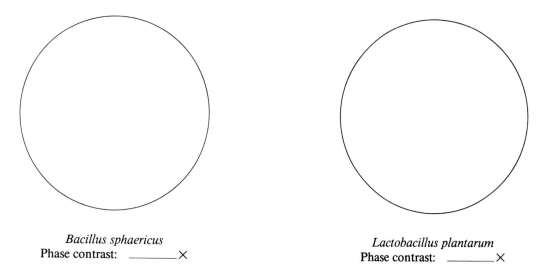

Bacillus sphaericus
Phase contrast: _____ ×

Lactobacillus plantarum
Phase contrast: _____ ×

Bacillus sphaericus is described by Bergey's Manual as forming spherical endospores, located in the terminal part of the cell, and usually distending the sporangium. How does this description compare with your observations?

CONCLUSIONS

RESULTS B: OBSERVING ENDOSPORES WITH SIMPLE AND GRAM STAINS

Use colored pencils to draw the appearances of cells from both cultures so that these cells appear larger than actually observed with the oil-immersion objective. Label the drawing to show endospores, free spores, vegetative cells, and sporangia.

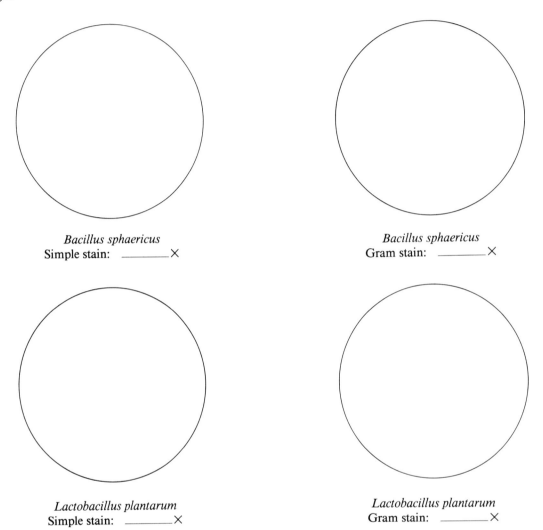

Bacillus sphaericus
Simple stain: _____×

Bacillus sphaericus
Gram stain: _____×

Lactobacillus plantarum
Simple stain: _____×

Lactobacillus plantarum
Gram stain: _____×

CONCLUSIONS

NAME _____

DATE _____

SECTION _____

RESULTS C: STRUCTURAL STAINING OF ENDOSPORES

Use colored pencils to draw the appearances of cells from both cultures so that the cells appear larger than actually observed with the oil-immersion objective. Label the drawing to show endospores, free spores, vegetative cells, and sporangia.

Bacillus sphaericus
Spore stain: _____ ×

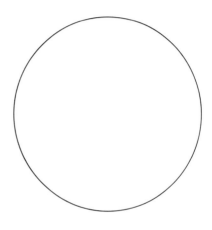

Lactobacillus plantarum
Spore stain: _____ ×

CONCLUSIONS

QUESTIONS

Completion

1. One weak basic dye that can be forced into mature endospores by applying heat is called _____.

2. That part of the cell surrounding an endospore is called the _____.

3. Two bacterial genera that form endospores are called _____ and _____.

4. Two types of antimicrobial chemicals that do not adversely affect endospores are _____ and _____.

True – False (correct all false statements)

1. All bacterial cells have the genetic ability to form an endospore. _____

2. Endospores resist the penetration of dyes at room temperature. _____

3. The endospore stain is an example of a differential stain. _____

4. A Gram stain does not allow detection of spores if they are present. _____

5. Spores cannot be detected with a simple stain. _____

6. Spores are resistant to autoclaving. _____

7. Spores are resistant to pasteurization. _____

10

Bacterial Flagella and Motility

The function of flagella for procaryotes is to propel the cell through its aqueous environment just as cilia and flagella do for eucaryotic cells. The movement caused by the action of flagella is called **motility**. Motility *toward* a favorable environment (such as one containing nutrients) is called **positive chemotaxis,** whereas motility *away* from an environment that is less favorable or possibly harmful (such as one containing toxic substances) is called **negative chemotaxis.** Microbiologists now know that this behavior is caused by specific proteins within the plasma membrane known as **chemoreceptors.** Thus cells that have flagella and chemoreceptors probably have an advantage over other microbes because the environment of most microbes is constantly changing.

Bacterial flagella are long, narrow strands that are coiled into *rigid* helices (spirals) that revolve around their points of attachment and push against the surrounding medium. Their length is often many times the length of the cell, but their diameters range from about 0.01 to 0.05 μm, about 100 times smaller than the width of most bacteria and about 10 times smaller than the *limit of resolution* for the light microscope. Therefore, these structures are too small to be seen with the light microscope.

In 1954, Einar Leifson published a staining method that made flagella visible in the light microscope. His stain uses **tannic acid,** which precipitates around the flagella and serves as a mordant for a red dye, basic **fuchsin.** The bound dye makes the flagella appear larger so that they can be resolved with the light microscope. The development and application of this stain greatly advanced the study of bacterial flagella and affected bacterial classification.

Leifson's publication inspired much research on the flagellation characteristics of bacteria. For a time, microbiologists believed that all bacteria fit into two categories: cells with flagella inserted only at the ends of a cell (called **polar flagellation**) and cells with flagella inserted all over the surface (called **peritrichous flagellation**). Much weight was placed on which type of flagellation a bacterium exhibited. For example, cells were placed in one major taxonomic group or another according to the arrangement of flagella. Although the type of flagellation exhibited by a bacterium is still considered somewhat important, recent evidence suggests greater variation in flagellar attachment in a single species than was once believed. This has led microbiologists to place less emphasis on flagellar arrangement in bacterial taxonomy.

Motility is an important microbial activity. Therefore, you should understand something of the nature of the flagellum and the consequence of its action.

LEARNING OBJECTIVES

• Understand the principles of staining and observing flagella with the light microscope.

• Be able to recognize stained flagella in the light microscope.

• Learn how to recognize motility in the phase-contrast or brightfield microscope.

• Know how to use soft-agar deeps to exhibit motility by facultatively anaerobic microorganisms.

A · Structural staining of flagella

Unless certain precautions are followed, this staining method can be difficult to perform. So many microbiologists have had trouble with Leifson's flagella-staining method in the past that it is known as the most difficult of all the stains to perform correctly. But, if the staining solution is prepared correctly and if proper care is taken with the staining procedure, Leifson's flagella stain can be accomplished by the beginning microbiology student.

Today, the electron microscope is more often relied on for determining the presence of flagella and their arrangements on cells than is Leifson's flagella stain. However, not every laboratory has access to an electron microscope; so the flagella stain may still be important in helping to classify unknown bacterial cultures.

Many things can go wrong with Leifson's procedure. For example, flagellar attachment is not very stable, and flagella are easily broken off. Detached flagella aggregate in clumps that look like locks of wavy hair and are often missed when the investigator examines the slide. Another problem lies in the chemical composition of the staining ingredients. For example, batches of dye may vary sufficiently to cause an altered affinity for the flagella, or slight errors in measuring ingredients may make the stain more or less effective. Another problem relates to the chemistry of the flagella themselves. All bacterial flagella are composed of a protein called **flagellin,** but not all molecules of this protein are exactly the same. For example, some contain different combinations or different arrangements of amino acids within the protein molecule. Each bacterium makes its own unique flagellin proteins, and some may have a greater affinity for the stain than others. Finally, the concentration of cells on the slide, the amount of stain remaining on the slide, and the cleanliness of the slide all must be perfect before the cells and their flagellar patterns can be seen.

On the other hand, we have confidence in your abilities, and we are sure that you can successfully stain bacterial flagella *if you follow these exact procedures.*

MATERIALS

Cultures   *Proteus mirabilis* (peritrichous flagella)
           Pseudomonas aeruginosa (polar flagella)

Media   None

Supplies   Leifson's flagella stain
           New microscope slides
           Wax marking pencils
           Pasteur (disposable) pipettes (nonsterile)
           Demonstration slides (if available)

PROCEDURE

1. **Select a new microscope slide from the box and handle it only by its edges.** Note that slides for this stain must be absolutely clean. This is because any organic material, including your fingerprints, will take up the dye. Your laboratory instructor may suggest that you clean this slide even though it is new.

2. **Heat one side with the bunsen burner, and use a wax pencil to draw lines on the same side of the slide.** Make sure to leave about one inch on one end of the slide so that you will have a place to hold on to the slide. The purpose of the wax pencil lines is to form an impenetrable barrier that will retain the stain.

    Your laboratory instructor will have previously, and *very gently,* added sterile distilled water to a slant containing the flagellated culture. This allows the motile, and thus only the flagellated, organisms to swim off the slant into the water.

3. **Touch the tip of a nonsterile pasteur pipette to the surface of the water, and allow this cell suspension to be drawn up into the pipette by capillary action.** This gentle method avoids the shear forces inherent in normal pipetting and will possibly avoid detaching flagella from the cells.

4. **Hold the slide at a 30-degree angle so that the waxed lines are upward and position *A* is raised.**

5. **Place a drop of cell suspension onto position *A*,** and let it run down the slide to position *B*. Keep this slide at the same angle until the film between *A* and *B* has air dried. (You may lean the slide against the side of a test-tube rack to keep the upper part of the film thin to facilitate drying.)

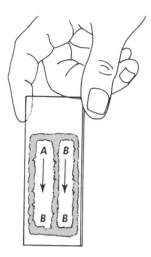

6. **Remove excess fluid from position *B*** by absorbing it with a torn piece of paper towel. When the excess fluid is in the toweling, saturate this piece of towel with disinfectant and discard it with other contaminated materials as directed.

7. **Air dry this slide for at least 10 minutes.** Do something else while you are waiting. *Do not heat fix this slide.*

8. **Add Leifson's stain to the wax-enclosed areas** so that the stain beads up on the surface. The hydrophobic wax line should contain the stain. If it does not, it is probably for one of three reasons: The wax lines were not heavy enough, the wax did not stick to the glass well enough, or the lines were not continuously drawn across the glass surface.

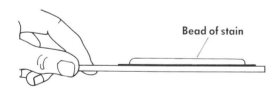

9. **Stain for 10 minutes.** Add stain if necessary to keep the stain beaded up on the slide. As time passes, you will note the formation of a greenish sheen on the surface of the stain. This is a precipitate that forms on the surface as the stain evaporates. *This precipitate must not come in contact with the surface of the slide.* If this happens, it will cover cells and flagella with precipitated stain, which cannot be washed off the slide; your preparation is ruined.

10. **Gently add water with the wash bottle so that the slide is slowly flooded.** The goal is to gently *raise the precipitate* until it falls off the edge of the slide, instead of letting it come in contact with the surface of the slide. Continue washing the surface for two to three minutes; then also wash the bottom of the microscope slide.

11. **Gently remove excess water by blotting, and air dry.**

12. **Examine the stained slide with the oil-immersion objective.** Start in the center of the area that contains cells and work toward the end that contains the least stained background material between the cells. *Spend lots of time with this slide.* It will be worth the effort.

    You will note that some parts of the slide contain cells with no apparent flagella, and other parts of the slide are loaded with flagellated cells. The reason for this is not known.

TECHNICAL NOTE

Leifson's flagella stain contains ammonium sulfate, tannic acid, and basic fuchsin dissolved in ethanol. It can be stored in the refrigerator for 24 to 48 hours, but it is best when freshly made. Basic fuchsin is a dye that you have previously used for a counterstain. It is a strong basic dye, and most bacteria are stained red by the basic fuchsin in Leifson's stain. The precipitate that builds up around the flagella (tannic acid plus basic fuchsin) will also appear red. If the slides are completely free of organic matter, the background around the cells should be unstained or slightly pink with small granules of stain. If the background is uniformly stained a dark red, you probably will not be able to see the flagella because they are buried in the stained organic matter surrounding the cell.

---

13. **Discard the slide in the special container provided for these slides.** It is very difficult to clean these slides because of the wax. Special procedures must be used, and it helps greatly if you keep these slides separate from all others.

14. **Observe demonstration slides of stained flagella** (if available).

B · Observing motility with the phase-contrast or brightfield microscope

Just because bacterial cells have flagella does not mean that they are motile! For example, an old culture may contain flagellated cells that are not motile, either because the cells are dead or because they can no longer produce enough energy from the environment to support motility.

Conversely, bacteria may be actively producing flagella, but the flagella may become detached from the cells by rough handling. In that case, cells may be alive and motile in culture but will not show motility under the microscope.

In addition, not all bacteria exhibit flagella and motility under the same cultural conditions. For example, some bacteria form flagella and exhibit motility best on solid media, whereas other bacteria show flagella and motility better when grown on liquid media. The ability to form flagella is dependent upon growth temperature in some bacteria.

For a bacterial culture to show motility, (1) its cells must have the genetic capacity to produce flagella, (2) it must be young and actively growing in the correct cultural conditions, and (3) it must be treated in such a way that the cells retain their flagella. The easiest and fastest way to observe culture motility is direct examination with the brightfield or the phase-contrast microscope.

MATERIALS

Cultures Live *Rhodospirillum rubrum* or *Spirillum volutans* (spiral-shaped cells; polar flagella)

Live *Proteus mirabilis* (small rods; peritricheous flagella)

Dead *Proteus mirabilis* (killed with formalin; nonmotile)

Media None

Supplies Phase-contrast microscope (preferred)
Brightfield microscope (acceptable)
Microscope slides and cover slips

PROCEDURE

1. **Prepare three clean microscope slides and three cover slips.**
2. **Prepare wet mounts** of either the live *Rhodospirillum* or the *Spirillum* culture and of the live *Proteus mirabilis* culture as described in Exercise 6. If possible, observe these preparations with a phase-contrast microscope.
3. **Prepare a wet mount** of the dead *Proteus mirabilis* culture. If possible, observe this preparation with a phase-contrast microscope.
4. **Describe your observations** in the Results section.
5. **Discard these slides and coverslips and clean the microscope** as directed.

C · Observing the consequence of motility on plates and in soft-agar deeps

You can observe motility indirectly, by watching the mass movement of cells in a culture growing on plates or in semisolid media. For example, a very actively motile bacterial culture *may* rapidly spread itself across the surface of agar plates. This type of culture is called a **spreader**. Species in the genus *Proteus* commonly show this characteristic. These cultures are often a problem for the microbiologist working with mixed cultures because they quickly overgrow other types of colonies on an agar plate, making it very difficult to isolate the other microorganisms. Fortunately, spreading is not characteristic of many microorganisms.

A more common way to observe motility indirectly is to use **soft-agar deeps** (see Exercise 13). Test tubes are partially filled with media that contain about 0.5

percent agar, a common solidifying agent. This results in a semisolid gel that is easily penetrated by actively motile cells. You **stab** the soft-agar deep with an inoculation loop or needle and then examine the culture after an appropriate incubation period. If the microbe is actively motile, you will see turbidity radiating outward from the stabbed line. If it is not actively motile and if it can grow with little or no air, the culture will grow only within the stabbed line. Note, however, that this procedure works only if the bacteria do not require lots of oxygen. The stabbed agar quickly flows together after inoculation. Cells that require lots of oxygen quickly use up all that is available, making further metabolism and growth impossible. Such cells will no longer be motile nor will they grow.

Therefore, stabbed soft-agar deeps serve as a check for motility only if the inoculated microbe is **facultatively anaerobic** (see Exercise 25), and then only if the culture is very actively motile.

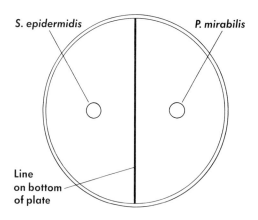

MATERIALS

Cultures *Proteus mirabilis* (motile, facultatively anaerobic)
Staphylococcus epidermidis (nonmotile, facultatively anaerobic)

Media Trypticase-soy agar plates
Trypticase-soy soft-agar deeps

Supplies Inoculating loops and/or needles

PROCEDURE: FIRST LABORATORY PERIOD

1. After reading or reviewing Exercise 16 on how to inoculate agar deeps, **stab inoculate one soft-agar deep** with *Proteus mirabilis*.

2. **Stab inoculate another soft-agar deep** with *Staphylococcus epidermidis*.

3. **Ask your instructor to "sham" inoculate several soft-agar deeps** to serve as uninoculated control tubes. This is done by stabbing sterile soft-agar deeps with a sterile loop or needle. This type of control tube will help you to determine the difference between disturbed agar and growth of nonmotile cultures after inoculation.

4. **Divide a plate into two equal parts,** by drawing a line on the bottom of the plate with a marking pen, and proceed as follows:

   a. **Label each half of the plate** with the name of the organism to be inoculated.
   b. **Inoculate *Proteus mirabilis* in a spot** about 5 mm (1/4 inch) in diameter near the center of one half of the plate.
   c. **Inoculate *Staphylococcus epidermidis* in a spot** about 5 mm (1/4 inch) in diameter near the center of the other half of the plate.

5. **Incubate the tubes and plate at 30°C for 48 hours.**

PROCEDURE: SECOND LABORATORY PERIOD

1. **Describe your observations in** the Results section.

2. **Discard plates and tubes** as directed.

NAME _____

DATE _____

SECTION _____

RESULTS A: STRUCTURAL STAINING OF FLAGELLA

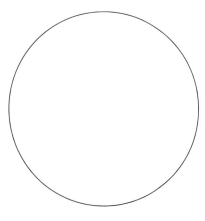

Leifson's flagella stain
Proteus mirabilis
Magnification: _____ ×

CONCLUSIONS

RESULTS B: OBSERVING MOTILITY WITH A MICROSCOPE

1. Verbal description of either the live *Rhodospirillum* or the live *Spirillum* culture:

2. Verbal description of the live *Proteus mirabilis* culture:

3. Verbal description of the dead *Proteus mirabilis* culture:

CONCLUSIONS

10 BACTERIAL FLAGELLA AND MOTILITY

NAME _____

DATE _____

SECTION _____

RESULTS C: OBSERVING THE CONSEQUENCE OF MOTILITY

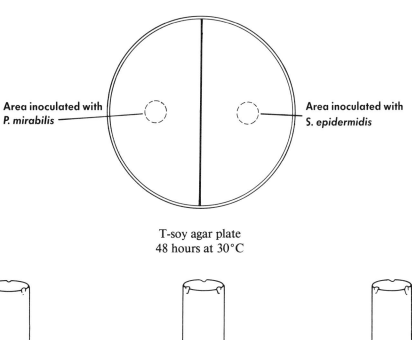

T-soy agar plate
48 hours at 30°C

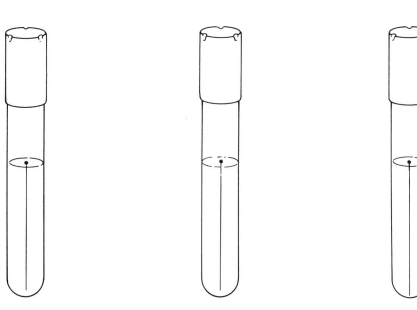

Proteus mirabilis
Soft-agar deep
48 hours at 30°C

Staphylococcus epidermidis
Soft-agar deep
48 hours at 30°C

Sham inoculated
Soft-agar deep
48 hours at 30°C

CONCLUSIONS

QUESTIONS

Completion

1. Simply stated, microbial flagella function to _____.
   Movement of the cell closer to a favorable environment is called _____
   _____. Movement of the cell away from an unfavorable environment is called
   _____.

2. The diameter of a flagellum is about _____ times smaller than the width of the average bacterial cell.

3. _____ is the person credited for discovering the flagella stain.

4. If a microbe has one or more flagella at the ends of the cell only, this is called _____ flagellation. If flagella are randomly distributed around the cell's entire circumference, this is called _____ flagellation.

5. Bacterial flagella are chemically composed of pure _____ (a type of polymer), and the specific name given to that polymer is _____.

6. How should you properly describe the physical appearance of stained flagella as seen in the light microscope?

7. After the flagella-staining procedure, the flagella are colored _____ and the cells are colored _____.

8. Why must great care be taken in handling cultures prior to attempting to stain their flagella?

9. _____ is a word that describes the distribution of flagella on motile *Proteus* cells.

True–False (correct all false statements)

1. All cells have the genetic ability to form flagella. _____

2. Flagella are visible after you stain them because the flagella absorb basic fuchsin. _____

3. The flagella stain is an example of a differential stain. _____

4. A Gram stain does not allow detection of flagella if they are present. _____

5. Flagella cannot be detected with a simple stain. _____

6. Unstained flagella can be seen with the phase-contrast microscope. _____

7. Almost all bacterial motility is a consequence of flagellar activity. Therefore, one could say correctly that almost all motile bacteria have flagella, but not that all flagellated bacteria are motile. _____

Bacterial Capsules: Indirect Observation with India Ink

Many bacteria produce **extracellular polymers** that accumulate around the outer surface of the cell wall. Most of these are *polysaccharides* (long chains of covalently bound sugar molecules). A few bacteria produce extracellular polymers made up entirely of *protein* (long chains of covalently bound amino acids).

Unfortunately, microbiologists use a variety of terms for these extracellular polymers. For example, some use the term **glycocalyx** to refer to a structured mass of extracellular polysaccharide that tightly adheres to the cell wall. Others use the term **capsule** to refer to this same structure. If the extracellular polysaccharide is less organized and not so closely bound to the cell surface (or even released into the growth medium), this may be called a **slime layer.** Other authorities believe that capsules, slime layers, and glycocalyces are too similar to warrant separate names.

Since the term *capsule* is the oldest and apparently the most commonly used, it will be used to refer collectively to all extracellular polymers that are produced by the cell and accumulate (at least in part) around its outer surface.

Not all bacteria have the ability to form capsules. Even when the required genes are present, environmental factors influence the production of capsules. For example, the types of available nutrients can determine how much capsular material a bacterium makes or whether it makes any. Therefore, to examine capsules in the laboratory, you must grow capsule-producing bacteria in appropriate environmental conditions.

Because of their chemical characteristics, capsules are usually not stainable by basic dyes. They are best observed by a procedure called **negative staining.** Two forms of negative staining are used for capsule detection: The first uses India ink and the wet-mount technique; the second uses either India ink or a dye called *nigrosin,* followed by smearing and drying on the slide.

Of these two methods, the wet-mount technique using India ink is preferred because it keeps the cell and capsule well hydrated throughout observation. Most capsules are about 99 percent water; so their structures tend to collapse when they are dried.

India ink is composed of fine carbon particles (less than 0.1 μm in diameter) that are suspended in water and that form a true **colloid** (do not settle out of suspension). These particles are too large to penetrate the gellike matrix of the capsule. When a drop of India ink is mixed with a drop of broth culture, and a wet mount is prepared and observed with the light microscope, the capsules and cells stand out as transparent zones of differing density that are surrounded by a dark (India ink) background. Thus, the capsule and cell are observed indirectly because they exclude the carbon particles (Figure 11-1). However, the wet mount must be *exceedingly thin* for this method to work well.

There are problems with this technique. For example, the thin wet mounts dry very quickly; so you must seal the edges of the coverslip to retard evaporation. Also, when using the brightfield microscope, you will find it hard to distinguish the cells from the capsules that surround them. This latter problem can be alleviated by using phase-contrast optics, in which the cells appear dark and the capsules transparent (Figure 11-2).

The negative stain technique that uses smearing and drying is less desirable because of the drying step. In this procedure, you mix a drop of culture with a

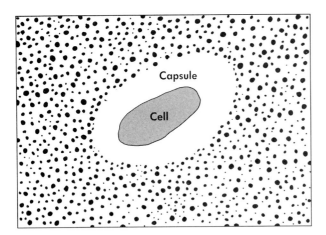

FIGURE 11-1 Schematic appearance of cell, capsule, and India-ink particles (black dots) when brightfield optics are used to observe very thin wet mounts. When phase-contrast optics are used, the cell looks much darker than the capsule that surrounds it.

drop of either India ink or nigrosin and spread the mixture across a glass slide to create a thin smear. After the smear dries, you counterstain it with crystal violet or a similar basic dye and examine it with conventional brightfield optics. The cells stand out because they are stained with crystal violet, whereas the capsules are transparent and easily distinguished from the dark background of ink or nigrosin.

Nigrosin produces results similar to those seen with India ink but for different reasons. **Nigrosin** is a black acidic dye that is dissolved in a solvent before use. Like all acidic dyes, it is repelled by the negatively charged cell and capsule surfaces. As the nigrosin dries, it creates a dark background outside the capsule.

Dried-smear preparations are more permanent than those made with the wet-mount method, and the cells are easier to recognize. However, judging the actual sizes of the capsules is difficult since they may shrink during drying. The cell may also shrink away from the ink during drying, a change that produces false "capsules" around unencapsulated cells. You should include positive and negative controls (that is, known capsulated and unencapsulated strains) in dried-smear capsule stains to help you detect these problems.

Capsules perform a variety of functions for the bacteria that produce them. They may protect the cell from desiccation and act as buffers or osmotic barriers between the cell and its environment. Capsules are also thought to retard the loss of nutrients from the cell and to help scavenge or concentrate nutrients from the environment. Some types of capsules protect the cell from attack by viruses (bacteriophages) and other predators.

Perhaps the most significant and best-documented function of bacterial capsules is to help cells attach to

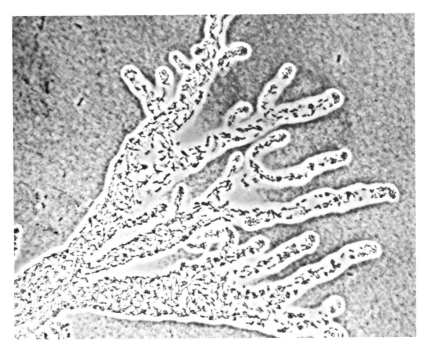

FIGURE 11-2 Phase-contrast light micrograph of a wet mount of *Zoogloea ramigera*, prepared with India ink and the wet-mount method. The capsule material is visible as a trasparent halo around the cells. The dark background is cased by the India ink particles. (Courtesy R. F. Unz, *International J. of Systematic Bacteriology* 21:91–99, 1971.

surfaces. Most bacteria in oceans, lakes, rivers, soil, and similar places attach to solid surfaces (for instance, to rocks in a stream) rather than drift freely in the surrounding water or air. Attachment probably helps these bacteria to survive; nutrients are concentrated at surfaces, and the chemical conditions (ionic composition, pH, and so on) may be more favorable or protective there than in other parts of the environment.

Capsules also facilitate the attachment of bacteria to surfaces on or in plants, animals (including humans), and other eucaryotic organisms. For example, *Streptococcus mutans,* a normal inhabitant of the human mouth, produces a type of polymer (called a **dextran**), with which it can attach to tooth enamel. The attachment of *S. mutans* to the enamel is the first step in the process of plaque formation, the process that eventually leads to **tooth decay.** (Tooth decay occurs later, when *S. mutans* and other microorganisms in the plaque produce acids that erode the tooth enamel.) *Streptococcus mutans* produces dextrans when sucrose (table sugar) is present in its environment. Sucrose is a disaccharide (two sugars covalently bonded) composed of one glucose molecule and one fructose molecule. *Streptococcus mutans* breaks sucrose molecules in half and uses the energy stored in the bond between the two sugars to link the released glucose molecules into a long chain (the dextran). (The fructose half of the sucrose molecule is used for other purposes.) Thus, *Streptococcus mutans* need not expend its own energy to make a dextran, nor does it make dextrans when it grows on carbon sources other than sucrose. For example, artificial sweeteners are not used as an energy source nor as a substrate for dextran capsule formation; therefore, foods that are artificially sweetened do not contribute to plaque formation or tooth decay.

Capsules are also important from a medical standpoint because they enhance the **virulence** of some bacteria; that is, they enhance the ability of these organisms to invade and produce disease in the human body. The capsule of *Streptococcus pneumoniae* (the primary cause of bacterial pneumonia) protects it from **phagocytosis,** the process by which some types of white blood cells engulf and destroy invading microorganisms. Because this organism can overcome this important human defense mechanism, *S. pneumoniae* easily establishes itself in the lungs, where it produces disease. By comparison, *S. pneumoniae* strains that have lost the ability to make their capsule (through mutation) are easily overcome by the body before they can produce pneumonia symptoms.

Bacterial capsule polymers can be used for a number of industrial purposes, especially as emulsifiers in foods, paints, and other products. **Xanthans,** a class of polysaccharide polymers made by the genus *Xanthomonas,* are used to make dripless paints and to enhance the recovery of oil from shales or other petroleum-bearing geological formations. In the pharmaceutical industry, purified capsular material is used to make certain types of vaccines.

Bacteria that produce capsules can also cause problems in industry. For example, *Alcaligenes viscolactis* produces a slimy polymer in milk, which causes a type of milk spoilage called *ropiness. Bacillus subtilis,* a spore-forming bacterium, can cause a similar condition in bread.

Thus, capsules are important for the survival of bacteria that form them because they retard drying, help cells avoid engulfment by other cells, attract and retain nutrients, and enable cells to stick to favorable environments. Bacterial capsules are important for humans too because some capsules have commercial benefit, others cause undesirable textures in foods, and still others enhance the ability of the cell to cause infections. Therefore, the ability to detect microbial capsule formation can be an important technique for the microbiologist.

LEARNING OBJECTIVES

• Learn microscopic methods for visualizing bacterial capsules and how to interpret the results of these methods.

• Understand how negative staining works and why it is used to examine bacterial capsules.

• Learn about the chemical properties and staining characteristics of capsules.

• Learn about the functions that capsules perform for the bacterial cell.

• Appreciate the significance of capsules in the environment, in human health, and in industry.

MATERIALS

Cultures *Alcaligenes viscolactis,* a bacterium that produces prominent capsules (7-day-old slant culture)
Gluconobacter oxydans, a bacterium that does not produce capsules (slant culture)
Unknown bacterium that may or may not produce capsules (slant culture)

Supplies Glass microscope slides
Cover slips
India ink (dropper bottle)
Crystal violet stain
Bibulous paper (or filter paper)
Toothpicks or wooden applicators
Phase-contrast microscope for demonstration slides

Two methods for staining capsules are provided in this exercise. Your instructor will ask you either to perform both methods or to perform one and observe the other as a demonstration.

A · Wet-mount method

Perform the following procedures and observations for the three microbial cultures listed. For the best results, you should prepare the slides one at a time and observe each slide immediately after preparing it.

PROCEDURES

Preparing the slides

1. **Select a glass microscope slide and clean it thoroughly** (see Exercise 5). This procedure will work best if the slide is completely free of oil.
2. **Place a loopful of tap water on the slide.**
3. **Pick up a small amount of culture from the slant** with a sterile loop or needle, and mix it thoroughly with the drop of water on the slide. Use only enough culture to make the droplet of water *slightly* turbid.
4. **Add one drop of India ink to the cell suspension, and mix** thoroughly with a loop or needle.
5. **Gently place a glass cover slip** on top of the inked cell suspension.
6. **Place the slide on a piece of bibulous paper,** and then place a second piece of bibulous paper on top of the slide.
7. **Press down hard on the paper** over the coverslip with your thumb, rolling your thumb across it. This step forces most of the cell suspension out from under the coverslip, leaving only a thin layer of material behind. The bibulous paper should absorb the fluid that is forced out. *If the coverslip moves* while you are pressing on it, start over with a new slide.
8. **Place the used bibulous paper in the contaminated waste container,** and wash your hands with disinfectant soap.
9. **Use a toothpick** (or wooden applicator stick) **to apply a thin layer of immersion oil around the edge of the coverslip.** If properly applied, the oil retards evaporation of the sample under the coverslip.

Observing the slides

1. **Examine the slide with the high-dry lens** (40×). Focus carefully, and look for the tiny granules of India ink. The cells are difficult to see at first because they are not stained; look for small, transparent areas in the film of India ink particles. The cells will have a very faint blue-green color that changes as you move in and out of focus. Capsules, if present, are visible as completely transparent areas surrounding the cells. Ask your instructor for help if you cannot find the cells.

   Note: The density of the India ink background will vary from one part of the slide to another. The thinnest part of the slide will probably be best for viewing cells and detecting capsules.

2. After you locate and recognize the cells on your slide, **place a drop of oil on the coverslip,** and **observe the preparation** with the oil-immersion (100×) lens. Close the condenser diaphragm to increase the contrast of the cells.

3. **Record your observations** in the Results section.

4. **If you have access to phase-contrast optics, observe your slide** with the oil-immersion phase-contrast lens. Otherwise, examine the demonstration slides with the phase-contrast microscope. The cells stand out clearly because, with phase-contrast optics, they appear dark even though they are not stained. However, the capsules are transparent under these conditions (see Figure 11-2).

B · Dried-smear method with counterstain

Perform the following procedures and observations with all three microbial cultures. You can save time by staining the three slides simultaneously.

PROCEDURES

Staining the slides

1. **Select a glass microscope slide, and clean it thoroughly** (see Exercise 5). This procedure works best if the slide is completely free of oil.
2. **Place a drop of India ink near one end of the slide.**
3. **With a sterile loop or needle, pick up a small amount of culture from the slant, and mix it** thoroughly with the drop of India ink on the slide.
4. **Place the short edge of a second clean microscope slide on the cell–ink mixture, and, holding it at an acute angle, push the mixture** across the first slide with a rapid, smooth motion.
5. **Air-dry the smear on the first slide.** *Do not heat-fix the smear.*
6. **Stain** the dried smear with crystal violet for 1 minute.
7. **Gently rinse** the slide with water. It is normal for some of the smear to be lost during rinsing because it was not heat-fixed.
8. **Air-dry the slide.** (Do not blot the slide dry.)

Observing the slides

1. **Place a drop of oil on the slide and examine the smear** with the oil-immersion objective (100×). Search the smear for an area with a medium-to-dark-gray India ink background. The cells will be purple. Capsules, if present, will appear as transparent zones around the cells. Ask your instructor for help if you cannot find the cells.
2. **Record your observations** in the Results section.

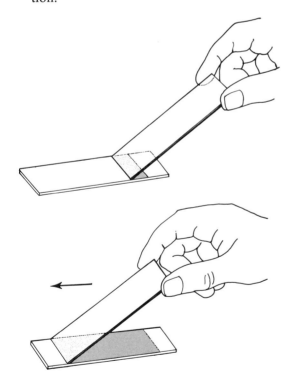

11 BACTERIAL CAPSULES: INDIRECT OBSERVATION

NAME _____

DATE _____

SECTION _____

RESULTS

1. **Culture texture.** Use words to describe the consistency (texture) of each culture, as observed upon probing with a loop or needle.

   a. *Alcaligenes viscolactis:*

   b. *Gluconobacter oxydans:*

   c. Unknown organism:

2. **Wet-mount method.** Draw a few representative cells from each wet-mount slide, as seen with the oil-immersion lens:

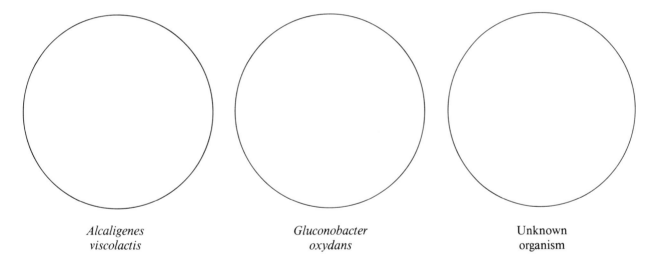

3. **Dried-smear method.** Draw a few representative cells from each dried-smear slide, as seen with the oil-immersion lens:

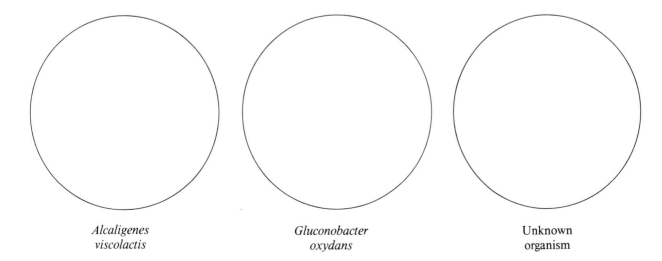

CONCLUSIONS

1. Does your unknown microorganism have capsules? Explain your conclusion.

2. Which of the methods for observing capsules do you think is more effective? Why?

3. Do you have any reason to think that artifactual results were obtained with the dried-smear method? Why or why not?

QUESTIONS

Completion

1. Most bacterial extracellular polymers are composed of _____.

2. *Streptococcus mutans* produces a specific type of capsule polymer called a(n) _____, which enables this organism to become attached to _____.

3. _____ is an example of a bacterium with a capsule that protects its from phagocytosis in the human body.

4. If a bacterial capsule is composed of polysaccharide, it may also be called a(n) _____.

5. Some capsule-producing bacteria cause a type of milk spoilage called _____.

True–False (correct all false statements)

1. Most capsules are easily stained with common basic dyes. _____

2. The wet-mount technique for staining capsules is more likely to produce artifacts than the dried-smear method. _____

3. Bacterial capsules are hard to preserve for microscopy because they contain very little water. _____

4. Capsules enable bacterial cells to attach to surfaces in many natural environments. _____

IV

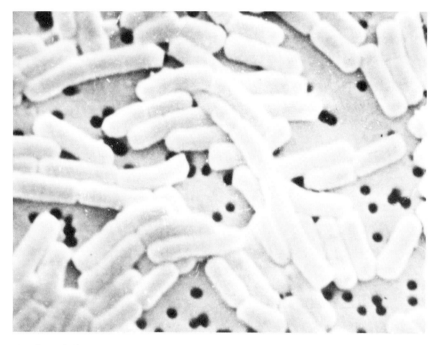

Bacillus subtilis on a 0.4 μm polycarbonate membrane filter.

All things require energy to grow; primarily because they are making order out of disorder.

—Sinstrom

STERILIZATION PRINCIPLES AND CULTURE MEDIA PREPARATION

The purposes of Part IV are to emphasize the need for adequate sterilization procedures in the microbiology laboratory, to explain how media and materials are sterilized, and to show you how a few simple media are prepared and made ready for growing microorganisms.

12

Sterilization Principles and Methods

In microbiology, **sterilization** is defined as the destruction or removal of all living organisms. An object that has been freed of living organisms is said to be **sterile.** You should understand that microbiological sterility is an absolute state: An object either is sterile or it is not; it cannot be almost sterile or half sterile.

Sterilization is one of several approaches used to control microorganisms. You can control microorganisms by killing them or by inhibiting their growth. Either of these methods can facilitate treatment of disease, prevent microorganisms from infecting the human body, and prevent contamination of pure cultures.

Sterilization differs from other methods of controlling microorganisms because it destroys or removes *all* living organisms. Other methods of microbial control may eliminate many organisms, but they do not eliminate all of them; therefore, these other methods do not make an object sterile. You will learn more about the uses of alternative methods for controlling microorganisms in Exercises 43 and 44.

Sterilization is important in microbiology laboratories because most experiments are performed with **pure cultures** (cultures that contain only one species or strain of microorganism). Microbiologists frequently use pure cultures because doing so allows them to study one type of microorganism without having to worry about how other types might affect their experimental results. To work with pure cultures, however, you must first sterilize the nutrient media and the containers in which the organism of interest will be grown, as well as any implements with which it will be handled. If this is not done, unwanted microorganisms (contaminants) may grow and produce invalid results. The concept of pure cultures is presented more fully in Exercises 14 and 15.

Sterilization is also important outside of the microbiology laboratory. In hospitals, surgical instruments must be sterilized so that microorganisms on them do not infect patients during surgery. Similarly, a drug company producing a vaccine must be certain that its product is sterile so that dangerous microorganisms are not introduced into the body when the vaccine is injected. The food industry also uses sterilization extensively, primarily in the canning process.

If we consider only those sterilization techniques that *destroy* (rather than remove) microorganisms, then we can state generally that destructive sterilization procedures are designed for the *most resistant* microbes. If a procedure destroys the most resistant form, then one can reason that the more sensitive forms are also destroyed.

Bacterial **endospores,** produced by species of bacteria such as those in the genera *Bacillus* and *Clostridium,* are the most heat-resistant living things known; some of them can survive in boiling water for several hours. They are also resistant to drying, starvation, degradative enzymes, toxic chemicals, and radiation. Because of this remarkable resistance, the physical parameters of most sterilization procedures are designed specifically to destroy all bacterial endospores. In fact, purified endospore suspensions are used to test the effectiveness of most sterilization methods.

Several sterilization methods are available to the microbiologist because no single procedure is suitable for all of the various materials that must be sterilized. For example, a method that uses high temperatures cannot be used to sterilize a heat-sensitive vitamin solution. Thus, one must select a method that is appropriate for each application. The most commonly used methods for sterilization are described in the following paragraphs. These methods and their applications are summarized in Table 12-1.

Sterilizing with heat

All living organisms have a **maximum growth temperature** above which they cannot survive. This temperature varies from one species to another and is determined by an organism's genetic traits. When the temperature of its environment rises above an organism's maximum growth temperature, the heat damages sensitive molecules such as proteins and nucleic acids. The organism eventually dies because its cell(s) cannot function without these molecules. Microbiologists apply this principle in three common sterilization methods: direct flame, dry heat, and autoclaving (see Table 12-1).

TABLE 12-1 Summary of sterilization methods and typical applications

Method	Application
Direct flame (incineration)	Small laboratory implements (inoculating loops, etc.) Contaminated wound dressings Disposable operating gowns Paper cups and plates Diseased animal carcasses
Dry heat (hot-air oven)	Empty glassware Dry powders (not heat labile) Laboratory equipment Surgical instruments Syringes (glass) and needles Oils (boiling point above 170°C)
Autoclaving	Microbiological growth media Laboratory equipment Discarded microbial cultures Heat-stable solutions Glassware Surgical instruments Hospital linen
Filtration	Gases (air in laboratories, etc.) Solutions containing heat-sensitive chemicals
Chemical (ethylene oxide)	Laboratory equipment Disposable plasticware and plastic tubing Surgical instruments Surgical supplies (sponges, etc.) Optical equipment Artificial heart valves
Ionizing radiation	Powders Medical supplies (lancets, pads, bandages, surgical sutures) Disposable plasticware Mattresses and bedding Surgical gowns Pharmaceutical products Foods

The *rate* at which microorganisms die when exposed to elevated temperatures is a function of the nature of the medium in which the heating takes place, the length of exposure time, and the extent of temperature elevation. Aspects of the medium that have an effect on heat sterilization include the pH, the salt concentration, and the type of nutrients available.

In general, the higher the temperature is raised above the microorganism's maximum growth temperature, the faster that organism dies. For example, *Mycobacterium tuberculosis* (the bacterium that causes tuberculosis) dies in about 30 minutes at 58°C, in about 2 minutes at 65°C, and in only a few seconds at 72°C. Microbiologists have defined two important parameters based on the relationship between time and temperature as microorganisms are killed by heat: (1) the **thermal death time (TDT),** which is the length of time required to kill all of the organisms in a liquid suspension at a designated temperature; and (2) the **thermal death point (TDP),** which is the lowest temperature at which all of the organisms in a liquid suspension are killed within 10 minutes.

Because the goal of sterilization is to kill all types of microorganisms, heat-sterilization procedures use TDTs and TDPs sufficient to kill all bacterial endospores (the most heat-resistant forms). A large safety factor is always built into the procedure so that the chance of having even a single survivor is less than one in a million.

Direct flame (incineration) The simplest heat-sterilization procedure is the **direct-flame** (or **incineration**) method, in which the object to be sterilized is exposed directly to an open flame. The very high temperatures within the flame burn up and quickly kill all living organisms on the object.

In the microbiology laboratory, the direct-flame method is used on a small scale to sterilize inoculating loops, forceps, open ends of test tubes, and so on. You learned how flame-sterilization is used in Exercise 2, and you will use this technique frequently in future exercises. Direct-flame sterilization is also used on a larger scale; for example, some hospitals incinerate contaminated wound dressings and other disposable items. The direct-flame approach is fast and effective but is obviously not suitable for sterilizing heat-sensitive objects that you want to reuse.

Dry heat (hot-air oven) Sterilization with **dry heat** is a less drastic alternative to the direct-flame method. The objects to be sterilized are placed in a **hot-air oven** (similar to a kitchen oven) and baked until all living organisms are destroyed (for example, at 170°C for 2 hours). Dry heat is thought to kill the organisms by oxidizing important cellular structures and macromolecules.

Dry heat is used in laboratories and hospitals to sterilize empty glassware (pipettes, flasks, and so on) and other solid objects (see Table 12-1). Dry heat cannot be used to sterilize liquids (such as microbiological growth media) because most liquids are aqueous solutions that will boil at approximately 100°C. As long as these solutions continue to boil, their temperatures do not rise above 100°C. Boiling is not a form of sterilization because some viruses (for example, the hepatitis virus) and many bacterial endospores can survive for hours at 100°C.

Autoclaving (moist heat) The primary limitation of dry-heat sterilization, its inability to sterilize liquids, is overcome by **autoclaving,** which is a sterilization process that uses moist heat and pressure so that all parts of the material to be sterilized reach *121°C for 15 minutes*. Autoclaving is probably the most versatile and most frequently used sterilization procedure. Many of the microbiological growth media and supplies that you will use in this course will be sterilized by autoclaving.

An **autoclave** (Figure 12-1) is essentially a large pressure-cooker, that is, a chamber that can be sealed off from the surrounding air. Materials to be sterilized are placed in the chamber, the door is closed and sealed tightly, pressurized steam is forced into the chamber until all of the cooler air is driven out through an exhaust valve, and then the exhaust valve is closed. Steam continues to be forced into the chamber until the pressure reaches 15 pounds per square inch (psi) above atmospheric pressure. High pressure is required only to achieve a sufficiently high temperature to kill the microorganisms; the pressure itself is not directly involved in the killing process. At sea level, a pressure of 15 psi pushes the temperature in the chamber to 121.5°C. At higher terrestrial elevations, somewhat more pressure is required to reach that temperature. In the autoclave, aqueous solutions do not boil at this temperature because of the high pressure. Larger volumes require more time than 15 minutes because it takes longer for the center of a large fluid volume to reach 121.5°C. After sterilization, the steam pressure is decreased slowly to atmospheric pressure (the liquids would boil over if it decreased suddenly). The door can then be opened to remove the sterile objects. Your instructor will show you the autoclave used to prepare media and supplies for your laboratory and will demonstrate its operation.

Autoclaves sterilize at lower temperatures and in less time (121°C; 15 minutes) than hot-air ovens (170°C; 120 minutes) because moist heat is used.

Moist heat (steam) kills cells more effectively than dry heat, because molecules (such as proteins) are more active when they are moist than when they are dry. This increased activity disrupts the hydrogen bonds that maintain the three-dimensional structure of the molecule. As a result, enzymes are denatured and protein-dependent cell structures such as membranes are destroyed. Moist heat also penetrates more effectively than dry heat because it is conducted through the cell at a faster rate.

Autoclaving has a variety of applications because it is suitable for both liquids and solids (see Table 12-1). In the microbiology laboratory, it is used primarily to sterilize culture media and to dispose of unwanted cultures safely. Hospitals use autoclaving to sterilize bedding, surgical instruments, and other items; and the food industry uses it in the canning process. Although autoclaving is suitable for many purposes, it should not be used on objects that are sensitive to heat or moisture.

FIGURE 12-1 Typical modern autoclave. (Courtesy of American Sterilizer Co., Erie, Penn.)

Sterilizing without heat

Filtration For heat-sensitive *liquids* or *gases*, the most commonly used sterilization method is **filtration**, which differs from heat sterilization in that it *removes* organisms instead of destroying (killing) them. Liquids or gases to be sterilized are simply passed through filters that have openings (pores) so small that microbial cells cannot pass through them.

Several kinds of filters have been used for sterilization since the early 1900s, but **membrane filters** are used most often today. Membrane filters are thin pieces of synthetic material (usually cellulose acetate or polycarbonate) that contain very small openings or pores. All membrane filters and the equipment with which they are used (Figure 12-2) must be sterilized by other methods (often by autoclaving) before the filter apparatus can be used to sterilize a fluid. Your instructor will show you some examples of membrane-filter devices and explain how they are used.

Membrane filters are available in a range of pore sizes, but few of the filters commonly used for sterilization can remove *viruses* because viruses are small enough to pass through the pores. If one considers viruses to be living organisms (as many microbiologists do), then membrane filters do not really sterilize the liquids that pass through them; rather, they remove cellular forms of microorganisms. This problem is not serious in many microbiological experiments because viruses do not grow in microbiological media; viruses grow only inside other cells. Therefore, the presence of viruses does not affect experiments using bacteria unless the viruses infect those bacteria. Filters (with very small pores) for removing viruses from liquids are available but are difficult to use because they clog rapidly. There is also some disagreement as to whether these filters really do remove the smallest viruses and bring about true sterilization.

Filtration is used in many hospitals and microbiology laboratories to sterilize solutions containing a

FIGURE 12-2 Typical small filter-sterilization apparatus. The liquid to be sterilized is poured into top of the apparatus (A) and accumulates in the bottom (B) as it passes through the filter (C). The apparatus must be connected to a vacuum source (D), which pulls the liquid being sterilized through the filter. The entire apparatus must be sterilized by autoclaving before it is used to sterilize a liquid. A separate membrane filter is shown at E. (Courtesy David L. Balkwill.)

wide range of heat-sensitive chemicals such as vitamins or antibiotics and other chemotherapeutic drugs. Some companies now use filtration to remove microorganisms from wine and beer; this approach avoids the heat-induced loss of volatile, flavor-producing chemicals that occurs with pasteurization. Filtration can also be used to sterilize gases (for example, the air pumped into a vessel containing a pure culture). High-efficiency particulate air (HEPA) filters are used instead of membrane filters when large volumes of air are filtered, such as that needed to supply clean rooms or laminar-flow hoods used for pathogen transfers. Filtration, as a method of sterilization, is obviously limited to gases and liquids.

Chemical sterilization **Chemical sterilization** is a popular method for sterilizing heat-sensitive *solid* objects. This method kills by exposing organisms to a toxic chemical. The actual mechanism of killing varies, depending upon the chemical used. Chemicals used for sterilization must be able to kill all living organisms and, therefore, should not be confused with disinfectants, antiseptics, and chemotherapeutic agents (drugs), which are useful for controlling microorganisms in various ways but cannot sterilize.

The most widely used chemical for sterilization is **ethylene oxide** (EtO). (Alternatives include propylene oxide and beta-propriolactone.) Ethylene oxide is an *alkylating agent* that reacts with proteins and nucleic acids. It reacts primarily with hydroxyl, sulfhydryl, and other chemical groups, as shown in Figure 12-3. Because proteins and nucleic acids no longer function after reacting with EtO, cells exposed to EtO die rapidly.

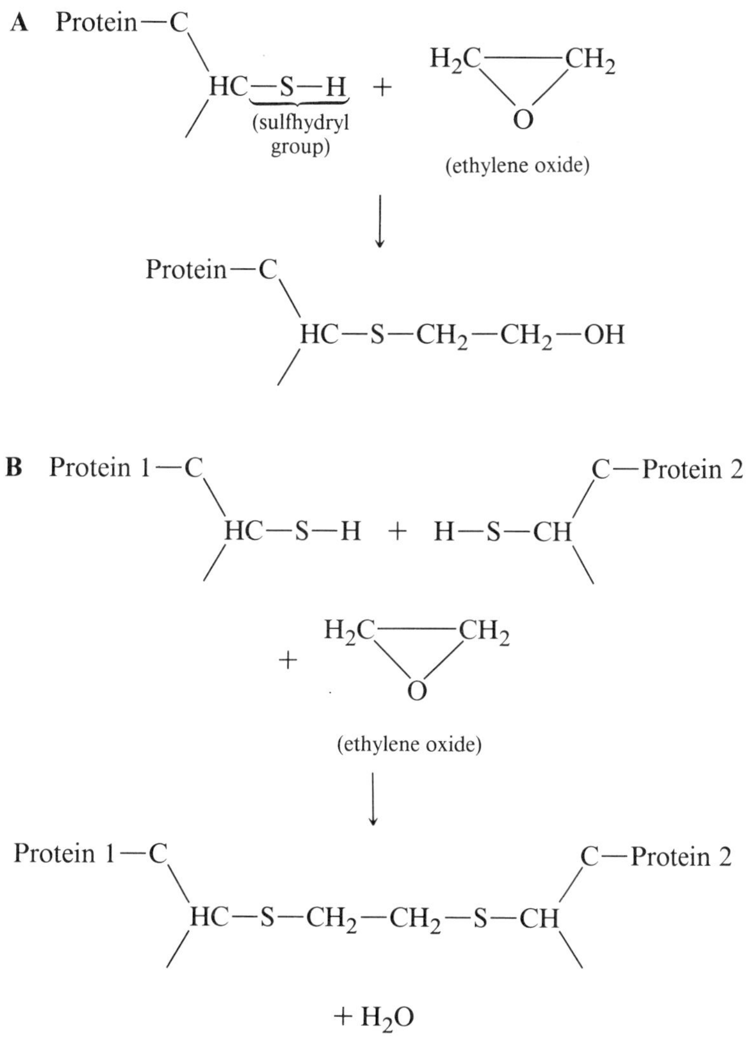

FIGURE 12-3 Examples of reactions between ethylene oxide and protein molecules. (A) Addition of an alkal group ($-CH_2-CH_2-OH$) to a sulfhydryl group ($-SH$). (B) Cross-linking of two protein molecules.

Ethylene oxide penetrates quickly and sterilizes effectively because it is a gas at room temperature. It must be handled carefully, however, because it is toxic to humans. It is also explosive, a problem that can be avoided if it is mixed with an inert gas such as freon or CO_2. Because of its dangerous properties, EtO sterilization is done only in special chambers that keep the gas from exploding or coming into contact with humans. Objects being sterilized must be exposed to EtO for about 4 hours, after which they must be flushed with inert gas for 8 to 12 hours. Flushing is done to ensure complete removal of the gas, since residual EtO can interfere with experiments, poison humans, and cause burns to the skin and other tissues.

Ethylene oxide is used to sterilize many materials that are sensitive to heat and/or moisture (see Table 12-1), especially disposable plasticware such as Petri dishes, pipettes, and syringes, which melt in an autoclave. Ethylene oxide is also used to sterilize satellites and space capsules sent to other planets by the National Aeronautics and Space Administration (NASA). The main disadvantage of EtO sterilization is the relatively high cost of the sterilizing chambers, a drawback that limits its use to larger laboratories and hospitals.

Ionizing radiation **Ionizing radiation** can be used instead of chemicals to sterilize heat-sensitive *solid* objects. Gamma radiation emitted by a radioactive cobalt source is used most often. When gamma rays enter a cell, they interact with water molecules to produce ions (such as OH^-) and free radicals (such as $\cdot OH$). The free radicals are so reactive that they can alter or destroy almost any type of molecule in the cell. The cell eventually dies as a result of the damage done to critical molecules such as enzymes and nucleic acids.

Sterilization with gamma rays is accomplished simply by exposure of objects to the source of radioactivity. A dose of 2.5 million rads (Mrad) is generally sufficient to kill all living cells, viruses, and spores. Treating solid materials with gamma irradiation does not render the materials radioactive, although it can bring about undesired *chemical* changes.

Gamma radiation is used to sterilize plastics (some companies now prefer it to EtO), powdered pharmaceuticals, and other heat-sensitive materials (see Table 12-1). Because of its power to penetrate solids, it can also be used to treat foods to retard spoilage. It is not yet being used to treat food on a large scale (except in Europe) because the radiation sometimes alters the food and because the public is suspicious of anything exposed to radiation. Gamma radiation is also so expensive (elaborate equipment is required to contain the radioactive source safely) that it is limited to large-scale industrial applications.

12 STERILIZATION PRINCIPLES AND METHODS

NAME _____

DATE _____

SECTION _____

QUESTIONS

Completion

1. Microbial cultures that contain only one type (one species or strain) of microorganisms are called _____.

2. _____ are usually the most resistant forms of life during sterilization.

3. Ionizing radiation kills microorganisms by producing _____ and _____ that reacts with and destroys important cell components such as proteins and nucleic acids.

4. _____ is a gaseous chemical that is often used to sterilize heat-sensitive solid objects.

True – False (correct all false statements)

1. Sterilization is the destruction or removal of pathogenic microorganisms. _____

2. The thermal death point is the lowest temperature at which all of the organisms in a liquid suspension will be killed within 10 minutes. _____

3. The physical parameters used in sterilization procedures are always designed specifically to kill the least resistant microbial form that is likely to occur in the object being sterilized. _____

4. Most of the membrane filters that are commonly used for sterilization do not really sterilize the liquids that pass through them. _____

Multiple Choice (circle all that apply)

1. Which of the following can be sterilized effectively in an autoclave?
   a. Typical microbiological growth media
   b. Empty glassware
   c. Plastic Petri dishes
   d. Dry powders

2. Which of the following methods is (are) suitable for sterilizing metal or glass surgical instruments?
   a. Chemical sterilization (EtO)
   b. Filtration
   c. Autoclaving
   d. Dry heat (hot-air oven)

3. Which of the following can be sterilized effectively by filtration?
    a. Heat-sensitive liquids
    b. Liquids that are not heat sensitive
    c. Syringe needles
    d. Air that is pumped into a laboratory

4. Which of the following methods is (are) suitable for sterilizing liquid solutions that are not heat sensitive?
    a. Direct flame
    b. Filtration
    c. Autoclaving
    d. Dry heat (hot-air oven)

13

Preparing Culture Media

Microbial nutrition

To study a microorganism thoroughly, you must be able to grow it. To a microbiologist, **growth** means an *increase in cell number.* Microbial growth is possible only if the proper physical and chemical conditions are provided.

An example of a *physical condition* that must be provided is heat. Each type of microorganism grows best at a certain temperature. The effect of temperature on microbial growth is developed in Exercise 23.

The *chemical conditions* for growth are provided by the components in the grown medium. For example, microbial growth requires free (unbound) water, the proper hydrogen-ion concentration (pH), essential minerals, and appropriate sources of carbon, energy, and nitrogen. In addition, some microbes require growth factors, which are complex organic compounds, such as vitamins and amino acids, that the cell cannot make; they must be provided in the growth medium. The minerals, growth factors, and sources of carbon, nitrogen, and energy necessary to support growth of a microorganism are collectively called **essential nutrients.**

The nutrients that are essential for growth vary widely among microorganisms. The more growth factors that a microorganism requires, the more **fastidious** it is said to be. The lactic acid bacteria are very fastidious, needing many growth factors; other bacteria, such as *Escherichia coli,* are able to grow on a medium containing only a simple sugar, ammonium phosphate, and a few mineral salts. Thus, *E. coli* is not considered nutritionally fastidious.

Many types of media (plural, *medium)* are used in microbiology. From a nutritional standpoint, all media types fit into two categories: complex and defined. **Complex media** are rich in the type and concentration of nutrients available. They often contain water-soluble extracts from partially degraded plant or animal tissue; these extracts are abundant sources of amino acids, sugars, vitamins, and minerals. However, since the exact types and quantities of nutrients in such media are unknown, they are called *complex.* Typical components of complex media are **peptones** and **tryptones** (enzymatically digested animal protein) and **yeast extract** (a water-soluble extract from broken yeast cells). These materials contain many minerals and organic nutrients. Frequently, a sugar such as glucose is added to the tryptone, peptone, or yeast extract to serve as the main carbon and energy source in the complex medium.

Defined media are composed of pure chemicals in carefully measured (known) concentrations. For example, *E. coli* grows well in a medium containing the following components (given in grams per liter): glucose (1.0), ammonium sulfate (1.0), dibasic potassium phosphate (7.0), monobasic potassium phosphate (3.0), sodium citrate (0.5), and magnesium sulfate (0.1). These compounds are dissolved in double-distilled water. Note that the exact chemical composition of this medium is known. Some bacteria, *E. coli* for instance, make all of their polysaccharides, proteins, nucleic acids, and lipids needed for growth from a simple sugar (glucose) as its carbon and energy source, an inorganic nitrogen source (ammonium sulfate), and various mineral salts. More fastidious bacteria require the addition of known quantities of purified amino acids, vitamins, and purines or pyrimidines (growth factors); then they too would grow in a defined medium.

Liquid and solid growth media

Microbiologists use culture media in liquid, solid, and semisolid forms. **Liquid media,** also called **broths,** contain only nutrients dissolved in water. Microorganisms grow while submerged in these broths (see Exercise 17). **Solid media** contain nutrients dissolved in water, plus a solidifying (gelling) agent to make the medium rigid. Microorganisms are often inoculated onto the surfaces of solid media, where they grow in *colonies* (see Exercise 14). **Semisolid media** are similar to solid media except that a lower concentration of solidifying agent is used, which gives them a jellylike consistency. Semisolid media are used for motility or gelatin-hydrolysis experiments (see Exercises 10 and 28).

Solidifying agents for media

The most common solidifying agent used in microbiological culture media is **agar**. This is a complex polysaccharide composed of galactose and galacturonic acid. It is found in many marine plants and is commercially extracted from certain red marine algae such as *Eucheuma, Gelidium, Gracilaria,* and *Rhodophyta*. Agar, often added to media at a concentration of 1.5 percent weight per volume (w/v), is commercially supplied as dry granules or powder. This solid material rapidly swells in cold water. As the water temperature approaches *100°C* (boiling), the swollen polymer turns into a transparent, colloidal **sol** (a fluid colloidal system). Since the sol is transparent, it looks as if the granules have dissolved, but this is *not* the case. When you allow the agar-containing liquid to cool, the viscosity of the sol increases until a gel forms, at around *38° to 40°C* (comfortably warm to the touch). As the medium continues to cool to room temperature (about 25°C), the gel becomes firmer and more opaque, until the resulting solid medium has the consistency of very firm Jello. Once solidified, the agar does not again become a fluid and transparent sol until you raise the temperature to 100°C.

The ability of agar to remain as a solid gel at temperatures up to 100°C is an important property of agar that is not shared by many other gels. For example, a protein called **gelatin** turns into a fluid, transparent sol when the temperature reaches 37°C (body temperature). Many microbes grow best at 37°C or higher; so gelatin is not commonly used as a solidifying agent for microbiological media.

Resistance to enzymatic breakdown is another important property of agar. Only a very few microorganisms produce and excrete enzymes that break down (hydrolyze) agar. These few microbial types are found mostly in marine environments; so many microbiologists never encounter them. Since most microorganisms that are cultivated in laboratories do not hydrolyze agar, nearly all microbiologists use it as a solidifying agent in culture media.

If you need to cultivate those few microorganisms that hydrolyze agar, **silica gel** can be substituted for agar as a solidifying agent. Silica gel is made from *silicic acid* (H_2SiO_3), and the gel formed from this acid is completely resistant to microbial breakdown. As you can see, silicic acid is an inorganic chemical. For this reason, silica gel is also used as a solidifying agent when one wants to cultivate organisms on solid media that are devoid of organic matter.

Sterilization of media

Before you can use liquid or solid media to grow microbial cultures, you must sterilize the medium so that it will contain only those cells that you transfer into it (see Exercise 12).

The function of the following procedures is to introduce you to the methods by which microbiologists make, sterilize, and dispense liquid and solid media into closed containers. Regardless of whether or not you are asked to perform these experiments, you should understand the basic methods and principles involved.

LEARNING OBJECTIVES

- Learn more about the components of complex media.

- Learn how microbiological culture media are prepared, sterilized, and stored.

- Learn about the melting and solidifying properties of agar.

- Learn how to prepare agar deeps, slants, and plates.

A · Preparing liquid media (broths)

Not long ago, if you wanted to grow microorganisms, you first had to kill an animal or plant, remove the appropriate parts, and extract the nutrients from the tissue. Later, several companies began to supply dehydrated forms of animal- or plant-tissue extracts in either paste or powdered form; and the microbiologist had to mix several different types together to prepare an appropriate complex medium. Today, most companies supply commonly used complex media in a premixed form. One need only add the correct amount of water to these premixed dehydrated media to dissolve the nutrients in the proper concentration.

For example, **nutrient broth** is a common type of complex medium. As supplied by the manufacturer, dehydrated nutrient broth is a mixture of powdered beef extract and peptone. *Beef extract* is prepared by boiling lean beef in water and evaporating the water from the dissolved nutrients. Thus, beef extract contains all of the water-soluble nutrients that are present in this tissue, including vitamins and minerals. *Peptone* refers to an enzymatically digested (hydrolyzed) protein from an animal source (for instance, meat or milk) or a vegetable source (for instance, soybeans or cottonseed). Peptones are complex mixtures of amino acids, polypeptides, carbohydrates, vitamins, and mineral salts whose exact chemical composition is not known. Thus, nutrient broth is a complex medium.

Trypticase-soy broth is another complex medium and is an excellent medium for growing a variety of microbes commonly used in the introductory microbiology laboratory. Trypticase-soy broth contains (in grams per liter): trypticase peptone (17.0), phytone peptone (3.0), sodium chloride (5.0), dipotassium phosphate (2.5), and glucose (2.5). *Trypticase peptone* is a powder prepared from enzymatically digested (hydrolyzed) milk protein (casein). *Phytone* is a powder prepared from enzymatically digested soybean meal. All of the ingredients are separately dehydrated and then mixed in the proper proportions, in such a way that the correct concentrations are achieved when 30 grams of powdered mixture are added to 1 liter of distilled water.

Once a broth is prepared, it is dispensed into proper containers, capped or plugged, and then sterilized.

MATERIALS

Cultures None

Media Trypticase-soy broth (premixed powder)

Supplies Weighing paper, spatulas, balances
Beaker, 250 ml
Graduated cylinder
Test tubes with caps or plugs and test tube rack
Pipettes, 10 ml (nonsterile)
Cart marked *To Be Sterilized*
Optional: Hotplate and magnetic-stirrer combination with stirring bar

PROCEDURE

1. **Follow the instructions on the bottle** of dehydrated trypticase-soy broth. Students should prepare 100 ml of this medium.

2. **Weigh the powder on weighing paper.** Use a clean spatula. Transfer the powder to a 250-ml beaker.

3. **Dissolve the powder in 100 ml of distilled water.** Stir until the powder is completely dissolved. Use a magnetic stirring bar and a magnetic stirrer if it is available. If not, use a clean glass stirring rod. It may be necessary to warm this mixture slightly to completely dissolve the powder (depending upon the water temperature, medium manufacturer, and manufacturer's lot number).

4. **Dispense 5 ml of broth into each of 20 clean test tubes. Cover each tube** with a plastic cap or insert a foam plug. If neither of these is available, have your instructor demonstrate how to make cotton plugs to fit these tubes.

5. **Label your tubes** with your name and the name of the medium. You may place a label on each tube or attach a single label to the rack holding the tubes. Place these tubes on the cart marked *To Be Sterilized.*

Your laboratory instructor will demonstrate the use of the autoclave when the class has prepared all of its media.

6. **Autoclave these tubes so that the medium is held at 121°C for 15 minutes.** If the autoclave has automatic controls, make sure it is set for *slow exhaust;* if it has manual controls, set them for slow exhaust according to your instructor's directions at the end of the sterilization time.

7. **Remove your tubes of sterile media and allow them to cool** to room temperature. Either use or store according to your instructor's directions.

PREPARATION NOTES

Record the notes you take during your instructor's demonstrations:

Record problems you encounter and noteworthy observations you make during preparation of the media:

B · Preparing solid media: Deeps, slants, and plates

Before you begin, you should review the first few pages of this exercise, especially those sections dealing with microbial nutrition, solidifying agents for media, and the material commonly used to prepare complex media.

One way to make a solid medium is to prepare a broth and then add about 1.5 percent of powdered agar. For example, you could dissolve the proper quantity of commercially prepared, dehydrated trypticase-soy broth in water and then add 1.5 percent (weight/volume) of agar.

Another way to make the same solid medium is to use commercially prepared, dehydrated trypticase-soy agar that is made of agar already mixed with all the ingredients normally found in the broth.

Regardless of which way you choose or which type of medium you choose, special procedures are necessary in preparing solid media because agar is not soluble in water. Therefore, particles of agar quickly settle out after you mix it with water. If you then pipette this mixture into test tubes, you will get unequal amounts of suspended agar particles in each tube. Here is the way to prepare a solid medium and dispense it into tubes:

First, add water to the dehydrated medium that contains powdered agar. This dissolves all water-soluble materials, and the agar particles begin to swell (become hydrated).

Second, heat this mixture to 100°C. This turns the agar into a fluid sol. Remember that it does not dissolve but instead becomes transparent. The fluid sol easily mixes with the dissolved nutrients and stays evenly mixed as long as it is in the fluid state.

Third, dispense this hot mixture into test tubes or other containers. The medium stays in the fluid state as long as the temperature is above 40°C. As soon as it drops much below this temperature, the medium becomes viscous and begins to form a gel; the temperature must be raised to 100°C before the gel will change back into the fluid state.

Fourth, cap or plug the test tubes, and then sterilize them.

Perhaps the three most common ways to use solid media in the microbiology laboratory are in deeps, slants, or plates (Figure 13-1).

You prepare **agar deeps** by dispensing heated and mixed agar media into test tubes. After the tubes are

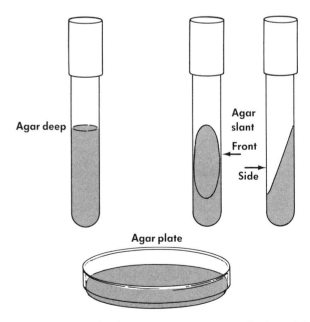

FIGURE 13-1 The three most common ways of using solid media in the microbiology laboratory: agar deep, agar slant, and agar plate.

capped (or plugged) and autoclaved, you allow them to cool upright in the test-tube rack. You inoculate agar deeps by pushing an inoculating needle (or loop) containing microbes down through the center of the agar, a method called **stabbing**. Once you have inoculated the agar deeps, you incubate them under appropriate conditions for growth so that the accumulated cells can be seen with the unaided eye. Inoculated agar deeps are primarily used to determine the oxygen requirements of pure cultures, as demonstrated in Exercise 25.

You prepare **agar slants** in a similar way, except that you tilt the tubes to one side while the agar cools. This produces a tube of solid agar with a large surface area. You inoculate the surfaces of slants by rubbing an inoculating needle (or loop or cotton swab) containing microbes in a zig-zag pattern across the slanted agar surface. Once you have inoculated the agar slants, you incubate them to stimulate growth. Agar slants are commonly used to grow and store pure cultures, as described in Exercise 16.

You prepare **agar plates** by first autoclaving the medium in a container (such as an Erlenmeyer flask), then thoroughly mixing the sterile medium, and finally pouring the fluid agar into sterile *Petri dishes* and allowing it to cool. The Petri dish containing a solid medium is called a *plate.* The main advantage of a plate is that it has a very large surface area for inoculation. You inoculate plates by **streaking** the agar surface with a loop, needle, or cotton swab; this technique spreads out the microbial cells so that the resulting colonies can be examined after incubation. This procedure is detailed in Exercise 14.

In this exercise, you will use one type of solid medium to prepare agar deeps, agar slants, and agar plates. Note this common usage for the word *agar.* *Agar* literally means the insoluble polysaccharide extracted from marine algae, but it also may refer to any solid nutrient medium that contains agar. For example, a microbiologist who refers to an *agar plate* is talking about a Petri dish containing any solid nutrient medium for growing microorganisms and not about a Petri dish containing only agar. You would need to inquire further to determine the exact chemical nature of the nutrients in that plate.

MATERIALS

Cultures   None

Media   Trypticase-soy agar (premixed powder)

Supplies   Weighing paper, spatula, balance
           500-ml Erlenmeyer flask
           250-ml graduated cylinder
           Nonabsorbent cotton, cheese cloth, scissors
           10-ml pipette (nonsterile)
           Heavy gloves for handling hot flask
           Cart marked *To Be Sterilized*
           Optional: Hotplate and magnetic-stirrer combination with stirring bar

PROCEDURES

Preparing trypticase-soy agar

1. **Follow the instructions** on the bottle. Prepare 250 ml of this medium.

2. **Weigh the powder** on the weighing paper. Use a clean spatula. Transfer the powder to a 500-ml Erlenmeyer flask; then add the necessary amount of water. Mix the powder well, without making it foam, and try to suspend the powder. Remember that this powder contains agar and that agar does not change to a fluid sol at room temperature.

3. **Heat the mixture carefully.** It boils over very easily!

   a. **If you use a magnetic-stirrer/hotplate,** add a magnetic stirring bar to the flask, and center the flask on the stirrer/hotplate. Move the flask around so that the stirring bar is centered on the bottom of the flask; then recenter the flask on the hotplate surface. Turn the stir-bar control to a speed at which the fluid is swirling but not splashing and the stirring bar stays in the center of the flask. Turn the heat control to *high,* and leave it at that setting until the medium begins to clear (becomes more transparent). When this happens, the medium is about ready to boil. Now, turn off the stirring bar for a moment, and see if bubbles are forming on the bottom. If so, turn the heat-control to *low,* and stir until the medium is completely transparent and many bubbles are rising from the bottom. Put on the heavy gloves, and remove the flask from the heat.

   b. **If you use a bunsen burner,** use a low heat and swirl the flask frequently. Never let agar particles settle onto the bottom of the flask. If this happens, they will burn and stick to the flask. This not only removes agar from the medium but also makes the flask very hard to clean. Use the heavy gloves to handle the hot flask. Watch the medium for signs of boiling, as described in step 3a. As soon as the medium is fully transparent and begins to boil, remove the flask from the heat.

4. **You must transfer this medium** to test tubes for deeps or slants before the temperature reaches 40°C. Otherwise, the fluid medium will solidify in the container used to heat it. The part of the medium to be used for plates should remain in the flask.

Preparing agar deeps

1. **Dispense** 10 ml of the hot, fluid medium into each of three test tubes.
2. **Rinse the pipette with hot tap water (>45°C)** as soon as all pipette transfers are completed. If you do not do this, the medium will solidify inside the pipette, from which it is *very* difficult to remove.
3. **Cap or plug each tube but keep screw caps loose while autoclaving.** Otherwise, the media inside the tubes may not reach sterilization temperatures; also, tubes may explode during the slow exhaust.
4. **Label each tube** with your initials and the name of the medium, and place the tubes in a rack on the cart labeled *To Be Sterilized*.
5. **Sterilize the tubes of agar** as instructed (see "Sterilizing the agar medium," below).
6. **Remove tubes from the autoclave.**
7. **Keep tubes upright in the rack until cooled to room temperature.** Note that agar in the fluid-sol state is *transparent* but that agar that has cooled to room temperature and gelled is *opaque* (translucent but not transparent).
8. **Store these tubes** as instructed. If screw caps are used, you should tighten them before storage to prevent the medium from drying. You may be asked to use these tubes for Exercise 16.

Preparing agar slants

1. **Dispense 5 ml of the hot medium into each of three test tubes.**
2. **Rinse the pipette with hot water (>45°C)** as soon as all pipette transfers are completed. If you do not do this, the medium will solidify inside the pipette, from which it is *very* difficult to remove.
3. **Cap or plug each tube but keep screw caps loose while autoclaving.** Otherwise, the media inside the tubes may not reach sterilization temperatures; also tubes may explode during the decompression cycle.
4. **Label each tube** with your initials and the name of the medium, and place the tubes in a rack on the cart labeled *To Be Sterilized*.
5. **Sterilize these tubes of agar** as instructed (see "Sterilizing the agar medium," below).
6. **Remove tubes from the autoclave.**
7. **Slant the tubes as follows:** Place your well-rinsed 10-ml pipette on the bench top, and lean the three tubes of melted agar slants on the pipette so that the notch between the tube and its loose cap catches on the pipette. If this is done correctly, the slanted surfaces of the medium should extend almost to the bottoms of the tubes. Do not move these tubes until the agar has cooled to room temperature and looks *opaque*.

An entire rack of tubes can be slanted as follows: Place a 10-ml pipette on the bench top. Tip the test-tube rack on its side so that the pipette supports the top edge of the rack. Leave the tubes in this position until the medium has cooled to room temperature and looks *opaque*.

8. **Store these tubes** as instructed. If screw caps are used, you should tighten them before storage to prevent the medium from drying. You may be asked to use these tubes for Exercise 16.

Preparing agar plates

1. **Prepare a cotton plug** for the Erlenmeyer flask that contains the remaining melted agar. Your instructor will demonstrate how to make cotton plugs. If you used a hotplate/stirrer to melt the agar, take it out with a stirring-bar retriever before inserting the cotton plug.
2. **Place your name and the name of the medium on this flask;** then place the flask on a cart marked *To Be Sterilized*.
3. **Your laboratory instructor will demonstrate the use of the autoclave** when the class has prepared all of its media.

TECHNIQUE NOTES

Cotton plugs usually have several layers of cheese cloth on the outside to prevent the cotton from sticking to the neck of the flask after autoclaving. A properly made cotton plug has these features: (1) Enough cotton protrudes above the flask so that you can easily grasp and remove the plug; (2) the plug should not be too loose because it may permit the entrance of contaminated dust; and (3) it should not be too tight because the flask may not easily decompress during the slow-exhaust cycle, and the plug may then pop out of the flask. An inserted cotton plug is too tight if you cannot freely turn it in the tube or flask; it is too loose if it will not support the weight of the empty tube or flask.

4. **Sterilize the flask of agar** as instructed (see "Sterilizing the agar medium," below).
5. **Remove flask from the autoclave and mix** the contents well with vigorous swirling motions.
6. **Cool this flask of sterile agar in a 45°C waterbath.** This process is called **tempering** the agar. To determine when the sterile agar has reached 45°C, feel the temperature of the water in the bath with your hand. Now, remove the flask from the waterbath, and hold your hand on the bottom for about 15 seconds while you gently swirl the melted agar. If the flask seems to be the same temperature as the water, then the agar is tempered and can now be poured into Petri dishes. **Before removing this flask from the waterbath, complete step 7.**

TECHNIQUE NOTES

Tempering the agar before placing it in the Petri dish is important. *If the agar is poured too hot,* steam condenses inside on the top of the plate before the agar solidifies. Later, when you are handling these plates, the water may drop to the agar surface. This water may spread cells of a pure culture all over the plate so that individual colonies will not develop; instead, the entire plate surface will be covered with microbial growth. This is called **confluent growth**. Water can also disturb the physical separation of cells from a mixed culture, which can lead to single colonies that have mixed populations of cell types. *If the agar is too cool,* it may solidify in the flask before it is poured, or it may begin to gel as it is being poured into the plate, producing an uneven surface that is difficult to streak.

7. **Label five Petri dishes.** Each Petri dish has two parts. The top half is larger than the bottom half. Place each dish so that its bottom (the smaller half) is facing up. Place your initials and the name of the medium on the bottom of each Petri dish; then turn it over. *Always write with small letters on the bottom* of the Petri dish because (1) inoculated plates are always incubated upside down and the labels can be read if you have written on the plates' bottoms, and (2) if you knock over a stack of plates, the labels will not be separated from their plates.
8. **Organize the five Petri dishes along the edge of your laboratory bench** in a row and about 2 in. apart with the top (larger half) up.
9. **Remove the tempered agar from the waterbath,** and wipe the water off the flask. If you do not do this, the water may drip into your plates as you pour them and contaminate the sterile medium.
10. **Pour each plate as follows:**
    a. Open the hot-water valve until hot (>45°C) water flows from the faucet; then turn the water off. This assures that hot water will be immediately accessible to you for step h.
    b. Grasp the neck of the Erlenmeyer flask with the hand you write with.
    c. Turn your other hand palm side up, and clamp the cotton plug between two fingers.
    d. Remove the plug, and flame the lip of the Erlenmeyer flask.
    e. Use the hand holding the cotton plug to lift the lid of the Petri dish. Now, **pour until the bottom is about two-thirds covered** with agar. Hold the lid so that it partially covers the bottom of the dish as you pour. This helps prevent microbes on airborne dust particles from dropping into your sterile plate and contaminating it.

f. Immediately replace the lid, and slide the dish gently in a circular pattern on the bench top until the fluid agar entirely covers the bottom of the dish.

Your plate should now contain a continuous layer of liquid agar about 4 mm (0.25 in.) thick. It takes about 20 ml of fluid agar to pour a plate in this way.

g. Flame the lip of the flask, and pour the next plate in the same way.

h. Immediately after pouring all five plates, dilute the remaining agar with hot tap water ($>45°C$), and discard it in the sink. Rinse the flask several times with hot tap water before setting it down.

i. If bubbles occur on the surface of the fluid agar, break them aseptically by quickly passing the bunsen-burner flame over the surface. If you hold the burner too long, the flame may melt the edge of the plastic Petri dish.

11. **Do not disturb these plates until they have cooled to room temperature** and the agar is opaque.

12. **Store these plates** as instructed. You may be asked to use them for Exercise 16.

Sterilizing the agar medium

1. **Autoclave** the capped or plugged tubes and flasks as instructed, so that all portions of the medium reach 121°C and remain at that temperature for 15 minutes.

2. **Slow exhaust** the pressure to prevent the liquid from boiling over inside the autoclave.

3. **Carefully remove** the tubes and flasks as instructed, and use as described in the preceding sections.

PREPARATION NOTES

Record the notes you take during your instructor's demonstrations:

Record any problems you encounter or any noteworthy observations you make during preparation of the media:

13 PREPARING CULTURE MEDIA

NAME _____

DATE _____

SECTION _____

QUESTIONS

Preparing liquid media

Completion

1. Microbiologists use the term *growth* to refer to _____.

2. Another term (or descriptive phrase) used for vitamins, amino acids, and other complex organic compounds required for growth is _____.

3. A microbe that requires many preformed organic compounds for growth is said to be _____.

4. The thing that differentiates complex media from defined media is that the exact chemical composition of a complex medium is _____.

5. Two common components of complex media are _____ and _____.

6. _____ is the name of a sugar that is often added as an energy source to both complex and defined media.

7. If a defined medium contains glucose, ammonium sulfate, potassium phosphates, manganese sulfate, and magnesium sulfate, the cell must be using _____ as its carbon and energy source.

8. Another name for a liquid medium is a _____.

9. *Sterilization* is defined as _____.

True-False (correct all false statements)

1. Most common media used by microbiologists today are supplied as premixed powders that need only to be rehydrated and sterilized before use. _____

2. *Peptone* is a general term that refers to enzymatically hydrolyzed protein, regardless of its source. _____

3. The protein in milk is called *trypticase*. _____

4. *Trypticase peptone* specifically refers to enzymatically hydrolyzed casein. _____

5. Tubes and flasks should be capped or plugged immediately after autoclaving to prevent airborne contaminants from growing in the sterile medium. _____

Preparing solid media

Completion

1. Probably the most commonly used solidifying agent used by microbiologists is called _____. Chemically speaking, this is composed of a type of polymer called _____.

2. If one wants to have a completely inorganic solid medium or if the growing microbes break down organic solidifying agents, then _____ can be used to solidify the medium.

3. Once liquified, mixed, and dispensed into tubes, agar media should be capped or plugged because _____.

4. Agar slants are primarily used for _____ and _____ of pure cultures.

5. Agar plates should not be poured when the agar is too hot because _____.

6. The process of cooling agar until it is the correct temperature for pouring into plates is called _____.

True – False (correct all false statements)

1. Agar does not go into the fluid state in water until the temperature reaches 100°F, and it does not gel until the temperature cools to about 4°C. _____

2. Once agar has cooled and formed a gel, it does not return to the liquid state until the temperature is raised to 45°C. _____

3. Agar is a complex protein that is not soluble in water. _____

4. Agar exists as a colloid at low temperatures, but when the temperature is sufficiently raised, it turns into a transparent state called a fluid sol. _____

V

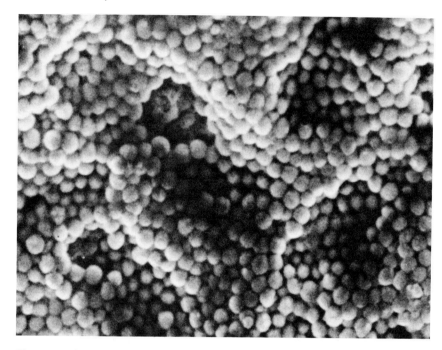

The center of an intact *Staphylococcus aureus* colony.

For every man the world is as fresh as it was the first day, and as full of untold novelties for him who has the eyes to see them.

— *Thomas Huxley*

TECHNIQUES FOR DETERMINING CULTURE PURITY

By definition, a pure culture contains a single species of microorganism. Most natural habitats contain many species, each interacting with its environment in its own characteristic way. As microbiologists, we are frequently faced with the task of isolating a single species so that it can be characterized. Therefore, we need a simple, inexpensive method for separating microbes.

In the late 1800s, Robert Koch observed isolated masses of growth (colonies) on decaying potato slices and noted that they frequently had different sizes, colors, and shapes. Using his microscope, he saw that all microorganisms in one colony were always identical. His work made it evident that pure cultures could be isolated on solid media. Many of the techniques that we now use to obtain pure cultures were developed in Robert Koch's laboratory.

The most practical method for obtaining pure cultures is the streak plate (see Exercise 14). With this method, you may start with a mixed culture and spread it over the surface of a solid medium until each cell is separated from all others. The colony that develops from one cell is a pure culture by definition because all cells in that colony were derived from that one cell.

However, not all colonies begin with a single cell. For example, the cells of many microorganisms cling tightly to each other after cell division, and these groups of progeny cells do not always come apart when you streak the culture on the solid nutrient surface. Yet the colony that arises from this group of cells is still a pure culture because all cells in it are the same species.

Some types of microbes excrete sticky capsules that bind them tightly to anything they touch. When these cells stick to other types of cells, the mixed-cell cluster can also give rise to a colony, which would be a mixed culture. Therefore, one should not believe that a colony always begins with a single cell or that a single colony always contains a pure culture. From a practical standpoint then, *how can you know when you have a pure culture?*

Assume that you have some well-isolated, identical looking colonies on a streak plate. You need two additional types of evidence to convince yourself (and other microbiologists) that you have a pure culture. Starting with a single colony, you make more streak plates; then, (1) each streak plate must show only one colony type, and (2) all cells from that colony type must have the same size and shape (see Exercise 15) and the same Gram reaction (see Exercise 7). Even meeting these criteria is no absolute guarantee of culture purity, but such evidence makes it highly probable. It is wise to check your cultures periodically to make sure that contaminants are not present.

The main objectives of Part V are to introduce you to the theory and the use of streak plates in separating microbial types, to further introduce you to the concept of pure cultures, and to give you an opportunity to separate and purify different strains from a mixed culture.

14

Separating Microbes on Streak Plates

If you were to examine an animal or plant, a bit of soil, water, or food, or any other part of your environment, you would find many types of microorganisms. Rarely would you find a single species. Nevertheless, we must be able to separate and grow each type of microorganism individually to study it. If microorganisms could not be separated, we could not determine the effect each has on our environment. For example, we could not tell which one causes a disease, or which one causes the decay of a stored food, or which one produces a desired antibiotic.

Consider for a moment the problem of trying to sort various types of cells that can be seen only with a microscope. The cells are too small to be picked out singly, except by tedious and complicated mechanical methods using expensive equipment. However, if you can spread cells on a streak plate so that you separate cells from each other by a few millimeters or more, each cell that grows will give rise to a **colony.** Since all cells in a colony come from one parent, it is unnecessary to study individual cells because all cells have identical genetic characteristics.

It is fortunate that microorganisms with different genetic characteristics often have different colony types. Colonies can vary greatly in their size, which you measure with a metric ruler, and their color, which you describe in conventional terms. One must be careful in using colony size as a criterion for purity because colonies that grow very close to one another are smaller than those that are spread farther apart on the agar surface. Crowded colonies rapidly deplete nutrients and do not grow as large as those with an unlimited nutrient supply. If you want to use colony size as the sole criterion for differentiating two microbial types, make sure that this size difference occurs between well-isolated colonies (>5 mm from the closest neighbor).

To a less obvious extent, colony types also vary in texture and shape. These differences often require close examination and the use of the terms found in Table 14-1. Note the variety of colony types evident in Figure 14-1.

If you are working with a pure culture, all colonies are identical in every respect; if you have a mixed culture, the colonies will usually vary in appearance. If the numbers of each type of microbe are almost equal in a mixed culture, then each colony type that develops from that mixture will be apparent on a well-prepared streak plate. By definition, a good streak plate is one that gives good separation (>2 mm) between adjacent colonies so that they can be examined separately and easily removed for further cultivation.

Many techniques exist for obtaining a good streak plate. Note, however, that one technique may not achieve the same result every time. The more cells that you have on your loop, the more you need to spread them across the agar surface. If the deposited cells are too close together, the colonies grow together **(confluent growth),** and microbial types cannot be separated. Many microbiologists prefer to overstreak rather than to not streak enough.

One further note on terms. Dr. R. J. Petri invented the Petri dish while working in Dr. Robert Koch's laboratory during the late 1800s. Tradition dictates that, as long as this dish is empty, we refer to it as a **Petri dish;** but when sterile agar has been added, it is referred to as an **agar plate.** In this exercise, you will be using trypticase-soy agar plates.

The following techniques are common to all methods of preparing streak plates.

1. Sterilize your inoculating loop (or needle), and let it cool before it touches the culture.

TABLE 14-1 Common variations in colony appearance

Texture when probed	Colony edge	Surface appearance	Elevation (side view)
Dry (breaks up easily)	Circular ("entire")	Glistening ("smooth")	Flat
Hard (difficult to break)	Irregular	Dull ("rough")	Convex
Creamy (like butter)	Rhizoid	Wrinkled	Pulvinate ("gumdrop")
Mucoid (like mucous)	Filamentous		Umbonate

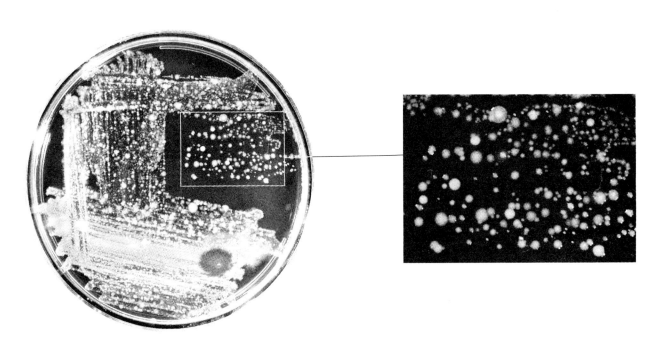

FIGURE 14-1 Streak-plate (left) inoculated with a mixed culture; a number of different colony types are evident. Enlargement shows differences in colony sizes, edges, apparent textures, and shapes.

2. Gently touch the loop to the agar surface during streaking; otherwise, it may penetrate the agar surface, and the cells will not form surface colonies.

3. Position yourself so you can see the faint scratches left by the loop as it glides across the surface. In this way, you can be sure where your loop has been during the last streak and can better position your loop for the next part of the streak.

4. Always sterilize your loop between streaking one area of the plate and the next. Then, subsequent streaks will spread only those cells that have been previously streaked onto the agar surface.

5. To prevent contaminate-laden dust from falling onto the agar surface while you streak the plate, hold the Petri-dish cover so that it partially covers the open plate.

The three methods for streaking plates described in Figures 14-2 through 14-4 seem to be those most commonly used by microbiologists. People tend to develop one or two methods that work best for them.

LEARNING OBJECTIVES

• Understand the purpose and theory of the streak-plate method.

• Achieve good colony separation using one or more methods, regardless of the cell density of the inoculum.

• Practice good aseptic technique.

• Gain experience in recognizing colony types.

MATERIALS

Cultures	Mixture of *Staphylococcus epidermidis* and *Escherichia coli*
Media	Trypticase-soy agar plates
Supplies	Inoculating loop Bunsen burner
Demonstration	Streak plates containing pure cultures of *S. epidermidis* and of *E. coli* (second laboratory period only)

PROCEDURE: FIRST LABORATORY PERIOD

For the purpose of describing this exercise, it is assumed that you have prepoured plates. If this is not the case, you will first need to prepare agar plates (see Exercise 13). You also should review Exercise 2 to refresh your memory on how to remove cells aseptically from a culture tube.

1. **Label one plate** along the bottom edge before you begin. Write small and include *your last name, date, culture used,* and *exercise number.* Always label on the bottom because the top may become separated from the bottom.

2. **Draw two heavy lines in the plate bottom** to divide the plate into three unequal parts (see Figure 14-2).

3. **Inoculate the plate with the mixed culture,** using the following general procedure for the *T-streak method:*

    a. **Aseptically remove a loopful of broth** from one culture tube, using the procedure outlined in Exercise 2. (Set this culture aside. It is no longer needed.)

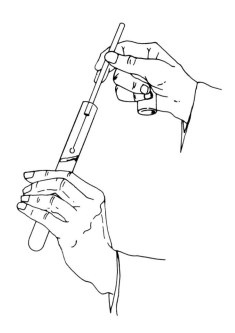

    b. **Lift the lid and gently streak area 1 on the plate** (see Figure 14-2). Take care not to touch the loop to anything other than the agar surface.

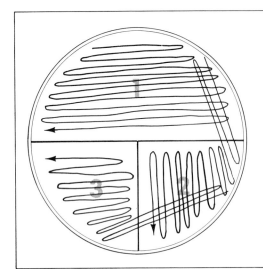

FIGURE 14-2

THE T-STREAK

1. Draw two lines on the bottom of your plate in the shape of a *T*, as shown.
2. Streak area 1, using one continuous motion (back and forth about 15 times).
3. Sterilize your inoculating loop, and allow it to cool.
4. Turn the plate about 90°. Draw your loop across area 1 a few times; then streak area 2.
5. Sterilize your loop, and allow it to cool.
6. Turn the plate again about 90° in the same direction. Draw your loop across area 2 a few times; then streak area 3.
7. Sterilize your loop before setting it down.

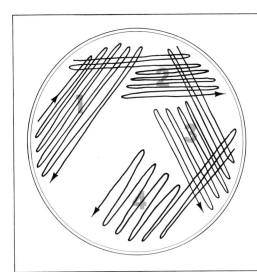

FIGURE 14-3

THE QUADRANT STREAK

1. Spread (streak) the culture over a small area near the edge of the plate (area 1), using one continuous motion.
2. Sterilize your inoculating loop, and allow it to cool.
3. Turn the plate somewhat. Draw your loop back and forth across area 1 a few times; then streak area 2.
4. Sterilize your loop, and allow it to cool.
5. Turn the plate again in the same direction. Draw the loop across area 2 a few times; then streak area 3.
6. Sterilize your loop, and allow it to cool.
7. Turn again in the same direction. Draw your loop across area 3 a few times; then streak area 4.
8. Sterilize your loop before setting it down.

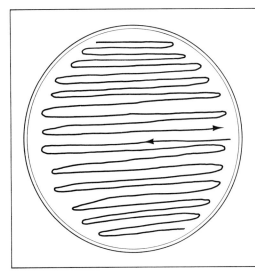

FIGURE 14-4

THE CONTINUOUS STREAK

1. Spread the culture over one-half of the plate in one continuous motion.
2. Turn the plate 180°.
3. Continue streaking the second half of the plate *without flaming the loop*. You must continue to use the same part of the loop, however.
4. Sterilize the loop before setting it down.

Note that this method allows you to continuously spread out those cells that are on your loop. This technique often does not work well unless there are relatively few cells on your loop.

14 SEPARATING MICROBES ON STREAK PLATES

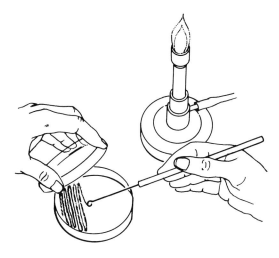

The purpose of the following steps is to spread a few of those cells that you already streaked in area 1.

d. Streak some cells from area 1 into area 2.

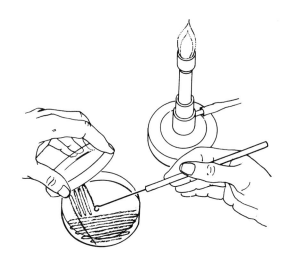

c. **Sterilize your loop** until the entire wire has been heated to redness and **cool the loop** for about 10 seconds. *Do not put your loop back into the culture before you streak area 2.*

e. **Sterilize your loop.**

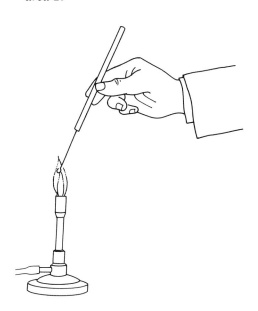

TECHNIQUE NOTE

After you streak area 1 and flame your loop or needle, you need to decide how many times to bring your loop (needle) across the lines streaked in area 1. This determines how many cells will be transferred from area 1 to area 2. In the beginning, consider these guidelines: If you transfer cells from a slant or a very turbid broth, slide the loop (needle) across area 1 only once. If there are only a few cells on the slant, or if the broth culture shows only slight turbidity, slide the loop (needle) across area 1 two or three times; because there are fewer cells in area 1, more passes are needed to collect enough cells to spread over area 2. *The more cells you transferred to area 1, the fewer times you need to pass the loop (or the shorter your single stroke should be) to transfer enough cells to area 2.* (The same principles are used when transferring cells from area 2 to area 3.)

f. Streak some cells from area 2 into area 3.

g. Sterilize your loop before setting it down.

4. **Use your sterilized loop to sham-inoculate a second plate.** Label this plate. This plate should remain sterile if you practice good aseptic technique.

EXPERIMENTAL DESIGN NOTE

The sham-inoculated plate is called a *control*. It will show you whether or not you sterilized your loop or needle and whether or not the medium was sterile. If you have adequately sterilized your loop then no microbes will grow along the streak after incubation. This also implies that any cells growing on inoculated plates are those you have transferred from cultures and not contaminates left on an unsterile loop. Some would call this sham-inoculated plate a *negative control* because there should be no signs of growth (a negative response). *All well-designed experiments have controls, one of which should be a negative control.*

---

5. **Stack the plates upside down.** Tape the plates together as directed by your laboratory instructor, and write your last name on the tape at the top of this inverted stack.

6. **Incubate at 30°C for 48 hours.**

TECHNIQUE NOTE

There is a good reason for inverting inoculated plates prior to incubation or storage. Solid media is >95 percent water, making the interior of an agar plate very humid. If these plates are transferred to a cool incubator, or (after incubating) to a refrigerator, water vapor inside the plate condenses. This condensation forms on the plate cover (because the cover cools faster than the bottom half of the plate which has an insulating layer of agar). If the plates are not inverted, condensation may fall from the cover onto the agar surface, spreading the cells; and confluent growth will result. Alternatively, condensation may stay on the cover during colony development but fall on the agar surface when the cover is removed. Either way, your plate may be ruined. *Always incubate and store plates inverted.*

PROCEDURE: SECOND LABORATORY PERIOD

1. **Observe the T-streak plate prepared from the mixed culture.** Sketch the appearance of this plate in your Results section.

2. **Identify each colony type** using the demonstration plates containing pure cultures. Use the terms given in Table 14-1 to help you *describe* the colony characteristics of each culture. **Record your description** in the Results section.

3. **Discuss your streak-plate technique** with your instructor. Mastery of the streak-plate technique is *essential* for much of your later work.

4. **Observe your sham-inoculated plate.** Can your aseptic technique be improved? Discuss problems with your instructor. If this plate is free of growth, you might want to leave the cover off for 5 to 10 minutes and then incubate the plate to see what kind of microorganisms are in the air. Alternatively, you might use it in the following steps.

5. **Label four additional plates,** as follows:
    a. Label all four on the bottom with *your last name,* the *date,* and *exercise number.*

14 SEPARATING MICROBES ON STREAK PLATES

   b. Label the first two on the bottom as *Staphylococcus epidermidis.*

   c. Label the second two on the bottom as *Escherichia coli.*

6. **Inoculate two plates with** *Staphylococcus epidermidis* that you isolated on your T-streak plate (use the demonstration plates as a guide to identification).

   a. Use the T-streak method for one plate.

   b. Use either the quadrant or continuous method for the other plate.

   c. Label each plate with the method used.

7. **Inoculate two plates with** *Escherichia coli* that you isolated on your T-streak plate (use the demonstration plates as a guide to identification).

   a. Use the T-streak method for one plate.

   b. Use the method not chosen in step 6b for the other plate.

   c. Label each plate with the method used.

8. **Invert, stack, and tape these plates.**

9. **Incubate at 30°C for 48 hours.**

PROCEDURE: THIRD LABORATORY PERIOD

1. **Observe all four streak plates.**

2. **Record your observations** in the Results section.

3. **Discuss your streak-plate technique** with your instructor. Also discuss why the continuous streak method might give good separation when very dilute broth cultures are transferred but might not give separate colonies when cells are transferred from turbid broth cultures or from colonies on solid media.

4. **Properly discard your plates.**

14 SEPARATING MICROBES ON STREAK PLATES

NAME _____

DATE _____

SECTION _____

RESULTS: FIRST LABORATORY PERIOD

1. Sketch the appearance of your streak plates in the space provided below.

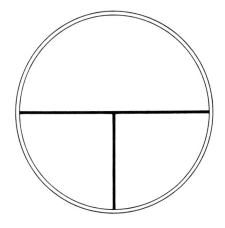

T-streak method
(indicate where separation occurred)

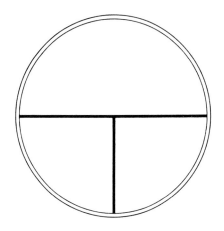

Sham inoculated
(with sterilized loop only)

2. Record the description of each colony type in the table given below (use the demonstration plates to help you identify each colony type).

Colony characteristic	*Staphylococcus epidermidis*	*Escherichia coli*
Size		
Color		
Elevation		
Surface		
Edge		

RESULTS: SECOND LABORATORY PERIOD

Sketch the appearance of your streak plates in the space provided below. Indicate where separation occurred on each plate.

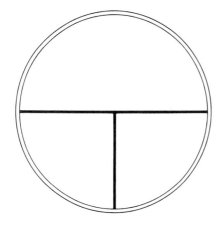

Staphylococcus epidermidis
Method: T-streak

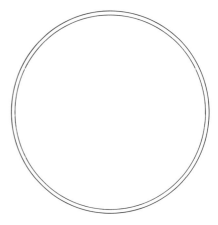

Staphylococcus epidermidis
Method:

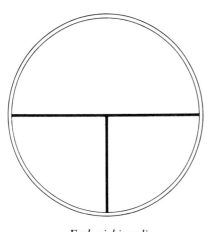

Escherichia coli
Method: T-streak

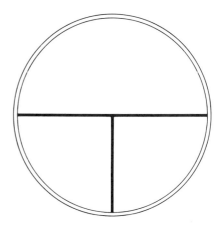

Escherichia coli
Method:

CONCLUSIONS

14 SEPARATING MICROBES ON STREAK PLATES

NAME _____

DATE _____

SECTION _____

QUESTIONS

1. Do you have contamination on any of your plates? How can you tell?

2. It is possible to streak a pure culture and observe that all colonies have the same color and texture but different sizes. How should this be explained?

3. Why is the streak plate so important to the concept of pure cultures?

4. Briefly state the purpose of the streak plate.

5. Of the methods described for streaking plates, which one(s) would give you good colony separation if you were transferring cells from

  a. a colony growing on an agar plate?

  b. a broth culture with barely visible turbidity?

  c. a broth culture with extremely high turbidity?

  What basic principle should you have used in making the above decisions?

6. What kind of control should you use to determine if you were actually using sterile plates for this experiment?

15

Determining Culture Purity

The first part of this exercise (Testing the purity of a colony) should be performed after you have prepared a **primary streak plate** from a mixed culture (see Exercise 14).

The second part of this exercise (Separating cultures from a unknown mixture) should be performed only after you have developed skills in preparing streak plates, in detecting different colony types, and in performing the Gram stain.

A · Testing the purity of a colony

In Exercise 14, you learned that all colonies arising from a single cell are pure, and that not all colonies begin with a single cell. The surfaces of some cells are sticky. If such a cell sticks to other cells from the same parent as it comes to rest on the surface of a streak or spread plate, the resulting cell cluster will form a **pure colony**. On the other hand, if that sticky cell adheres to cells from other species as it is deposited on the plate surface, the cell cluster will form a **mixed (contaminated) colony.**

Sometimes, you can easily differentiate mixed (contaminated) colonies and pure colonies by appearance. For example, some mixed clusters form **segmented colonies**, in which one part of the colony looks very different from the rest. Segmented colonies often appear that way because the two species produce different pigments and/or grow at different rates. Sometimes, a mixed colony appears homogeneous (not segmented) but quite different from the pure colonies of each species that it contains; this may happen when the mixture contains about equal quantities of cells from each species where each exhibits the same color and grows at a similar rate. At times, a mixed colony may look exactly like a pure colony; this happens when the contaminating cells are present in low numbers.

The important point to remember is that you cannot depend upon appearance to determine with certainty that a colony is pure. Therefore, you must continue testing the colonies that are isolated on a **primary** streak plate. To do this, you should (1) transfer cells from one colony on the primary plate to a **secondary** streak plate, (2) incubate the secondary streak plate, (3) closely examine the resulting colonies for consistency, and (4) microscopically examine cells from representative colonies for their appearance after they are Gram stained.

From a practical standpoint, a culture is considered pure if, starting from a single colony, *all subsequent streak plates show only one colony type,* and *cells from that colony type always show the same size, shape, and Gram reaction.* It is very important that your aseptic technique be good enough to prevent contamination of media and cultures while you are performing these tests. It is also important that you have good skills in preparing streak plates and in making and interpreting Gram-stained smears.

In this exercise, you may be asked to use the primary streak plates prepared for Exercise 14. You will examine Gram stains prepared from each colony type and transfer cells from each colony type to a secondary streak plate. The Gram stain will give you preliminary indication about the purity of each colony you examine. If the colony from the primary streak plate contains only one type of microorganism, then all colonies formed on the secondary streak plate should appear the same as those on the primary plate, and all colonies on the secondary streak plate should appear identical. Finally, each colony from the secondary plate should contain cells having the same size, shape, and Gram-staining characteristics.

LEARNING OBJECTIVES

- Understand the tests required for practical assurance of culture purity.
- Gain experience in accurate observation of various colony types.
- Obtain skill in preparing streak plates.
- Practice good aseptic technique.
- Obtain more experience in performing the Gram stain and in interpreting the results.

MATERIALS

Cultures   Various types of colonies on a streak plate (may use primary streak plates prepared for Exercise 14)
One known gram-positive culture and one known gram-negative culture (standard for proper interpretation of the Gram-staining technique)

Media   Trypticase-soy agar plates

Supplies   Inoculating loop
Bunsen burner and striker
Gram-staining reagents

PROCEDURE: FIRST LABORATORY PERIOD

Remember to use good aseptic technique.

1. **Describe the appearance of two colony types** found on your primary streak plate. Call one type colony A and the other colony B. **Record your descriptions** in the Results section for part A of this exercise. If the plate used is the same as that prepared for Exercise 14, you may copy the descriptions from the Results section of Exercise 14.

2. **Prepare one slide that contains separate, well-labeled smears of four cell types:** one smear from colony A, one from colony B, one from the known gram-positive culture, and one from the known gram-negative culture. *Use only very thin smears.* If smears are too thick, it will be hard to decolorize them and hard to microscopically observe individual cells. Make sure that each smear is labeled both before and after staining.

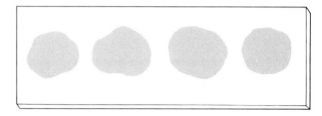

Gram (+) standard   Colony A   Colony B   Gram (−) standard

Gram stain this slide so that all smears are treated identically.

Microscopically examine at least 100 cells from colony A and 100 from colony B before you decide whether the colonies contain cells of the same size, shape, and Gram reaction. You should recall from Exercise 7 that gram-positive cultures (or colonies) that are past the exponential-growth phase may appear gram variable.

Record your observations in the Results section for part A of this exercise.

3. **Select one method for streaking plates** from those used in Exercise 14 (see Figures 14-2, 14-3, and 14-4). Carefully review each step of the selected streaking method.

4. **Use the selected streaking method to separately streak cells from colony A onto one sterile plate and cells from colony B onto another sterile plate.** These are called *secondary* streak plates because it is the second time that each cell type has had an opportunity to form pure colonies.

5. **Label each plate** with your *name, section, date* of inoculation, and *colony type*.

6. **Invert both secondary streak plates, and tape them together** with masking tape.

7. **Incubate both plates at 30°C for 48 hours.**

PROCEDURE: SECOND LABORATORY PERIOD

1. **Separately examine the secondary streak plates** prepared from colonies A and B. **Sketch the appearance** of each plate in the Results section for part A of this exercise.

   When you examine each plate, consider the following questions. Does your plate exhibit many well-separated colonies? (The answer to that question reflects your streaking technique.) Is there more than one colony type present? If you have only one colony type, does it look exactly like the colony transferred during the first laboratory period? If there are two colony types present, do they both appear like those on the primary streak plate? If there are two colonies present, and both look like those on the primary streak plate, what should you conclude about the colony used as an inoculum for streaking the secondary streak plate?

   Use these questions to help develop your conclusions, and place these conclusions in the Results section for part A of this exercise.

2. **Prepare one slide that contains separate, well-labeled smears of four cell types:** one smear from colony-type A, one from colony-type B, one from your known gram-positive culture, and one from your known gram-negative culture.

   Gram stain this slide so that all smears are treated identically. **Microscopically examine** at least 100 cells each from colony types A and B.

   Record your results in the Results section for part A of this exercise.

3. **Discuss your results with your laboratory instructor.** It is important that you learn how to prepare streak plates and to test for culture (or colony) purity because these techniques are essential for your later laboratory work. Your instructor will help you to analyze your results so that you can improve your techniques.

B · Separating cultures from an unknown mixture

Microbiologists must often purify unknown mixtures of microorganisms from infected plant or animal tissue or from complex environments such as soil and water. They often accomplish this by streaking the mixed culture on an appropriate medium to yield colonies each of which may be composed of a pure culture.

Many physiological and morphological tests are used to identify an unknown microorganism. However, if microbiologists are to rely on the test results, it is absolutely essential that they conduct those tests on a pure culture. Therefore, preparing one or more streak plates is frequently the first step in identifying an unknown bacterium. *If you do not prepare these streak plates correctly and do not achieve a pure culture, all hope for accurate identification is lost.* Please remember that statement. It cannot be emphasized strongly enough.

Near the end of a course such as this, students are often asked to identify unknown microorganisms. Therefore, it is important that you learn how to separate microorganisms on streak plates and know when you have a pure culture.

In this part of Exercise 15, you can test your skill at isolating pure cultures early in the academic term. If you find that you need more skill, you will have sufficient time to acquire it before you must depend upon the streak-plate technique for isolating and identifying pure cultures.

LEARNING OBJECTIVES

- Obtain practice in isolating pure cultures from mixtures of microorganisms by the streak-plate technique.

- Develop more skill and confidence in the proper interpretation of streak plates and Gram stains made from isolated colonies.

MATERIALS

Cultures *First period:* For each student, one broth culture containing approximately equal numbers of different species of microorganisms; each tube should contain a unique number

Both periods: One known gram-positive culture and one known gram-negative culture (standards for proper interpretation of the Gram-staining technique)

Media   Trypticase-soy agar plates

Supplies   Inoculating loop
Bunsen burner and striker
Gram-staining reagents

PROCEDURE: FIRST LABORATORY PERIOD

Review streak-plate procedures (Exercise 14, Figures 14-2, 14-3, and 14-4) and your notes from part A of this exercise before you begin this procedure.

1. **Record the number on the tube containing your unknown mixture** in the Results section for part B of this exercise.

2. **Thoroughly mix this culture; then streak two plates,** using any technique that will assure you the best separation of this mixed culture.

3. **Simultaneously Gram stain both your mixed culture and your Gram-stain standard cultures.** The appearance of the mixed culture will tell you what types of microorganisms you should have in the separated pure colonies on your streak plates. Mix the culture well before sampling to make smears.

4. **Properly label and refrigerate your mixed broth culture** until the next laboratory period.

5. **Incubate your streak plates at 30°C for about 24 hours.**

    You will be responsible for checking colony development on your plates and removing them from the incubator when colonies are sufficiently large to distinguish different colony types.

    In most cases, you should not incubate beyond 18 to 24 hours, because you will need to Gram stain each colony type during the second laboratory period. Only young cultures should be Gram stained.

    In some cases, however, more than 24 hours incubation will be needed. Some microbial types form only very small (pinpoint) colonies after 24 hours incubation.

Therefore, examine your plates carefully after 24 hours to determine if you have different colony types. If these are not evident, incubate longer. Use your judgment in this matter.

6. **After sufficient incubation, refrigerate these streak plates** until your next laboratory period.

PROCEDURE: SECOND LABORATORY PERIOD

1. **Select your best streak plate, and determine the morphology and Gram-staining characteristics of cells from each colony type.** Simultaneously stain separate smears of the known gram-positive and gram-negative standards when you stain cells taken from each colony. Prepare all smears on one slide, as shown on page 132.

2. **Record your observations** in the Results section for part B of this exericse.

3. **Write your name on your best slide** in the upper right-hand corner (over the gram-negative standard). Carefully *blot off the excess oil* with a paper towel.

4. **Tape this slide** to the bottom of the Results section. Use one small piece of masking tape at each end of the slide. This is done so that your instructor can check your results microscopically.

5. **Carefully examine the colony types** on your best streak plate, and **record your observations** in the Results section.

NAME _____

DATE _____

SECTION _____

RESULTS A: TESTING THE PURITY OF A COLONY

1. Describe colonies on the *primary* streak plate. Include age, size, and color (use descriptive terms from Table 14-1).

   Colony A:

   Colony B:

   Microscopic appearance of cells obtained from typical colony types A and B from the *primary* streak plate:

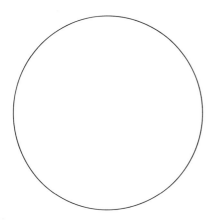
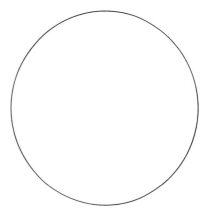

Cells from colony A, Gram stain
Magnification: _____ ×

Cells from colony B, Gram stain
Magnification: _____ ×

2. Describe colonies on the *secondary* streak plates.

   Colony A:

   Colony B:

   Microscopic appearance of cells obtained from typical colony types A and B from the *secondary*-streak plates:

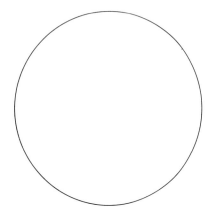
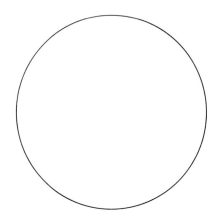

Cells from colony A, Gram stain
Magnification: _____ ×

Cells from colony B, Gram stain
Magnification: _____ ×

CONCLUSIONS

Were your colonies pure or contaminated? State the reasons to support each of your conclusions. Also consider the questions asked in the second-period procedure, step 1.

Primary Streak Plate:

Secondary Streak Plate:

NAME _____

DATE _____

SECTION _____

RESULTS B: SEPARATING CULTURES FROM AN UNKNOWN MIXTURE

Unknown identification number: _____

First Laboratory Period

Describe Gram-stained broth culture.

Second Laboratory Period

Observation	Culture A	Culture B	Culture C
Colony characteristics (include age, size, color; see Table 14-1 for descriptive terms)			
Cell shape			
Gram reaction			

Comments:

For instructor's use only

Streak-plate technique			

Tape blotted slide here:

15 DETERMINING CULTURE PURITY

NAME _____

DATE _____

SECTION _____

QUESTIONS A·TESTING THE PURITY OF A COLONY

Completion

1. A culture is considered pure if (a) all cells taken from one colony type and streaked on a plate show _____, (b) cells from that colony examined with a light microscope all appear _____, and (c) all subsequent streak plates and Gram-stained smears show these same characteristics.

2. A pure culture contains cells of one type. In other words, all cells are of the same _____ _____.

True–False (correct all false statements)

1. Only colonies starting from a single cell are pure colonies. _____

2. Any colony beginning from more than one cell should be considered a mixed culture. _____

3. Difference in colony size is a good indicator of a mixed culture. _____

4. Broth cultures should routinely be used to help determine whether a colony is a pure culture. _____

5. If a culture shows only short gram-negative rods, it should be considered a pure culture without need for further work. _____

Multiple Choice (circle all that apply)

1. A mixed culture may be indicated by
   a. both gram-positive and gram-negative cells.
   b. various colored colonies.
   c. various sized colonies.
   d. pulvinate and umbonate colonies having the same size and color.
   e. rhizoid-shaped colonies containing gram-positive rods and spheres.

2. You can be sure you have a pure culture if
   a. colonies contain only gram-negative rods.
   b. all colonies have the same color.
   c. you use cells from only one colony as an inoculum source.
   d. all colonies are convex.
   e. the inoculum contained only gram-positive cocci.

VI

Proteus mirabilis swarming on nutrient agar.

when you cannot measure . . . it in numbers, your knowledge is of a meager and unsatisfactory kind; . . . it may be the beginning of knowledge, but you have scarcely . . . advanced to the stage of science . . .

—*Lord Kelvin, 1883*

CULTURING AND QUANTITATING MICROORGANISMS

Part VI introduces you to characteristic patterns of growth of pure cultures in liquid media and to ways in which solid media are used for growing and storing pure cultures. You will also see how changes in turbidity can be used to determine the growth rates of cultures and to demonstrate when these cultures shift from exponential growth to the stationary phase. You will use cultural methods to determine the numbers of live cells in broth cultures and to distinguish the growth of various types of microorganisms on specialized media. You will also learn how cultural techniques can be used to enrich the growth of certain species when their numbers are few and they are mixed with a large number of other types of microorganisms.

16

Agar-Slant and Agar-Deep Cultures

Once you isolate a microbial species (see Exercises 14 and 15), you can use **agar slants** (see Exercise 13) to grow and then store (maintain) the pure culture. Inoculated slants should be incubated only until growth is barely visible on the slant surface. If you then store these slant cultures in a cold environment, you can maintain them for long periods of time in a suspended state because the cold temperature of a household refrigerator (about 4°C) will sufficiently slow the cells' metabolism. To prevent media from drying out during storage, you often prepare and inoculate slants in screw-capped test tubes; then you screw down the caps tightly for storage. Stored in this way, many microbial cultures stay relatively inactive but alive for 4 to 6 months.

Refrigerated slant cultures are examples of **stock cultures.** At any time, you can remove a stock culture from the refrigerator, use it as a source of inoculum for other cultures, and then return it to the refrigerator. This is often the way stock cultures are maintained for class use.

Slant cultures are not used to characterize and identify microorganisms. Instead, they are used for the growth and the refrigerated storage of pure microbial cultures.

Agar deeps are made in either a solid (1.5 to 2.0 percent agar) or semisolid (0.5 to 0.7 percent agar) form (see Exercise 13). You inoculate either form by carefully inserting a culture-laden loop, or (preferably) needle, down the center to the bottom of the deep. After incubation, this **stab culture** exhibits growth either along the entire length of the stab or only at certain places, depending upon the microbes' need for or tolerance to oxygen (see Exercise 25). Because there is not much oxygen inside an agar deep, and because the little that is present is quickly consumed by the growing microbes, you can use stab cultures to determine a microbe's ability to grow in environments where oxygen is partially or wholly depleted. You also can use stab cultures to determine whether the microbes excrete certain metabolic end-products (see Exercise 33). Although it is not usually done, you can store stab cultures in the refrigerator and use them as stock cultures.

LEARNING OBJECTIVES

• Learn how to use slant cultures as sources of inoculum.

• Learn how to inoculate sterile agar slants and agar deeps.

• Learn to recognize the appearance of microbial growth on an agar slant and in an agar deep.

MATERIALS

Cultures	*Bacillus subtilis*
	Escherichia coli
	Saccharomyces cerevisiae
Media	Trypticase-soy agar slants
	Trypticase-soy agar deeps
	Sabouraud-agar slants
Supplies	Bunsen burner
	Inoculating loop or needle

PROCEDURE: FIRST LABORATORY PERIOD

For inoculating agar slants, the following procedure uses the *single-tube transfer* method. If you prefer, you may use the multiple-tube method described in Exercise 17. Regardless of which method you select, use good aseptic technique.

1. **Collect needed materials and label tubes** before you begin. Write your *name*, today's *date*, the *culture name*, and *exercise number* on each tube.

2. **Aseptically enter the culture tube, and remove some cells with a sterile loop.** Gently touch the growth on the agar surface. Do not scrape off a large quantity of cells from the agar surface. Aseptically close the tube, and return it to the rack.

3. **Flame the lip** of the culture tube, **replace the cap,** and **return** the culture tube to the test-tube rack.

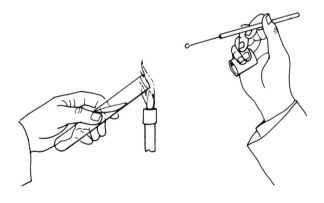

4. **Aseptically open the sterile slant tube, and transfer (inoculate) these cells onto the surface of a sterile slant** with a zig-zag motion that starts at the bottom of the slant and ends at the top. Note that you must inoculate the sterile agar with the same side of the loop that you used to touch the culture.

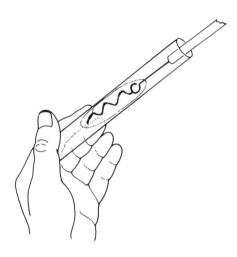

5. **Flame the lip** of the freshly inoculated slant, cap it, and sterilize your loop.

6. **Inoculate each bacterial species** (*Bacillus subtilis* and *Escherichia coli*) onto a separate trypticase-soy agar slant.
    Inoculate the yeast (*Saccharomyces cerevisiae*) onto both a Sabouraud-agar slant and a trypticase-agar slant.

To inoculate *Bacillus subtilis* and *Escherichia coli* into separate T-soy *agar deeps,* perform the following steps with an inoculating needle (not a loop).

7. **Straighten the needle.** It is easier to insert a straight needle than a crooked needle down the center of an agar deep. Also, a straight needle will form a small slit in the agar, which will close more tightly than a large slit.

8. **Flame sterilize the entire needle** because all of it will contact the sterile agar during inoculation.

9. **If you inoculate from a broth culture, immerse the needle in the broth,** and then aseptically transfer this inoculum to the sterile deep.
    If you inoculate from a slant culture, touch the end of the needle to the slant surface, and then aseptically transfer to the sterile deep.

10. **Carefully insert the culture-laden needle down the center of the deep** so that the needle is entirely inserted into the sterile deep.

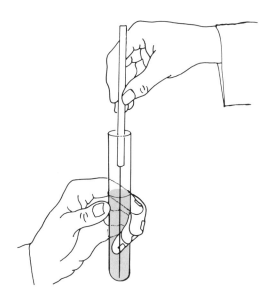

11. **Remove the needle from the inoculated deep, and flame sterilize the needle.**

12. **Place all inoculated slants and deeps in a small container** along with a scrap of paper that contains your name, so that your cultures can be easily identified in the incubator.

13. **Incubate at 30°C for 48 hours.**

14. **Ask your instructor** to put aside several uninoculated slants and sham-inoculated deeps. These will serve as uninoculated controls that you can use for comparison with your inoculated tubes.

PROCEDURE: SECOND LABORATORY PERIOD

1. **Observe the types of growth on your inoculated agar slants and deeps.** If growth is not heavy and is spread across the agar surface or all along the stab line, you may have difficulty seeing it. In that case, it may help to compare your inoculated slant or deep with uninoculated or sham-inoculated controls. It may help to use both reflected and transmitted light.

2. **Use a sterile loop to examine the texture** of each surface culture. Use proper aseptic technique.

3. **Record your observations.**

4. **Return sterile control tubes** to a rack at the front of the room.

5. **Remove all markings on your culture tubes, and discard the tubes as directed.** Do this only after all persons in your group have finished their observations.

16 AGAR-SLANT AND AGAR-DEEP CULTURES

NAME _____

DATE _____

SECTION _____

RESULTS

1. Illustrate the growth on each *slant culture* compared with the uninoculated control. Probe the surface growth on each slant with a sterile loop, and then use the terms for texture given in Table 14-1 to describe each culture.

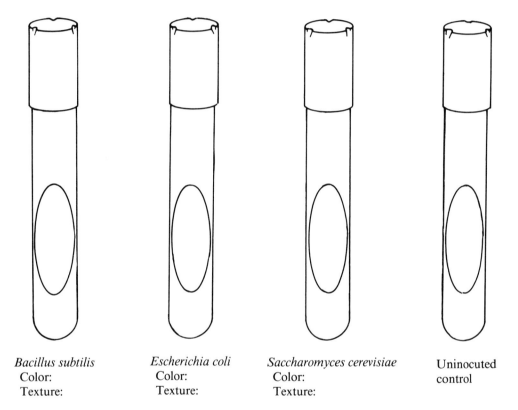

Bacillus subtilis
Color:
Texture:

Escherichia coli
Color:
Texture:

Saccharomyces cerevisiae
Color:
Texture:

Uninocuted control

2. Illustrate the growth of each *stab culture* (inoculated agar deep) compared with the uninoculated control. Do your results suggest whether these bacteria require oxygen or grow in the absence of oxygen?

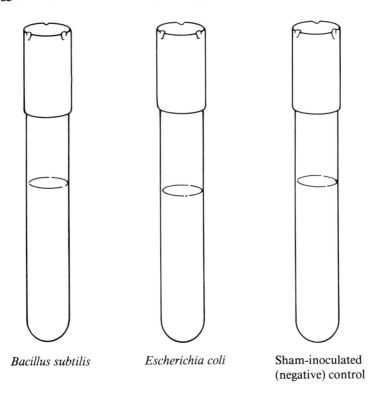

Bacillus subtilis *Escherichia coli* Sham-inoculated (negative) control

CONCLUSIONS

16   AGAR-SLANT AND AGAR-DEEP CULTURES

NAME _____

DATE _____

SECTION _____

QUESTIONS

Completion

1. The main functions of agar-slant cultures are to _____ pure cultures and to facilitate long-term _____ of these cultures.

2. Agar slants are frequently used to prepare stock cultures. After incubation, they are usually stored in _____.

3. Of the three microbes used in this exercise, _____ is the name of a eucaryotic microorganism that is a type of yeast.

True–False (correct all false statements)

1. Stock cultures prepared on agar slants are commonly stored for several years without noticeable change in the culture. _____

2. When you use an agar slant as a source of inoculum, you should remove a large quantity of cells from the agar surface. _____

3. Colony characteristics are commonly observed on agar slants. _____

4. Agar-slant cultures are commonly used to help identify a microorganism because growth on the slant surface shows characteristic patterns for different types of microbes. _____

17

Broth Cultures

Any liquid medium used for growing microorganisms is called a **broth,** a term that dates back to the days when the most common liquid media were soups derived from cooked animal and vegetable matter. The volumes of broth used to culture microorganisms vary from a few millimeters in a test tube to thousands of liters in an industrial fermentor.

To grow microorganisms in a broth, you must provide the appropriate chemical and physical environments. For example, the liquid medium called **trypticase-soy** broth contains the following chemical components: glucose, enzymatically digested milk protein, enzymatically digested soybean meal, and a small amount of sodium chloride. These chemicals provide the necessary sources of energy and nitrogen, plus the vitamins and minerals that are necessary for growing many microorganisms.

Growing cultures require a certain amount of heat in their environment, which is supplied by an incubator. Many cultures also require oxygen, a certain amount of which is absorbed into the surfaces of broth in test tubes sitting in an undisturbed rack on the incubator shelf; this condition is referred to as **static incubation.** Some microbes require more oxygen than can be absorbed at the surfaces of static broth tubes, in which case, you can place your culture tubes at an angle (to increase the surface area) and rotate them on a **roller drum apparatus** (Figure 17-1). This is one method of providing oxygen to broth cultures; a process known as **aeration.** For larger volumes of broth, air is filter sterilized and pumped into large broth-filled vessels called **fermentors.**

Before you can use a broth to grow cultures, you must sterilize it. Otherwise, organisms present in a freshly prepared broth will contaminate the culture you inoculate into that broth. An autoclave is used to sterilize most types of broth (see Exercise 13).

When you inoculate a microorganism into an appropriate broth, it grows in a way that is often characteristic for that microorganism. For example, some microbes uniformly increase the **turbidity** (milkiness or cloudiness) of the broth as their numbers increase. Most of the microbes you will work with in your laboratory this academic term have that type of growth.

Other species of microorganisms grow only at the surface of a static broth culture, forming a mat of cells called a **pellicle.** These cells usually require considerable oxygen for their growth and excrete capsular material that holds clumps of cells together at the surface.

Other microbial species may settle to the bottom of the tube to form a **sediment** or **button** of cells that stick together. Such cells do not usually need much, if any, oxygen. These cultures should be thoroughly stirred before they are used as sources of inoculum.

Other microorganisms form intertwined **mycelia** (singular, *mycelium*) during growth; these may form clumps of filamentous cells having the appearance of small cotton balls floating in otherwise clear broth. Clumps of mycelia suspended within a broth are characteristic of the filamentous fungi.

In this exercise, you will examine several characteristic ways in which microorganisms grow in broth. Note, however, that this appearance is not a very important diagnostic characteristic for identification of microbial species. The main use of broth cultures is for cultivating large numbers of microorganisms so that they can be used or studied. For example, most antibiotics are produced by microbes in broth culture, and baker's yeast is mass produced in broth culture.

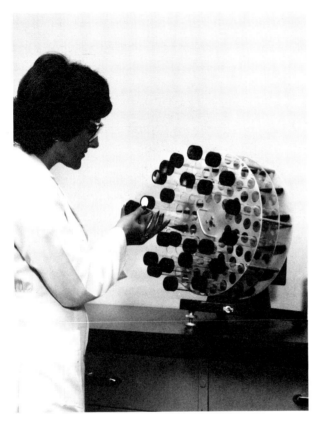

FIGURE 17-1 A rollerdrum apparatus used for improving aeration of broth cultures in test tubes. The inclined tubes have increased surface areas for greater oxygen absorption. Rotation of these inclined test tubes provides continuous circulation of the broth below the enlarged surface and, therefore, an increased amount of oxygen absorption. (Courtesy of New Brunswick Corp., Model TC.)

LEARNING OBJECTIVES

- Practice culture transfer and aseptic technique.
- Become familiar with the characteristics of microbial growth in static broth cultures.
- Learn the terms associated with microbial growth in broth.

MATERIALS

Cultures	*Bacillus subtilis* (a bacterium)
	Escherichia coli (a bacterium)
	Saccharomyces cerevisiae (a single-celled fungus; called a yeast)
	Penicillium notatum (a filamentous fungus; called a mold)
Media	Trypticase-soy broth in test tubes
Supplies	Bunsen burner
	Inoculating loop

PROCEDURE: FIRST LABORATORY PERIOD

In the following procedure, the *multiple-tube transfer* method is described; but if you feel uncomfortable with this procedure, you may handle only one tube at a time (see *single-tube transfer* method described in Exercise 16). Use good aseptic technique.

1. **Collect needed materials and label your test tubes** before you begin. Label the test tubes with your *name, date* of inoculation, *culture* used, and *exercise number*. Write small and only on the glass.

2. **Thoroughly mix the culture** if a broth culture is used as an inoculum. Thump the bottom of the tube to create a vortex for mixing.

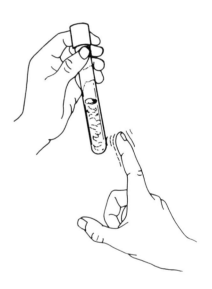

3. **Hold both the culture and the tube of sterile broth in one hand.**

17 BROTH CULTURES

4. **Sterilize your loop,** and then **remove both caps** (or plugs) from these tubes by grasping them between the fingers of the hand holding the loop.

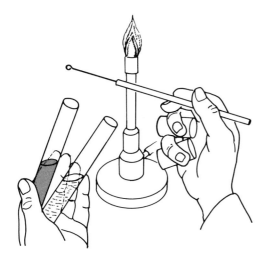

5. **Alternately flame the lips of both tubes.**

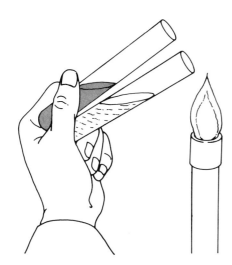

6. **With the sterile loop, transfer a loopful of culture into the sterile broth.** After this transfer, touch the loop to the inner surface of the tube to remove excess broth.

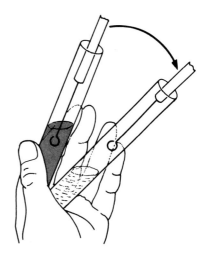

7. **Alternately flame the lips of both tubes, and replace both caps** (plugs). Make sure that each cap is returned to the correct tube.

8. **Flame sterilize your inoculating loop** before returning it to its holder.

9. **Inoculate each organism into a separate tube** of sterile trypticase-soy broth. Place all tubes in an appropriate container.

10. **Ask your instructor** to put aside several uninoculated tubes to serve as controls that you can use for comparison with your inoculated tubes.

11. **Incubate all tubes at 30°C for 24 hours.**

PROCEDURE: 24 HOURS LATER

1. **Carefully remove your cultures** from the containers. *Do not shake* or otherwise disturb the broth cultures until you have observed the effects of growth.
2. **Record the undisturbed appearance of growth in all tubes.** Each culture should be visually compared with uninoculated control tubes.
3. **Return these tubes to the incubator** after recording your observations.

PROCEDURE: SECOND LABORATORY PERIOD

1. **Carefully retrieve your cultures** from the incubator without disturbing them.
2. **Record any change in appearance** from those observed 24 hours after incubation (and the number of days of incubation) on your Results page.
3. **Vigorously mix each tube** by thumping the bottom of it against a finger, and record how easily the sediment (if any) was suspended, or the pellicle disturbed, and how homogeneous the suspension appeared both before and after resuspension.
4. **Return sterile control tubes** to a specially marked rack provided by your instructor.
5. **Remove all markings on the tubes, and discard** these cultures as directed. Discard only after everyone has recorded the results and compared them with the results of others.

CHARACTERISTIC MICROBIAL GROWTH PATTERNS

Turbidity Once the microbial cell numbers reach about 1×10^6 cells/ml, you will notice a slight turbidity. A turbid culture looks cloudy or milky because the cells are scattering the light that would normally pass through the clear medium. (You might hold a printed page up to the tube to see how clearly you can see the printing through the medium.) Uniform turbidity during growth in broth is a characteristic of many species of bacteria. As this type of culture ages, the cells may settle to the bottom of the tube; you can see this sediment or button by examining the tube from the bottom and then swirling the tube gently to partially disperse the cells into the medium.

Pellicle Cultures that require a lot of oxygen and have the ability to stick together may be found in a surface film. These pellicles vary from very thin, filmy, and almost invisible to thick and so heavy that they sink into the medium. Usually, when a culture forms a pellicle, the rest of the medium is not very turbid, or not turbid at all.

Button Some cultures grow rapidly and settle to the bottom of the tube into a cell mass called a *sediment button*. This button of cells may or may not be easily resuspended into the broth after you disturb the culture. The formation of buttons is characteristic of some yeast cultures. Although yeast cells may consume oxygen, they do not require it to grow. Also, some yeasts form sticky capsular material, and these cells will stick together and settle out of suspension during growth.

Filamentous fungi The filamentous fungi grow in long strands called *hyphae* (plural) that tangle together during growth. When this mass gets large, it can take a number of forms. Some fungi form pellicles at the surface, referred to as *hyphal mats*. Other filamentous fungi form submerged masses of hyphae that look like fuzzy balls of cotton within in the medium.

17 BROTH CULTURES

NAME _____

DATE _____

SECTION _____

RESULTS

1. Appearance of cultures *after 24 hours of incubation.* (Draw and label with descriptive terms; refer to Procedure: 24 hours later):

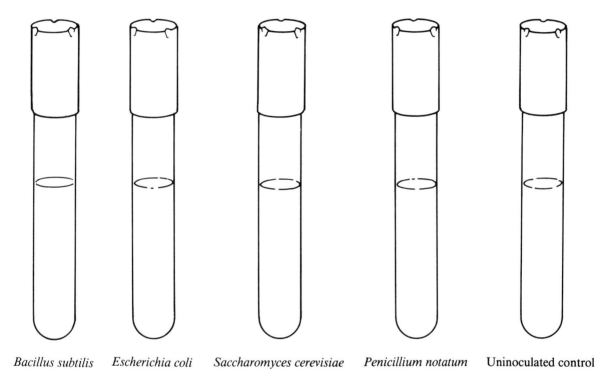

Bacillus subtilis   *Escherichia coli*   *Saccharomyces cerevisiae*   *Penicillium notatum*   Uninoculated control

2. Appearance of cultures *after* _____ *days of incubation* (number of days between first and second laboratory periods):

Culture	Changes noted from 24 h observation	Reaction of culture to rapid mixing
B. subtilis		
E. coli		
S. cerevisiae		
P. notatum		
Uninoculated control		

CONCLUSIONS

17 BROTH CULTURES

NAME _____

DATE _____

SECTION _____

QUESTIONS

Completion

1. A broth is defined as _____.

2. Another term for cloudiness or milkiness in a culture is _____.

3. The thick film of cells covering the surface of a broth is called a _____.

4. The primary purpose of broth cultures is to _____.

True – False (correct all false statements)

1. Broth tubes are always capped or plugged before sterilization. _____

2. Trypticase-soy broth need not always be sterilized before it is used to grow pure cultures. _____

3. Cells that require much oxygen may form a pellet during active growth in statically incubated broth tubes. _____

4. It is necessary to sterilize only the tip of the inoculating loop before it is used to enter a tube. _____

Multiple Choice (circle all that apply):

1. Each of your culture tubes should be labeled with
   a. your name or initials.
   b. culture name.
   c. date of inoculation.
   d. experiment number.
   e. manual page number.
   f. social security number.

2. Trypticase-soy broth contains
   a. partially degraded milk protein.
   b. beef extract.
   c. sodium cyanide.
   d. partially degraded soybean meal.
   e. sucrose.

18

Culturing Anaerobes: Thioglycolate Use and the Anaerobe Jar

Microorganisms that grow only in the presence of oxygen (air) are called **aerobes,** and those capable of growth only in the absence of oxygen (air) are called **anaerobes.** No sharp boundary exists between aerobic and anaerobic environments, and there is no sharp distinction between the microbes that exist in these environments. Some microbes grow in either the presence or absence of oxygen; these are usually called **facultatives** or **facultative anaerobes** (see Exercise 25 for more detail).

A microbe's ability to grow in either the presence or the absence of oxygen is directly related to its metabolism. Strict aerobes are incapable of fermentation and can respire only by using oxygen as their terminal electron acceptor (see Exercises 36 and 57). Facultatives, on the other hand, are capable of either respiratory or fermentative metabolism (see Exercises 31 and 32); they shift their metabolism depending upon the availability of oxygen. Anaerobes, unable to respire, have only a fermentative type of metabolism.

Anaerobes have a wide range of oxygen tolerance. Oxygen is very **toxic** for the strict (stringent or obligate) anaerobes, which can survive and grow only where oxygen is completely excluded. It is believed that dissolved oxygen in the medium forms toxic free radicals and hydrogen peroxide, which strict anaerobes are incapable of adequately detoxifying.

Nonstringent anaerobes can tolerate some oxygen and can even grow slightly in the presence of small amounts. Therefore, these microbes are called **aerotolerant anaerobes.** However, oxygen is not used in their metabolism, and their growth is more luxuriant when oxygen is totally absent. Thus, it is believed that aerotolerant anaerobes can detoxify oxygen, but not to the same extent as the facultatives or the strict aerobes.

Oxygen has at least two other effects on culture media in addition to forming free radicals. First, molecular oxygen is absorbed into media and exists there as dissolved oxygen (O_2). Dissolved oxygen is used by many microorganisms and higher forms of life (see Exercise 25). As with all gases, the quantity absorbed or dissolved in a liquid depends upon temperature.

Second, oxygen affects the O/R potential of culture media. Consider what is meant by the term **O/R potential.** At any one time, some organic molecules in a medium are in the oxidized (O) state, and others are in the reduced (R) state. The proportion of oxidized to reduced molecules (O/R) is used to describe the relative state of oxidation or reduction of the medium. This overall oxidation-reduction state can be measured with instruments, and its numerical value is the O/R potential. For example, when oxygen dissolves in a medium, organic compounds become more oxidized, and the medium exhibits a *positive* O/R potential. As organisms consume oxygen or heat drives it off, or reducing agents (chemicals that contribute hydrogen) reduce the medium components, the medium shows a *negative* O/R potential. Strict (stringent) anaerobes do not grow unless the O/R potential of the medium is very low (negative).

Several ways exist for lowering the O/R potential of a culture medium and keeping it low after oxygen is removed. For example, you can add reducing agents such as cysteine or sodium thioglycolate to the medium. **Cysteine** [$HS-CH_2CH(NH_2)COOH$] is an amino acid that contains a sulfhydryl group ($-SH$) that easily donates its hydrogen to other compounds.

Thus, cysteine is able to reduce other organic compounds in the medium. **Sodium thioglycolate** (HS—CH_2COONa) also contains a sulfhydryl group and also is a good organic reducing agent. Both chemicals are used for culturing anaerobes in liquid media. Thioglycolate is the more commonly used.

In summary, the cultivation of strict (stringent) anaerobes requires employment of a method for removing oxygen from the atmosphere above the medium and for establishing and maintaining a low O/R potential.

Many ways have been devised for removing gaseous oxygen dissolved in a liquid medium or present in the atmosphere above the medium. The most satisfactory method depends on the type of anaerobe being cultured and its tolerance to oxygen.

Nonstringent anaerobes are easy to grow in liquid media without special equipment. For example, you may completely fill a sterile, screw-capped test tube with a freshly prepared broth (to eliminate an atmosphere above the medium), autoclave it (to sterilize the medium and drive off dissolved O_2), and then cool, inoculate, and seal it again to prevent entry of O_2. A reducing agent is often added to the medium prior to autoclaving to reduce the medium and keep its O/R potential low. Note, however, that this method should be used only for incubation of non-gas-producing anaerobes because voluminous gas production will make the tightly sealed tube explode, and gas production is common during growth of anaerobes.

For growing gas-producing nonstringent anaerobes, you can prepare media in non-screw-capped tubes, autoclave, cool, and inoculate them, and then layer melted Vaseline or mineral oil on the surface of the broth. This keeps the inoculated liquid from absorbing atmospheric oxygen, yet the gases produced during fermentation either push up the Vaseline plug or bubble through the mineral oil, thereby releasing pressure without breaking the tube.

The most efficient way to remove oxygen from liquid media is to aseptically flush chemically reduced media with a sterile, oxygen-free gas (such as carbon dioxide or nitrogen) during inoculation. Then you tightly reseal the tube with its sterile rubber stopper. Usually, you use heavy-walled tubes sealed with clamped stoppers to prevent explosions caused by microbially produced gases.

A modern way to grow nonstringent anaerobes on solid media is with an **anaerobe jar** containing a gas-generating system (Figure 18-1). This technique was developed in 1966 by Brewer and Allgeier. Equipment to support this technique is now commercially produced. The gas-generating system is contained in

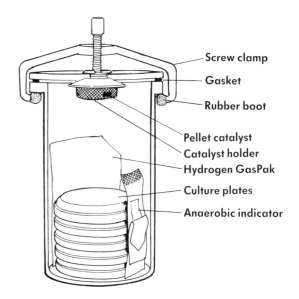

FIGURE 18-1 Cross-sectional diagram of a sealed anaerobe jar containing an activated gas-generating system package (hydrogen GasPak) and anaerobic indicator strip. Note placement of the pellet catalyst inside the jar. (Courtesy of Baltimore Biological Laboratories, Baltimore, Md.)

one small package and produces both *hydrogen* and *carbon dioxide*. Each package contains one sodium borohydride tablet, one sodium bicarbonate tablet, and powdered citric acid. When you add water to the foil package, these reagents combine and generate gaseous hydrogen (H_2) and carbon dioxide (CO_2). In the presence of a solid catalyst, the hydrogen combines with atmospheric oxygen to produce water ($2H_2 + O_2 \rightarrow 2H_2O$), thus removing oxygen from the jar and also providing a reducing atmosphere above the medium. The carbon dioxide is helpful because growth of many anaerobes is stimulated by extra CO_2. In this exercise, you will use an anaerobe jar and a gas-generating system to produce a proper anaerobic environment for growth of nonstringent anaerobes.

In most of the techniques described above, removing oxygen from the atmosphere and the growth medium requires considerable time. Although such procedures are very satisfactory for aerotolerant anaerobes, they are not suitable for growing strict anaerobes because some oxygen always remains. Strict anaerobes must be transferred and inoculated in a completely anaerobic environment, which can be accomplished in an anaerobic chamber similar to the one shown in Figure 18-2. However, these chambers are not commonly found in microbiology laboratories because they are very expensive.

Anaerobic microorganisms contribute to the environment by degrading many complex organic mole-

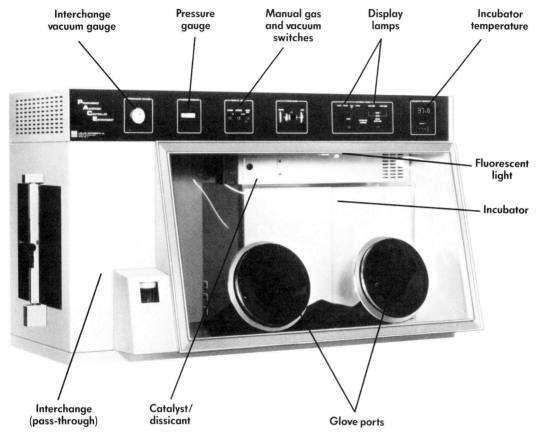

FIGURE 18-2 A modern chamber used for transferring and incubating strict anaerobes in an oxygen-free environment. (Courtesy of Lab-Line Equipment Co., Model 6500 PACE.)

cules that are found in decaying plant and animal tissue. They also occur in huge numbers in the animal intestinal tract. They are responsible for some severe human infections (such as tetanus) and food intoxications (such as botulism). Therefore, it is important that microbiologists know how to isolate and cultivate anaerobic microorganisms.

LEARNING OBJECTIVES

• Become familiar with atmospheric and medium conditions necessary for cultivation of anaerobes.

• Be aware of transfer and cultivation requirements for both aerotolerant and strict anaerobes.

• Learn how to operate the anaerobe jar and the gas-generating system for removing atmospheric oxygen.

• Compare the effects of aerobic and anaerobic incubation on growth of a strict aerobe, a facultative anaerobe, and an aerotolerant anaerobe.

MATERIALS

Cultures   *Micrococcus luteus*
           Clostridium sporogenes
           Escherichia coli

Media   Trypticase-soy agar plates

Supplies   Anaerobic jars
           Indicator strips
           Hydrogen and carbon-dioxide generator packages
           Fresh catalyst pellets
           10-ml pipettes (nonsterile)

PROCEDURE: FIRST LABORATORY PERIOD

1. **Draw lines on the bottoms of two plates to divide each plate into three equal sections,** as shown in Figure 18-3.

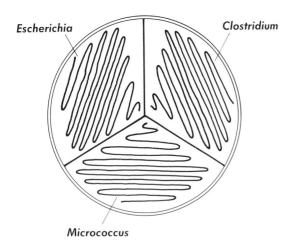

FIGURE 18-3 Method for marking on the bottom of the plate to divide it into three sections. Each section should be inoculated with a different microorganism, as shown.

2. **Label each of the three sections,** on each plate, with the name of the microbe to be inoculated on that section. Also add your *name* and the *date* of inoculation on the bottom of both plates. Make sure that you mark which plate will be incubated *anaerobically.*

3. **Streak each microbe on both plates,** as illustrated in Figure 18-3. Make sure each microbe is inoculated only onto the section containing its name. Work quickly so that your anaerobic culture will not be exposed to air any longer than necessary. Use good aseptic technique!

4. **Invert both plates, and place one in the anaerobe jar.** Note the color of the exposed indicator strip after being exposed to the air for a few minutes.

TECHNICAL NOTE

The foil package contains a filter-paper strip moistened with methylene blue (structure given in Exercise 4). *This dye is colorless when reduced and blue when fully oxidized.* The foil package is sealed under anaerobic conditions; so the white filter-paper strip is reduced and colorless when the package is first opened. After the dye is exposed to air, it quickly oxidizes and turns blue. Once the jar is sealed, the GasPack activated, and the atmospheric oxygen depleted, the dye is reduced to its colorless state. Therefore, this strip provides a visual check on the atmosphere inside the jar. It takes about 6 hours for the strip to be fully reduced (lose its blue color) at room temperature.

---

5. **Your instructor will start the hydrogen and carbon-dioxide generator and seal the jar.** The gas generator is started by adding 10 ml of water through the cut opening in the envelope. The jar is sealed and placed in the incubator. The water produced by the catalyzed reaction between the hydrogen gas and atmospheric oxygen will appear as condensation on the inner walls of the jar. This will be evident about 30 minutes after sealing the jar.

    Note that many anaerobes grow better when provided with carbon-dioxide concentrations higher than the amounts normally found in air. Concentrations inside these jars normally reach about 7 percent shortly after the hydrogen and carbon-dioxide generator is activated and the jar sealed.

6. **Incubate the second inoculated plate aerobically.**

7. **Incubate both plates at 30°C for 2 to 5 days.** This extended incubation allows these cultures to achieve maximum growth under each atmospheric condition.

PROCEDURE: SECOND LABORATORY PERIOD

1. **Examine the anaerobe jar before it is opened.** Record the color of the indicator strip. What does this color tell you about the internal atmosphere?

2. **Open the jar, and obtain your plate.** Record any odors coming from the jar.

3. **Record the extent of growth of all microbes incubated aerobically and all those incubated anaerobically.** Since growth on an agar plate cannot easily be quantitated, simply illustrate the comparative amount of growth observed. Note also the presence of pigments where observed.

4. **Properly discard your plates.**

NAME _____

DATE _____

SECTION _____

RESULTS

Color of the indicator strip before opening jar:

Odor of atmosphere upon opening jar:

For which of the three bacteria inoculated is this odor characteristic?

Record the appearance of each plate below:

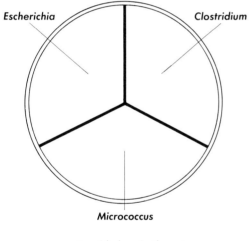

Aerobic incubation at
30°C for _____ days

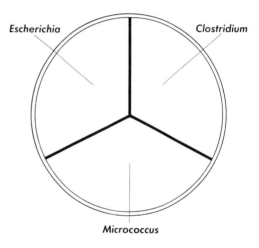

Anaerobic incubation at
30°C for _____ days

Note here how these results compare with those from Exercise 25:

CONCLUSIONS

QUESTIONS

Completion

1. Microbes that grow only in the presence of oxygen are called _____. Those that grow only in the complete absence of oxygen are called _____ (two words needed). Those that are capable of growth in both the presence and the absence of oxygen are called _____ (one word) or _____ (two words).

2. Anaerobes prefer or require a medium having a _____ O/R potential. This value represents the extent to which the medium has been chemically _____. The most commonly used compound for adjusting the O/R potential in this manner is _____.

3. One way to eliminate oxygen from the atmosphere above liquid media in test tubes is to _____ _____ prior to incubating the culture.

4. The type of metabolism characteristic of strict anaerobes is called _____, whereas _____ is the type that is characteristic of strict aerobes.

5. The anaerobe jar and gas-generator package are used to provide a suitable environment for cultivation of _____ anaerobes.

6. The gas-generator package generates two gases. These are _____ and _____. The first of these is important because it _____, and the second gas is important because it _____.

True – False (correct all false statements)

1. Considering the three bacteria used in this experiment, *Escherichia coli* is the only one that grew like a strict aerobe. _____

2. Screw-capped tubes that are filled to capacity, autoclaved, and then inoculated and sealed immediately after cooling are satisfactory for cultivating gas-producing facultative anaerobes. _____

3. It takes about 6 minutes for all oxygen to be consumed inside the anaerobe jar after activating the gas-generating package and sealing the jar. _____

4. The indicator strip used inside the anaerobe jar is blue when fully oxidized and white (colorless) when reduced. _____

Multiple Choice (circle all correct answers)

1. You might find a strict anaerobe growing in which of the following environments?
   a. A thioglycolate-containing medium, purged with oxygen-free nitrogen, and sealed off with a sterile rubber stopper
   b. A thioglycolate-containing medium sealed off with a layer of Vaseline
   c. Petri plates incubated inside an anaerobe jar containing an activated gas generator
   d. Petri plates and thioglycolate-containing liquid media, transferred and incubated inside an anaerobic chamber
   e. Your own intestines

NAME _____

DATE _____

SECTION _____

2. You might find facultative anaerobes growing in which of the following environments?
   a. A thioglycolate-containing medium, purged with oxygen-free nitrogen, and sealed off with a sterile rubber stopper
   b. A thioglycolate-containing medium sealed off with a layer of Vaseline
   c. Petri plates incubated inside an anaerobe jar containing an activated gas-generating package
   d. Petri plates and thioglycolate-containing liquid media transferred and incubated inside an anaerobic chamber
   e. Your own intestines

19

Measuring Turbidity Changes during Growth of Broth Cultures

If you hold a printed page behind a tube of sterile broth, you can see the print clearly through the tube. If you add a few drops of milk to the broth, the print becomes difficult if not impossible to read. This happens because very small particles of phosphoprotein (casein) are suspended in the milk and act as a **colloid** to scatter the light as it passes through the tube. **Turbidity** is the effect of light scattering by a colloidial suspension. The more milk that you add to the broth, the more particles present, the more light scattered, and the more turbid the broth appears.

Many growing microorganisms act like colloids: They do not settle out of suspension, and they scatter light, creating turbidity. Microbiologists use this behavior to measure microbial growth. As cultures grow, the number of cells indirectly increases, accompanied by a corresponding increase in turbidity. In general, the amount of light scattered is directly proportional to the number of cells in a culture. However, you should remember that this is true only for microbes that do not form clumps or long filaments in the growth medium.

Microbiologists usually measure changes in turbidity with an instrument called a **spectrophotometer**, which measures the amount of light that passes through a liquid medium (Figure 19-1). The lamp produces light that is split into various wavelengths by the diffraction grating. A small part of the diffracted light passes through a **cuvette** (special test tube) that contains the liquid sample. The light that passes through the sample (the transmitted light) is collected by a phototube, and the quantity entering the phototube is shown on the galvanometer. Measurements are usually begun by inserting into the instrument a cuvette that contains *only* sterile broth and adjusting the galvanometer to read 100 percent transmittance. Then a second, optically matched cuvette is filled with a growing culture (in the same type of broth) and placed in the light path. The turbid culture, which scatters more light than the sterile broth, transmits less light to the phototube and thus produces a lower reading on the galvanometer. The more turbid the culture, the lower the percent transmittance. In other words, there is an *inverse relationship between cell numbers and percent transmittance.*

Growth is not usually reported as loss in transmittance with time, however. Instead, it is traditional to record growth as the change in **optical density (OD)** with time. This is easier to deal with because *OD is directly proportional to cell concentration.* As cell numbers increase, so does the optical density.

Let us assume that you want to follow the growth of a broth culture in a large fermentor. You could periodically remove culture samples and compare their OD with that of the sterile medium. You would find that OD *increases exponentially (logarithmically) with time* during active growth. You would also discover that the rise in OD stops when active growth stops. If something should get into your culture and destroy (lyse) the cells, the OD would drastically decrease. Therefore, the primary purpose of spectrophotometry is to show relative changes in population density. Note that spectrophotometry alone does not tell you the exact number of cells in your culture.

Mathematically speaking, **OD** $= 2 - \log(\% \, T)$. The relationship between OD and percent transmittance (% T) can be seen by examining the galvanometer on most spectrophotometers. Both scales are usually present. However, OD is referred to as *absorbance* because these instruments are often used to measure

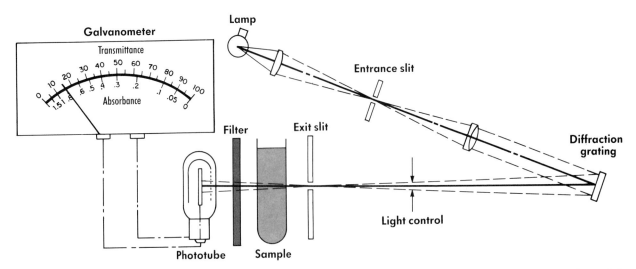

FIGURE 19-1 Schematic diagram of the light path in a Bausch & Lomb Spectronic-20 spectrophotometer. (Courtesy of Bausch & Lomb.)

concentration of color-absorbing solutes in a solution. The relationship between % T and absorbance is exactly the same as the relationship between % T and OD. Thus, you can read the values on the log scale (Abs) and consider these to be OD values.

There is one problem with measuring increases in turbidity (growth) with the conventional cuvette: Samples must be transferred from the culture vessel to the cuvette. If you take many measurements, you will rapidly deplete the culture volume, which can alter the microorganism's growth characteristics. One way to avoid this is to use a **growth flask,** as shown in Figure 19-2. Instead of transferring samples from the flask to a cuvette, you simply tip the culture into the attached cuvette and insert the cuvette into the spectrophotometer. A black cloth bag placed over the flask prevents stray light from interfering with measurements.

LEARNING OBJECTIVES

- Learn the relationship between changes in microbial numbers in a broth culture and changes in optical density.

- Learn how to operate a spectrophotometer to measure cell density in broth cultures.

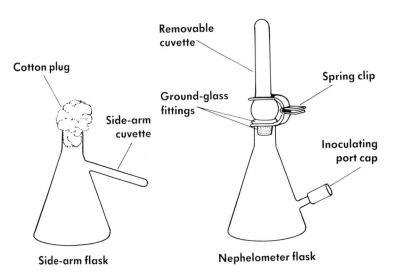

FIGURE 19-2 Two types of modified Erlenmeyer flasks (growth flasks) used for measuring turbidity of broth cultures without removing samples of broth.

19 TURBIDITY OF BROTH CULTURES

MATERIALS

Cultures	Samples of an *Escherichia coli* culture taken and chemically fixed at intervals during growth and then placed into matched cuvettes and sealed with parafilm
Media	Sterile trypticase-soy broth in one cuvette to be used as a blank to zero the spectrophotometer
Supplies	Bausch & Lomb Spectronic-20 spectrophotometer set at 620 nm Matched cuvettes Kimwipes
Demonstration	Example of growth flask (if available)

PROCEDURE

Preparing the spectrophotometer

Refer to Figure 19-3.

1. **Turn the power switch clockwise** until a distinct click is heard. In some older models, the unit is turned on by plugging the power cord into a 110-V outlet.
2. **Check the wavelength control** to make sure that the unit is adjusted for the required wavelength of incident light (620 nm).
3. **Allow at least 10 minutes for the unit to warm up** before you take readings.

Operating the spectrophotometer

Your instructor has already matched the cuvettes to make sure that they are optically identical. When this is so, then spectrophotometric differences must be due to the contents of the cuvettes rather than to different optical properties of the cuvettes. Ask your instructor how the cuvettes were matched.

1. **Turn the power switch clockwise** to adjust the galvanometer to zero percent transmittance (% T). This is considered the dark-current adjustment. The galvanometer now indicates that no light is being transmitted to the phototube.
2. One of the matched cuvettes contains sterile T-soy broth. This is called the **blank. Wipe the outside of this cuvette with a clean paper tissue, and insert the cuvette into the sample holder.** Make sure that the vertical line marked on the tube is exactly opposite the raised line under the sample-holder lid. Close the sample-holder lid.

   Adjust the 100%-T control until the galvanometer shows 100% T. The galvanometer now indicates that 100% of the light leaving the light source is being received by the phototube.
3. **Select the cuvette labeled T_0** (the sample taken immediately after inoculation). Gently **invert the parafilm-sealed cuvette several times** to uniformly mix the cell suspension without introducing bubbles. **Place this cuvette into the sample holder** of the adjusted

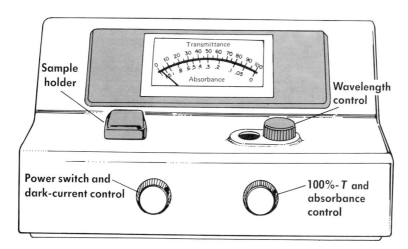

FIGURE 19-3 Front view of the Bausch & Lomb Spectronic-20 spectrophotometer. (Courtesy of Bausch & Lomb.)

spectrophotometer. **Close the sample-holder lid,** and, without touching the control knobs, **read the optical density (absorbance)** on the logarithmic scale. If the sample is turbid, the OD should be greater than zero.

4. **Remove the cuvette from the sample holder. Repeat procedure 3 for all remaining samples in numerical order** (samples T_1 to T_8) for samples taken from one to eight hours after inoculation.

    Adjust the dark current to zero before determining the OD value of each sample.

    Each time you take a reading, you must first wipe fingerprints from the outside of the cuvette and make sure the vertical line on the cuvette is aligned with the raised line on the holder.

5. **Record** your results.

19 TURBIDITY OF BROTH CULTURES

NAME _____

DATE _____

SECTION _____

RESULTS

1. Record the optical density value for each sample in the following table:

Culture (hours of incubation)	Optical density (620 nm)

2. Plot these optical density values on the semilog paper provided on the following page.

CONCLUSIONS

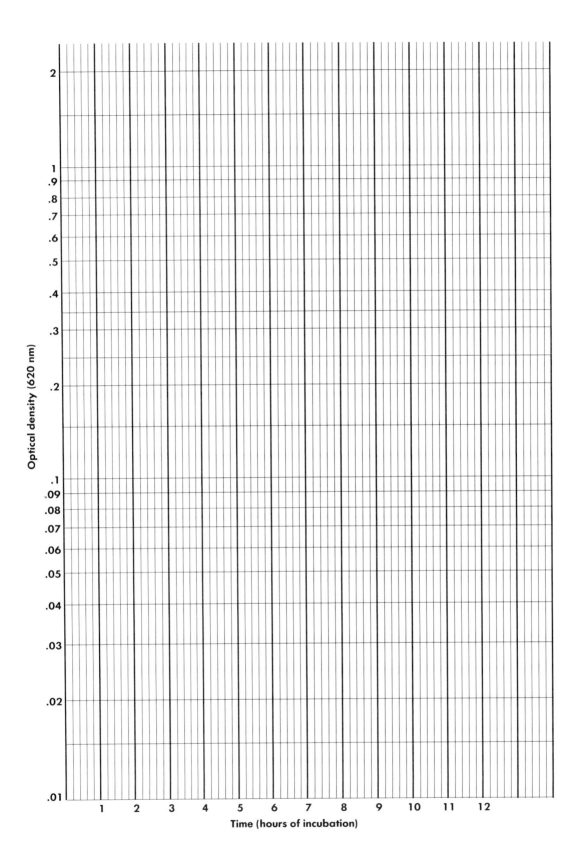

19 TURBIDITY OF BROTH CULTURES

NAME _____

DATE _____

SECTION _____

QUESTIONS

Completion

1. _____ and _____ are two examples of particles that act like colloids when suspended in a liquid.

2. When a liquid contains colloidal particles in suspension, its appearance is said to be _____.

3. Light scattering is _____ to the number of microorganisms suspended in a liquid.

4. An instrument that measures the quantity of light scattered by a colloidal suspension is called a _____.

5. The tube used to hold a sample in the instrument for measuring scattered light is called a _____.

6. A formula showing the relationship between OD and percent transmittance is _____.

True – False (correct all false statements)

1. One test of a colloid is that the particles quickly settle out of a static liquid. _____

2. Growth of filamentous fungi and bacteria that clump are commonly followed by light-scattering measurements. _____

3. Light passing through a sample in the spectrophotometer is collected by a galvanometer, and the amount of light collected is shown on the instrument's phototube. _____

4. As a microbial culture grows, its optical density increases. _____

5. The term *optical density* is used to indicate light scattering by colloidial particles, whereas *absorbance* refers to light absorbed by solutes in solution. Both are mathematically related to the percent transmittance in the same way. _____

6. The primary purpose of spectrophotometry is to tell the microbiologist the exact number of cells that are in a broth suspension. _____

Multiple Choice (circle all that apply)

1. A sterile (blank) broth sample is used in the spectrophotometer to
   a. adjust the dark current.
   b. determine the turbidity of the uninoculated broth.
   c. calibrate it for 100-percent transmittance (zero OD).
   d. convert absorbance to optical density.

2. Growth of a well-suspended microbial culture is accompanied by
    a. a decrease in percent transmittance.
    b. an increase in optical density.
    c. an increase in turbidity.
    d. an increase in broth volume.

20

Counting Viable Cells: Serial Dilution and Spread Plates or Pour Plates

In many microbiological studies it is important to know how many live microorganisms are present in a culture. But how does one determine whether such a tiny creature is alive? This question is not easy to answer, but microbiologists routinely use a simple definition: Microbes are alive (or viable) if they are capable of forming colonies on a suitable solid medium. Therefore, a **viable number** is that number of cells capable of division on a solid medium. Note that this is a definition of convenience, but it is used because there is virtually no other reliable way to count living microorganisms.

One might ask whether a cell can be alive if it is not capable of division. The answer is yes, especially if the medium does not provide the essential nutrients for division. If one begins with a mixed culture from the environment, it is impossible to provide one medium that supports the growth of all possible microbial cells. It is also conceivable that some older cells have lost the capacity for division, yet are still capable of metabolism and other activities that we normally attribute to living cells. It is for reasons such as these that microbiologists often refer to **colony-forming units (CFU)** rather than to viable numbers when they describe cultures.

Serial dilution and plating is the best method available for determining viable numbers, even though it does have certain limitations. For example, it works well for cells that separate shortly after division but not for cells that stick together after division. The reason is that, although many cell divisions may have contributed to one chain or clump, each chain or clump of cells forms only one colony on the agar surface. (This phenomenon is discussed in Exercise 14.)

Choosing a suitable solid medium for plating live cells is not always easy because many microorganisms have very specific nutritional needs. Especially with mixed cultures, you cannot always be sure that you have provided an appropriate medium for growing all of the cell types present. If you are working with a pure culture, however, the literature usually can suggest a suitable medium for growth. Remember that the medium chosen must promote cell multiplication.

Other potential problems with serial dilution and plating are sampling error and technical error. **Sampling error** usually occurs because of an unequal distribution of cells in the culture or dilution fluid; the goal is to distribute the cells evenly by thorough mixing and then to obtain a representative sample in the pipette. **Technical error** is most often due to some inaccuracy in preparing dilution blanks or in pipetting. You can minimize these errors by precise measurements in preparing blanks and by accurate pipetting technique. It is impossible to avoid all error, however. When microbiologists examine exponentially growing cultures, they often make about six separate transfers between dilution blanks before the diluted culture is plated. Each pipette transfer into separate dilution blanks introduces the possibility of sampling and technical errors. At best, this method results in about 10 percent error. You can best observe these errors by plating many replicates from many identical serial dilutions; the errors show up as variations in the number of colonies on presumably identical plates.

In summary, serial dilution and plating is, at best, an estimate of the number of live microorganisms present in a sample. This method frequently produces a low estimate because some cells may be alive but incapable of division, and some may exist in pairs, chains, or clumps. Nevertheless, serial dilution and plating is currently the best method available for determining viable numbers of microorganisms in a liquid sample.

Principles of serial dilution

Most laboratory cultures and some environmental samples contain many thousands or even hundreds of millions of cells per milliliter. Therefore, you must dilute these cultures or samples before plating them. **Serial dilution** is the method of sequentially diluting a culture through a series of sterile dilution blanks. The fluid in the dilution blanks **(diluent)** may be sterilized distilled water, saline, or an appropriate (but chilled) liquid growth medium. It is helpful to know something about the microbes before the diluent is selected; otherwise, cell viability may be lost during the serial dilution process. For example, if you are diluting an aquatic microorganism from a lake sample, sterile distilled water or sterile lake water may be an appropriate diluent. But if you take cells from a nutrient-rich environment, such as animal tissue, and dilute them in distilled water, cellular damage and loss in viability may result because of rapid changes in **osmolarity** (solute concentration) of the fluid surrounding the cells (see Exercise 24). In general, one should *dilute microorganisms in a fluid that is osmotically similar* to the environment from which they were transferred. Care must be taken, however, not to let the cells grow (increase in numbers) during the serial dilution process; cold sterile diluents are sometimes used to avoid that possibility.

Examine the dilution scheme in Figure 20-1, and *note how microbiologists refer to dilutions.* If you mix one part of culture with one part of diluent, you obtain a 1 : 2 (one-to-two) dilution. One part of culture placed in nine parts diluent is called a 1 : 10 (or 1/10 or 10^{-1}) dilution. If you take one part of the 10^{-1} (1 : 10) dilution and place it in nine parts of diluent, you obtain an additional one-to-ten dilution. This is because the final (total) dilution is a multiple of all dilutions. In the preceding example, the final dilution is $1/10 \times 1/10 = 1/100$ (or 1 : 100 or 10^{-2}). The denominator of the final dilution (100 or 10^2) is called the **dilution factor.** *To determine the final dilution, multiply all dilution factors.* For example, if you place one part of a 1 : 100 (10^{-2}) dilution into nine parts of diluent (a further 1 : 10 dilution), the resulting dilution factor is $10 \times 100 = 1000$ or 10^3.

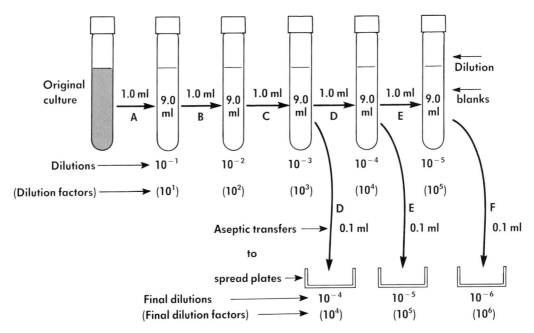

FIGURE 20-1 Serial-dilution and SPREAD-PLATING scheme for the culture used in this exercise. Note how much volume is aseptically transferred (arrows). Capital letters (A – F) refer to sterile pipettes and how they should be used. For example, *if 1.0-ml pipettes are used,* make two transfers with pipette D. First, aseptically withdraw 1.0 ml of sample from the 10^{-3} dilution, and transfer this volume to the sterile 10^{-4} blank. Second, use the same pipette to remove fluid from the 10^{-3} dilution, and transfer 0.1 ml to each of one or more spread plates labeled 10^{-4}. *If 1.1-ml pipettes are used,* withdraw 1.1 ml from the 10^{-3} blank with pipette D, transfer 0.1 ml to one 10^{-4} spread plate, and then place the remaining 1.0 ml to the 10^{-4} blank. *If more than one plate is to be prepared for each dilution,* the same pipette should be used to transfer another 0.1 ml to the second spread plate. Use proper aseptic technique for each transfer.

Now, let us explore how **dilution problems** are calculated. The goal of serial dilution and plating is to determine how many viable cells (CFU) are in each milliliter of the *original culture*.

Note in Figure 20-1 that 0.1 ml plated from a 10^{-3} dilution blank is labeled 10^{-4}. Let us examine why. Pretend that the 10^{-3} blank shown in Figure 20-1 contains 1000 cells in 1 milliliter (cells/ml). When this cell suspension is further diluted 1:10, and 1.0 ml of this 10^{-4} dilution is plated, there will be 100 colonies on that plate. If, instead, we simply plate 0.1 ml from the 10^{-3} blank, we also would have 100 colonies form on the plate.

Look at this another way. If we plate 0.1 ml from the 10^{-3} blank, we would have exactly the same number of colonies as if we had plated 1.0 ml of the 10^{-4} dilution. Therefore, we label plates that contain 0.1 ml of a dilution the same as those that contain 1.0 ml of the next higher (10×) dilution.

There are several advantages to transferring only 0.1 ml to each plate. You eliminate one transfer with a pipette; you use one less dilution blank; and, with the spread-plate technique, you also avoid the need to spread 1.0 ml on the surface of an agar plate (an amount that many plates cannot easily absorb). A rule of thumb for calculating dilution schemes and performing serial dilutions is to *always spread 0.1 ml on each plate and consider this the same as an extra 1:10 dilution.*

If you cannot remember how to handle exponential numbers, now would be a good time to review the rules of exponents in an algebra book. Get used to expressing dilutions as exponents of ten! Then, make sure you understand how to work dilution problems. *The ability to work dilution problems is a basic skill, which must be mastered in every beginning microbiology course.*

Accurate pipetting technique

Accurate pipetting is essential for many quantitative measurements in microbiology, but beginning students are often confused by the types of pipettes and the methods of using them.

Two types of pipettes are found in most microbiology laboratories today (Figure 20-2). **Serological pipettes** deliver a measured volume and are calibrated down to the delivery tip. If you see two concentric rings etched near the top, it is a serological pipette, which delivers the exact amount shown if you blow the last drop out of the pipette. Therefore, a serological pipette that is marked in this way is commonly called a **blow-out pipette**. Another way to identify these pipettes is to examine the tips. If the pipette is graduated all the way to the tip, it is a serological, or blow-out, pipette.

The second type of pipette is the **measuring pipette**, which delivers a measured volume but is not calibrated all the way to the tip. These pipettes do not have etched rings near the top. If you are asked to use the measuring pipette shown in Figure 20-2 to transfer 1.0 ml, you should draw the fluid up to the 0

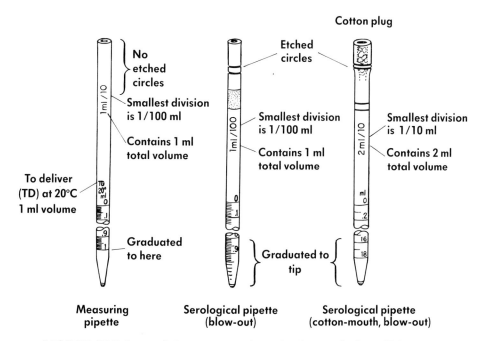

FIGURE 20-2 Types of pipettes commonly used in the microbiological laboratory.

line and transfer fluid until the fluid **meniscus** is at the 1 line. You can see from the figure that some fluid will remain inside the pipette near the delivery tip after you deliver the milliliter of fluid. If you use this type of pipette to transfer microbial cultures, the remaining fluid should be discarded into a jar containing liquid disinfectant.

Both the serological and the measuring pipettes are usually calibrated **to deliver (TD)** the stated volume of nonviscous liquid at 20°C. Both types also exhibit a whole number near the top of the pipette that represents the total volume that is held in that pipette (see Figure 20-2). Both pipettes also show a fractional number that states the smallest graduated division on that pipette.

Some pipettes have a chamber at the upper end in which a cotton plug is placed to safeguard the user when working with microbial pathogens. This plug prevents the accidental passage of culture fluids into your mouth or into a mechanical pipetting device. This type of pipette is usually called a *cotton-mouth* or *cotton-plugged* pipette. (Many microbiologists feel that microbial cultures should never be transferred by mouth. They believe that you should use a mechanical pipetter, sometimes referred to as a **propipette**. Ask your laboratory instructor for the policy in your classroom.)

If you are asked to *mouth pipette,* you should control the column of fluid in the pipette with your *index finger,* not your thumb, regardless of whether you are right- or left-handed (Figure 20-3). There are two reasons for this rule. First, with a little practice, you can achieve more precise control with your index finger than with your thumb. This is just the way most people are built. Second, you can hold a tube or bottle cap between your small finger and the palm of the same hand if you pipette with your index finger. You should become very proficient with this technique since you need to pipette many times during this academic term. **Ask your instructor to demonstrate this technique for you,** and then you should practice with nonsterile pipettes.

If you are right-handed, you should hold the pipette between your thumb and central fingers of your right hand. Place your index finger over the top of the pipette (see Figure 20-3). Pick up the dilution blank or culture tube with your left hand. Remove its cap with the little finger and pad of your right hand. Flame the lip of the culture tube or dilution blank, and then place the sterile tip of the pipette about 10 cm below the surface of the liquid. After you have withdrawn the appropriate amount of fluid, flame the lip again, replace the cap, and set the tube down. Remember to *use good aseptic technique.*

If you are left-handed, the procedure is the same except that you hold the pipette in your left hand and the tube or dilution blanks in your right hand.

To avoid pipetting errors when measuring fluid volumes, position your eyes so that they are horizontal with the top of the fluid column in the pipette (see correct position in Figure 20-4). The term *meniscus* refers to the curved air-liquid interface at the top of the fluid in the pipette. Proper eye alignment avoids parallax error. **Parallax error** comes from holding the pipette too high or too low to align the meniscus with the mark on the pipette, as shown in Figure 20-4. The volume in the pipette is correctly read when the bottom of the meniscus is positioned directly opposite the line on the pipette.

Smooth control of the liquid column in the pipette depends in part upon the condition of the tip of your index finger. If your skin is too dry, it cannot hold the liquid column, and the fluid will leak out of the pipette. If your finger tip is too moist, the fluid column is easily held, but you will not be able to control its slow release from the pipette. Slow and precise pipetting is something that you can achieve only with practice. Ask your instructor to set up a practice station so that

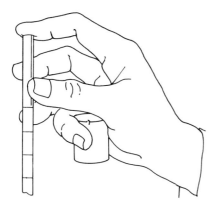

FIGURE 20-3 Technique for holding the pipette and controlling liquid flow with the index finger while also holding the culture-tube cap.

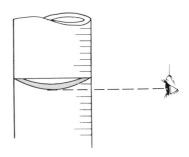

FIGURE 20-4 Correct positioning of the eyes in relation to the meniscus to avoid parallax errors in pipetting.

Spread-plate versus pour-plate techniques

After dilution, you need to **plate** the cells in some way so that you can determine their viable numbers (colony-forming units, or CFU). The two accepted methods for plating these dilutions are the spread-plate method and the pour-plate method.

The **spread-plate method** involves spreading a small, known volume (usually 0.1 ml) of cell suspension onto the surface of a prepoured agar plate. If you spread the cell suspension evenly over the agar surface, the colonies that develop after incubation should be evenly distributed. The technique for accomplishing this is explained in the procedure section of this exercise. This method has two main advantages. The first and most important is that distinct colonies develop that can be characterized according to size, shape, color, texture, and so on. You can easily transfer cells from these colonies to other media for study. The second advantage is that only this method is satisfactory for growing strict (obligate) aerobes. Since all cells are spread on and grow at the agar surface, there is plenty of oxygen to support the growth of strict aerobes. The major disadvantage of this technique is that it takes a little more time and involves more manipulations than the pour-plate technique. *The spread-plate technique is preferred* if you want to determine viable numbers of strict aerobes, to study the proportions of cell types in a mixed culture, or to check for culture purity.

In the **pour-plate method,** a known volume of cell suspension is inoculated into a *tempered* agar deep. Agar deeps (see Exercise 13) are tubes containing solid (unslanted) media. When an agar deep is changed into a liquid sol and then cooled and held at a temperature just above that at which the agar solidifies (about 40°C), the medium is said to be **tempered.** A dilution of the culture to be plated is added to the tempered agar deep; most microorganisms can withstand from 40° to 45°C for a short period of time. The agar-cell suspension is then thoroughly mixed and immediately poured into a sterile Petri dish where it is allowed to harden (Figure 20-5). Note that the cells are now thoroughly distributed in the hardened agar. Upon incubation, these cells will form colonies that develop throughout the agar. Colonies developing *on* the agar surface are characteristic of the microbial type (see Table 14-1), whereas all colonies that develop *below* the surface have a **lenticular shape,** regardless of the type of microbe that forms that colony. Lenticular colonies look like a biconvex lens; in other words, they are convex on both sides. This technique has two main advantages. It is easy to perform because it requires few manipulations and no special equipment. Also, this technique is satisfactory for growing either the **facultative** or the **microaerophilic** bacteria [those that do not grow in the absence of oxygen or in the presence of 20 percent oxygen (air)]. The major disadvantages of the pour-plate method are (1) strict aerobes do not grow below the agar surface, (2) you cannot detect contaminants and you cannot study colony types of a mixed culture because you cannot characterize the type of colony growing below the agar surface, (3) it is more difficult to isolate cells from subsurface colonies, (4) it is often more difficult to count colonies embedded in a pour plate because they may be stacked on top of one another, and (5) some cells may be damaged or killed when held for even a short

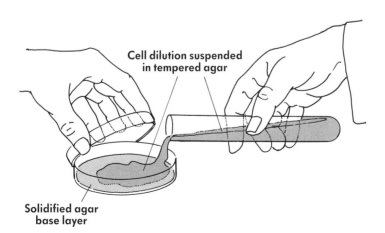

FIGURE 20-5 Correct positioning of the Petri-dish cover while pouring the tempered overlay agar that contains a diluted sample of culture.

time at 40° to 45°C. However, you may prefer the pour-plate technique if you want only to determine the viable number of cells in a pure culture in which the cells are not strict aerobes (facultatives or microaerophiles).

At this point, beginning students often ask why you could not use a streak plate to determine viable numbers. There are two reasons. First, it is important that a *known volume* be transferred from the dilution blanks to the agar plate, and a fragile wire loop cannot be depended upon to consistently transfer a known volume of culture. Second, the volume of cells transferred to plates must be *evenly distributed* on the surface or throughout the agar so that each colony can be counted. From your results in Exercise 14, you should recall that the beginning of each streak plate showed confluent growth. The cells on this part of the streak plate were so close together that, following incubation, you could not distinguish separate colonies. Thus, streak plates cannot be used for counting the viable microorganisms in a culture fluid.

LEARNING OBJECTIVES

- Learn the principles and limitations involved in determining viable numbers.
- Learn how to plan and carry out serial dilutions.
- Learn how to prepare spread plates.
- Learn how to prepare pour plates.
- Know how to calculate viable numbers from plate counts.

MATERIALS

Cultures	*Escherichia coli*
Media	Trypticase-soy agar plates (for spread plates)
	Trypticase-soy agar deeps (for pour plates)
Supplies (spread plates)	1.0- or 1.1-ml pipettes (sterile)
	9.0-ml distilled water dilution blanks (sterile, screw-capped tubes)
	Spreader wheels, glass rods, and 95 percent ethanol
Supplies (pour plates)	1.0- or 1.1-ml pipettes (sterile)
	99.0-ml distilled water dilution blanks (sterile, dilution bottles)
	9.0-ml distilled water dilution blanks (sterile, screw-capped tubes)
	Petri dishes (sterile)
	Water bath set at 45°C (to temper the liquified agar deeps)
Supplies (practice station)	1.0- or 1.1 ml pipettes (nonsterile)
	99.0-ml distilled water dilution blanks
	9.0-ml distilled water dilution blanks

PROCEDURE: FIRST LABORATORY PERIOD

Before you begin your dilutions, you should do the following:

Observe your instructor's demonstration of aseptic methods for transferring cultures with a pipette.

Practice pipetting using good aseptic technique. Do this even though you have used pipettes in other laboratories. Practice before you begin preparing your serial dilutions.

Label one set of dilution blanks and plates with the appropriate dilution factors. If you are using the *spread-plate method,* use Figure 20-1 as a guide. If you are using the *pour-plate method,* use Figure 20-6 as a guide.

Mark each plate with your *name, date* of inoculation, and the name of the *microorganism* used. Place these identifying marks on the *bottom* of the plate so that plates can be identified even if the lids are accidently removed. Write small and only around the edge.

Preparing spread plates

Follow the dilution scheme given in Figure 20-1 during each of the following steps.

1. **Use pipette A for the first transfer** of 1.0 ml of sample to the first 9.0-ml dilution blank. This is a one-to-ten (1:10, 1/10, or 10^{-1}) dilution of the original sample

2. **Properly discard pipette A** after you make the transfer.

SAFETY NOTE

Pipettes used for culture transfer should always be placed into a container of liquid disinfectant immediately after use. Never put the pipette down or let it touch anything. Avoiding contamination in this way is part of good aseptic technique.

3. **Mix this 10^{-1} dilution blank thoroughly** after adding the diluted culture. This is best done by tightening the screw cap and gently inverting the tube about ten times.

SAFETY NOTE

Violent shaking will create an aerosol inside the tube. (An aerosol is composed of very small droplets of liquid suspended in air.) An aerosol created inside the culture tube will contain microorganisms. When you open the tube, the aerosol may be released into the room where it can be inhaled or can contaminate materials in the laboratory. Since you will not be using virulent pathogens for your experiments, this is not of great concern. However, a microbiologist knows how aerosols are created and takes steps to avoid their creation. *Invert culture tubes gently.*

4. **Use pipette B for the second transfer** of 1.0 ml of the cell suspension from the 10^{-1} tube to the dilution blank marked 10^{-2}. This is a 1/10 dilution of a 1/10 dilution. In other words, this is a 1/100 (10^{-2}) dilution of the original sample.

5. **Thoroughly mix** the 10^{-2} tube to distribute the cells in the liquid.

6. **Use pipette C** to transfer 1.0 ml from the 10^{-2} tube to the tube labeled 10^{-3}. This is a 1/1000 or a 10^{-3} dilution of the original sample.

7. **Use pipette D for two transfers** from the suspension in the tube marked 10^{-3}. *First,* transfer 1.0 ml of the 10^{-3} suspension to the dilution blank labeled 10^{-4}. *Second,* transfer 0.1 ml of the 10^{-3} suspension to the *center* of one or more prepoured agar plates marked 10^{-4}. This center placement makes it easier to spread the sample and to keep the sample away from the edge of the plate. *Your instructor will tell you how many replicate plates to prepare* from each dilution tube.

8. **Spread replicate plates for each dilution immediately** after all 0.1-ml samples are transferred to the plates. If this is not done, the agar may absorb too much of the sample fluid, and an unusually high concentration of cells will develop into colonies on that part of the plate.

9. **Spread each plate in the following way:**
   a. Center the plate on the spreading wheel.
   b. Remove the spreading rod from the ethanol, touch the rod to the edge of the plate to drain off the excess ethanol, and quickly pass the rod through the flame to ignite the ethanol. Wait a few seconds for the flame to go out. *Take care that the flaming ethanol does not drip on something combustible* (like you!).

   Start the wheel turning with your other hand, and then lift the plate cover with that hand. Hold the lifted plate cover over the plate so that dust does not drop onto the exposed agar surface while you spread the culture.

SAFETY NOTE

Be very careful with burning ethanol! You will be supplied with 95 percent ethanol in a glass Petri dish. This should be placed at least one foot away from the bunsen burner. Remember that ethanol burns with an *almost invisible* flame. Do not allow the flaming ethanol to drip on your skin, your clothing, or in the Petri dish filled with ethanol. Set the Petri dish cover *next* to the ethanol-filled Petri-dish bottom. This way, you can quickly cover the dish to put out the fire if the ethanol is accidentally ignited. *Work with care!*

c. Touch the flame-sterilized spreading rod to the center of the agar. The combined action of the rod and the rotating plate should spread the fluid evenly over the agar surface. It is only necessary to spend a few seconds to spread the fluid. It is not necessary to continue spreading until the fluid is absorbed into the agar. (Keep the fluid away from the edge of the plate. Colonies developing here are very hard to see and count.)
d. Replace the cover, and return the spreading rod to the alcohol.
e. Leave each completed spread plate right-side up for about 5 minutes to give the fluid time to absorb into the agar.

10. **After each plating, thoroughly mix the most recently prepared culture dilution,** and continue with the dilution and plating scheme.

11. **Invert your labeled plates, tape them together, and incubate them at 30°C for 24 hours.**

12. **Refrigerate plates after 24 hours of incubation.** Refrigeration stops growth and prevents colonies from growing larger and fusing. This will enable you to count these plates more easily during your next laboratory period. *Check with your instructor* to find out who will refrigerate the plates.

Preparing pour plates

Before you begin, remember that tempered agar is held at a temperature just above the point at which it begins to solidify (gel). Therefore, you need to leave the tempered-agar deeps in the water bath until just

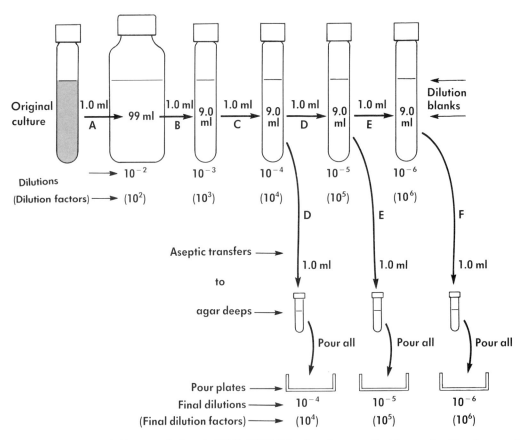

FIGURE 20-6 Serial-dilution and POUR-PLATING SCHEME for the culture used in this exercise. Note how much volume is aseptically transferred (arrows). Capital letters (A–F) refer to the sterile 1.0-ml pipettes and how they should be used. For example, three transfers are made with pipette D. First, aseptically withdraw 1.0 ml of sample from the 10^{-4} dilution, and transfer this volume to the sterile blank labeled 10^{-5}. Second, use the same pipette to remove 1.0 ml from the 10^{-4} dilution, and transfer this to one tempered-agar deep labeled 10^{-4}. If you are to prepare more than one pour plate, repeat this last step with the same pipette. Use proper aseptic technique for each transfer.

before you are ready to add cells, thoroughly mix them with the tempered agar, and pour that plate. Plates should be poured one at a time. If you work in groups, your instructor will give you suggestions on how to coordinate your work.

Follow Figure 20-6 when preparing your dilutions, and read the safety notes in the left-hand column on page 181. Prepare your serial dilutions only through the 10^{-4} dilution in the following way.

1. **Use pipette A for the first transfer** of 1.0 ml of the original culture to the 99-ml dilution blank. This is a 1 to 100 (1:100, 1/100, or 10^{-2}) dilution of the original sample. *Thoroughly mix* this suspension before making the next transfer.

2. **Use pipette B for the second transfer** of 1.0 ml of cell suspension from the 10^{-2} bottle to the dillution blank marked 10^{-3}. This is a 1/10 dilution of a 1/100 dilution. In other words, this is a 1/1000 (10^{-3}) dilution of the original sample. *Thoroughly mix* this suspension before making the next transfer.

3. **Use pipette C for the third transfer** of 1.0 ml of cell suspension from the 10^{-3} tube to the blank marked 10^{-4}. This will achieve a 1/10,000 or 10^{-4} dilution. *Thoroughly mix* this suspension before making the next transfer.

4. **Use pipette D for the fourth transfer** of 1.0 ml from the mixed 10^{-4} dilution to the blank labeled 10^{-5}. Keep this pipette uncontaminated so that you can use it for step 6.

5. **Remove one 9.0-ml tempered agar deep from the waterbath.**

6. **With the same pipette (D),** aseptically transfer 1.0 ml from the 10^{-4} dilution to the tempered agar deep just taken from the waterbath. Screw the cap on tightly after this transfer is made.

7. **Mix these cells thoroughly in the tempered deep,** using about ten gentle inversions. Mixing should be gentle enough not to make bubbles but rapid enough so that the agar does not harden.

8. **Aseptically pour this entire mixture into one sterile Petri dish** labeled 10^{-4}. Gently swirl the plate so that the agar is evenly spread over the bottom of the plate. Set this plate aside to harden before you place it in the incubator.

   If you are asked to prepare more than one pour plate per dilution, repeat steps 5, 6, 7, and 8 to obtain additional 10^{-4} pour plates. Remember to discard each pipette into the disinfectant solution after you are through making all aseptic transfers from one dilution blank.

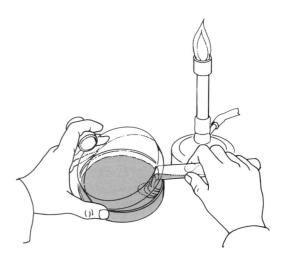

CONSERVATION NOTE

Do not discard tempered-agar deeps that hardened before you add culture. Place them in a rack beside the waterbath marked *Sterile Deeps*. If your agar deep hardened *after* you added culture but before it could be poured, discard this deep as you would any other culture or contaminated material. If a large blob of gel falls into the plate as you pour your inoculated deep, inoculate a new tempered deep and repeat the procedure. These difficulties occur (1) if you take too much time between the waterbath and pouring the plate, (2) if you do not mix the inoculated deep quickly enough after inoculation, or (3) if the waterbath is not keeping the agar deeps hot enough. If any of these problems occur, please determine the cause and ask your instructor to correct it as soon as possible.

9. **After each plating, thoroughly mix the newly prepared dilution,** and continue as before until all dilutions are appropriately plated (see Figure 20-6).

10. **Invert your labeled plates, tape them together, and incubate them at 30°C for 24 hours.**

11. **Refrigerate plates after 24 hours of incubation.** Refrigeration stops growth and prevents colonies from growing larger and fusing. Stopping growth while the colonies are small

will allow you to count these plates more easily during your next laboratory period. *Check with your instructor* to find out who will refrigerate the plates.

PROCEDURE: SECOND LABORATORY PERIOD

1. **Count the colonies on each plate** that you prepared during the first laboratory period.
   a. You might want to use a *Quebec colony counter,* which provides an indirect (oblique) source of light, a magnifying glass, and grid markings to help in the systematic examination of the plate. If you use a colony counter, you should turn the plate upside down on the counter.

   These colonies may be large enough to count easily with the unaided eye. In that case, hold the plate up so that it is between your eyes and the ceiling light. Look through the bottom of the plate to count the colonies.
   b. Place marks on the bottom of the Petri dish with your marking pen so that a mark appears to cover each colony as you count it. This will help to prevent you from counting the same colony twice.
   c. Write the number of colonies on the bottom of each plate, and circle this number for easy identification.
   d. Check the very edge of the agar for hard-to-see colonies. Double check to see that each colony is marked.
   e. To estimate the number on plates containing many colonies, draw two perpendicular lines on the bottom of the plate that divide the plate into four roughly equal parts (quadrants). One quadrant may be counted and its count multiplied by four to estimate the total number of colonies on the plate.
2. **Record the total (or estimated) number of colonies per plate** in the Results section.

The next step is to calculate the number of viable microbes per milliliter of the original culture using **the 30-to-300 rule.** Statisticians tell us to use only those plates containing between 30 and 300 colonies in this calculation, because samples containing fewer than 30 cells per 0.1 ml are subject to large fluctuations in numbers of viable cells (sampling errors) and plates containing more than 300 colonies cause human errors in counting. Statistitions also tell us that, for greatest accuracy, we should plate each dilution three times and use the average of these counts. Therefore, **triplicate plating** of each dilution is used in all research laboratories, but this procedure is usually too expensive for large classes.

Plates with more than 300 colonies are considered **too numerous to count (TNTC)** accurately. Therefore, they should not be used in calculating the plate count. Beginning students, however, should estimate the number of colonies on these plates by counting one quadrant to make sure that the plates do contain more than 300 colonies. (The term *TNTC* sometimes means that people were too lazy to count more than a few colonies on the plate. This is TLTC, not TNTC.)

3. **Multiply the number of colonies per plate by the dilution factor** [see Figures 20-5 (spread plates) and/or 20-6 (pour plates)] **to obtain the number of viable cells (CFU) per milliliter of the original culture.** Remember that the dilution factor is the total number of times the culture was diluted. This is the same thing as the denominator in the fraction that expresses the total dilution. For example, if you diluted the original sample 10,000 times (a 1/10,000 or 10^{-4} dilution), then the dilution factor is 10,000 or 10^4.

Let us use an example to show how microbiologists use plate counts to calculate the number of viable cells in a sample. Assume that you find an average of 223 colonies on the 10^{-4} (1/10,000 dilution) plates. Then the correct viable-number calculation would be

$$\frac{\text{CFU}}{\text{ml}} = \left(\frac{\text{average number of colonies}}{\text{plate}} \right) \times \text{dilution factor}$$

that is, $223 \times 10,000$ (or 223×10^4 or 2.23×10^6) = 2,230,000 viable cells (CFU) per milliliter of original sample. Note, however, that the numeral 3 in this number has little significance because all plate counts have an error of at least 10 percent. Therefore, *microbiologists traditionally round off viable counts to two significant figures.* Therefore, 223×10^4 should be expressed as 220×10^4. It is also traditional to *express viable numbers as exponents of numbers between 1 and 10 (to the nearest one-tenth).* Thus, this plate count should be expressed as 2.2×10^6.

Another way to calculate the number of viable cells in the original culture is to use the following equation:

$$\frac{CFU}{ml} = \frac{(\text{average number of colonies/plate})}{\left(\begin{array}{c}\text{dilution}\\\text{plated}\end{array}\right)\left(\begin{array}{c}\text{volume plated}\\\text{in milliliters}\end{array}\right)}$$

For example, assume that you found 223 colonies on the plate resulting from 0.1 ml transferred from the 10^{-3} dilution tube. Placing these numbers into the equation, we find

$$\frac{CFU}{ml} = \frac{223}{(10^{-3})(0.1)} = \frac{223}{10^{-4}}$$
$$= 223 \times 10^4 = 2.2 \times 10^6$$

4. **Record your calculated plate counts** in the Results section.

5. **Also place your results on the chalk board** so that everyone can examine the class results.

6. **Copy the class results in your Results section,** and calculate the class average.

"I'm afraid that's not the way to bridge the communications gap between scientists and nonscientists."

20 VIABLE NUMBERS: SPREAD OR POUR PLATES

NAME _____

DATE _____

SECTION _____

RESULTS

Spread-plate counts

Your results from spread plates

Dilution counted (e.g., 10^{-4})	Total number of colonies per replicate plate (mark estimates with an asterisk*)	Calculated viable number of microbes in original culture (circle the numbers in column 2 used in calculations) (CFU/ml)

Show your calculations here:

Give your conclusions here:

Class results from spread plates

Student	Dilution with 30–300 colonies	Number of colonies per plate	Viable numbers in original Sample (calculated CFU/ml)
1			
2			
3			
4			
5			
6			
7			
8			
9			
10			
11			
12			
13			
14			
15			
16			
17			
18			
19			
20			
21			
22			
23			
24			
25			
26			
27			
28			
29			
30			
	Class average =		

Give your conclusions here:

20 VIABLE NUMBERS: SPREAD OR POUR PLATES

NAME _____

DATE _____

SECTION _____

Pour-plate counts

Your results from pour plates

Dilution counted (e.g., 10^{-4})	Total number of colonies per replicate plate (mark estimates with an asterisk*)	Calculated viable number of microbes in original culture (circle the numbers in column 2 used in calculations) (CFU/ml)

Show your calculations here:

Give your conclusions here:

Class results from pour plates

Student	Dilution with 30–300 colonies	Number of colonies per plate	Viable numbers in original Sample (calculated CFU/ml)
1			
2			
3			
4			
5			
6			
7			
8			
9			
10			
11			
12			
13			
14			
15			
16			
17			
18			
19			
20			
21			
22			
23			
24			
25			
26			
27			
28			
29			
30			
	Class average =		

Give your conclusions here:

20 VIABLE NUMBERS: SPREAD OR POUR PLATES

NAME _____

DATE _____

SECTION _____

QUESTIONS

Completion

1. Using 99-ml dilution blanks (only), draw a scheme that would yield between 30 and 300 colonies from an original sample that contains 4.7×10^8 cells/ml.

2. What is the minimum number of pipettes you need to achieve one plate with between 30 and 300 colonies in that dilution scheme?

3. Why are plates containing between 30 and 300 colonies the only ones counted?

4. Why must pipettes be changed between dilutions?

5. Name two methods commonly used to plate serial dilutions: _____ and _____.

6. You observed that a high dilution of a young culture was plated on a medium known to support growth, but after 48 hours of incubation, the plate showed no growth. Give *two* possible explanations for these results.

Dilution Problems

1. What is the final dilution obtained after each dilution series shown below? (Express each answer in exponential numbers.)

    a. 1:10, 1:10, and 1:100: _____

    b. 10^{-1}, 10^{-1}, and 10^{-2}: _____

    c. 10^{-3} and 10^{-5}: _____

2. Determine the number of cells per milliliter of the original culture when the total number of colonies and the total dilutions are as follows:

Plate count	Final dilution	Viable no. in original culture (CFU/ml)
37	1:100	
88	1:1,000,000	
147	10^{-5}	
221	10^{-7}	

Show *one sample calculation* here:

NAME _____

DATE _____

SECTION _____

3. In the following dilution scheme, write the correct dilution factor for each plate and the number of colonies that you should have on each plate.

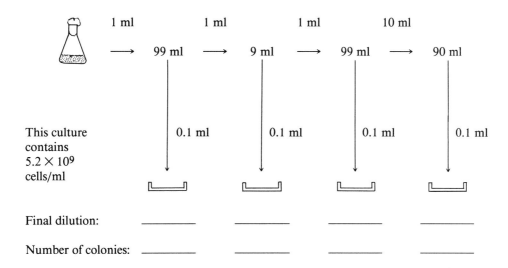

Final dilution: _____ _____ _____ _____

Number of colonies: _____ _____ _____ _____

4. Draw a dilution scheme that would produce countable plates if the original culture has 4,730,000 viable cells/ml.

21

Selective and Differential Media

Many of the growth media used in microbiology are **general** or **all-purpose media,** which typically contain a variety of nutrients and can therefore support the growth of a variety of microorganisms with different nutrient requirements. All-purpose media are used for maintaining pure stock cultures of many types of microorganisms and for growing these microbes for laboratory experiments. Most of the pure cultures you work with in this course are grown on all-purpose media.

All-purpose media are also used in some situations that involve mixed cultures. For example, you might want to estimate the number of viable microorganisms in a soil sample by plating serial dilutions (see Exercise 20) of the soil on all-purpose media. However, because most soils contain a great variety of microorganisms and because your plate count will detect only the types of microorganisms that grow on the medium used, you should plate the soil on an all-purpose medium that will permit the widest possible variety of microorganisms to grow.

All-purpose media are *not* appropriate for all situations. Suppose you want to isolate, from a mixed-culture sample such as soil, an organism that is present in very **low numbers** (say 10^2 cells/g) compared with the rest of the microbial population (say 10^7 cells/g). You could try to isolate this organism by making a streak plate (see Exercise 14), but streaking usually yields isolated colonies of only the most numerous organisms in a mixed culture. Therefore, the organism you want to isolate would probably remain hidden in the first portion of the streak, where the colonies run together into a confluent area of growth. You could also attempt to isolate the organism by plating serial dilutions of the sample, but this approach is also likely to fail. The organisms that are present in low numbers will grow only on plates from the lowest dilutions, and these plates will almost certainly be overgrown by the microbes that are present in higher numbers. Thus, the chances of obtaining an isolated colony of the desired organism using these methods will be poor.

Such a situation can often be dealt with by using a **selective medium** for streaking or plating, instead of an all-purpose medium. Selective media contain at least one ingredient that can *inhibit* the growth of unwanted (more numerous) microorganisms without preventing the growth of the desired type (less numerous). Ingredients that inhibit the growth of unwanted organisms are called **selective agents.** Commonly used selective agents include antibiotics, inhibitory chemicals, and certain types of dyes. To isolate a microorganism present in low numbers, you have to construct a selective medium in which the selective agent inhibits the more numerous organisms in the sample without affecting the one you want to isolate. It would then be easy to isolate the desired organism by streaking or plating on this medium because the desired organism would no longer be overgrown by the other forms.

In a very strict sense, all media are at least slightly selective. Microorganisms are so diverse in their nutritional and environmental requirements that no medium or set of growth conditions can support optimum growth of all microbial types. For example, most bacteria are inhibited by high concentrations of salt, but some bacteria *require* high salt concentrations for survival and growth. Obviously, such different organisms cannot be grown on the same medium. Even the most general all-purpose medium, then, is selective against those few organisms that cannot grow on it; however, we usually think of such media as being nonselective because their slight selectivity is unintentional. Selective media, then, are those that you *deliberately* make selective to accomplish some specific purpose.

We have seen that nonselective all-purpose media are not very useful for isolating organisms present in relatively low numbers. These media are also of little use if one wants to *distinguish* one type of microorganism from another. Suppose that you need to determine the number of **coliform bacteria** in a sample of drinking water. (Coliform bacteria are residents of the human intestine, and their presence in water indicates that the water is contaminated with fecal material; see Exercise 53.) If you plate the water on most suitable growth media, you will not be able to count the coliforms because their colonies are not visibly different from those of noncoliform bacteria. You could isolate cells from every colony on the plates and then test each isolated culture for coliform characteristics, but this approach would be tedious and time-consuming.

One can better approach the problem by plating the water sample on a differential medium. **Differential media** are designed to distinguish one type of microorganism from all others in a mixed culture. Differential media contain a special ingredient that changes during the growth of only certain types of microorganisms, causing their colonies or the agar around their colonies to take on a distinctive appearance. For example, when counting coliforms mixed with other types of bacteria, you would use a differential medium that contains a chemical used or changed only by coliform bacteria. You could then count the distinctive coliform colonies directly, without having to perform further tests. (In fact, coliforms are routinely counted this way; see Exercise 53.)

Differential media can also be selective or nonselective. **Nonselective** differential media are similar to all-purpose media because they permit the growth of many microbial types; however, the differential properties of nonselective differential media only enable you to recognize a specific microbial type among the many forms that grow on the plates. You can make differential media **selective** by adding a selective agent that will inhibit the growth of most other microbial types. In this case, the differential properties of the selective medium permit one to distinguish microbial types within the limited group of organisms that can grow. For example, the medium used to count coliforms is both selective and differential (see Exercise 53).

Selective and differential media are important in microbiology because they have many uses, such as (1) the isolation, detection, and enumeration of microbial types in mixed-culture samples, (2) routine sampling of products for quality-control analysis, and (3) elimination of potential contaminants during the transfer and growth of pure cultures. Selective and differential media are especially important in **medical microbiology,** in which the quick isolation and identification of microorganisms can hasten disease treatment. Many of the standard testing procedures found in hospital laboratories make use of selective and differential media.

In this exercise, you will compare the growth of four pure cultures and one mixed culture on three types of media.

The first medium is **nutrient agar,** a typical all-purpose medium that is both nonselective and nondifferential. All of the microorganisms you use in this exercise should grow on this medium.

The second medium is **high-salt agar.** This is a *selective* medium because it contains 7.5 percent sodium chloride in addition to the usual components of an all-purpose medium. The high concentration of sodium chloride inhibits most bacteria but does not affect salt-tolerant bacteria, such as members of the genus *Staphylococcus.* Thus, high-salt agar selects for organisms like staphylococci but does not differentiate among the salt-tolerant types of bacteria.

The third medium is **mannitol-salt agar,** a medium that is differential and selective. Mannitol-salt agar is *selective* because it contains 7.5 percent sodium chloride (its selectivity is identical to that of high-salt agar) and is *differential* because it contains both mannitol (a six-carbon sugar alcohol) and phenol red (a pH indicator). When a salt-tolerant organism ferments the mannitol to acid, the release of acid from its cells causes the phenol red in the agar around its colonies to turn from red to yellow. (See Exercise 31, which explains fermentation in greater detail.) Thus, you can distinguish between salt-tolerant organisms that ferment mannitol and those that do not. (Mannitol-salt agar contains other sources of carbon and energy on which non-mannitol-fermenting, salt-tolerant organisms can grow without producing a color change in the medium.) With mannitol-salt agar, you can distinguish between fermenting and nonfermenting staphylococcci; this distinction is important in clinical microbiology because most nonfermenting staphyloccci (such as *S. epidermidis*) are relatively harmless residents on human skin whereas most mannitol-fermenters (such as *S. aureus*) are opportunistic pathogens (see Exercise 47).

21 SELECTIVE AND DIFFERENTIAL MEDIA

LEARNING OBJECTIVES

- Learn what differential and selective media are and understand how they work.
- Understand how differential and selective media differ from each other and from all-purpose (general growth) media.
- Appreciate the various applications of differential and selective media.
- Become familiar with mannitol-salt agar and its significance in clinical microbiology.

MATERIALS

Cultures *Escherichia coli*
  Pseudomonas aeruginosa
  Staphylococcus epidermidis
  Staphylococcus aureus
  Mixed culture containing a very large proportion of *E. coli* and a very small proportion of *S. aureus* or *S. epidermidis*

Media Nutrient agar plates
  High-salt agar plates
  Mannitol-salt agar plates

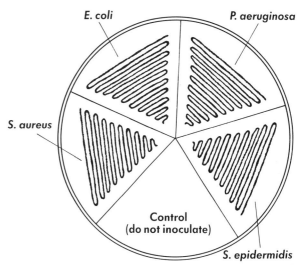

FIGURE 21-1 Scheme for inoculating plates with pure cultures.

PROCEDURE: FIRST LABORATORY PERIOD

1. **Label the following** with your *name*, the *date*, and the name of the *medium* each contains.
   a. Two nutrient agar plates
   b. Two high-salt agar plates
   c. Two mannitol-salt agar plates

2. **Perform the following procedure with the pure cultures. Use one nutrient agar plate, one high-salt agar plate, and one mannitol-salt agar plate** (Figure 21-1).
   a. Divide the bottom of each plate into five equal sectors with your marking pen.
   b. Label each of four of the sectors of the plate with the name of a different microorganism (listed above). Label the fifth sector of the plate with the word *control*.
   c. Inoculate each of the four sectors of the plate with the appropriate organism. Transfer a loopful of culture to the plate, and streak it throughout the sector. *Do not inoculate the control sector.*

3. **Prepare a streak plate with the mixed culture on one nutrient agar plate, one high-salt agar plate, and one mannitol-salt agar plate,** just as you would for colony isolation (see Exercise 14).

4. **Invert, tape, and incubate all six plates** (three from step 2 and three from step 3) **at 30°C for 48 hours.**

PROCEDURE: SECOND LABORATORY PERIOD

1. **Examine your plates and record your observations in the Results section** (see the Interpretation Note on page 198). A color change from red to yellow around the area of growth on standard mannitol agar plates indicates that the cells have fermented mannitol and have excreted an acid end-product.

   If an organism produces a large amount of acid, some of it may diffuse through the agar into the neighboring sectors of the plate and cause parts of those sectors to turn yellow. Do not confuse this phenomenon with a true positive reaction. Ask your instructor for help if you are not sure how to interpret your plates.

2. **Prepare Gram stains of the organisms growing on the plates that you inoculated with pure cultures** (see Exercise 7).

3. **Properly discard all plates.**

INTERPRETATION NOTE

Phenol red (phenolsulfonphthalein) is a dye (structure given in Exercise 31, Figure 31-2) that is commonly used as an acid-base indicator in microbiological growth media. The dye is *red above pH 8.4*, changes from red to yellow between pH 8.4 and pH 6.8 (pK 7.9), and is *yellow below pH 6.8*. If the medium under a colony changes completely from red to yellow, this means that sufficient acid has been produced by cells in that colony to lower the pH of the medium below the colony from above 8.5 to below 6.8. Since colonies grow on the plate surface, they must be able to grow under aerobic conditions; yet fermentation is characteristic of an anaerobic environment (see Exercise 31). The fermentation of mannitol that occurs on mannitol-salt agar spread plates probably occurs deep within the colony where the cells have used up all the oxygen and are faced with relatively anaerobic conditions. Here, the cells ferment mannitol and excrete acid fermentation products that diffuse through the colony and into the surrounding medium. Metabolism and growth under both aerobic and anaerobic conditions is characteristic of *facultatively anaerobic* bacteria (see Exercise 25).

21 SELECTIVE AND DIFFERENTIAL MEDIA

NAME _____

DATE _____

SECTION _____

RESULTS

1. Record the results from the plates that you inoculated with *pure cultures* in the table:

Organism	Growth (+ or −)			Mannitol fermentation [mannitol-salt agar] (+ or −)
	On nutrient agar	On high-salt agar	On mannitol-salt agar	
E. coli				
P. aeruginosa				
S. epidermidis				
S. aureus				

2. Differences in types and numbers of colonies formed upon inoculation of the *mixed culture* on each of the following media.

   a. Nutrient agar:

   b. High-salt agar:

   c. Mannitol-salt agar:

3. Record the results from Gram stains of cells from colonies on plates inoculated with *pure cultures* in the table:

Organism	Gram reaction (+ or −) and morphology (cell shape)		
	On nutrient agar	On high-salt agar	On mannitol-salt agar
E. coli			
P. aeruginosa			
S. epidermidis			
S. aureus			

4. Did you obtain isolated colonies of the *Staphylococcus* species on the nutrient agar plate streaked with the mixed culture? If not, explain your results.

5. Was the *Staphylococcus* species in your mixed culture *S. aureus* or *S. epidermidis* (choose one or the other)? How do you know?

CONCLUSIONS

21 SELECTIVE AND DIFFERENTIAL MEDIA

NAME _____

DATE _____

SECTION _____

QUESTIONS

Completion

1. Media that contain a wide variety of nutrients and that support the growth of many different kinds of microorganisms are called _____ media.

2. Selective media are especially useful for isolating microorganisms that are present in relatively _____ numbers in a mixed culture.

3. An investigator can enumerate coliform bacteria easily by plating a sample on a(n) _____ medium.

4. Sodium chloride serves as a(n) _____ in mannitol-salt agar.

5. The pH indicator in mannitol-salt agar is _____, which turns _____ at low pH.

True – False (correct all false statements)

1. Media can be selective or differential, but they cannot be both selective and differential. _____

2. Some media support the growth of all known types of microorganisms. _____

3. An all-purpose medium is more appropriate than a selective medium for counting the total number of viable microorganisms in a mixed-culture sample. _____

4. All media are at least slightly selective, but not all media are differential. _____

5. All-purpose media are used frequently for routine pure-culture tasks, such as growing organisms for use in laboratory experiments. _____

22

Enrichment Techniques

Most environments contain a variety of microbial types. To understand how these different organisms interact and how they affect their environment, one must know their metabolic and genetic characteristics. The most efficient way to obtain such information is to isolate the various microbial types and study them in the laboratory as pure cultures. You learned in Exercise 21 that organisms occurring in very low numbers cannot be isolated by streaking or plating on all-purpose media because they are overgrown by the more numerous forms. This problem can be addressed by using either a selective medium (as described in Exercise 21) or an enrichment culture technique.

Enrichment-culture technique makes use of conditions designed to enable a particular type of microorganism to grow faster than all others in the sample; thus, the *number* of this type is enriched when the mixed culture is inoculated into a suitable medium. To isolate an organism that breaks down cellulose in soil, for example, one could introduce the soil into a liquid medium containing cellulose as the *only* carbon source. The cellulose-degrading organisms would have abundant carbon under these conditions, but other microbial types would have only the low concentrations of carbon-containing nutrients contributed by the soil. Because the cellulose degraders are able to grow faster than other microbial types (especially after alternative nutrients in the soil are used up), they eventually become the most numerous forms in the enrichment culture. It is then easy to isolate them by streak- or spread-plating.

Sometimes it is necessary to carry out several sequential transfers and incubations in a liquid enrichment medium before streaking or plating on solid media. In the case of the cellulose-degrader enrichment, it might be that other organisms in the soil will grow faster and more extensively before they run out of nutrients. If that happens, the cellulose degraders will not become one of the more numerous forms before all growth stops. But, if you transfer a small amount of the liquid from the first enrichment culture to a second, identical, sterile, enrichment medium, the cellulose degraders may grow faster if the other carbon sources are now depleted. Further transfers may be required, but the cellulose degraders will eventually become one of the most numerous forms.

Enrichment cultures and selective media accomplish their purpose in similar but slightly different ways. *Selective media* are designed to inhibit the growth of undesired organisms so that only the desired type can grow. *Enrichment media,* on the other hand, are designed to favor the growth of the desired organism without deliberately repressing other forms. In the latter case, undesired types may or may not grow, depending on the composition of the medium and the conditions of incubation. Although not designed to repress undesired forms, most enrichment media do exert a selective effect on a mixed culture. For example, a medium that contains fructose as a sole carbon source obviously selects against organisms that cannot degrade fructose, but *no inhibitory ingredient is added to most enrichment media.* Some authors now recognize the selectivity of enrichment media by referring to the enrichment culture approach as **enrichment selection.** These authors refer to the use of *selective* media as **repression selection.**

Conditions that favor the growth of a particular microbial type in an enrichment culture can be created with chemical, physical, and/or biological strategies. **Chemical strategies** include techniques such as the use of specific carbon sources (such as cellulose) to enrich for organisms with specific degradative abilities, or the use of an acidic pH to enrich for fungi. (Fungi grow faster than most bacteria do at low pH; see Exercise 41.) **Physical strategies** include methods such as incubation at high temperatures to enrich for thermophilic organisms or incubation in the absence of oxygen to enrich for anaerobes. An example of a **biological strategy** (these are used less frequently than chemical or physical approaches) is the addition of live host cells to enrich for a specific type of bacterial virus (bacteriophage). By using combinations of these strategies, one can design enrichment media for nearly any conceivable need.

Besides their usefulness in isolating relatively scarce microbial types from mixed-culture environments (for example, soils, muds, lakes, and oceans), enrichment cultures can reveal the existence of *new microbial types.* Several interesting groups of bacteria were first observed by early enrichment-culture pioneers such as Martinus Beijerinck and Sergei Winogradsky.

Enrichment cultures provide a convenient way to study the effects of environmental changes on mixed cultures. To determine which organisms in a soil are stimulated by a particular nutrient, for example, one can introduce a sample of the soil into a medium that is enriched in that nutrient. Use of enrichment cultures is also an effective way to seek microorganisms with particular enzymatic capabilities. In recent years, microorganisms that can break down pollutants have been isolated by enrichment of samples of soil and other materials in media that contain pollutants as their sole sources of carbon. These pollutant-degrading organisms may someday (possibly after genetic modification to increase their efficiency) be used to remove dangerous pollutants from our environment.

In this exercise, you will see how culture conditions can be varied to enrich for three types of microorganisms in a soil sample. The first of these microbial types is a group of organisms known as **sulfate-reducing bacteria;** they play a role in the recycling of sulfur in the environment, and they can adversely influence soil fertility by affecting the amount of sulfate available to plants.

Sulfate-reducing bacteria are anaerobic and chemoheterotrophic organisms that oxidize organic compounds and pass the resulting high-energy electrons to the electron transport system; this results in energy and low-energy electrons, which are used to reduce sulfate to hydrogen sulfide (see Exercise 33). Because sulfate serves in place of oxygen as a terminal electron acceptor in respiration, this type of metabolism is an example of **anaerobic respiration.** Some typical H_2S-producing sulfate reducers are *Desulfovibrio, Desulfomonas, Desulfococcus,* and *Desulfosarcina* — all gram-negative bacteria that differ in cell shape and size.

TABLE 22-1 Chemical composition of some enrichment media

Microorganism to be enriched	Name of medium	Medium components and concentration (grams per liter distilled H_2O)	
Sulfate-reducing bacteria	Van Delden's	Sodium lactate	5.0
		$MgSO_4 \cdot 7H_2O$	2.0
		Asparagine	1.0
		$CaSO_4$	1.0
		K_2HPO_4	0.5
		$FeSO_4 \cdot 7H_2O$	0.1
Algae and cyanobacteria	Bristol's	$NaNO_3$	0.250
		$MgSO_4 \cdot 7H_2O$	0.075
		K_2HPO_4	0.075
		$FeCl_3$	0.050
		$CaCl_2$	0.025
		NaCl	0.025
		KH_2PO_4	0.018
Cyanobacteria	Chu's	$MgSO_4 \cdot 7H_2O$	0.025
		Na_2SiO_3	0.025
		$NaCO_3$	0.020
		K_2HPO_4	0.010
		Ferric citrate	0.003
		Citric acid	0.003

Sulfate-reducing bacteria are present in relatively low numbers in soil (typically, less than 10^3 cells per gram) unless the soil becomes waterlogged (for example, by excess rainfall) and thus remains anaerobic for a long period of time; therefore, they cannot be isolated easily unless their numbers are increased by enrichment. **Van Delden's medium** (Table 22-1), an effective enrichment medium for sulfate-reducing bacteria, uses *three selective factors* that favor the growth of these organisms: anaerobic conditions, the presence of carbon and nitrogen sources (sodium lactate and asparagine, respectively) that are preferred by sulfate reducers, and the toxic effects of the H_2S that is produced by sulfate reducers. The growth of sulfate reducers in Van Delden's medium produces a black precipitate because some of the microbially produced H_2S combines with iron to produce insoluble iron sulfide:

$$H_2S + FeSO_4 \longrightarrow H_2SO_4 + FeS\downarrow$$

Thus, the formation of this black precipitate tells the microbiologist that the enrichment is working without the need to open the culture tube, which would risk the introduction of oxygen into the medium. A similar reaction is used to specifically test for H_2S production by bacteria (see Exercise 33).

22 ENRICHMENT TECHNIQUES

You will also carry out an enrichment for **algae** and **cyanobacteria** in this exercise. These are *photosynthetic* microorganisms (eukaryotic and prokaryotic, respectively). They usually grow as chemoautotrophs, using energy and electrons generated by photosynthesis to fix carbon dioxide. In other words, these organisms use CO_2 as a carbon source by reducing it and incorporating it into certain organic molecules inside their cells. Their ability to use CO_2 as a carbon source enables them to grow in the absence of organic carbon compounds, something that very few other microorganisms can do.

Some cyanobacteria can *fix* (reduce and incorporate) both CO_2 and atmospheric nitrogen (N_2). In fixing nitrogen (see Exercise 54), these bacteria use energy and electrons produced by photosynthesis to reduce nitrogen to ammonia, which they can incorporate into cellular compounds such as amino acids. These cyanobacteria are probably the most self-sufficient of all microorganisms because they obtain both carbon and nitrogen from the atmosphere. Such organisms are typically multicellular and carry out nitrogen fixation in special cells called **heterocysts,** which are produced only when no reduced nitrogen is available in the environment. They provide reduced nitrogen to **vegetative cells** (nonheterocyst cells that cannot fix nitrogen); in return, the vegetative cells supply the heterocysts with carbon. Most of these cyanobacteria produce long filaments of cells in which heterocysts periodically occur after every few vegetative cells along the filament. Typical cyanobacteria of this type include the genus *Anabaena* and the genus *Nostoc.*

Algae and cyanobacteria play important roles in recycling carbon in the environment, and some of the cyanobacteria are important in recycling nitrogen. These organisms are present in most soils (near the surface where light is available), but their numbers are usually very low. You can use **Bristol's medium** to enrich for algae and cyanobacteria that live in soils and other natural environments. Chemoheterotrophic microorganisms do not grow in this medium because it lacks a reduced source of organic carbon. Only the chemoautotrophs grow after residual organic compounds provided by the soil are used up. Incubating the cultures under bright light gives photosynthetic chemoautotrophs an advantage over nonphotosynthetic chemoautotrophs because the light provides them with an unlimited source of energy. (Nonphotosynthetic autotrophic microorganisms derive their energy from inorganic chemical compounds.)

Biologists often refer to **fixed nitrogen,** meaning any compound that contains nitrogen in either the reduced form (like ammonium chloride) or the oxidized form (like sodium nitrate). Bristol's medium contains *sodium nitrate* as a nutritional source of nitrogen. Fixed forms of nitrogen are used by both the eucaryotic algae and the procaryotic cyanobacteria; therefore, Bristol's medium enriches for both the algae and the cyanobacteria.

On the other hand, *only the cyanobacteria use atmospheric nitrogen* (N_2) as a sole nitrogen source for synthesis and growth. This is the principle upon which **Chu's medium** is based (see Table 22–1); it lacks fixed nitrogen so that cells must first fix atmospheric nitrogen to grow. Chu's medium is used to specifically enrich for the cyanobacteria, and enrichment cultures using this medium should contain few if any algae.

In this exercise, you will introduce identical soil samples into Bristol's medium and Chu's medium so that you can directly compare the enrichment populations obtained with these two media. You will also inoculate soil into nutrient broth, an all-purpose medium rather than an enrichment medium. Consequently, only those microbes that can grow the fastest and are initially present in the greatest numbers will grow and quickly use all available nutrients. These will probably be chemoheterotrophic bacteria, which are predominant in most soils. Growth of soil microbes in the nutrient broth should provide an interesting comparison with the growth that occurs in the more selective enrichment media.

LEARNING OBJECTIVES

• Understand the principle of the enrichment culture technique.

• Understand how media used for enrichment culture differ from selective media and all-purpose media.

• Learn about the varied uses of enrichment cultures.

• Learn how variations in the enrichment conditions affect the outcome of an enrichment culture.

MATERIALS

First laboratory period

Media   Van Delden's broth (steamed and cooled just prior to use)
Bristol's broth
Chu's broth
Nutrient broth

Supplies	Topsoil sample
	Small spatula
	Ethanol (95 percent for flaming spatula blades)
	Bright light (artificial; for Bristol- and Chu-broth incubation)
Demonstrations	1-g soil sample (for volume estimation)
	How to flame spatula blades

After 4.5 weeks

Media	Nutrient broth
Supplies	Same soil sample

After 5 weeks

Supplies	Methylene blue stain
	Microscope slides and coverslips
	Phase-contrast microscopes (if available)
Demonstrations	Uninoculated Van Delden's broth, Bristol's broth, Chu's broth, and nutrient broth (for comparison with cultures inoculated during the first laboratory period)

PROCEDURE: FIRST LABORATORY PERIOD

Enrichment for sulfate-reducing bacteria

1. **Flame sterilize the blade of a spatula with 95 percent ethanol** according to the method demonstrated by your instructor.

2. **Use the sterile spatula blade to transfer about 1 g of soil to a bottle of Van Delden's medium.** The exact quantity is not critical. Ask your instructor to weigh out 1 g of soil for you to use as a volume approximation.

   Avoid introducing oxygen to the bottle. Remove the stopper, add the soil, and tightly replace the stopper as quickly as possible. The enrichment for sulfate reducers will work only if the medium is anaerobic at the time of inoculation and remains anaerobic during incubation. To provide an anaerobic medium for inoculation, bottles of Van Delden's medium will be steamed to drive off oxygen and then cooled to room temperature shortly before your use.

3. Incubate at room temperature for 5 weeks.

Enrichment for algae and cyanobacteria

1. Flame sterilize a spatula blade using 95 percent ethanol, and use it to transfer about 1 g of soil to one bottle of Bristol's medium (broth).

2. Flame sterilize a spatula blade using 95 percent ethanol, and use it to transfer about 1 g of soil to one bottle of Chu's medium (broth).

3. Incubate both broths under a bright light at room temperature for 5 weeks.

PROCEDURE: AFTER 4.5 WEEKS

The following steps should be performed 48 hours prior to examining Van Delden's, Bristol's, and Chu's enrichment media after 5 weeks of incubation.

1. Flame sterilize a spatula blade using 95 percent ethanol, and use it to transfer about 1 g of soil to one bottle of nutrient broth.

2. Incubate at room temperature for 48 hours. Then refrigerate if necessary to hold the culture until the next laboratory period.

The purpose of using nutrient broth is to provide an all-purpose medium that will support growth of the more numerous and faster growing microorganisms that are present in soil. The nutrient-broth culture will be used during the next laboratory period (after 5 weeks) for comparison with the types of microbes found in the enrichment media.

PROCEDURE: AFTER 5 WEEKS

You should now have three enrichment cultures inoculated with soil and incubated at room temperature for 5 weeks: Van Delden's broth (enrichment for sulfate-reducing bacteria), Bristol's broth (enrichment for algae and cyanobacteria), and Chu's medium (enrichment for cyanobacteria). You should also have one nonenrichment culture (nutrient broth) inoculated with soil from the same sample and incubated at room temperature for 48 hours.

1. **Without disturbing the bottles, examine the appearance of each of the four cultures,** and record their appearance in the Results section. Compare cultures with uninoculated media.

2. **Gently shake each bottle to distribute the microorganisms throughout the medium.** Wait about 60 seconds for the largest soil particles to settle before sampling.

3. **Simple stain each culture with methylene blue** (see Exercise 5 for a review of Simple Staining). Use a sterile inoculating loop to transfer material from the culture bottles to the slides.

4. **Prepare a wet mount of each culture** (see Exercise 6 for a review of this method).

5. **Observe your stained slides and wet mounts** with the low-power, high-dry, and oil-immersion objectives.

6. **Record your observations** in the Results section.

7. **Properly dispose of your cultures and other materials.**

NAME _____

DATE _____

SECTION _____

RESULTS

1. **General appearance of each culture.** Be sure to include information on color and turbidity. Your instructor will provide *uninoculated* bottles of media for comparison. If the appearance of the medium varies from one part of the bottle to another, describe this variation.

   a. Van Delden's medium:

   b. Bristol's medium:

   c. Chu's medium:

   d. Nutrient broth:

2. **Type of microorganisms in each culture.** Describe those that you observed most frequently in the *simple stains* and *wet mounts*. Include a sketch of each microbial type that you describe.

   a. Van Delden's medium:

   b. Bristol's medium:

   c. Chu's medium:

   d. Nutrient broth:

3. Does the Van Delden's medium have an unpleasant odor? If so, how do you explain this?

4. Briefly summarize the major differences between the organisms you saw in Bristol's medium and those you saw in Chu's medium. Explain these differences.

5. Briefly summarize the major differences between the organisms you saw in nutrient broth and those you saw in Bristol's medium or Chu's medium. Explain these differences.

QUESTIONS

Completion

1. Enrichment media make use of conditions that are designed to _____ the growth of a particular microbial type.

2. Effective enrichment conditions can be created with chemical, _____, and _____ strategies.

3. Some authorities now recognize the selectivity of enrichment media by calling the enrichment culture technique _____.

4. Sulfate-reducing bacteria grow under anaerobic conditions and reduce sulfate to _____.

5. Algae and cyanobacteria can use _____ as a source of carbon.

True – False (correct all false statements)

1. A medium designed to enrich specifically for nitrogen-fixing cyanobacteria should contain sodium nitrate. _____

2. Enrichment media and selective media work in exactly the same way. _____

3. Enrichment media can be used to search for and isolate new microorganisms that degrade pollutants. _____

4. Bristol's medium makes use of two conditions to achieve an enrichment for algae and cyanobacteria: (1) the absence of reduced organic carbon and (2) the presence of bright light. _____

5. Van Delden's medium makes use of three selective factors to favor the growth of sulfate-reducing bacteria. _____

VII

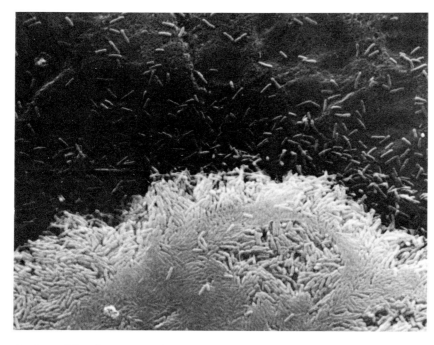

A colony of *Pseudomonas aeruginosa*.

> ... *everything that we discover about microorganisms applies also to elephants, only more so!*
>
> —*F. Jacob*

THE EFFECT OF ENVIRONMENT ON MICROBIAL GROWTH AND VIABILITY

Part VII introduces you to the effect of environment on the structure (Exercise 24) and metabolism (Exercise 25) of microbial cells and their rates of growth (Exercise 23). However, much of the field of microbiology deals with interactions between microbes and their environment; therefore, other parts of this text also deal with the effect of environment on microorganisms. For example, Exercises 18 and 25 show the effects of either the presence or the absence of oxygen on microbial growth and the methods used to achieve anaerobic growth. Exercises 21 and 22 demonstrate how humans can use selective chemicals and/or selective nutritional environments to favor or inhibit the growth of certain microorganisms. Exercises 38 and 39 deal with the effect of soluble mutagens and ultraviolet light on microorganisms, and Exercise 51 shows how a selective environment can be provided to enhance the growth of desirable bacteria during cabbage fermentation to form sauerkraut.

Every kind of microorganism has one type of environment in which it will optimally prosper and, perhaps, even grow. The diversity of favorable environments occupied by microorganisms is very broad. Anything that deviates from the optimum will adversely affect the microbe that occupies that environment. We can apply our knowledge of optimum environmental conditions to either cultivate or selectively inhibit the growth of specific types of microorganisms.

23

The Effects of Temperature on Growth

When microbiologists refer to **growth** they usually mean an increase in cell number. If you inoculate bacteria into a broth and incubate it at a constant temperature, the relationship between change in cell number and time is called a **growth curve** (Figure 23-1). The part of the growth curve at which cell number is doubling at a constant exponential (logarithmic) rate is called the **exponential phase** of growth.

Consider what would happen if you inoculated the same type of bacteria into the same kind of sterile broth but incubated at a temperature 5 degrees lower than before. The resulting growth curve would probably be similar, except that the rate at which the cells doubled (rate of exponential growth) would be altered. If you grew the same cells in the same broth at varied temperatures, the varying rates of growth might be similar to those shown in Figure 23-2. That temperature at which the growth rate is the highest is called the **optimum growth temperature.** The optimum temperature for growth of a microbial species is closely correlated to the temperature of that organism's habitat. For example, the optimum temperature for growth of microbes pathogenic for humans is near body temperature (37°C). You should expect microorganisms from a mountain stream to have lower optimum growth temperature than that for human pathogens.

Most known microorganisms have optimum growth rates between 28° and 38°C; these are called **mesophilic** (middle loving). Some species have optimum growth rates at temperatures lower than 16°C; these are **psychrophilic** (cold loving). A few types have optimum growth rates around 60°C; these are **thermophilic** (heat loving). Note that these temperatures are only for *optimum growth rates;* they do not indicate where survival is possible.

Every microorganism has an optimum, a minimum, and a maximum growth temperature (Figure 23-3). For example, the hypothetical data in Figure 23-2 suggest that 30°C is the optimum growth temperature. **Maximum growth temperature** refers to the highest temperature at which growth occurs: somewhere between 35° and 40°C for the microbe plotted in Figure 23-2. The **minimum growth temperature** is the lowest temperature at which growth occurs. For the culture plotted in Figure 23-2, this would probably be somewhere below 15°C. Note again that these temperatures refer to growth rates and not to survival.

To measure growth rates, you would have to periodically measure increases in viable cell numbers (see Exercise 20), total cell numbers, or changes in turbidity (see Exercise 19). You will not have to be that precise in this exercise; instead, you will roughly compare the maximum *extent of growth* achieved by a microbe incubated at three temperatures over an incubation period of 2 to 7 days. Your results should

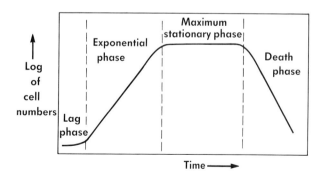

FIGURE 23-1 Typical growth curve of an unicellular microorganism inoculated into a tube of sterile broth. The phases of growth are separated by dotted lines. Note how cell numbers change with time.

213

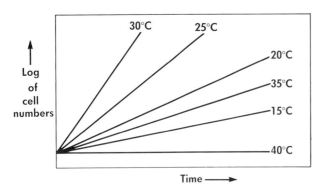

FIGURE 23-2 Effect of temperature on exponential rates of cell growth of a culture having an *optimum* growth temperature of 30°C.

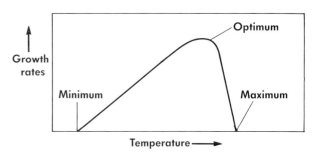

FIGURE 23-3 Relationship between a microorganism's growth rate and incubation temperature. As temperature rises, so does growth rate (rate of doubling). As temperature rises above the optimum, rates of growth fall rapidly until the maximum growth temperature is reached. Temperatures above the maximum usually result in cell *death*. Temperatures below the minimum slow or stop growth, but *cells do not die*. As a matter of fact, most microorganisms survive for long periods of time at cold temperatures and will quickly begin to grow as soon as the temperature rises.

give you an idea of whether the inoculated microbes are psychrophilic, mesophilic, or thermophilic (Figure 23-4).

LEARNING OBJECTIVES

• Learn how to measure microbial growth rates and how to determine optimum growth temperatures.

• Learn the characteristic optimum growth temperatures for psychrophiles, mesophiles, and thermophiles.

MATERIALS

Cultures	*Escherichia coli*
	Bacillus stearothermophilus
	Micrococcus cryophilis
Media	Trypticase-soy broth tubes
Supplies	4°C incubator
	30°C incubator
	55°C incubator

PROCEDURE: FIRST LABORATORY PERIOD

1. **Inoculate each microbe into three broth tubes.** One tube will be incubated at 4°C, one at 30°C, and one at 55°C.

2. **Label each tube** with your *initials,* the name of the *organism, date* inoculated, and *temperature* of incubation. Write only on the glass, not on the cap.

3. **Separate these tubes according to incubation temperature.** Your class will probably be provided with one test tube rack clearly marked for each incubation temperature. Place each inoculated tube in the appropriate rack.

4. **Incubate all tubes for 48 hours.**

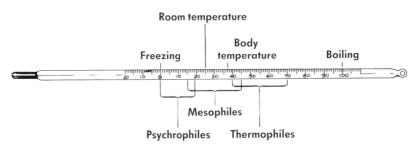

FIGURE 23-4 Commonly encountered Celsius temperatures and approximate ranges for growth of psychrophiles, mesophiles, and thermophiles.

23 THE EFFECTS OF TEMPERATURE ON GROWTH

PROCEDURE: SECOND LABORATORY PERIOD

1. **Organize your tubes according to microbial species.** Note that some cells may have settled out of suspension during incubation.

2. **Mix each tube thoroughly** by thumping the end of the tube against your finger, as in Exercise 17, Procedure 2.

3. **Examine the growth of *E. coli* at 30°C. Use this tube as your standard** for determining the amount of growth in the other tubes. Call this amount of growth ++ (heavy). After thoroughly mixing all other tubes, rate the amount of growth for each culture on a scale from 0 (no growth) to ++ (same as *E. coli*).

INTERPRETATION NOTE

Psychrophilic microorganisms often grow *slowly,* even at temperatures for which they show *optimum* growth rates. Therefore, you may find that 48 hours is not enough incubation time to see growth of any of your inoculated cultures at 4°C, including psychrophiles. If that happens, incubate all cultures for an additional period of time. Five to seven days (total) incubation at 4°C is often necessary to selectively show growth of psychrophiles.

---

4. **Record your results.**

5. **If further incubation is needed** for one or more cultures, incubate each culture as before. **If no further incubation is needed,** remove the markings from all tubes, and discard as instructed.

23 THE EFFECTS OF TEMPERATURE ON GROWTH

NAME _____

DATE _____

SECTION _____

RESULTS

Effect of temperature on relative microbial growth. Use 0 for no growth and ++ for heavy growth. For each organism, record the *total* incubation time in *days* at the time your observation was made.

Incubation temperature (°C)	Growth (0 to ++)		
	E. coli (Incubation time: ____)	*M. cryophilis* (Incubation time: ____)	*B. stearotherm.* (Incubation time: ____)
4			
30			
55			

CONCLUSIONS

Give the best temperature for supporting the growth of each microbe and the probable classification according to maximum growth (that is, psychrophile, mesophile, or thermophile) in the following table.

Microbe	Temperature of best growth (°C)	Probable classification
E. coli		
B. stearothermophilus		
M. cryophilis		

QUESTIONS

Completion

1. When microbiologists speak of growth, they usually are referring to an increase in _____.

2. The phase of growth at which cells are doubling at a constant rate is called the _____.

3. The temperature at which the growth *rate* is at its maximum is called the _____ growth temperature.

4. There are three terms that classify a microorganism's preference for temperature (that is, the temperature at which its growth rate is optimum). Of these three terms, most types of microorganisms are _____.

5. Where might you find a thermophile in the natural environment? Where might you find a psychrophile?

6. How do refrigeration temperatures affect mesophiles?

True – False (correct all false statements)

1. Psychrophilic microbes are those that exhibit an optimum growth rate around 55°C. _____

2. Cultures show a minimum growth rate at their optimum growth temperature. _____

3. Growth rate measurements must show an increase in numbers or turbidity. _____

4. Pathogens that grow in living human tissue are usually thermophilic. _____

24

The Effects of Elevated Sugar and Sodium Chloride Concentrations on Growth

Any solution is composed of two parts: that which is dissolved (the **solute**) and the fluid in which the solute is dissolved (the **solvent**). The solute concentration is related to a phenomenon known as **osmotic pressure.** If you were to place a nonliving semipermeable membrane between two sides of a water-tight box filled with distilled water, you could demonstrate how solute concentration affects osmotic pressure. If you dissolved sugar in the water on one side of the membrane, making the solute concentration higher on this side, water would flow from the low solute side to the high solute side, causing an imbalance in pressure. The membrane would then bulge toward the compartment with the low solute concentration. In this way, you would demonstrate that higher solute concentrations create higher osmotic pressures.

Inside a microorganism are many dissolved materials. In fact, the solute concentrations are much higher inside the cell than outside; therefore, the osmotic pressure inside the cell is higher than it is outside. This internal pressure pushes the cell's plasma membrane against the cell wall. This is easy to remember if you think about water flowing across a membrane from the side with lower solute concentration to the side with higher solute concentration, creating more pressure on the high-solute side of the membrane.

The solute concentration of blood and other body fluids is sometimes used as an osmotic standard in research with biological systems. **Tonicity** refers to the normal osmotic pressure exerted by body fluids, which amounts to about the same as **physiological saline,** a 0.9 percent (w/v) sodium chloride solution. Physiological saline is said to be **isotonic** with tissue fluids or blood, and it is often used as a diluent when working with microbes isolated from humans and other mammals. Fluids having a lower solute concentration are said to be **hypotonic** (less than tonic), and those with greater solute concentrations are called **hypertonic** (more than tonic).

In most natural environments and in laboratory media, the bacterial cell wall has sufficient strength to withstand the difference in pressure between the inside and the outside of the cell (Figure 24-1A). However, if you rapidly *decrease* the osmotic pressure outside some bacteria (for example, by plunging cells into distilled water), water rushes into the cell, creating a very large pressure difference. The cell literally explodes because the cell wall is not strong enough to withstand the high internal pressure. This phenomenon is called **plasmoptysis.** It does not occur with all bacteria but is lethal (bactericidal) for those in which it does occur.

An opposite effect is observed when one *increases* the exterior solute concentration above that found inside the microbial cell. In that case, there is a net flow of water out of the cell, the protoplasmic constituents shrink, and the plasma membrane pulls away from the wall (Figure 24-1). This condition is called **plasmolysis.** As cells lose water, metabolism slows down to such a low level that cell division (growth) stops. The cells do not usually die but enter a static state, a condition called **bacteriostatic.**

Certain methods of food preservation use these principles. For example, if you dry (remove water from) any plant or animal tissue, solutes become more concentrated, causing plasmolysis of the bacteria that are always present on these tissues; thus these foods are preserved because decay (microbial degradation) is inhibited. Another example is the addition

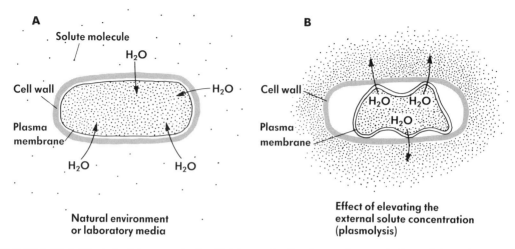

FIGURE 24-1 Effect of solute concentration (dots) on water flow (arrows) across the bacterial membrane and on osmotic pressure. In the normal environment that favors growth of bacterial cells (A), the solute concentration of the cell's interior is greater than that of the surrounding medium; thus the interior pressure keeps the plasma membrane pressed against the cell wall. When the solute concentration outside the bacterial cell is greatly increased (B), water flows out of the cell, and the cell's interior shrinks. This effect is called *plasmolysis* and is usually *bacteriostatic*.

of sugar to fruits to make jams and jellies, which also increases the solute concentration and accomplishes the same bacteriostatic effect.

Most bacteria are sensitive to high osmotic pressure, which is why you do not find them growing in jellies (about 65 percent sugar) or other high-sugar foods or in dried plant or animal tissue. Yeast and molds, however, are more tolerant to high osmotic pressure and can be found growing in or on the surface of materials such as honey, jams, canned and dried fruit, dried meats, and even clothing, shower stalls, and shoes (if the humidity of the surrounding air is high enough). Yeasts and molds (fungi), therefore, are said to be **osmophilic** (osmotic-pressure loving) and sometimes also **saccharophilic** (sugar loving).

Sodium chloride (NaCl, table salt) may be used to preserve meats if it is used in large amounts. High concentrations of salt dehydrate or pull water out of tissue, thus increasing the solute concentration inside the tissue and creating an osmotic effect upon contaminating microorganisms (plasmolysis). Salt is also toxic to many microorganisms if its concentration is from 7 to 9 percent or higher. The reason for this toxicity is not known. One theory is that, because NaCl dissociates when placed in solution, the ions provide a greater amount of osmotic pressure than the same concentration of sugars or other organic molecules. Another theory is that too many anions or cations interfere with the normal functioning of the microbial plasma membrane. Regardless of the reason, you should remember that too much NaCl may have both an *osmotic* and a *toxic* effect.

A few bacteria, called **halophilic** (salt loving), actually require high sodium chloride concentrations for optimum growth rates. Halophilic bacteria can often be isolated from natural salt lakes and other environments where salts are in high concentration.

Perhaps the most important point to remember is that preservation of plant or animal tissue by drying or by adding either sugar or salt has primarily an osmotic effect on microbial cells whose presence would otherwise cause decay of that tissue. Preservation of the food results from the prevention of normal microbial activity. It does not mean that the microorganisms have been killed. On the contrary! The plant or animal tissue undoubtably contains many viable microorganisms that will grow actively if the osmotic pressure is decreased. Plasmolysis is *bacteriostatic* not bacteriocidal. This type of food preservation is our most ancient method.

LEARNING OBJECTIVES

• Become aware of the relationships between solute concentration and osmotic pressure.

• Learn how increased osmotic pressure differentially affects the bacterial and fungal cell.

• Understand the practical role played by increased osmotic pressure in food preservation.

24 EFFECTS OF SUGAR AND NaCl CONCENTRATIONS

MATERIALS

Cultures
Escherichia coli
Penicillium notatum
Saccharomyces cerevisiae
Staphylococcus aureus

Media
Trypticase-soy agar plates containing each of the following sodium chloride (table salt) concentrations in percent (w/v): 0.5, 5.0, and 20.0
Trypticase-soy agar plates containing each of the following sucrose (table sugar) concentrations in percent (w/v): 0.5, 15.0, and 60.0

Supplies
Metric ruler (second period only)

PROCEDURE: FIRST LABORATORY PERIOD

1. Select a 0.5 percent NaCl plate, and write its solute and concentration on the bottom of the plate.
2. Repeat step one with all the other NaCl concentrations. Label each plate as you pick it up.
3. Select and mark the three plates containing different sucrose concentrations. Label each plate as you pick it up.
4. Divide each plate into four equal sections with your marking pen and label each section with the name of one microorganism.

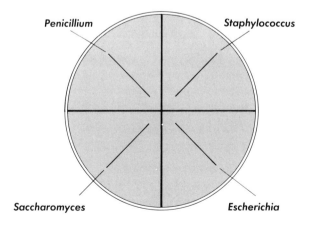

5. Use your inoculating loop to transfer the appropriate microorganism to its section on all three NaCl plates and all three sucrose plates. Make sure that you transfer the microbe only to that section containing its name. Make a separate aseptic transfer for each section on each plate. Streak the agar with *one straight line,* as shown in step 4. It is important to inoculate in this way because later you will estimate the extent of growth by measuring the *width* of growth in each single streak.

When you finish inoculating, you should have three NaCl plates, one for each salt concentration, and each should be inoculated with four different microorganisms. You should also have three sucrose plates, one for each sugar concentration, and each should be inoculated with all four microorganisms.

6. **Incubate all plates at 30°C for 48 hours.** If the next laboratory period is more than 48 hours away, either you or your instructor should refrigerate these plates after the 48-hour incubation. Make sure that you know who will take care of this before you leave the laboratory.

PROCEDURE: SECOND LABORATORY PERIOD

1. **Arrange your plates** in front of you so that they are lined up in order of their increasing NaCl or sucrose concentration.
2. **Measure the average width of each inoculated streak, and record the relative density of growth of that microorganism.** Let the *E. coli* streak on the 0.5 percent sucrose plates be your standard for very dense (++) growth. Record the growth as zero (0) if you detect no growth.
3. **Record your results, and properly discard your plates.**

NAME _____

DATE _____

SECTION _____

RESULTS

Relative growth of four cultures on NaCl and sucrose plates of varying concentrations. Record the relative density of growth from 0 (zero) for no growth to ++ for heavy growth (like that of *E. coli* on 0.5 percent sucrose plates).

Solute concentration (%)	*Escherichia*		*Saccharomyces*		*Penicillium*		*Staphylococcus*	
	Width (mm)	Density	Width (mm)	Density	Width (mm)	Density	Width (mm)	Density
NaCl 0.5								
5.0								
20.0								
Sucrose 0.5								
15.0								
60.0								

CONCLUSIONS

QUESTIONS

Completion

1. In any solution, the dissolved material is called the _____, and the fluid in which it is dissolved is called the _____.

2. If you artificially increase the solute concentration outside the cell, this will _____ the osmotic pressure on the outside of the cell's plasma membrane; this phenomenon is called _____.

3. If the osmotic pressure outside the cell is rapidly decreased, this may cause the inside of the cell to _____, resulting in a condition known as _____.

4. Isotonic saline has a NaCl concentration of about _____%. If the NaCl concentration is increased, this salt solution is considered to be _____.

5. Any physical or chemical factor that stops cellular growth without killing the cells is called _____. A similar term that refers to materials causing cell death is _____.

6. In this exercise, the microorganism(s) used to represent the *fungi* is(are) called _____ and that(those) used to represent the bacteria is(are) _____.

True – False (correct all false statements)

1. Sodium chloride exerts osmotic effects, but it may also kill the cell at higher than isotonic concentrations. _____

2. When you dry fruit, vegetables, or meat, you are creating an osmotic-pressure effect on the bacteria present, and this results in sterilization of the food. _____

3. Water flows across a membrane toward the side that contains the highest solute concentration. _____

4. Extreme plasmolysis initially creates a bacteriostatic effect on most bacterial cells. _____

5. Yeast and molds usually tolerate higher osmotic pressures than bacteria. _____

Multiple Choice (circle the best answer)

1. If you want to isolate a saccharophilic yeast, you might have success examining the following substances.
   a. Apple juice
   b. Burlap sacks
   c. Honey
   d. Concentrated orange juice
   e. Brine pickles

2. Where might you expect to find halophiles?
   a. Natural salt lakes
   b. Mountain streams
   c. Jellies and jams
   d. Oceans
   e. Hail stones

25

The Effects of Free-Oxygen Concentration on Growth: Agar-Deep Cultures

Microorganisms vary widely in their requirements for oxygen. **Strict (or obligate) aerobes** grow only when oxygen is present, and they are commonly found on dust, on plant surfaces, and in the uppermost layers of soil. These organisms are capable only of respiratory metabolism, and they can use only oxygen as the final electron acceptor in their electron transport chain. (The introduction to Exercise 36 gives further background on the metabolism of respiration.)

Strict (obligate) anaerobes require no oxygen for growth; in fact, oxygen is toxic to these bacteria. Strict anaerobes are commonly found in marsh and bog muds, stomachs of ruminant animals, intestines of humans and other nonruminant animals, and deep-penetration tissue wounds. The metabolism of these organisms is commonly **fermentative.** They usually lack an electron transport chain and are, therefore, incapable of respiring. Those strict fermenters that do not find oxygen toxic but cannot use it in their metabolism are called **aerotolerant anaerobes.** (The introduction to Exercise 31 gives more information about fermentative metabolism.)

Most known microorganisms fall somewhere between the two extremes of strict aerobes and aerotolerant anaerobes. These **facultative anaerobes** can grow either in the presence or the absence of oxygen. They are usually found where oxygen sometimes is present and sometimes is not; they respire in the presence of oxygen and shift to fermentation when the oxygen supply is depleted. They obtain more energy from respiration so they grow faster in aerobic than in anaerobic conditions.

If a facultative anaerobe is inoculated into broth or tempered agar and incubated, these cells respire until all of the oxygen is consumed, and then they shift to fermentation and continue to grow at a slower rate. This phenomenon happens frequently in aquatic environments during the decay of plant and animal material and in controlled fermentations that produce foods and beverages.

Knowledge of how oxygen affects the metabolism of a microorganism should also suggest the best way to grow this culture and how to classify it.

In this exercise, you will examine the growth characteristics of three cultures suspended in a sterile, semisolid medium. These sterile **agar deeps** are prepared as shown in Exercise 16. Prior to class time, they will be heated to 100°C to liquify the agar and then tempered to 45°C until you are ready to inoculate them. You will inoculate the deeps, thoroughly mix the cells in the medium, allow the medium to solidify, and incubate the resulting cultures.

Heating the agar deeps not only liquifies the agar but also drives off all the oxygen. After inoculation and cooling, the semisolid-agar column forms an oxygen gradient, with the greatest amount of oxygen at the top of the tube and the least amount (or no oxygen) at the bottom (Figure 25-1).

In the past, some microbiologists referred to inoculated agar deeps, in which cells are evenly suspended in the solid medium, as *shake cultures,* but this term is misleading because these tubes are not actually shaken before incubation; such action would mix air into the medium. The use of the word *shake* in this context refers merely to the act of mixing the cells in tempered-agar deeps before the agar is allowed to solidify.

The nature of microbial growth in agar deeps reflects the cells' relative need for oxygen or an oxygen-free environment. Strict aerobes grow only at or near

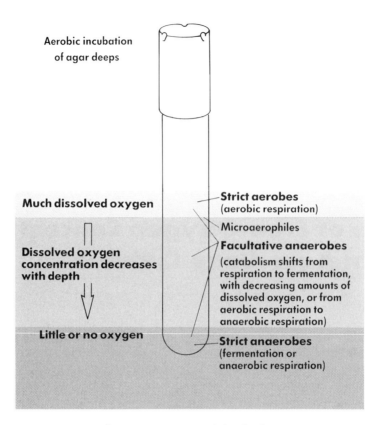

FIGURE 25-1 Relative concentration of dissolved oxygen in an agar-deep culture and its relationship to growth of aerobes, microaerophiles, facultative anaerobes, aerotolerant anaerobes, and strict anaerobes. The heating of agar deeps to 100°C drives off all dissolved oxygen. Once the agar deep has been cooled (tempered), inoculated, and allowed to solidify, these agar deeps should begin absorbing oxygen and form an oxygen-concentration gradient from the top to the bottom of the tube.

the surface of the agar in the tube. Facultative anaerobes should grow from the top to the bottom of the tube. Aerotolerant anaerobes grow throughout the tube but probably better near the bottom. The growth of strict anaerobes in agar deeps depends on how far oxygen diffuses into the solid agar medium and how sensitive the cells are to the toxic forms of oxygen (such as hydroxyl ions and superoxide radicals). Anaerobes that are extremely sensitive to an oxygen-containing environment may not grow at all, but others will be found growing at the bottom of the tube.

Microbiologists often use the growth characteristics of a pure culture in agar deeps to help them describe the microbe's need for or tolerance to oxygen. This can help the microbiologist to identify the microorganism, but perhaps more importantly, this knowledge helps understand how best to grow the microbe for further study.

LEARNING OBJECTIVES

• Observe how agar-deep cultures can be used to determine differences in the way in which microorganisms respond to oxygen.

• Understand that there is a relationship between how a microorganism responds to the presence or concentration of oxygen and whether it is capable of respiration or fermentation and how sensitive the cells are to the toxic forms of oxygen.

• Develop skill in inoculating agar deeps.

MATERIALS

Cultures   *Clostridium sporogenes*
            Micrococcus luteus
            Escherichia coli

25 EFFECTS OF OXYGEN: AGAR-DEEP CULTURES

Media   Trypticase-soy agar deeps (melted and tempered)

Supplies   Waterbath set at 45°C

PROCEDURE: FIRST LABORATORY PERIOD

1. **Remove one tube at a time from the waterbath for inoculation.** If you remove all tubes at once, chances are good that one or more will solidify before you can inoculate them. If your tube solidifies before you inoculate it, please put it in an appropriately marked rack outside the waterbath. This sterile medium will be saved for later use.

> **TECHNICAL NOTE**
>
> When the polysaccharide called *agar* is placed with water and heated, it changes from a suspended solid to a *fluid colloid*. That fluid colloid is called a *sol* or *liquid sol*. For convenience and in keeping with common usage, however, the liquid–sol state of agar will be referred to as a *liquid* in this exercise. For more information on the properties of agar, see Exercise 13.

2. **Dry the outside of the tube with a paper towel, and label the tube** (only on the glass) with *your name*, the *microbe's name*, and *date* inoculated.

3. **Inoculate by aseptically transferring one loopful of one culture to the bottom of the liquified-agar deep.**

4. **Stir the liquified agar with the loop as you withdraw it.** This is done to mix the cells from the bottom to the top of the tube. To make this an adequate test, cells must be distributed throughout the tube so that they can metabolize and grow where the oxygen concentration or lack of oxygen meets their requirements.

   This medium was heated to drive off dissolved oxygen; therefore, do not shake the tubes after inoculation to mix cells in the medium because shaking will reintroduce oxygen into the medium. Remember that the purpose of using agar deeps is to allow oxygen to diffuse from the surface of the solidified agar deep, so that a decreasing oxygen-concentration gradient is formed from the top to the bottom of the tube.

5. **Place the inoculated agar deeps in a rack at room temperature.** Do not move these tubes until they have completely solidified.

> **TECHNICAL NOTE**
>
> You can tell when the agar has completely solidified by the change in its appearance. Liquified agar is transparent. When the temperature drops below about 40°C, the agar solidifies and simultaneously loses its transparency. Solidified agar deeps have an opaque (cloudy) appearance.

6. **Label and inoculate other liquified agar deeps using the two remaining microbes and the same methods.**

7. **Ask your instructor to select a few sterile, agar deeps for the entire class to use as uninoculated controls.** Incubate these tubes with the inoculated tubes until the next laboratory period. These tubes will be used as *negative controls*. (What is the purpose of a negative control in an experiment?)

8. **Incubate all tubes at 30°C for 48 hours.** If your next laboratory period is more than 48 hours away, either you or your instructor should place the incubated tubes in the refrigerator after the 48-hour incubation. Make sure you know who is responsible for this before you leave the laboratory.

PROCEDURE: SECOND LABORATORY PERIOD

1. **Examine the inoculated tubes, and compare their appearances with the uninoculated control.** Look for evidence of microbial growth and its position in the tube.

   Growth may appear either as increased turbidity (if the inoculated cell concentration was heavy) or as isolated colonies embedded in the agar (if the inoculum was light). Do not mistake condensation on the outside of the tube as growth if your tubes were refrigerated. Does the agar appear to be breaking up? If so, this may be the result of the hydrogen and carbon dioxide gases given off during fermentaion of nutrients. Does the presence of gas correspond with growth of only the anaerobe?

2. Record your results.

3. Return the negative-control tubes to your instructor.

4. Remove markings from all inoculated tubes, and properly discard them.

NAME _____

DATE _____

SECTION _____

RESULTS

Illustrate the location and general appearance of the growth (if any) for each tube in the drawing. Give the microbe's *name,* the appropriate *term* that describes its need for oxygen, and *other observations* (if appropriate).

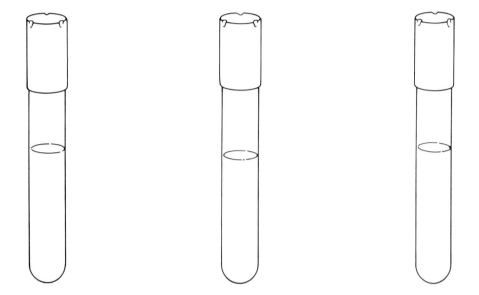

CONCLUSIONS

QUESTIONS

Completion

1. Microbes that are capable of either fermentation or respiration and that can grow in either the presence or the absence of oxygen are called _____.

2. Metabolism in which oxygen is the terminal electron acceptor is called _____. Cells that have only this type of metabolism are called _____.

3. Microbes that grow only in the complete absence of oxygen are called _____ anaerobes. If they tolerate some oxygen but do not use it, they are called _____ anaerobes.

4. Judging from your experimental results, how far into the tube has oxygen penetrated? _____ mm. Give your reasoning for this conclusion.

5. Of the three genera inoculated, which would you expect to have the greatest ability to adapt to an environment in which oxygen availability varies from day to day?

True – False (correct all false statements)

1. The primary function of oxygen in respiration is to serve as the final electron acceptor in electron transport. _____

2. Strict aerobes grow only just below the surface of inoculated agar deeps. _____

3. Heat increases the ability of gases like oxygen to be absorbed into growth media. _____

4. Agar becomes a liquid sol at about 100°C and solidifies at about 38° to 40°C. _____

5. The negative control for this experiment consisted of one tube that was handled exactly like all the rest except that it was not inoculated. Therefore, by comparison, the negative control shows the appearance of microbial growth in the tube. _____

Multiple Choice (circle all correct answers)

1. You might find a strict anaerobe growing in which of the following environments?
    a. A cow's rumen
    b. Infected muscle wounds
    c. Marsh mud
    d. Your own intestine
    e. The surfaces of plants
    f. Canned foods

NAME _____

DATE _____

SECTION _____

2. You should expect an aerotolerant anaerobe to
   a. die in the presence of oxygen.
   b. be capable of respiratory metabolism.
   c. survive but not grow in the presence of oxygen.
   d. be capable of both fermentation and respiration.
   e. use oxygen as the terminal electron acceptor in the electron transport chain.

3. You should expect a strict aerobe to
   a. die in the presence of oxygen.
   b. be capable of respiratory metabolism.
   c. survive but not grow in the presence of oxygen.
   d. be capable of both fermentation and respiration.
   e. use only oxygen as the terminal electron acceptor in the electron transport chain.

4. You should expect a facultative anaerobe to
   a. die in the presence of oxygen.
   b. be capable of respiratory metabolism.
   c. survive but not grow in the presence of oxygen.
   d. be capable of both fermentation and respiration.
   e. use oxygen as the terminal electron acceptor in the electron transport chain.

VIII

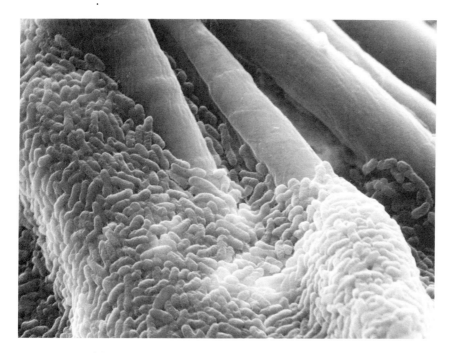

Zymomonas mobilis cells immobilized on cotton fibers for ethanol manufacture.

> . . . *there is a difference between bacterial physiology and bacterial biochemistry. Biochemistry is concerned with the chemical composition and the chemical reactions of living organisms; physiology is concerned, in addition, with the* functions *of these reactions in the life of the organism.*
>
> —E. L. Oginsky and
> W. W. Umbreit

PHYSIOLOGICAL CHARACTERISTICS OF MICROORGANISMS

By definition, physiology is the study of function. There is probably no topic as broad as that dealing with how a living cell functions. An understanding of its function can come only from studies of its structure, its metabolism, its genetic composition, how it responds to its environment, and how all of these things interrelate. Thus, a microbial physiologist should be a student of cytology, biochemistry, genetics, and the environment.

According to E. L. Oginsky and W. W. Umbreit (1954), "So rich and varied is the field of bacterial physiology that selection of what are the basic principles . . . and how they are to be illustrated becomes a matter of personal judgment." And so it is with Part VIII. Actually, this entire text deals with an introduction to how the microbial cell functions.

In the minds of many microbiologists, however, the field of bacterial physiology has a more limited scope and deals primarily with how cells alter their immediate environment, how nutrients get into the cell, the metabolism of the cell, and the metabolic products that the cells excrete. Therefore, for tradition's sake, Part VIII is organized into three sections:

Hydrolysis and use of large extracellular materials (Exercises 26 through 29)

Transport and use of small molecules (Exercise 30)

Measurement of products and effects of metabolism (Exercises 31 through 36)

Emphasis here is on bacterial function, that is, on function with respect to bacteria (not the function of bacteria with respect to humans, as stated by Oginsky and Umbreit). For example, you will learn in Exercise 34 what *indol* is, how it is formed, and the value of this metabolic pathway for the cell. But you should also be aware that indol formation (or the failure to form it) is often of value in identifying certain microorganisms, especially those that are pathogenic for humans. For this reason, indol formation and other such characteristics are often placed in the medical section of other introductory microbiology laboratory manuals. For that matter, most of the exercises in this part could have been placed in Part XI (Medical Microbiology). However, the purpose of this manual is to give you an understanding of *how* and *why* microbes function as they do. If you choose to do so, you may later study more thoroughly how this knowledge can be put to use in fields such as medicine, food, industry, or environmental microbiology.

26

Introduction to Extracellular Degradation

For a microorganism to grow, it must have a source of energy. Some microbes use sunlight, whereas others use inorganic or organic molecules as energy sources; but whatever the form, the energy source must get into the cell. For example, before photosynthetic cells can use sunlight, the light must penetrate the surrounding medium and the cell surface. And before cells can use inorganic or organic molecules, these must be transported into the cells. But many organic molecules are too large to enter microbial cells.

Does that mean that microorganisms cannot use large molecules? If the answer to that question were yes, then the people on this planet would be in deep trouble because polymers from dead plant and animal tissue would not decompose. Without microbial decomposition, there would be no nutrient recycling, and life could not continue as we know it. Some microorganisms break down large molecules, such as proteins, nucleic acids, polysaccharides, and lipids, and use them for energy. How do microorganisms do this?

Some microorganisms make and then release enzymes into the surrounding medium. Because they are released by the cell, these are called **extracellular enzymes.** (A few people call them *exoenzymes;* but that is the same term biochemists use for enzymes that attack the end of a large molecule, and not all extracellular enzymes are exoenzymes). Some extracellular enzymes (hydrolytic, water-breaking, enzymes) can break down large molecules in the environment surrounding the cells. Most extracellular enzymes produced by microbes are hydrolytic, but not all hydrolytic enzymes are extracellular.

Hydrolytic enzymes are so named because of the way they break up large molecules (polymers). When such an enzyme comes in contact with a polymer (Figure 26-1), the enzyme first binds a water molecule and breaks the bond between the hydrogen and hydroxyl ions; then the enzyme binds the polymer and breaks a bond that joins two of its parts; and finally, the enzyme adds the hydrogen ion from the water to one part of the broken polymer and adds the hydroxyl ion of the water to the other part, at the site of the broken bond. In summary, a water molecule is broken, and its parts are added to either side of a broken polymer molecule.

Extracellular hydrolytic enzymes are roughly classified according to the type of large molecule they attack.

Glycosidases break apart *polysaccharides* (large molecules of sugars covalently bonded together).

Proteinases (proteases) break apart *proteins* (large molecules of amino acids covalently bonded together).

Esterases break apart *fats* and *lipids* (large molecules of long, fatty-acid chains often covalently bonded to a molecule of glycerol).

In each case, the name of the enzyme type refers to the type of bond holding the large molecules together. Microorganisms that produce and release these types of enzymes are examined in the exercises that follow.

Just because a large molecule is broken into smaller parts does not mean that these smaller parts can get into the cell. One function of the microbial cell's **plasma membrane** is to select which small molecules it binds and transfers into the cell. Thus, the cell's plasma membrane is **selectively permeable;** that is, it

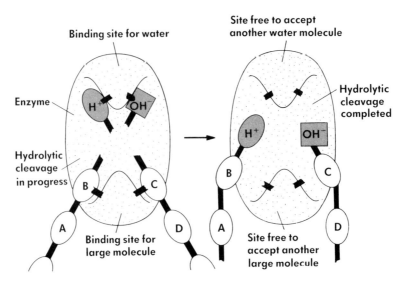

FIGURE 26-1 Schematic representation of how a hydrolytic enzyme breaks one bond within a large molecule (polymer), splits a water molecule, and inserts the hydroxyl and hydrogen ions, before releasing the two parts of the broken polymer.

allows only certain kinds of small molecules into the cell. To get in, the small molecule must bind to a specific **binding protein** at the outer surface of the plasma membrane (Figure 26-2). The binding proteins, which are usually selective, dictate the kind of small molecules that enter the cell. If a cell lacks the specific binding protein for one type of small molecule, then that molecule will not enter the cell, and the cell cannot use it as a source of energy or as a building block for growth and reproduction.

After a small molecule is attached to a specific binding protein, it must then be moved from the outside to the inside of the membrane (into the cell). This movement of molecules across the membrane is called **translocation**. Although not fully understood, it appears that separate enzyme-like proteins in the

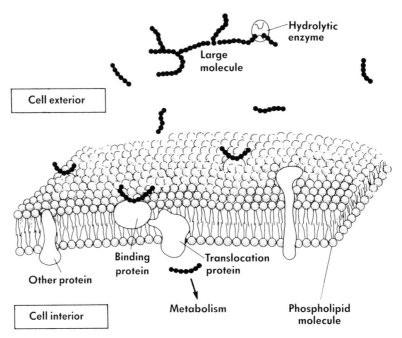

FIGURE 26-2 Model showing the molecular structure of the cell's plasma membrane and the specific binding proteins and translocation proteins needed to get small molecules inside the cell. Other types of proteins are shown inserted into and through the phospholipids of the plasma membrane.

26 EXTRACELLULAR DEGRADATION

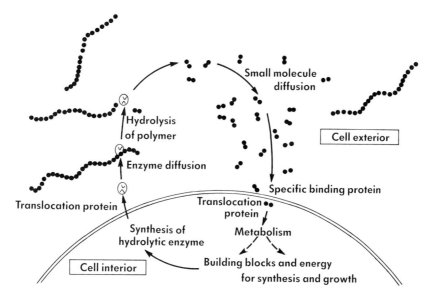

FIGURE 26-3 Overview of extracellular enzyme production and the many different types of proteins needed for breakdown and use of large molecules in the cell's environment.

plasma membrane move the small molecules across the membrane. Therefore, cells lacking the specific **translocation protein** are not able to move these small molecules across the membrane to the inside of the cell.

After the small molecules are bound and translocated, they can be metabolically broken down by the cell and used as sources of energy or used directly as building blocks for synthesis. But this happens only if the cell makes all of the enzymes needed for metabolizing that type of small molecule. Thus, cells must make many other types of enzymes besides extracellular enzymes in order to use the large molecules found in their environment (Figure 26-3).

Only certain kinds of microorganisms have the ability to produce and release extracellular enzymes for the hydrolysis of specific large molecules. These microbes are found in the soil where active decomposition of organic material normally occurs.

The three exercises that follow (27, 28, and 29) deal with microbial degradation of polysaccharides, proteins, and fats (respectively).

26 EXTRACELLULAR DEGRADATION

NAME _____

DATE _____

SECTION _____

QUESTIONS

Completion

1. The term *hydrolytic* literally means "_____," and the result of this type of enzymatic activity is to _____.

2. Glycosidases are hydrolytic enzymes that break up large molecules collectively called _____.

3. Once large molecules outside the cell are broken down and the smaller molecules have penetrated the cell's plasma membrane, the smaller molecules can be used as a source of either _____ or _____ for the cell.

4. The microbe's plasma membrane is said to be *selectively permeable*. This means that _____.

5. The process by which microbes produce and release extracellular enzymes that break down large organic molecules in the environment is part of a widespread process generally known as microbial _____.

True – False (correct all false statements)

1. Most extracellular enzymes are hydrolytic, but not all hydrolytic enzymes are extracellular. _____

2. Once extracellular enzymes break up large molecules into sufficiently small molecules, these smaller molecules can readily diffuse through the cell's plasma membrane. _____

3. Esterases enzymatically cleave fats and lipids into two components: fatty acids and the molecule these fatty acids are bonded to (usually glucose). _____

Microbial Degradation of Polysaccharides (Starch)

Carbohydrates, as the name implies, are compounds of carbon, hydrogen, and oxygen. They can be divided into three broad categories.

Monosaccharides *(single sugars)* are the five- or six-carbon sugars (pentoses and hexoses, respectively). **Oligosaccharides** *(few sugars)* are larger molecules composed of two or more monosaccharide units joined by **glycosidic bonds.** For example, both sucrose (a disaccharide) and raffinose (a trisaccharide) are oligosaccharides. There is no clear distinction, however, between oligosaccharides and polysaccharides, the third group of carbohydrates. **Polysaccharides** *(many sugars)* are large polymers of monosaccharide units joined together by glycosidic bonds; they have large molecular weights. Polysaccharides are usually classified into structural and nutrient polysaccharides.

Many types of **structural polysaccharides** exist. Among them are cellulose (found in all plants), lignin (found in higher plants), mannan (found in yeasts), pectin (mostly in fruits), hyaluronic acid (the "cement" holding animal cells together), and chitin (shells of invertebrates such as lobsters and crabs).

Cellulose is composed of long, unbranched chains of glucose units connected by $\beta(1 \rightarrow 4)$ glycosidic bonds (Figure 27-1). Each chain may have a molecular weight of about 35,000 daltons, and many parallel chains make up the cellulose fibers of plants. Cellulose is thought to be the most prevalent structural polysaccharide; in fact, it is the most abundant organic compound on earth. Because cellulose is so abundant, the ability of microorganisms to hydrolyze it is also very important in the recycling of dead plant material. Cellulose is insoluble in water; and since it is such a large molecule, only microorganisms that produce and release the enzyme **cellulase** can break it down. Only a few bacteria, such as myxobacteria, clostridia, and actinomycetes, produce cellulase. However, many fungi produce cellulase, and they are thought to be predominately responsible for the aerobic cellulose decomposition that occurs in our environment. Anaerobic cellulose digestion is accomplished primarily by a few clostridia; this occurs in lake sediments, systems for anaerobic sewage digestion, and the rumen. Few mammals can digest cellulose. Ruminant animals, however, harbor cellulase-producing microbes in their rumen that break down the cellulose for them.

Nutrient polysaccharides are formed by both animal and plant cells, and they are used as reserve food supplies by the cells that form them. **Glycogen** is the reserve polysaccharide of animal cells; and **starch,** a chemically similar polysaccharide, serves a similar function for plant cells. There are two forms of starch. **Amylose** is a straight chain made only of glucose molecules connected by $\alpha(1 \rightarrow 4)$ glycosidic bonds (Figure 27-2). **Amylopectin** also contains only glucose units connected by identical bonds, but its structure is branched, with one branch for about every 30 glucose units (see Figure 27-2). The name for the enzyme (glycosidase) that hydrolyses starch is **amylase.** There are several types of amylase, all giving **maltose** (a disaccharide composed of two glucose units) as the smallest product of starch breakdown. As you might expect, maltose is very soluble and is used as an energy source by many microorganisms.

Starch is hydrolyzed by the amylases produced and released by many molds (filamentous fungi) and bacteria. These amylases diffuse outward from the cells and break down starch. Since enzymes are catalysts and are not consumed in chemical reactions, a few

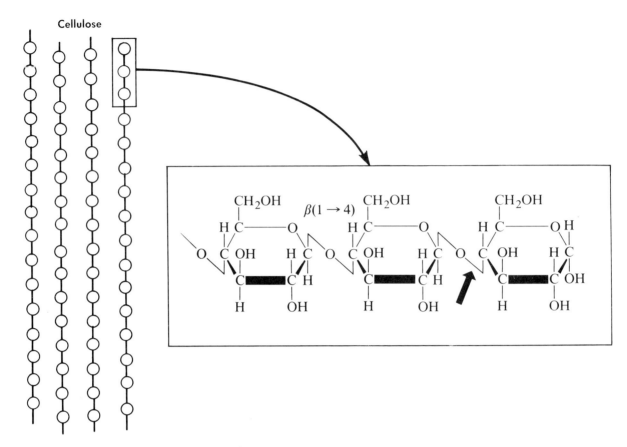

FIGURE 27-1 Schematic illustration of cellulose and the type of glycosidic bonds that join the monosaccharide (glucose) units together to form this polysaccharide. The large arrow (inset) indicates where this molecule is cleaved by cellulase.

molecules of enzyme are used over and over until a great deal of starch is broken down. When a mixture of starch and a growth medium containing agar is sterilized, poured into plates, and allowed to cool, the starch becomes translucent and almost impossible to see. Therefore, the extent of hydrolysis of starch cannot be seen unless the plate is treated to make the starch visible. This is done by applying an *iodine solution* to the plate. The iodine reacts with the starch and forms a characteristic blue or red-brown color. If the starch plate is inoculated, incubated to allow for microbial growth, and then treated with an iodine solution, there will be a clear zone (no color in the agar) around the microbes that produce extracellular amylases. This clear zone, called the **zone of hydrolysis,** results from the enzymatic breakdown of the insoluble starch into smaller oligosaccharides and the disaccharide called maltose. Both are fully soluble in the medium and do not react with iodine. One can assume that the microbe that produces amylase also has the necessary binding and translocating enzymes to translocate these carbohydrates into the cell and use them in obtaining energy and making intermediates for synthesis.

LEARNING OBJECTIVES

• Demonstrate the ability of microbes to produce amylase and the effect of this enzyme on starch.

• Become more aware of the importance of microbial degradation of plant polysaccharides in our biosphere.

MATERIALS

Cultures   *Escherichia coli*
           Bacillus cereus
           Streptococcus salivarius

27 HYDROLYSIS AND USE OF POLYSACCHARIDES

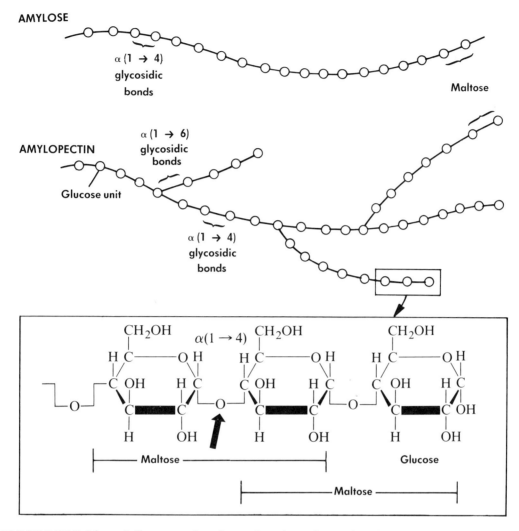

FIGURE 27-2 Schematic illustration of two forms of starch (amylose and amylopectin) and the type of glycosidic bonds that holds the glucose units together. Amylase (large inset arrow) is the hydrolytic glycosidase (enzyme) that attacks both forms of this polysaccharide. Maltose (a disaccharide) is the smallest product formed by the action of amylase on starch.

Media   Starch-agar plates

Supplies   Iodine solution (second period)

PROCEDURE: FIRST LABORATORY PERIOD

1. **Draw lines on the bottom of one starch-agar plate** with a marking pen, dividing it into three equal parts, and label the parts with the names of the three microorganisms to be used.

2. **Inoculate each microorganism in one straight streak** in the middle of its part of this plate. These inoculations should be sufficiently far apart so that a large zone of hydrolysis from one microbe does not overlap a smaller zone or a negative reaction from the second microbe.

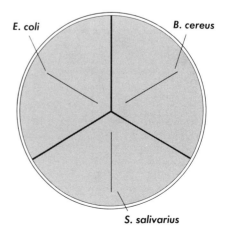

3. **Incubate at 30°C for 48 hours.** Longer incubation may enable one microbe to hydrolyze all of the starch on the plate. Before leaving the laboratory, make sure that you know who is responsible for refrigerating this plate after 48 hour's incubation.

PROCEDURE: SECOND LABORATORY PERIOD

1. **Examine the inoculated plate closely.** Can you see evidence of starch hydrolysis?
2. **Flood the plate with Gram's (Lugol's) iodine solution** (the same solution you use for the Gram stain).

INTERPRETATION NOTE

Amyloses give a deep-blue color and amylopectins give a red to brown color when they react with iodine. You may see one or a combination of these colors, depending on the source of starch used in this medium. For example, potato starch contains about 20 percent amylose and a much greater amount of amylopectin. Starches from other sources contain different proportions of these two polysaccharides; thus their color reactions with iodine are different from that of potato starch. Once starch is completely hydrolyzed, however, the oligo- and disaccharides do not form a color complex with the iodine. Remember that zones of hydrolysis are evident by the *lack* of color.

---

3. **Determine if a zone of hydrolysis is evident.**
4. **Record your results.**

STARCH HYDROLYSIS TEST SUMMARY

Test name	Starch hydrolysis
Cultures used	Many molds and bacteria hydrolyze starch.
Reagents and times	Inoculate starch-agar plate with single streak. Incubate at 30°C for 48 hours. Flood plate with iodine solution, and observe as soon as the starch reacts fully with the iodine.
Positive reaction	Clear (colorless) zone surrounds growth of colony or streak. Remainder of starch agar stains deep blue to red-brown.
Negative reaction	Entire plate stains deep blue or red-brown with no clear (colorless) zones.

27 HYDROLYSIS AND USE OF POLYSACCHARIDES

NAME _____

DATE _____

SECTION _____

RESULTS

Illustrate the appearance of your plate after testing for starch hydrolysis, and label your drawing.

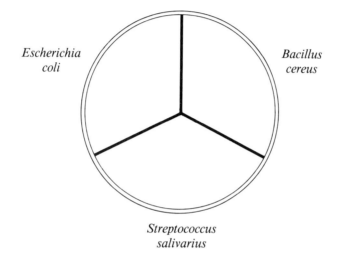

Reagent used:

Color of starch reaction:

Presence of hydrolysis zone:
- *Escherichia coli:* _____ (+/−)
- *Bacillus cereus:* _____ (+/−)
- *Streptococcus salivarius:* _____ (+/−)

CONCLUSIONS

QUESTIONS

Answers to some of the following questions will be found in Exercise 26; the remainder are located in this exercise.

Completion

1. Enzymes that are made by the cell and released into the surrounding medium are called _____.

2. Hydrolytic enzymes are so named because they insert _____ across the covalent bond in the polymer as it is broken.

3. Extracellular hydrolytic enzymes that break down proteins are called _____. Those that break down polysaccharides are called _____, and those enzymes that break down fats and lipids are called _____.

4. Assume that a polymer outside the cell has been broken down by the appropriate enzymes. Before the smaller molecules can get into the cell, at least two proteins in the plasma membrane must be present. These are _____ and _____ the proteins.

5. The type of covalent bond that connects adjacent sugars making up a polysaccharide is called a _____ bond.

6. One example of a structural polysaccharide is _____. An example of a nutrient polysaccharide is _____.

7. Amylose is an enzyme that attacks starch. The smallest product of this hydrolysis is called _____.

8. Starch hydrolysis is evident on starch plates by _____. This effect is produced because _____.

9. The chemical used to detect microbial starch hydrolysis on starch plates is _____.

True–False (correct all false statements)

1. A molecule must get into a cell before the cell can use it as a source of energy. _____

2. The ability to produce and release extracellular enzymes for hydrolyzing a specific type of polymer is a property shared by all microorgaisms. _____

3. Some small molecules transported into the cell may be used directly as building blocks for biosynthesis. _____

4. Carbohydrates are molecules composed of three types of atoms: carbon, oxygen, and nitrogen. _____

5. A positive starch-hydrolysis test is indicated by a blue zone surrounding the area of growth on a starch plate treated with merthiolate. _____

6. Of the two bacteria tested in this exercise, only the species of *Bacillus* gives a strong positive reaction. _____

28

Microbial Degradation of Proteins (Casein and Gelatin)

Proteins are long, complexly folded chains of **amino acids** covalently linked by **peptide bonds** (Figures 28-1 and 28-2). Large proteins in culture media cannot be used as nutrients by microbes because they are too large to be taken into cells; however, if they are broken into smaller molecules, they can be taken in and used as sources of carbon and/or energy or as building blocks for protein synthesis.

Some microorganisms produce and release extracellular enzymes, called **proteinases** (or **proteases**), that accomplish hydrolytic cleavage of protein molecules outside the cell (see Exercise 26). Proteinases are also called **proteolytic enzymes** because they break up (hydrolyze) large protein molecules into smaller chains of amino acids called **peptides** (Figure 28-3).

Once the large protein molecules are broken apart by the proteinases, the smaller peptides may be further attacked outside the cell by **extracellular peptidases** or taken directly into the cell and further broken down into amino acids by **intracellular peptidases** (see Figure 28-3). Peptides must be broken down to individual amino acids before the cell can use them in its metabolism.

Many fungi and fungilike bacteria *(the actinomycetes)* and some of the more typical bacteria produce extracellular proteinases. For example, pathogenic bacteria (such as species of *Staphylococcus* and *Streptococcus*), aquatic bacteria (such as the pseudomonads and some *Proteus* species), and soil bacteria (such as *Clostridium* and the *Bacillus* species) are known to produce extracellular proteinases.

As with the production of extracellular glycosidases (see Exercise 27), the microbial ability to produce extracellular proteinase is very important in the degradation of dead plant and animal tissue and the recycling of nutrients in our environment. For example, if milk is spilled on the ground, it is desirable that it be degraded so that the nutrients can be used by other life forms. But these same microorganisms may be undesirable in other environments. For example, the dairy farmer does not want to have these microbes breaking down the milk collected for human consumption. Therefore, we should understand how important these microbes are in our environment, but we should also understand how to control their growth and proteolytic activity. (The effect of proteolytic activity on milk will be examined again in Exercise 35).

LEARNING OBJECTIVES

• Understand how microorganisms produce and release extracellular enzymes called proteinases and how the products are transported and used by the microbe.

• Demonstrate the effect of proteinase on colloidally suspended particles of milk protein (casein) and another animal protein (gelatin).

• Become more aware of the importance of microbial degradation of animal and plant proteins in our biosphere.

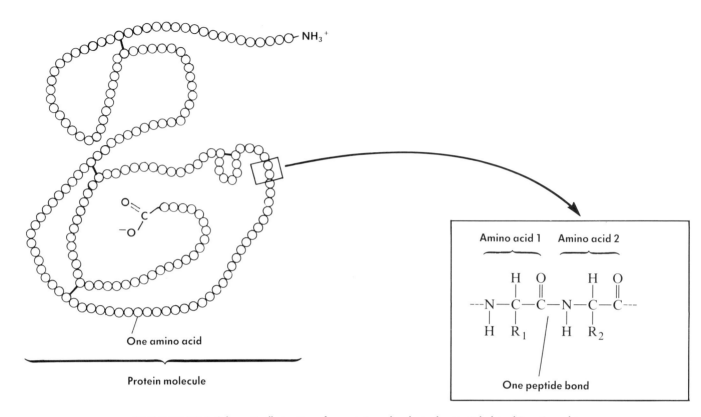

FIGURE 28-1 Schematic illustration of a protein molecule and a peptide bond (inset) used to covalently bond adjacent amino acids (O) to form a protein molecule. R_1 and R_2 represent the side chains that make each amino acid different from all others; in other words, these different side chains make glutamate different from lysine and make cysteine different from tryptophan.

A · Degradation of milk protein (casein)

Milk is an excellent medium for growing many microorganisms because it is rich in nutrients. Cow's milk contains about 4 percent fat, 5 percent lactose *(milk sugar)*, 3 percent protein, 1 percent vitamins and minerals, and 87 percent water. Milk that is secreted 3 to 5 weeks postpartum *(not* colostrum) contains three important proteins: casein, lactalbumin, and lactoglobulin. Casein comprises about 85 percent of the total protein in milk.

Casein in milk is probably not a homogeneous (pure) protein; however, most of it is a **phosphoprotein** (about 0.7 percent phosphorous) that binds to calcium forming a calcium salt called **calcium caseinate**. Molecules of calcium caseinate are very large and

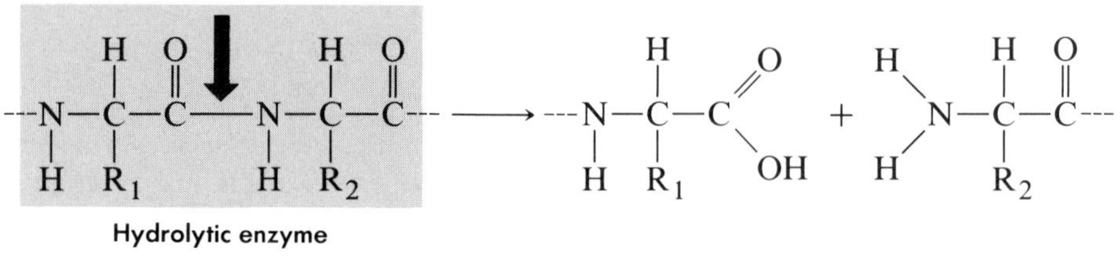

FIGURE 28-2 Hydrolysis of a peptide bond. An enzyme breaks the peptide bond (large arrow) and inserts water to complete the cleavage of the large molecule into two smaller parts. Proteinases and peptidases are examples of enzymes that accomplish hydrolytic cleavage of proteins (see Figure 28-3).

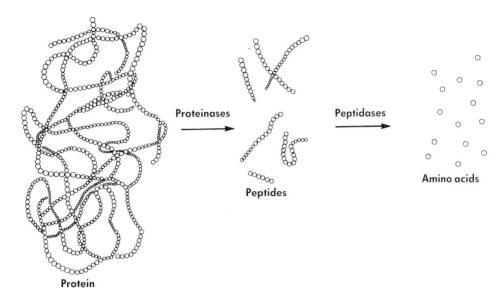

FIGURE 28-3 Schematic overview of the enzymatic breakdown (hydrolysis) of proteins into peptides and individual amino acids. Note that proteinases and peptidases are both hydrolytic enzymes and that both are needed in the extracellular environment (excreted by the cell) before proteins can be broken down into pieces small enough to be translocated into the cell across the plasma membrane.

are water *insoluble*, but they are not so large that they settle out of suspension. In other words, calcium caseinate exists as a true *colloid* in suspension, and it is these colloidally suspended particles that give milk its turbid appearance and white color.

Casein is used in this exercise because it is an insoluble protein that can be readily attacked and solubilized by microbially produced extracellular proteinases. Thus, casein is the **substrate** of this enzymatic reaction. Since casein is insoluble in water, it can be easily observed when suspended in solid culture media. The hydrolysis of casein by extracellular proteinases into soluble peptides and amino acids is evident by the disappearance of casein particles from the medium. In other words, there is a *clearing* of the agar surrounding the growing microbes, which are producing and releasing extracellular proteinases. The greater the enzyme activity or the more enzyme produced, the greater the zone of clearing around the microbes growing on plates containing casein.

The test for casein hydrolysis demonstrates the ability of microbes to degrade complex organic molecules and provides another test that helps microbiologists identify unknown microorganisms.

MATERIALS

Cultures *Escherichia coli*
 Bacillus cereus
 Streptococcus salivarius

Media Skim-milk agar plates

Supplies None

PROCEDURE: FIRST LABORATORY PERIOD

1. **Draw a line on the bottom of one skim-milk agar plate** with a marking pen, dividing the plate into three equal parts.

2. **Inoculate each microorganism with one straight streak** in the middle of its part of the plate.

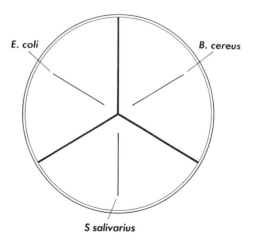

3. **Incubate at 30°C for 48 hours only.** Longer incubation may cause hydrolysis of all of the casein on the plate. If this happens, you will not know which organism hydrolyzed the casein or which one hydrolyzed casein faster. If your next laboratory period is more than 48 hours away, make sure that you know who is responsible for refrigerating your plate.

PROCEDURE: SECOND LABORATORY PERIOD

1. **Examine the inoculated plate for evidence of casein hydrolysis.** Clearing (solubilization) of the casein suspended in the agar surrounding growth of a microorganism is positive evidence for hydrolytic cleavage of the casein.

2. **Determine if zones of hydrolysis are evident.**

3. **Record your results.**

CASEIN HYDROLYSIS TEST SUMMARY	
Test name	Casein hydrolysis
Medium used	Skim-milk agar plates
Reagents and times	Inoculate with a straight streak on part of the plate. Incubate at 30°C for 48 hours.
Positive reaction	Clearing of the casein under and around the growing microorganisms
Negative reaction	No clearing of the casein under and around the growing microorganisms

B · Degradation of other animal proteins (gelatin)

After the microbiologist Schroeter observed in 1872 that various colonies of microorganisms grew on decaying potato slices, the great German bacteriologist Robert Koch set out in 1880 to develop a clear yet solid medium that could be prepared easily in the laboratory. Koch started with a liquid medium and solidified it by adding 5 to 10 percent *gelatin.* He sterilized this medium, placed it on a sterile microscope slide, and put this under a bell jar to keep the dust and accompanying microbes away from the sterile surface. Thus, gelatin provided a solid, transparent, sticky gel whose surface could be inoculated with a microbial culture.

There were several disadvantages to Koch's use of gelatin as a solidifying agent. Although the gel is solid when cold, it liquifies at temperatures above 28°C and, therefore, cannot be used for cultivating human pathogens at 37°C (body temperature). Also, gelatin is a protein, and many microorganisms produce and release extracellular proteinases that hydrolyze this particular protein. Thus, gelatin is the **substrate** of this enzymatic reaction. As enzymes hydrolyze gelatin, it changes from a solid to a liquid, thus destroying its use as a solidifying agent.

Gelatin is formed by boiling a protein called collagen in water or an acid solution (Figure 28-4). Most of the white connective tissue in animals is collagen, which is also found in yellow elastic tissue *(cartilage)* and in the organic matrix of bone *(ossein).* Although collagen is not water soluble, gelatin is, which is probably due to the way gelatin is formed.

When you boil collagen, you partially *denature* (unfold) this large molecule, making it soluble in water. It also becomes more susceptible to hydrolytic cleavage by proteinases, probably because these large enzymes can now more easily get into the various parts of the protein.

As Robert Koch found, many types of microorganisms produce and release proteinases that attack gelatin. Many fungi are particularly good at this, although comparatively few bacteria have this ability. For this reason, gelatin hydrolysis is used to distinguish among some genera and many species of bacteria, particularly among the enterics and pseudomonads, some gram-positive cocci, the gram-positive spore formers, and the strict and aerotolerant anaerobes (especially the clostridia).

When gelatin is enzymatically hydrolyzed into short peptides and individual amino acids, it loses its ability to gel even at low temperatures; therefore, many tests for gelatin hydrolysis are based upon **gelatin liquification.** Temperatures of incubation are important because some bacteria produce extracellular proteinases better between 20° and 26°C and others

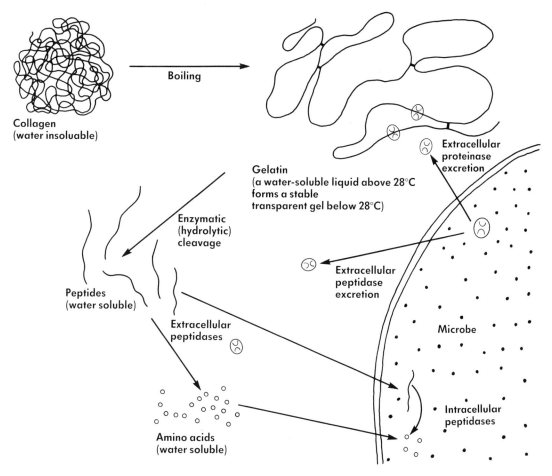

FIGURE 28-4 Overview of the production of gelatin from collagen and the effect of extracellular proteinases and peptidases on the hydrolysis of this protein.

at 37°C. At 37°C, untreated gelatin is always a liquid; so it is necessary to cool gelatin-containing media below 28°C to determine whether gelatin hydrolysis has taken place.

The **stab-inoculation method** for determining gelatin liquification has been used for many years but is an imprecise test for several reasons. For example, it is often difficult to determine partial hydrolysis of the gelatin. Some bacteria produce the right proteolytic enzymes, but they only partially liquify the gel, making it difficult to tell if gelatin hydrolysis has taken place. In addition, gelatin forms a stronger, more stable gel the longer it is stored at 22°C (about room temperature). Thus, the temperature and length of storage of the medium before inoculation can affect a microorganism's ability to liquify the gel.

Many methods have been employed over the years to make stab-inoculation of gelatin deeps more sensitive to partial hydrolysis of gelatin. One method uses *formalin-denatured gelatin.* Formalin alters the gelatin so that it no longer becomes a liquid at incubation temperatures above 28°C. (With this form of gelatin, you do not have to cool it below 28°C to detect hydrolysis.) Formalin-denatured gelatin is melted, mixed with sterile charcoal particles, and allowed to solidify. When pieces of this sterile gelatin (with the embedded charcoal) are added to sterile broth, even partial gelatin hydrolysis can be detected because some charcoal is released, which turns the broth black. The more gelatin is hydrolyzed, the blacker the medium appears. Today, modifications of this method are used as one of the tests in rapid detection (strip) kits in clinical laboratories to detect gelatin hydrolysis caused by suspected pathogenic bacteria.

Gelatin deeps are used in this exercise to demonstrate gelatin hydrolysis. This method was chosen because it demonstrates the principle of gelatin hydrolysis in the simplest way. To make sure that you can easily see positive gelatin hydrolysis, however, the selected microbial strains either lack the ability to hydrolyze gelatin or produce large amounts of the necessary proteinases. With many other microorganisms, it is not so easy to detect positive gelatin hydrolysis using agar deeps.

MATERIALS

Cultures *Escherichia coli*
Bacillus cereus
Proteus vulgaris

Media Trypticase-soy broth + 4 percent gelatin

Supplies Ice bucket (second and third periods only)

PROCEDURE: FIRST LABORATORY PERIOD

1. **Separately inoculate one gelatin deep with each microbe.** Use a heavy inoculum (3 to 4 loopsful of culture), and stab the deep all the way to the bottom of the tube in the same manner described in Exercise 16 for agar deeps.

2. **Incubate at 30°C for 48 hours.** If your next laboratory period is more than 48 hours away, continue the incubation until the next period. Your instructor should incubate a few uninoculated tubes to use as negative controls during the second and third laboratory periods.

PROCEDURE: SECOND LABORATORY PERIOD

1. **Examine all three tubes for evidence of growth.** Thoroughly resuspend any cells that have settled to the bottom before you record this growth. All three tubes should show heavy turbidity.

2. **Note the physical state of all three tubes.** (Remember that gelatin liquifies above 28°C and that these cultures were incubated at 30°C; so the liquid state of these tubes at this point in time *does not* demonstrate gelatin hydrolysis.)

3. **Stand all three cultures upright in an ice-filled bucket for at least 30 minutes.**

4. **Record your observations.**

5. **Reincubate your cultures and controls at 30°C until the next laboratory period.**

PROCEDURE: THIRD LABORATORY PERIOD

1. **Repeat the testing procedure from the second laboratory period** (steps 1, 2, and 3).

2. **Record your observations.**

3. **Properly discard your culture tubes.**

GELATIN HYDROLYSIS TEST SUMMARY	
Test name	Gelatin hydrolysis
Medium used	Gelatin deeps (T-soy broth +4 percent gelatin)
Reagents and times	Inoculate tubes with 3 or 4 loopsful. Incubate tubes at 30°C for at least 48 hours. Stand tubes upright in ice for at least 30 minutes to bring the temperature considerably below 28°C (melting point of gelatin). If negative, reincubate for up to 7 days.
Positive reaction	Gelatin deep shows evidence of growth and remains liquid after 30 minutes on ice.
Negative reaction	Gelatin deep shows evidence of growth but becomes a solid gel when left for 30 minutes on ice.

28 HYDROLYSIS AND USE OF PROTEINS

NAME _____

DATE _____

SECTION _____

RESULTS A: CASEIN DEGRADATION

Illustrate the appearance of your plate after testing for casein hydrolysis, and label your drawing.

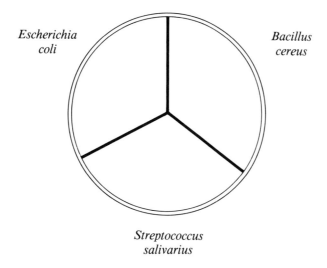

Reagents used:

Description of casein hydrolysis on plate:

Presence of hydrolysis zone:

  Escherichia coli: _____ (+/−)

  Bacillus cereus: _____ (+/−)

CONCLUSIONS

RESULTS B: GELATIN DEGRADATION

Record your results in the table below. Describe the relative amounts of growth (turbidity) in each tube immediately after taking them from the 30°C incubator, that is, while the medium is still in a liquid form.

| Microbe | Relative amount of growth | | Appearance of tube after 30 minutes on ice | | Gelatin hydrolyzed? |
	Second period (Incubation: _____ days)	Third period (Incubation: _____ days)	Second period (Incubation: _____ days)	Third period (Incubation: _____ days)	
Escherichia coli					
Bacillus cereus					
Proteus vulgaris					

CONCLUSIONS

28 HYDROLYSIS AND USE OF PROTEINS

NAME _____

DATE _____

SECTION _____

QUESTIONS

Casein degradation

Completion

1. From a chemical standpoint, casein is classified as a _____. From a physical standpoint, calcium caseinate forms a _____ when placed in a liquid. Calcium caseinate comes from _____.

2. Extracellular hydrolytic enzymes that break apart proteins are called _____. After a protein is broken down in that way, the smaller products are called _____.

3. Assume that a polymer outside the cell is broken down by the appropriate enzymes. Before the cell can take in these smaller molecules, at least two proteins must be present in the cell's plasma membrane. These are _____ and _____.

4. The type of covalent bond that connects adjacent amino acids making up a protein is called a _____ bond.

5. The substrate acted upon by the enzyme tested for in this part of Exercise 28 is _____.

6. Microbial degradation of proteins is important in our environment. Controlling the growth and metabolic activity of microbes that degrade proteins is of especial concern to the _____ industry.

7. Protein hydrolysis is evident on skim-milk agar plates by _____. This effect is produced because _____.

True–False (correct all false statements)

1. A positive proteolysis test is indicated by a blue zone surrounding the area of growth on skim-milk agar plates. _____

2. Of the three bacteria tested in this exercise, only the species of *Proteus* gave a strong positive reaction after the second laboratory period. _____

Gelatin degradation

Completion

1. Gelatin is chemically classified as a _____. Physically, when gelatin in water is above 28°C, it is a _____, but at temperatures below 28°C, it is a _____.

2. Extracellular hydrolytic enzymes that break gelatin apart are called _____.

3. The substrate acted upon by the enzyme tested for in this part of Exercise 28 is _____.

4. Gelatin is formed by boiling _____ in water or acid solutions. Boiling _____ this large molecule, and this probably allows easier entry of _____.

5. The material that we now almost universally substitute for gelatin as a solidifying agent for microbiological media is called _____.

6. Another name for the process of microbial hydrolysis of gelatin is _____. A positive test for this requires _____. A negative test is indicated by _____.

7. The test tube containing media for the gelatin hydrolysis test is called a _____.

True - False (correct all false statements)

1. Of the three bacteria tested for gelatin hydrolysis, *Bacillus cereus* was the only one that was strongly positive after the second laboratory period. _____

2. Denaturation of collagen is accomplished by heating, whereas microbially produced hydrolysis of gelatin is an enzymatic process. _____

29

Microbial Degradation of Lipids

Some bacteria can break down a variety of animal and vegetable lipids, especially a group of lipids called triglycerides. **Triglycerides** are composed of a *glycerol* molecule and *three fatty acids,* with one fatty acid attached to each carbon of the glycerol molecule by an **ester linkage,** as shown in Figure 29-1. The types of fatty acids attached to the glycerol vary from one kind of triglyceride to another.

Bacteria that degrade triglycerides produce enzymes called **lipases,** which hydrolyze (split with the addition of water) the ester linkages between glycerol and the fatty acid molecules (see Figure 29-1). Lipases are, of necessity, extracellular enzymes because triglycerides are too large to be transported into the cell. The glycerol and fatty acids that result from lipid hydrolysis *can* be transported into the bacterial cell, and the cell uses them either as building blocks for the synthesis of new cell structures or as sources of energy.

Triglyceride-degrading bacteria can grow in many environments that contain animal or vegetable lipids. They grow especially well in fatty foods, where they can cause rancidity. **Rancidity** is a form of food spoilage in which unpleasant (sometimes *very* unpleasant) odors and flavors develop because of the release of free fatty acids from triglycerides. Foods that are especially susceptible to this type of microbial spoilage include butter, margarine, tallow, vegetable oils, coconut oil, and fish with high lipid contents such as salmon and mackerel.

Microbiologists have developed specialized culture media for detecting triglyceride-degrading microorganisms. In this exercise, you will use **spirit blue agar** to test three bacteria for the ability to degrade triglycerides. Spirit blue agar is a conventional nutrient medium that contains a source of triglycerides and a pH indicator called spirit blue. When a triglyceride-degrading organism grows on this medium, excreted lipases hydrolyze triglycerides in the surrounding agar to glycerol and fatty acids. The release of fatty acids lowers the pH and causes the spirit blue indicator to turn from a pale lavender to a deep royal blue. A deep blue in the agar around the area of growth, then, is a positive indication that the test organism can break down triglycerides.

An alternative medium for detecting triglyceride degradation is **tributyrate agar.** This medium contains tributyrate, a triglyceride with three butyric acid molecules as its fatty acids. Because it is insoluble in water, tributyrate forms an opaque or cloudy emulsion when mixed with the rest of the medium. When a triglyceride-degrading organism grows on this medium, its lipases break down the insoluble tributyrate and produce a transparent zone around the area of growth (a process called **lipolysis**).

Some bacteria can degrade a class of lipids known as **phospholipids,** which are similar to triglycerides except that one of the three fatty acid residues is replaced by a non-fatty acid molecule that is linked to the glycerol through a phosphate group (Figure 29-2). Phospholipids are important molecules in all living organisms because they are one of the major chemical components of membranes. Bacteria that degrade these molecules produce extracellular enzymes called **phospholipases.** Some phospholipases remove only the fatty acid residues from a phospholipid (therefore, they are similar in function to the lipases already described), whereas others can remove the non-fatty acid group. In either case, the products of the reaction can be transported into the bacterial cell and used for synthesis of new cell materials or for generation of energy.

Bacteria that break down phospholipids are often of considerable medical interest because their phospholipases destroy membranes in the human body.

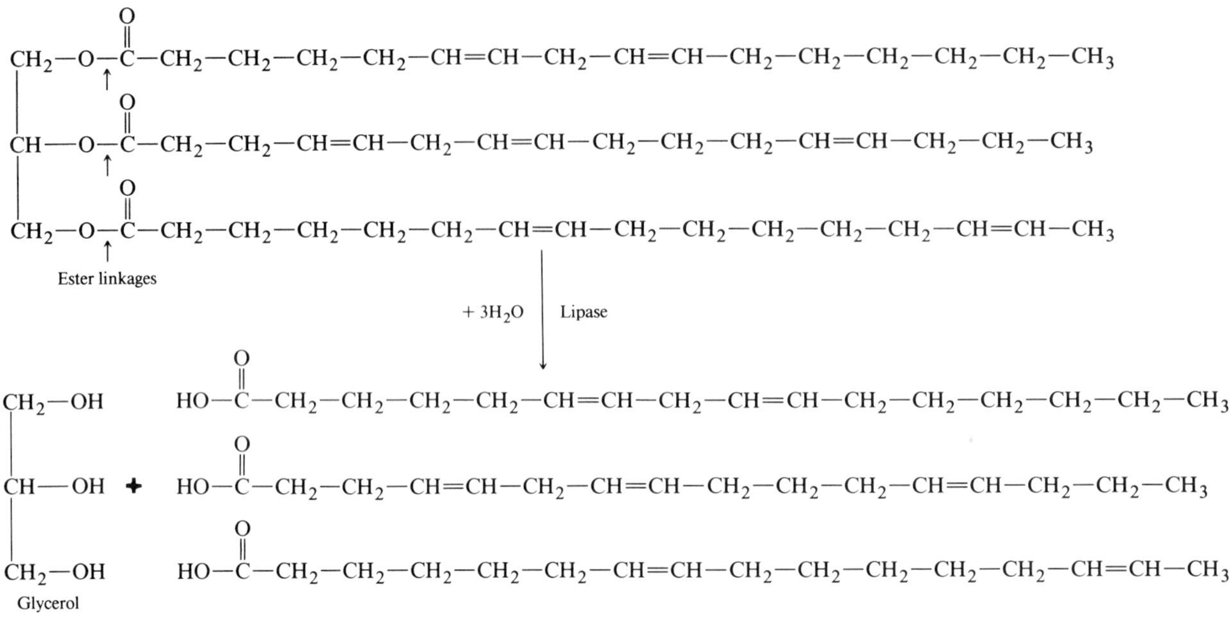

$$CH_2-O-\overset{\overset{O}{\|}}{C}-CH_2-CH_2-CH_2-CH_2-CH=CH-CH_2-CH=CH-CH_2-CH_2-CH_2-CH_2-CH_2-CH_3$$
$$CH-O-\overset{\overset{O}{\|}}{C}-CH_2-CH_2-CH=CH-CH_2-CH=CH-CH_2-CH_2-CH_2-CH=CH-CH_2-CH_2-CH_3$$
$$CH_2-O-\overset{\overset{O}{\|}}{C}-CH_2-CH_2-CH_2-CH_2-CH_2-CH=CH-CH_2-CH_2-CH_2-CH_2-CH_2-CH=CH-CH_3$$

Ester linkages

$+ 3H_2O$ | Lipase ↓

$$CH_2-OH \qquad HO-\overset{\overset{O}{\|}}{C}-CH_2-CH_2-CH_2-CH_2-CH=CH-CH_2-CH=CH-CH_2-CH_2-CH_2-CH_2-CH_2-CH_3$$
$$CH-OH \; + \; HO-\overset{\overset{O}{\|}}{C}-CH_2-CH_2-CH=CH-CH_2-CH=CH-CH_2-CH_2-CH_2-CH=CH-CH_2-CH_2-CH_3$$
$$CH_2-OH \qquad HO-\overset{\overset{O}{\|}}{C}-CH_2-CH_2-CH_2-CH_2-CH_2-CH=CH-CH_2-CH_2-CH_2-CH_2-CH_2-CH=CH-CH_3$$

Glycerol

FIGURE 29-1 Enzymatic hydrolysis of a triglyceride to glycerol and fatty acids. This reaction is catalyzed by an extracellular enzyme (called a *lipase*) that is excreted by some bacterial cells and hydrolyzes each ester linkage.

For example, some of the bacterial **hemolysins** that lyse red blood cells (see Exercise 46) are phospholipases that attack the red blood cell membrane. One of the phospholipases produced by *Clostridium perfringens* not only degrades an important lipid in the red blood cell membrane (called *lecithin* or *phosphatidylcholine;* see Figure 29-2) but also is partly responsible for triggering the gangrene process that occurs in some clostridial infections. Several other groups of pathogenic microorganisms also produce phospholipases.

The production of microbial phospholipases can be detected in the laboratory with **egg yolk agar,** another medium that you will use in this exercise. Egg yolk agar is an important medium in clinical laboratories because it is used for the identification of pathogenic species of *Clostridium* (its most common use) and of *Bacillus* and *Staphylococcus.* It is a nutrient agar that contains approximately 4 percent sterile egg yolk, which is a source of concentrated phospholipids. The phospholipids in the egg yolk are well dispersed in this medium because their non-fatty acid groups (such as choline and phosphate; see Figure 29-2) are polarly charged. When a microorganism grows on this medium and produces the right kind of phospholipases, however, the polar non-fatty acid groups are cleaved from the phospholipids by hydrolysis. The fatty acids have no polar groups; therefore, they are insoluble and separate from the aqueous medium to form an oily surface film that appears opalescent on and around the microbial growth. The appearance of this opalescent zone is positive evidence that the tested microorganism can break down phospholipids.

In summary, we have considered three types of **differential** (but not selective) media used to detect microbial production of extracellular lipases: spirit blue agar, tributyrate agar, and egg yolk agar. In this exercise, you will test two of these differential media: **spirit blue agar,** which shows the release of fatty acids that lower the pH and change the color of a pH indicator and **egg yolk agar,** which shows the hydrolysis of phospholipids by exhibiting the accumulation of the water-insoluble (nonpolar; uncharged) fatty acids. Egg yolk agar is commonly used for the clinical diagnosis of certain clostridium infections.

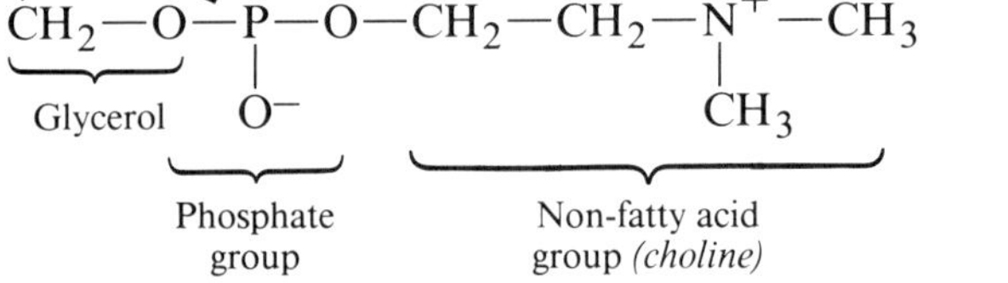

$$CH_2-O-\overset{\overset{O}{\|}}{C}-CH_2-(CH_2)_{15}-CH_3$$
$$CH-O-\overset{\overset{O}{\|}}{C}-CH_2-(CH_2)_{15}-CH_3$$
$$CH_2-O-\overset{\overset{O}{\|}}{\underset{O^-}{P}}-O-CH_2-CH_2-\overset{\overset{CH_3}{|}}{\underset{CH_3}{N^+}}-CH_3$$

Glycerol | Phosphate group | Non-fatty acid group *(choline)* | Fatty acids

FIGURE 29-2 Chemical structure of lecithin (phosphatidylcholine), a typical phospholipid. The arrow indicates where this molecule is cleaved by the extracellular phospholipase excreted by *Clostridium perfringens*. This enzyme and the microbe that excretes it are involved in the gangrene process.

29 HYDROLYSIS AND USE OF LIPIDS

LEARNING OBJECTIVES

- Learn laboratory methods for the detection of lipid (triglyceride and phospholipid) degradation by microorganisms.
- Learn about the chemical properties of lipids that are commonly degraded by microorganisms.
- Understand how microorganisms degrade lipids and how they use the products of lipid degradation.
- Understand the function and purpose of microbial extracellular enzymes.
- Appreciate the significance of lipid-degrading microorganisms in medicine and the food industry.

MATERIALS

Cultures *Bacillus subtilis*
Proteus mirabilis
Staphylococcus epidermidis

Media Spirit blue agar plate
Egg yolk agar plate

Supplies Metric ruler

PROCEDURE: FIRST LABORATORY PERIOD

1. **Perform** the following procedure on one spirit blue agar plate and one egg yolk agar plate (refer to Figure 29-3).
   a. **Divide** each plate into four equal sectors by marking on the bottom with your marking pen.
   b. **Label** each plate bottom with your *name*, the *date*, and the name of the *medium* it contains. Label each of three sectors with the name of a different *microorganism*, and label the fourth sector *control*.
   c. **Inoculate** each of three sectors of the plate with a different organism. To do this, transfer a loopful of culture to the center of the sector and spread it evenly over a small circular area that is about 5 to 10 mm in diameter. *Do not inoculate the control sector.*
2. **Invert and incubate the plates at 30°C for 48 hours.**

PROCEDURE: SECOND LABORATORY PERIOD

1. **Examine the spirit blue agar plate for hydrolysis of triglycerides.** A dark blue zone around the area of bacterial growth indicates that triglyceride hydrolysis has taken place. The agar will remain unchanged in appearance if the organism does not degrade triglycerides.
2. **Examine the egg yolk agar plate for hydrolysis of phospholipids.** A cloudy, opalescent zone on the agar surface around the area of bacterial growth indicates that phospholipid degradation has taken place. The agar will remain unchanged in appearance if the organism does not degrade phospholipids.

INTERPRETATION AND TECHNIQUE NOTES

Spirit blue is a combination of two basic dye salts (diphenyl-rosanilin chloride and triphenyl pararosanilin chloride) that together serve as a *pH indicator*. When fatty acids are released from triglycerides after hydrolysis by the extracellular enzymes, the pH becomes more acidic and changes the color of spirit blue from a pale lavender to a deep blue.

Egg yolk agar contains many polar phospholipids that are soluble in this aqueous environment. Enzymatic hydrolysis of the polar side chain by phospholipase changes the phospholipid into a nonpolar *insoluble* diacylglycerol (two fatty acids bound to glycerol). These molecules separate from the aqueous medium and form a surface film on and around cultures that produce the appropriate lipases. This film appears *opalescent* (having a milky iridescence or pearly sheen) on the culture surface and the surface of the surrounding agar when viewed with oblique light. This is frequently difficult to see unless the plate is uncovered and held at the correct oblique angle with respect to the light source.

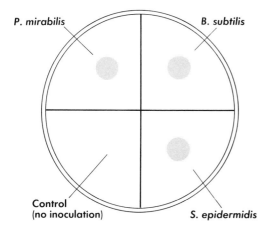

FIGURE 29-3 Method used for spot inoculating three sectors of the spirit blue agar and egg yolk agar plates. The inoculated areas should be about 5 to 10 mm in diameter.

3. **Record your observations** in the Results section.
4. **Properly discard the plates.**

TRIGLYCERIDE (LIPID) HYDROLYSIS TEST SUMMARY

Test name	Triglyceride hydrolysis
Medium used	Spirit blue agar
Reagents and times	Spot inoculate spirit blue agar plates with microbe. Incubate at 30°C for 48 hours.
Positive reaction	Microbial growth with a dark blue zone surrounding the growing culture on an otherwise pale lavender medium.
Negative reaction	Microbial growth, but no dark blue zones evident around microbial growth; entire medium is a pale lavender color.

PHOSPHOLIPID HYDROLYSIS TEST SUMMARY

Test name	Phospholipid hydrolysis
Medium used	Egg yolk agar
Reagents and times	Spot inoculate egg yolk agar plates with microbe. Incubate at 30°C for 48 hours.
Positive reaction	Microbial growth with an opalescent sheen on the culture surface and the agar surface immediately surrounding the microbes when the opened plate is observed with oblique light.
Negative reaction	Microbial growth, but no opalescent sheen on the culture surface or the agar surface when the opened plate is observed with oblique light.

29 HYDROLYSIS AND USE OF LIPIDS

NAME _____

DATE _____

SECTION _____

RESULTS

1. Record your observations in the following table.

Microbe	Spirit blue agar		Egg yolk agar	
	Zone size (mm)	Triglyceride hydrolysis (+ or −)	Zone size (mm)	Phospholipid hydrolysis (+ or −)
B. subtilis				
P. mirabilis				
S. epidermidis				

2. Did any of the organisms produce a positive result on spirit blue agar and a negative result on egg yolk agar? If so, how do you explain these results?

CONCLUSIONS

QUESTIONS

Completion

1. Enzymes that degrade triglycerides are called _____, whereas those that can degrade phospholipids are called _____.

2. Triglyceride-degrading bacteria sometimes grow in lipid-rich foods and cause a type of spoilage known as _____. The unpleasant flavors and odors associated with this type of spoilage are caused by _____ released during lipid degradation.

3. _____ agar is often used as an alternative to spirit blue agar for the detection of triglyceride hydrolysis.

4. *Clostridium perfringens* produces a lipid-degrading enzyme that attacks a lipid called _____, which is partly responsible for triggering the _____ process that sometimes occurs during clostridial infections.

True-False (correct all false statements)

1. Bacteria use extracellular enzymes to degrade triglycerides because they cannot transport free fatty acid molecules into their cells. _____

2. The color change that indicates a positive reaction on spirit blue agar is triggered by the release of glycerol during degradation of triglycerides. _____

3. Some of the bacterial hemolysins that attack red blood cells are actually phospholipases that attack red blood cell membrane lipids. _____

4. Egg yolk agar makes use of a pH indicator to determine whether a microorganism can degrade phospholipids. _____

5. Bacteria that degrade phospholipids are not very important in medicine because very few of them are pathogenic. _____

30

Differential Utilization of Citrate by Enteric Bacteria

The last few exercises have demonstrated that large molecules (such as polysaccharides, proteins, and lipids) must be broken into smaller pieces before they can be taken into a cell. Now you will learn that having a small size is no guarantee that a molecule can be taken into a cell or that, once inside, it can be used in the cell's metabolism.

Cells need to have specific binding proteins and translocation proteins in their plasma membranes, so that the small nutrient molecules can be bound to the plasma membrane and moved inside the cell. Therefore, microorganisms need genes to make these required proteins.

Assume that a cell has the necessary membrane proteins and that it has taken in small molecules. The cell must also make all of the enzymes necessary to convert these small molecules into energy or building blocks for cell growth. Each type of metabolic conversion requires many enzymes, and these enzymes are made only if the cell has the necessary genes on its chromosome (because it has evolved to use certain nutrients and not others). If a cell lacks any one gene, it will not make that enzyme; this will cause the entire metabolic pathway to fail and the corresponding nutrient will not be metabolized.

There are many reasons why microbes may not use certain nutrients. But the fact that some use one and others do not can be used by diagnostic microbiologists to help differentiate two microorganisms that are otherwise similar. For example, there is a large group of very similar gram-negative bacteria called the **enterics** because many are found in the intestines of animals. Many years ago, some researchers thought that one or more of the enterics could be used as indicators of fecal pollution; but, at that time, it was impossible to differentiate those enterics that are found only in animal intestines from those that are also found in other environments. *Escherichia coli* was selected as the bacterium whose presence would indicate fecal pollution. However, scientists then had to find a way to distinguish it from *Enterobacter aerogenes,* a similar bacterium that is found both in feces and in non-fecally polluted environments such as plant surfaces.

In 1923, S. A. Koser devised a synthetic, liquid growth medium that contained *ammonium phosphate* as the sole (inorganic) nitrogen source and *citrate* (an organic acid) as the sole carbon source. Koser found that the **aerogenes group** of enteric bacteria can use citric acid as a sole carbon source but the **coli group** cannot. His was the first study to use carbon-source utilization as a method to identify bacteria.

In 1926, J. S. Simmons added agar to Koser's medium and incorporated the pH indicator **bromothymol blue** to demonstrate an alkaline reaction. Bromothymol blue is green under acidic conditions (below pH 6.8) and dark blue under alkaline conditions (above pH 7.6). Simmons found that the medium turned blue when the aerogenes group of enteric bacteria grew on citrate as the sole carbon source. Since *Escherichia coli* did not grow on this medium, there was no color change. This medium is now called **Simmons' citrate agar.**

Currently, we do not clearly understand why the aerogenes group produces an alkaline reaction on Simmons' citrate agar. The predominant theory suggests that alkaline substances in the medium (such as the ammonium ion) cause the medium to become more alkaline as the cells take up and use citric acid. In other words, citrate use removes acid from the medium, thus increasing pH of the medium. In addition, the aerogenes group is thought to excrete alkaline products on this medium. Therefore, both the removal of the citric acid and the excretion of alkaline waste products may be responsible for raising the pH above 7.6.

Since the aerogenes group of enteric bacteria uses citrate as a sole carbon source for growth, one can assume that these bacteria make the proteins necessary for binding and translocating citrate across their plasma membranes. In addition, these bacteria also produce the intracellular enzymes necessary for metabolizing citrate.

When one experimentally manipulates the coli group of enteric bacteria in the laboratory to force citrate inside, one finds them to be fully capable of metabolizing citrate and producing quantities of alkaline excretion products. Therefore, the inability to transport citrate seems to be the barrier to the use of citrate by these bacteria.

Although Simmons' citrate agar is used here to demonstrate the ability or inability of bacteria to use small molecules as sole sources of carbon for metabolism and growth, one should not lose sight of the importance of citrate utilization for differentiating the coli and the aerogenes groups of bacteria. Such differentiation is particularly important when one wishes to establish whether or not an environment or foodstuff has been fecally contaminated or to detect the presence of enteropathogenic *Escherichia coli* in a clinical specimen.

Although in this exercise you will use the inability of *E. coli* to use citrate as the sole carbon and energy source, you should never use this criterion as the only test to identify a pure culture as *E. coli*, for two reasons: (1) Other types of bacteria are also unable to grow on citrate as the sole carbon source, and (2) occasionally, an atypical *E. coli* strain that is capable of citrate utilization has been isolated from agricultural and clinical environments. In every case, a **plasmid** (extrachromosomal genetic material) was found to carry the genes responsible for citrate utilization. These atypical *E. coli* are thought to have received these plasmids from other bacteria, and this receipt alters the normal **genotype** of *E. coli*. You will find further tests for differentiating these two groups of bacteria in Exercises 19, 32, and 34.

When you test for citrate utilization by bacteria, it is important to use a light inoculum to minimize carryover of nutrients from the medium used to grow the inoculum. Here you will add cells to sterile distilled water to control inoculum size and to dilute the growth medium. If you obtain questionable positive cultural reactions, nutrient carryover should be suspected and confirmed by transferring cells to a fresh citrate medium.

When you use a citrate medium with a pH indicator, you should record as positive only those pure cultures that change the medium to an alkaline pH. In some diagnostic laboratories, cultures are frequently incubated for 7 days before negative results are reported on Simmons' citrate agar.

LEARNING OBJECTIVES

• Demonstrate the ability of some cells to use citrate as a sole source of carbon for growth.

• Understand why citrate is used by some microbes and not others and how this fact can be used to differentiate the coli and aerogenes groups of enteric bacteria.

• Know how to differentiate a positive and a negative test, and know what the change of color means using Simmons' citrate agar.

MATERIALS

Cultures   *Escherichia coli*
             Enterobacter aerogenes

Media   Simmons' citrate agar slants with bromothymol blue

Supplies   Tubes containing 2 ml of sterile distilled water.

PROCEDURE: FIRST LABORATORY PERIOD

1. **Transfer a small amount of *E. coli* from a slant to a tube of sterile distilled water.** Continue to aseptically transfer cells from the slant to the distilled water until the water becomes only *slightly turbid*. Take care not to transfer any solid medium along with the cells. **Label this tube** so that you know it contains *E. coli*.

2. **Transfer *E. aerogenes* to a second tube of sterile distilled water** in the same manner.

TECHNIQUE NOTE

Cells are transferred from the slant to the distilled water to drastically reduce the concentration of nutrients around the cells. Remember that you are asking the cells to grow on a medum in which citrate is the only carbon source. If the inoculum you transfer from the slant also contains nutrients from the slant, the cells may grow on these nutrients (carryover) instead of on the citrate. Thus, to minimize nutrient carryover, you must rinse cells in distilled water before you transfer them to Simmons' citrate agar slants.

3. **Use these distilled-water suspensions to inoculate *Escherichia coli* and *Enterobacter aerogenes* onto separate sterile Simmons' citrate agar slants.** Use only one loopful of the distilled-water suspension for each inoculation, but **use a** *back-and-forth motion* to spread the inoculum over the entire surface of each slant.
4. **Ask your instructor to label several uninoculated tubes and to incubate them until the next laboratory period.** These tubes will serve as negative controls for this experiment.
5. **Incubate all tubes at 30°C for 48 hours** or until the next laboratory period.

PROCEDURE: SECOND LABORATORY PERIOD

1. **Examine the surfaces of both inoculated slants for evidence of growth and color change.** Compare the appearances of both slants with the negative (uninoculated) controls provided by your instructor.

INTERPRETATION NOTE

Recall that bromothymol blue is green when acidic (pH 6.8 and below) and blue when alkaline (pH 7.6 and higher). Only those cultures that demonstrate a pH shift of the medium to alkaline conditions should be considered positive.

2. **Record your results.**
3. **Properly discard all culture tubes.**
4. **Return all control tubes** to a specially marked test-tube rack as directed by your instructor.

CITRATE UTILIZATION TEST SUMMARY

Test name	Citrate utilization
Medium used	Simmons' citrate agar slants with bromothymol blue
Reagents and times	Suspend cells in sterile distilled water until suspension is slightly turbid; spread one loopful over the surface of a Simmons' citrate agar slant. Incubate for least 48 hours at 30°C.
Positive reaction	Only those pure cultures that exhibit an alkaline reaction (turn the pH indicator blue)
Negative reaction	Cultures that do not grow or that grow slightly but show no alkaline color reaction

30 DIFFERENTIAL UTILIZATION OF CITRATE BY ENTERICS

NAME _____

DATE _____

SECTION _____

RESULTS

Compare all inoculated tubes with a uninoculated (negative) control. Record *growth* as + (slight) or ++ (moderate to heavy). Record *no growth* as −.

Microbe	Growth	Color of agar	Positive or negative reaction
Escherichia coli			
Enterobacter aerogenes			
Uninoculated control			

CONCLUSIONS

QUESTIONS

Completion

1. To move a small molecule inside, a cell must have at least two types of proteins on or within its plasma membrane; these proteins are called _____ and _____.

2. The metabolically similar gram-negative bacteria that are found in mammalian intestines and other environments and that include the *Escherichia* and *Enterobacter* species are called the _____ group.

3. The name of the pH indicator found in Simmons' citrate agar is _____. This indicator is the color _____ when alkaline and _____ when acid.

4. Simmons' citrate agar is one test that helps differentiate the _____ group and the _____ group of bacteria. This test works because only the _____ group can grow and produce an alkaline reaction when citrate is the sole carbon source.

5. Before you inoculate pure cultures on Simmons' citrate agar, you transfer cells to sterile distilled water. The purpose of this transfer is to _____.

6. A citrate utilization test is positive when _____.

7. If all cultures inoculated onto this medium were positive, then a negative control would be essential to show _____.

True–False (correct all false statements)

1. Bacteria in the coli group are unable to use citrate as a carbon and energy source because they cannot translocate it into the cell. _____

2. Of the two bacteria tested in this exercise, only the species of *Escherichia* gave a strong positive reaction. _____

31

Acid and Gas Production from Sugar Fermentation

Microorganisms break down food to obtain energy in two ways: by fermentation (described in this exercise) and by respiration (introduced in Exercise 36). **Fermentation** has four basic features: (1) It can occur in the absence of oxygen; (2) energy-producing electron transport (like that found in respiration) is absent; (3) it is much less energy efficient than respiration; and (4) metabolic intermediates, often called **fermentation end products,** are excreted from the cell.

Fermentation end products are usually *acids* such as lactic and acetic, *neutral products* such as ethyl alcohol and butanediol, or *gases* such as carbon dioxide and hydrogen.

The types of fermentation end products are often characteristic of a genus or species of microorganism. For example, all species in the genera *Streptococcus* and *Pediococcus,* as well as some species of the genus *Lactobacillus,* are among those commonly called **homolactic acid bacteria** because they excrete lactic acid as the primary end product when they ferment glucose. Other species of *Lactobacillus,* as well as all members of the genus *Leuconostoc,* are called **heterolactic acid bacteria** because, although they make and excrete lactic acid when fermenting glucose, they also can make and excrete ethanol, glycerol, and acetic acid during glucose fermentation. Collectively, all of these genera are called **lactic acid bacteria.**

Let us consider the four distinctive characteristics of fermentation in more detail. Figure 31-1 shows what happens when glucose is fermented by homolactic acid bacteria in the genus *Streptococcus.* Note first that there is a net gain in ATP when one molecule of glucose is fermented to lactic acid by these bacteria (see inset table, Figure 31-1). The cell can later use the energy in each molecule of ATP for synthesizing new cell materials for growth.

First, oxygen is not used anywhere in this metabolic process. Each pair of hydrogen atoms removed from the glucose molecule during its fermentation is used to reduce one molecule of NAD^+, and the resulting $NADH_2$ molecule is later used to reduce a metabolic intermediate such as pyruvate. Therefore, a fermenting microorganism has no need for an electron transport mechanism to recycle NAD^+ molecules because a reduction reaction (such as pyruvate → lactate) serves this purpose.

Second, fermenting microbes lack an electron transport system. Since the cell is left with no excess hydrogens ($NADH_2$'s), it has no need of a cytochrome-linked electron transport system; indeed, strict fermentive microorganisms lack cytochromes. This is in contrast to the strictly respiring cells, in which oxygen is frequently used to accept the spent hydrogens (electrons) at the end of an electron transport system. Consequently, strictly fermenting microbes can ferment sugars in anaerobic environments.

Third, fermenting microbes obtain only a small amount of energy per molecule of sugar oxidized. Since these microorganisms do not use the electron transport system during fermentation, they cannot use electron transport as a mechanism to generate ATP. Therefore, they make ATP only by **substrate-level phosphorylation** during oxidation of the sugar. Figure 31-1 shows that *Streptococcus* species have a net gain of only two molecules of ATP per sugar molecule oxidized to lactic acid. This is about 18 times less than the net molecules of ATP produced per sugar molecule oxidized by a respiring microorganism. Be-

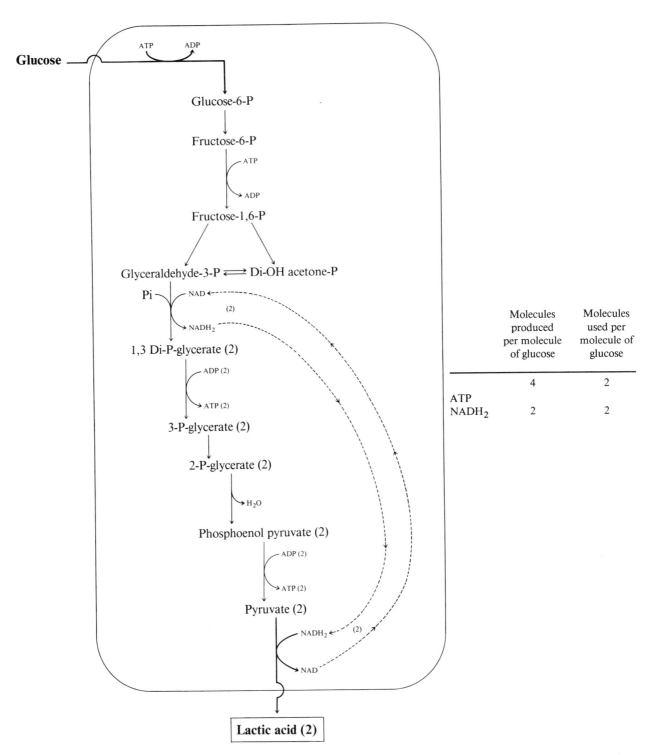

FIGURE 31-1 Homolactic acid fermentation of sugars by *Streptococcus* species. Note how this pathway represents all four characteristics common to all fermentations. This pathway is frequently called the Embden-Meyerhoff-Parnas (EMP) pathway after the three men who first described it.

cause fermenting cells derive so little energy, they grow more slowly and use much more sugar than respiring cells. In comparison with respiration, fermentation is very inefficient.

Fourth, fermenting microorganisms characteristically produce organic compounds that are released into their environment. Metabolic intermediates (organic compounds) resulting from sugar oxidation are reduced, and these reduced products are excreted from the cell. With homolactic acid bacteria, as an example, glucose is oxidized to pyruvate, and then pyruvate is reduced to lactic acid, which is released into the medium surrounding the cell.

When you want to identify an unknown microorganism, it is often helpful to know whether or not a pure culture of these microbes can ferment certain sugars. The production of acids (or other organic compounds) and gases is characteristic of certain sugar fermentations, and the detection of these products helps identify microorganisms. Therefore, it is important for you to know how to detect these fermentation products in culture media.

The detection of acid production in culture media in diagnostic microbiology is most often accomplished with **acid-base (pH) indicators** (Table 31-1). Although this exercise uses a classical (test-tube) method for detecting acid fermentation products, note that acid-base indicators are also used in the most modern and miniaturized diagnostic laboratory techniques.

Most organic compounds that act as acid-base indicators were first isolated from plants as potential dyes. Actually, they turned out to be very weak dyes, but their value as pH indicators was quickly noticed. Most of the useful acid-base indicators are in a class of aromatic compounds called *sulfonphthaleins,* and they change from one color to another with a change in pH (hydrogen-ion concentration).

These acid-base indicators are themselves either weak acids or bases. They do not change color abruptly at a single pH value, but do so gradually over a range of pH (see Table 31-1). The width of this band and its position on the pH scale depend on the strength of the indicator as an acid or as a base and on the relative color intensity of each of the two colored forms. In the simplest case, in which the intensities of the two colors are about the same, no change in color is detected by the eye until about 10 percent of the indicator is converted to its second form. The apparent color of the solution continues to change until about 90 percent of the indicator is in its second form. During that 10 to 90 percent transition, a dye that is

TABLE 31-1 Some pH indicators commonly used in microbiology

Acid-base indicator[a]	pH range[b]	Color change[c] (acid → alkaline)
Thymol blue (acid range) (0.04)[d]	1.2 ↔ 2.8	Red → yellow
Methyl orange (0.05)	3.1 ↔ 4.4	Red → yellow
Bromophenol blue (0.04)	3.1 ↔ 4.7	Yellow → blue
Bromocresol green (0.04)	3.8 ↔ 5.4	Yellow → blue
Methyl red (0.02)	4.2 ↔ 6.3	Red → yellow
Chlorophenol red (0.04)	5.1 ↔ 6.7	Yellow → red
Bromocresol purple (0.04)	5.4 ↔ 7.0	Yellow → purple
Bromothymol blue (0.04)	6.1 ↔ 7.7	Yellow → blue
Phenol red (0.02)	6.9 ↔ 8.5	Yellow → red
Cresol red (0.02)	7.4 ↔ 9.0	Yellow → red
Thymol blue (alkaline range) (0.04)	8.0 ↔ 9.6	Yellow → blue
Phenolphthalein (0.10)	8.3 ↔ 10.0	Colorless → red

SOURCE: M. Frobisher, R. D. Hinsdill, K. T. Crabtree, and C. R. Goodheart, *Fundamentals of Microbiology,* 9th ed. (Philadelphia: W. B. Saunders, 1974).
[a] Probably the most generally useful indicators in microbiology are methyl red, bromocresol purple, and phenol red.
[b] pH indicator ranges from Conn and Jennison (Eds.), *Manual of Microbiological Methods,* Society of American Bacteriologists (New York: McGraw-Hill, 1957). Note that these values may vary somewhat depending upon the literature reference used. The ranges given here indicate the pH values of the medium where the dye is changing from an alkaline to an acid color. Alkaline color occurs *above* the range, and acid color occurs *below* the range. Within each pH range, the dye will appear to be a combination of the acidic and the alkaline colors.
[c] Color change for each indicator is given from maximum acid color to maximum alkaline color.
[d] Numbers in parentheses are recommended percentage concentrations of dye in solution.

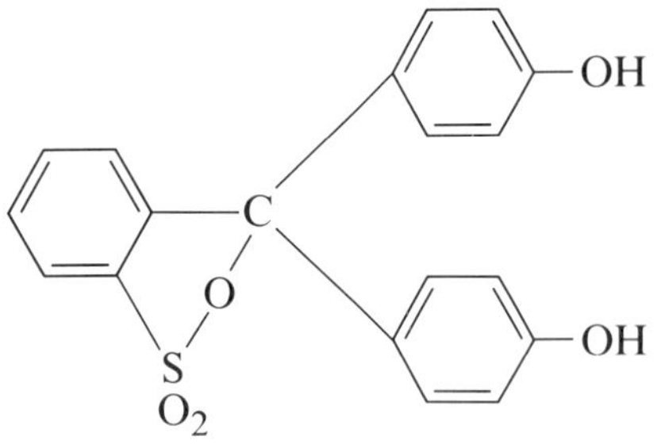

FIGURE 31-2 Chemical structure of the acid-base (pH) indicator called phenol red (phenolsulfonphthalein). Molecular weight = 354.4; dissociation constant (pK) = 7.9; pH range = 6.9 ↔ 8.5; acid color = yellow; alkaline color = red.

red at alkaline pH values (such as phenol red; Figure 31-2 varies through many shades (from yellowish-red through orange and pinkish-yellow) before the bright yellow acidic form is visible. On the other hand, if the color of the alkaline form is more intense than that of the acid, 20 to 30 percent of the acid form may have to be present before a color change can be detected with the eyes. Regardless, most monobasic acid-base indicators fully change color (from the alkaline to the acid form) over a pH range of about 2 pH units.

The position of the pH range over which an acid-base indicator changes color depends on the dissociation constant of the indicator. Indicators that are very weak acids change at higher pH values, and those that are strong acids change at lower values (see Table 31-1). The midpoint in the pH band over which the indicator appears to change color will agree with the midpoint of the titration curve (obtained potentiometrically) only if the intensities of both of the indicator's colors are about equal.

The color changes that occur during the transition from one color to the other commonly cause problems for the new microbiologist because people vary in their ability to detect color changes and in their description of the transition colors. Therefore, **standard diagnostic procedures** usually dictate that acid formation is not recorded as positive until *all* of the pH indicator is converted from the alkaline to the acid form. (Note that a standard tube for comparison can be prepared by adding excess acid to the sterile medium so that the maximum color change is achieved.)

Any color change that is less than complete, that is, any transition color, should be recorded as negative. This is not meant to be a test for the production of any specific amount of acid, but only a test for the production of enough acid to change the indicator completely from the alkaline to the acid form. Because these sulfonphthalein acid-base indicators are intensely colored, they are used in very low concentrations in culture media; therefore, only a small amount of acid is needed to effect a complete color change of the indicator.

Gas production by microorganisms during fermentation may be determined by collecting the gas inside a small, inverted vial that is submerged in a liquid growth medium. The vial is placed upside-down in a broth-filled test tube *before* autoclaving. During autoclaving, the heat expands the air inside the inverted vial and drives the air out. Upon cooling, the partial vacuum inside the vial allows the surrounding medium to rush into and completely fill the inverted vial. After inoculation, the gas produced during fermentation collects as a bubble at the top of the inverted vial. The evolved gas would otherwise escape into the atmosphere and not be detected. Therefore, bubble formation in the vial during incubation of microbial cultures is evidence for gas production during fermentation. These small, inverted, medium-filled vials are called *Durham tubes* (presumably named for microbiologist H. E. Durham).

Capped test tubes that contain a sterile broth, a pH indicator, and a Durham tube are traditionally called *fermentation tubes.* More modern (miniaturized) equipment employs these same basic principles to detect the release of fermentation products during microbial growth: use of a pH indicator and gas-entrapment.

LEARNING OBJECTIVES

• Understand the function of fermentation for a microbial cell.

• Know the four basic features characteristic of all fermentations.

• Know how pH indicators help detect the excretion of acidic and alkaline fermentation end products.

• Be able to recognize acid and gas production in sugar fermentation tubes containing phenol red.

MATERIALS

Cultures	*Escherichia coli* *Streptococcus faecalis* *Proteus mirabilis*
Media	Glucose fermentation tubes Lactose fermentation tubes Sucrose fermentation tubes
Supplies	None

PROCEDURE: FIRST LABORATORY PERIOD

1. **Separately inoculate each organism into each type of broth.** One loopful of broth or a small quantity of cells from a slant is sufficient. Too heavy an inoculum from a slant creates instant turbidity in the broth, and you may not be able to tell if growth has occurred after incubation.

   Do not vigorously dip the loop in and out of the broth. This action may introduce air into the inverted Durham tube.

2. **Check inverted Durham tubes for bubbles.** There should be none *before* incubation.

3. **Ask your instructor to label and incubate several uninoculated tubes of glucose broth as negative controls.**

4. **Incubate at 30°C for 48 hours.**

PROCEDURE: SECOND LABORATORY PERIOD

1. **Ask your instructor to demonstrate** the appearance of a tube of sterile phenol-red broth when excess acid is added to it.

2. **Carefully remove your cultures so that they are not shaken before you make your observations.**

3. **Observe the presence or absence of growth, acid production, and gas.** Checking for growth is important. It is possible that no fermentation occurred because no cells were present. Perhaps you forgot to inoculate one tube, or perhaps you touched the culture with a loop that was too hot and killed the cells. Or it may be that you inoculated live cells, but they cannot use this medium to support growth. No conclusions about fermentation products can be made from a sterile tube.

4. **Record your observations.**

5. **Remove the markings from your tubes,** and discard these tubes as directed.

INTERPRETATION NOTES

Growth If no growth is evident, record this only as *no growth*. Do not record this as a negative test for acid or gas production. Lack of growth may mean that these cells cannot use lactose, but it also can mean that you forgot to inoculate the broth or that your inoculating loop was too hot.

Acid production Phenol red is an acid-base indicator that is red above pH 8.5, yellow-red from pH 6.9 to 8.5, and yellow below pH 6.9. After phenol-red broth is first prepared and autoclaved, its pH is about 7.4; thus it is slightly alkaline and has a yellow-red color. Sometimes, when bacteria grow and ferment lactose in the anaerobic areas of the culture tube, they produce a small amount of acid that remains in the anaerobic areas of the tube; thus the yellow may appear only within the Durham tube or at the bottom of the culture tube. If enough acid is produced to lower the pH of the *entire tube* below 6.9, then the indicator in the entire tube appears yellow.

Tradition in diagnostic microbiology dictates that cells must produce enough acid to turn all of the medium yellow (below pH 6.9) before being scored as positive for acid production. This test is negative when there is evidence for growth but there is either no evidence of color change or yellow only in the bottom of the tube and/or inside the Durham tube.

Gas production When cells produce gas while fermenting a sugar, the gas first dissolves in the broth. After the broth becomes saturated with gas, however, continued gas excretion causes bubbles to form and rise through the broth. Bubbles formed in the inverted Durham tube rise and are trapped at the top. Active fermentation may produce a trapped bubble that is at least 2 to 3 mm long. Check closely! Sometimes the entire tube is filled with gas; this bubble might be missed without close inspection.

The test for gas production is positive when the inverted Durham tube contains a bubble of any size where none existed before inoculation. This test is negative if there is no gas accumulation in the inverted Durham tube.

SUGAR FERMENTATION TEST SUMMARY

Test name	Acid and gas production from sugars
Medium used	Phenol-red broth with one sugar added
Incubation	30°C for 48 hours
Substrate metabolized	Any sugar (tested separately)
Products formed	Any acid and/or any gas
Positive reactions	*Acid.* Turns phenol-red indicator yellow in the entire tube (pH at or below 6.9). *Gas.* Any bubble in the inverted Durham tube where none was present before incubation.
Negative reactions	Growth occurs, but media remains same color or yellow only in the Durham tube and/or in the bottom of the culture tube. Growth occurs, but no bubble in top of inverted Durham tube.
Purpose of test	To test for the production of acid and gas as fermentation products.

31 ACID AND GAS PRODUCTION

NAME _____

DATE _____

SECTION _____

RESULTS

Record your observations according to the directions given in the Interpretation Notes (see Procedure). Use the following abbreviations: growth (+ or −); acid (+ for the entire tube being yellow, or − for all other appearances); gas (+, and give the bubble size (mm) where there was none before inoculation; or − for no bubble inside Durham tube).

Organism	Sugar	Growth	Acid	Gas
Escherichia coli	Glucose			
	Lactose			
	Sucrose			
Streptococcus faecalis	Glucose			
	Lactose			
	Sucrose			
Proteus mirabilis	Glucose			
	Lactose			
	Sucrose			

CONCLUSIONS

QUESTIONS

Completion

1. Fermentation is one of two ways that microorganisms derive energy from nutrients in their environment. The other way is called _____.

2. In your own words, briefly list the four characteristics of fermentation.
   a.

   b.

   c.

   d.

3. Glucose fermentation by *Streptococcus* species proceeds along a metabolic pathway known as the _____ pathway. The major organic product excreted in this fermentation is _____.

4. Compared with respiration, fermentation is _____ energy efficient. When one examines the metabolic pathway, the efficiency is reflected in the net production of _____ per molecule of sugar fermented.

5. Why is it important for a microbiologist to know what sugars are fermented by a microbial species as well as the nature of the products excreted?

6. What is the function of sugar fermentation for the microbial cell?

7. The acid-base indicator used in this exercise is called _____. Its colors are _____ when alkaline (pH above 8.5), _____ around neutrality (pH 6.9 to 8.5), and _____ when acid (pH below 6.9).

True–False (correct all false statements)

1. Gas production in inoculated fermentation tubes is evident by gas bubbles continually rising in the broth. _____

2. A positive fermentation test requires that both acid and gas be produced. _____

3. Fermentation can take place in a totally anaerobic environment. _____

4. The negative control in this exercise functions as a sterility check and to determine if there is any nonbiological change in the pH indicator during incubation. _____

32

Acid and Neutral Products from Sugar Fermentation: Methyl-Red and Voges-Proskauer Tests

The methyl-red and Voges-Proskauer (MR/VP) tests are primarily used to differentiate two major types of facultatively anaerobic **enteric bacteria.** At first, these bacteria grow aerobically on glucose (in broth cultures) and rapidly use all of the oxygen via respiratory metabolism; then they commonly shift to one of two types of glucose fermentation: either the **mixed-acid** type or the **butylene glycol** type. The type of fermentation may be characteristic of the genus of enteric bacterium; therefore, identification of the fermentation end products may help a microbiologist to identify a pure culture of enteric bacteria. The genera that are considered enterics (family *Enterobacteriaceae*) and their fermentation types are shown in Table 32-1).

The enteric bacteria that accomplish *mixed-acid fermentation* of glucose excrete large amounts of formic, acetic, lactic, and succinic acids, and ethanol (Figure 32-1 and Table 32-2). The excreted acids cause a large decrease in pH (acidification), which microbiologists detect with methyl red, an indicator that does not fully change color until the pH drops to about 4.2. Therefore, the test for mixed-acid fermentation of glucose by enteric bacteria is called the **methyl-red test.**

The *butylene glycol (or butanediol) fermentation* of glucose produces only a fraction of the quantity of acids produced in the mixed-acid fermentation (see Figure 32-2 and Table 32-2). However, enteric bacteria that carry out this type of fermentation produce large amounts of **butanediol** and **ethanol,** neither of which alter the pH of the growth medium. **Acetoin** (pronounced "a-SET-owen"), also called *acetylmethylcarbinol,* is an intermediate in the synthesis of butanediol. Some acetoin is excreted into the culture medium during growth of these bacteria, although not shown in Table 32-2.

In 1898, O. Voges and B. Proskauer described a color reaction that we now know tests for acetoin. They found that the bacterial genera *Klebsiella* and *Enterobacter* tested positive, which is due to acetoin excretion during glucose fermentation; the bacterial genus *Escherichia* does not excrete acetoin. This test, now called **the Voges-Proskauer test,** is a major diagnostic tool for differentiating these bacteria. One drawback to the test is the 48-hour incubation needed to complete the test. Rapid tests have recently been developed, but the principles involved are the same as those used in this exercise.

LEARNING OBJECTIVES

- Realize that several types of acids and/or neutral products are characteristically formed and excreted by various microorganisms during glucose fermentation.

- Know how to perform the MR/VP test, so that you can use it to help differentiate certain enteric bacteria.

- Know what the abbreviations MR and VP stand for and what each test detects in a positive reaction.

TABLE 32-1 Types of Gram-negative facultative anaerobic rods (the enterics) and their methods of glucose fermantation

Genus	Fermentation method
Escherichia Salmonella Shigella	Mixed-acid
Enterobacter Serratia Erwinia Klebsiella	Butanediol
Edwardsiella Citrobacter Hafnia Proteus Yersinia Providencia Morganella	MR/VP not useful for genus identification

TABLE 32-2 Types and quantities of products formed that are characteristic of the mixed-acid and the butanediol fermentations

	Amount of product excreted (moles formed/100 moles glucose fermented)[a]	
Product formed	Mixed-acid type (*Escherichia coli*)	Butanediol type (*Enterobacter aerogenes*)
Formic acid	2	17
Acetic acid	37	1
Lactic acid	80	3
Succinic acid	11	0
Ethanol	50	70
2,3-butanediol	0	66
Carbon dioxide	88	172
Hydrogen	75	35

SOURCE: W. A. Wood, in I. C. Gunsalus and R. Y. Stanier, (Eds.) *The Bacteria* (New York: Academic Press, 1961). Vol. 2, pp. 59–149.
[a] All values reported are rounded off to nearest whole number.

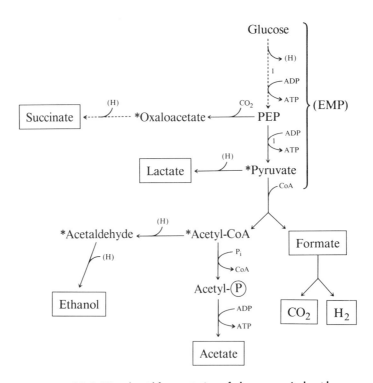

FIGURE 32-1 Mixed-acid fermentation of glucose carried out by certain genera of enterobacteria, such as *Escherichia*. Note especially the entire Embden-Meyerhoff-Parnas *(EMP)* pathway. Reaction sequences in which ATP is formed or hydrogens (electrons) are produced and consumed are indicated with the symbols *ATP* or *(H)*. Excretion products are shown in boxes, and the organic metabolic intermediates serving as terminal electron acceptors are marked with asterisks (*).

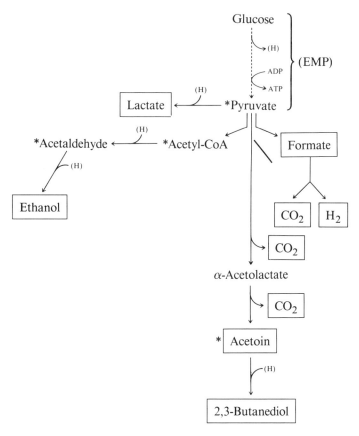

FIGURE 32-2 Butanediol (butylene glycol) fermentation of glucose carried out by certain enteric bacteria. Comments given in legend for Figure 32-2 apply here also. Note that acetoin (acetylmethylcarbinol) is an intermediate in the pathway that forms butanediol, the major excretion product in this type of fermentation; nevertheless, some acetoin is excreted by bacteria using this pathway.

MATERIALS

Cultures *Escherichia coli*
          Enterobacter aerogenes

Media   MR/VP broth

Supplies Clean test tubes
          5 percent alpha-naphthol reagent (w/v in absolute EtOH)
          40 percent KOH + 0.5 percent creatine
          0.02 percent methyl-red reagent

PROCEDURE: FIRST LABORATORY PERIOD

1. Inoculate each organism into a separate tube of MR/VP broth.
2. Set aside one additional tube of sterile MR/VP broth for an uninoculated control.
3. Incubate all three tubes at 30°C for 48 hours.

PROCEDURE: SECOND LABORATORY PERIOD

1. **Pour about one half of one culture into a clean test tube, and one half of the uninoculated broth (control) into another clean tube.** Make sure to label the tube before you make each transfer. You will use these tubes for the methyl-red test.

   Save the remaining half of each culture and the sterile broth for the Voges-Proskauer test.

Methyl-red test

2. **Add 5 to 6 drops of methyl-red reagent** to each of the two tubes that contain one-half of the cultures, and also to the tube that contains one-half of the *uninoculated control broth*. **Thoroughly mix the fluid in each tube.**

3. **Determine the results of this test immediately after mixing the methyl-red reagent.** The test is *positive* when the methyl red (acid-base indicator) turns the culture medium a bright red, showing complete conversion of the dye to the acid form (pH is 4.2 or lower).

INTERPRETATION NOTE

The uninoculated medium contains glucose and should have a neutral pH value (pH 7.0). When microbes ferment the glucose, releasing acids into the culture medium, the hydrogen-ion concentration increases (pH goes down). Solutions to which the acid-base indicator methyl red are added are yellow above pH 6.3; they are bright red only below pH 4.2. Therefore, cultures that make this indicator fully red have produced and excreted enough acid from glucose to decrease the pH from 7.0 to 4.2 or lower.

To see what a complete color conversion looks like, ask your instructor to add acid to a tube containing the same volume of sterile MR/VP broth plus 5 to 6 drops of methyl red.

The test is *negative* when the culture remains yellow after adding methyl red (pH is 6.3 or higher), or when the culture is some transition color between yellow and red (pH is between 4.2 and 6.3). For a review of acid-base indicators, see Exercise 31, Table 1.

METHYL-RED (MR) TEST SUMMARY

Test name	Methyl-red (MR) test
Medium used	MR/VP broth
Incubation	30°C for 48 hours
Substrate fermented	Glucose
Expected product	Mixture of acids (such as lactic, acetic, formic, and succinic)
Reagent and method	Methyl red. Read immediately after mixing.
Positive reaction	Red color throughout tube (complete conversion of pH indicator)
Negative reaction	Yellow color throughout tube or any yellowish-red (indicating incomplete conversion of indicator)

Voges-Proskauer test

4. **Add 0.6 ml of the alpha-naphthol reagent to the other tube containing one-half of one original culture and one-half of the uninoculated control, and mix thoroughly** by thumping the bottom of the tube with your finger to create a vortex, as described in Exercise 17.

5. **Add 0.2 ml of KOH (plus creatine) reagent to these same tubes, and mix vigorously for 30 seconds.** It is most important that you mix a lot of oxygen into the fluid, besides mixing the reagents.

TECHNICAL NOTE

If formed by the culture, acetoin (acetylmethylcarbinol) spontaneously oxidizes to diacetyl in the presence of oxygen and KOH. (Alpha-naphthol is used only to increase the sensitivity of this reaction.) Diacetyl then reacts with creatine to form a red complex whose chemistry is not known.

This procedure detects as little as 0.2 parts per million of diacetyl. The more acetoin that the culture forms, the more diacetyl is formed, and the more intense is the red color. However, remember that acetoin is not ultimately converted to a red color unless oxygen is present; so it is very important that excess oxygen be mixed into the fluid mixture. Thirty seconds of vigorous shaking should be sufficient to saturate this fluid with oxygen.

$$\underset{\text{Acetoin (acetylmethylcarbinol)}}{\begin{array}{c}CH_3\\|\\C=O\\|\\H-C-OH\\|\\CH_3\end{array}} \xrightarrow{\underset{KOH}{O_2}} \underset{\text{Diacetyl}}{\begin{array}{c}CH_3\\|\\C=O\\|\\C=O\\|\\CH_3\end{array}} \xrightarrow[(\alpha\text{-Naphthol})]{\text{Creatine}} \underset{\substack{\text{Red dye}\\\text{(Chemistry unknown)}}}{}$$

Creatine:
$$\begin{array}{c}COOH\\|\\CH_2\\|\\H_3C-N\\|\\H-C-NH_2\\|\\NH_2\end{array}$$

---

6. **Let these mixtures stand at room temperature for no less than 15 minutes.** Slanting the tubes during this 15-minute period increases the amount of surface area exposed to oxygen.

The test is *positive* if a red color (of any intensity) develops in the fluid. (If the fluid does not contain enough oxygen, the red color will be more intense at the surface where the dissolved oxygen concentration is highest.)

The test is *negative* if the mixture lacks red color. (Compare the color with the uninoculated control treated in the same way.)

7. **Repeat the methyl-red and the Voges-Proskauer tests with one-half of your second culture,** and compare your results with the appropriate controls.

8. **Record your results,** and properly discard all tubes.

VOGES-PROSKAUER (VP) TEST SUMMARY

Test name	Voges–Proskauer (VP) test
Medium used	MR/VP broth
Incubation	30°C for 48 hours
Substrate fermented	Glucose
Expected product	The neutral product acetoin (acetylmethylcarbinol)
Reagents and method	Oxygen, KOH, and creatine, plus alpha-naphthol to increase sensitivity. Mix vigorously for 30 seconds to oxygenate; let stand for 15 minutes.
Positive reaction	Red color (any intensity); may be most intense at the surface
Negative reaction	No red color (of any intensity; compare with uninoculated control)

NAME _____

DATE _____

SECTION _____

RESULTS

Record your results as either positive (+) or negative (−) for mixed-acid production (MR test) and for acetoin production (VP test) during glucose fermentation.

Microorganism	Methyl-red and Voges-Proskauer tests	
	Acid production (MR)	Acetoin production (VP)
Escherichia coli		
Enterobacter aerogenes		

CONCLUSIONS

QUESTIONS

Completion

1. The MR/VP test is important in differentiating two types of bacteria in a group known as the _____. This test distinguishes between the fermentation products produced by these two bacterial types. The two fermentation types are the _____ and the _____ fermentations.

2. The acid-base indicator used in this exercise is called _____. Its color is _____ when the pH is above 6.3 and _____ when the pH is below 4.2.

3. We use the Voges-Proskauer reagents to test for acetoin production, and this is a(n) _____ (acidic, basic, or neutral) excretion product.

4. The Voges-Proskauer test uses two reagents. Once both reagents are added to the culture, the tube should be _____ and then left to stand for _____ minutes before the results are determined. If color development occurs, it may be most evident in _____ [which part(s)] of the tube.

True-False (correct all false statements)

1. Acids are the only kind of organic molecules produced by microbial fermentation. _____

2. Species of both *Escherichia* and *Enterobacter* are considered mixed-acid fermentation types. _____

3. *Escherichia, Salmonella, Enterobacter,* and *Shigella* are all facultatively anaerobic gram-negative rods. _____

4. The methyl-red test can be read immediately after adding the pH indicator, and the color should be evident throughout the tube. _____

33

Hydrogen Sulfide (H₂S) Production from Thiosulfate and Sulfur-Containing Amino Acids

Microbial metabolism may produce hydrogen sulfide (H_2S) in two ways. (1) Some microorganisms metabolically oxidize organic sulfur-containing compounds to obtain energy for growth and excrete the sulfur as hydrogen sulfide; for example, the amino acid *cysteine* contains a *sulfhydryl (—SH) group* that is cleaved in the first step of its metabolic breakdown as a source of energy for the cell (Figure 33-1). (2) Other microorganisms produce hydrogen sulfide as an end product of **anaerobic respiration,** in a manner similar to the way water is produced from oxygen during aerobic respiration: Specifically, thiosulfate acts as the final electron acceptor at the end of an electron transport chain. In other words, these microbes can use either *sulfate* (SO_4^{2-}) or *thiosulfate* ($S_2O_3^{2-}$) instead of oxygen as the terminal electron acceptor (Figure 33-2), thus reducing the oxidized sulfur molecule to hydrogen sulfide (Figure 33-3). (Note that in *aerobic respiration,* oxygen is the final electron acceptor.)

Hydrogen sulfide is a colorless gas that has about the same toxicity as hydrogen cyanide, the gas used in execution chambers. However, its odor (largely responsible for the smell of rotten eggs) is so offensive and so easily detected at such low concentrations that accidental exposures to lethal doses of it are virtually unknown.

Another property of hydrogen-sulfide gas is its great solubility in water (437 gm/100 ml of water at 0°C). Because of its solubility, microbially produced hydrogen sulfide cannot be collected as a gas inside an inverted Durham tube. Therefore, other methods of detection are necessary. The most common is the formation of a black, insoluble precipitate, **ferrous sulfide** (FeS), when hydrogen sulfide (H_2S) reacts with ferrous ions (Fe^{2+}).

Modern culture media for the detection of hydrogen sulfide vary greatly, but all contain peptone and thiosulfate. **Peptone** is an enzymatically hydrolyzed animal protein that is rich in sulfur-containing amino acids (such as cysteine), which some microorganisms can metabolically oxidize to obtain carbon and energy. Because microbiologists have found that the amount of sulfur-containing compounds varies considerably among peptones, they also add **sodium thiosulfate** ($Na_2S_2O_3$) to media as an additional source of sulfur (see Figure 33-2) from which microbes can make hydrogen sulfide.

Microbes that accomplish this type of anaerobic respiration are called **sulfate-reducing bacteria.** The metabolic activity of the sulfate-reducing bacteria is particularly apparent in the mud at the bottoms of ponds and streams, in bogs, and along the seashore. Since seawater also contains a relatively high concentration of sulfate, microbial reduction of sulfate is an important process for the conversion of organic matter into inorganic minerals on the shallow ocean floors. Signs of this process in both fresh- and saltwater muds are the odor of hydrogen sulfide and the pitch-black color. The color of the mud is caused by accumulation of ferrous sulfide, the same insoluble iron salt used to detect hydrogen-sulfide production in laboratory culture media. Some coastal areas, where the accumulation of organic matter leads to a

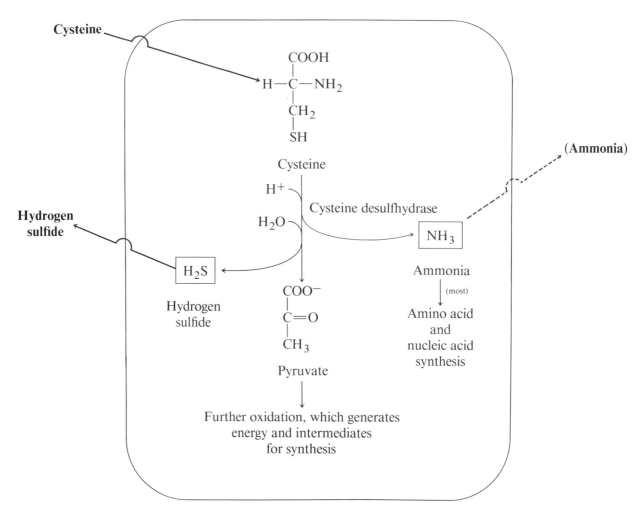

FIGURE 33-1 Microbial production and excretion of hydrogen sulfide during metabolic oxidation of the sulfur-containing amino acid cysteine. This amino acid comes from peptone in the medium; it is translocated into some cells and used as a source of energy. The cleavage products of the intracellular enzyme called cysteine desulfhydrase are hydrogen sulfide, ammonia, and pyruvate. Hydrogen sulfide is excreted from the cell. Some ammonia may also be excreted, but most is probably used for synthesis of nitrogen compounds needed for cell growth. Pyruvate is broken down further as a source of energy (for example, via the TCA cycle), and certain intermediates are also used as building blocks for synthesizing other things the cell needs to grow.

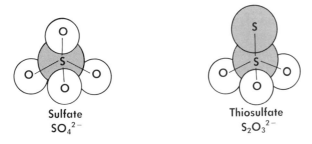

FIGURE 33-2 Comparison of the structures of the sulfate (SO_4^{2-}) and thiosulfate ($S_2O_3^{2-}$) ions. The prefix *thio-* indicates that a sulfur atom has been substituted for an oxygen in the molecule. Both of these compounds can be reduced by the spent electrons at the end of the electron transport chain during anaerobic respiration (see Figure 33-3).

particularly massive reduction of sulfate, are almost uninhabitable because of the odor and toxicity of hydrogen sulfide.

In the clinical laboratory, microbial production of hydrogen sulfide can be used as a test to help identify certain gram-negative **facultatively anaerobic rods,** many of which cause disease in humans. The test for hydrogen-sulfide production seems particularly helpful for differentiating species in the following genera: *Proteus, Citrobacter, Salmonella, Escherichia,* and *Yersinia.*

The media most commonly used in clinical laboratories for detecting hydrogen sulfide seems to be **Kligler's iron agar** and **triple-sugar-iron (TSI) agar.**

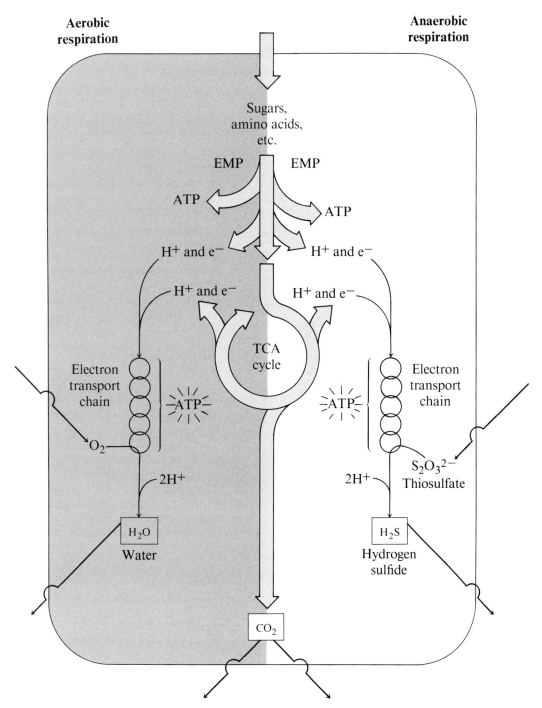

FIGURE 33-3 Comparison of metabolic oxidations linked to aerobic and to anaerobic respiration. Aerobic respiration reduces oxygen to water by means of the energy-spent electrons (plus protons) at the end of the electron transport system. Anaerobic respiration reduces thiosulfate (instead of oxygen) to hydrogen sulfide (instead of water) by means of the energy-spent electrons (plus protons) from the electron transport system.

Both contain peptone and sodium thiosulfate for microbes to use in forming hydrogen sulfide, but both also test for acid and gases from sugar fermentation. Therefore, this exercise uses peptone-iron agar to show hydrogen sulfide production because this dissolved gas is the only product detected in this differential medium. You should note, however, that peptone-iron agar is not necessarily the best medium for growing *all* hydrogen-sulfide producing bacteria.

LEARNING OBJECTIVES

• Realize that hydrogen-sulfide gas may be produced and excreted as an end product of microbial metabolism and that this gas is very water soluble.

• Understand that hydrogen-sulfide production is characteristic of only certain types of microorganisms.

• Know how to test for hydrogen-sulfide production so that this test may later be used in identifying unknown microorganisms.

MATERIALS

Cultures	*Proteus mirabilis*
	Escherichia coli
Media	Peptone-iron agar deeps
Supplies	Inoculating needles (not loops)

PROCEDURE: FIRST LABORATORY PERIOD

1. **Use a straight inoculating needle** that will cleanly penetrate the center of an agar deep.

2. **Inoculate by stabbing each culture into separate agar deeps.** Stab a heavy inoculum (lots of cells on the needle) into the center of the column of agar, and penetrate the agar at least two-thirds of the way toward the bottom.

3. **Ask your instructor to sham-inoculate and incubate several agar deeps** so that you may use these as uninoculated controls during the second laboratory period.

4. **Incubate at 30°C for 48 hours.**

PROCEDURE: SECOND LABORATORY PERIOD

1. **Observe deeps for evidence of growth.** *First,* hold the sham-inoculated deep so that light is transmitted through the tube. Note that uninoculated agar has a cloudy appearance; therefore, growth along the stab line in the inoculated tube may be difficult to see. *Next,* observe each stab-inoculated deep with transmitted light in the same manner. If you cannot detect growth along the stab line, check the surface of the agar deep; you may be able to see growth where the needle penetrated the agar.

2. **Observe deeps for evidence of hydrogen sulfide production.** A black color in the medium should be recorded as *positive* and the absence of a black color as *negative*.

INTERPRETATION NOTES

This medium turns black when hydrogen sulfide is produced because ferrous iron (Fe^{2+}) is present in the medium. Peptone-iron agar contains ferric ammonium citrate; when this is autoclaved, both ferrous (Fe^{2+}) and ferric (Fe^{3+}) ions are formed because of changes in the oxidation-reduction potential. Ferrous ions react with the hydrogen sulfide, excreted by the growing bacteria, to form ferrous sulfide (FeS), which is a black precipitate. Therefore, this black precipitate is present only in iron-containing media in which growing microbes produce hydrogen sulfide.

---

3. **Record your results,** and properly discard all tubes.

HYDROGEN-SULFIDE TEST SUMMARY

Purpose of test	Detect H_2S production
Medium used	Peptone-iron agar
Incubation	30°C for 48 hours
Substrate metabolized	Sulfur-containing amino acids or thiosulfate
Product formed	Hydrogen sulfide (H_2S)
Reagents used	None (ferrous iron already in medium serves as an indicator)
Positive reaction	Black precipitate (FeS) formed in agar
Negative reaction	No black precipitate formed

33 HYDROGEN SULFIDE (H$_2$S) PRODUCTION

NAME _____

DATE _____

SECTION _____

RESULTS

Record growth as either present (+) or absent (−). Describe the appearance of the medium after incubation. Record H$_2$S production as positive (+) or negative (−).

Organism	Growth	Medium appearance	H$_2$S production
Proteus mirabilis			
Escherichia coli			

CONCLUSIONS

QUESTIONS

Completion

1. Bacteria may excrete hydrogen sulfide by means of two metabolic mechanisms. One involves the breakdown of sulfur-containing _____. The second involves a process known as anaerobic respiration, in which either sulfate or thiosulfite is used as a terminal _____ _____ in place of oxygen. For which of these two mechanism(s) does stabbed peptone-iron agar test?

2. Peptone is used in all media for H_2S detection because _____ _____.

3. Although H_2S is very toxic, it is not dangerous in the laboratory because _____ _____.

4. A positive H_2S production test is indicated by the _____ color formed in agar deeps that contain the _____ ion. Chemically, this black precipitate is called _____ _____.

5. Sulfate and thiosulfate have a function in anaerobic respiration similar to that of _____ in aerobic respiration.

6. Why might you expect facultative anaerobes to accomplish anaerobic respiration in peptone-iron agar deeps when they are not incubated in an anaerobic chamber?

True–False (correct all false statements)

1. H_2S gas production in broth is best detected with an inverted Durham tube. _____

2. H_2S formed from the oxidation of sulfur-containing amino acids is not an indicator of either respiration or fermentation. _____

34

Indole Production from the Amino Acid Tryptophan and Catabolite Repression

The test for the ability to form indole from the amino acid tryptophan has long been used to help separate *Escherichia* from other genera in the family *Enterobacteriaceae,* especially *Klebsiella* and *Enterobacter* (see Figures 34-1 and 32-1). The indole test is now considered important in the identification of a wide variety of other organisms commonly isolated in the clinical laboratory.

The *Escherichia* and *Klebsiella-Enterobacter* groups of enteric bacteria are morphologically and physiologically similar. There is, however, a series of tests called **IMViC** that can distinguish between them. [In IMViC, *I* denotes the test for indole production; *M* denotes the methyl-red test for acid production during growth on a glucose medium (see Exercise 32); *V* denotes the Voges-Proskauer test for microbial production of acetoin (see Exercise 32); *i* simply makes the acronym easier to say; and *C* indicates that citrate can be the sole source of carbon and energy for the microbe (see Exercise 30)].

The IMViC tests are important in differentiating *Escherichia* from other types of enteric bacteria, primarily because the genus *Escherichia* is so common in the intestines of animals that its presence in the environment is used as an indicator of fecal pollution. From these tests, *Escherichia* gives (++--)—that is, positive for indole and methyl red, but negative for acetoin and citrate. Because you can find the *Klebsiella-Enterobacter* group of enterics in many places other than animal intestines, their presence does not necessarily indicate fecal pollution. Species of the genus *Enterobacter* give (--++) on the IMViC tests.

Bacteria form indole in the following way. The amino acid **tryptophan** is translocated from outside to inside the cell, where enzymes begin to break it. The first step is catalyzed by an enzyme called *tryptophanase,* which splits tryptophan into indole, ammonia, and pyruvate (see Figure 34-1). Then indole (and some ammonia) are excreted from the cell as waste products, while pyruvate is oxidized to provide energy for the cell. The oxidation of pyruvate in aerobically grown *Escherichia* occurs via the tricarboxylic acid (TCA) cycle. Since facultative anaerobes such as *Escherichia* produce more indole when incubated aerobically, indole should *not* be considered a product of tryptophan fermentation.

For bacteria that utilize tryptophan in the above manner, you should not use a growth medium that contains glucose because glucose inhibits indole production. Cells obtain much more energy from respiration on glucose than they do during respiration on tryptophan. If glucose is present, the microbe will use glucose preferentially over tryptophan; thus, little or no tryptophan is cleaved, and detectable levels of indole are not produced. The process whereby glucose shuts down tryptophan utilization is an example of **catabolite repression.** An unknown catabolite, derived from glucose catabolism, appears to repress the activity of genes that code for synthesis of the enzymes needed for tryptophan utilization.

To favor indole production, you should incubate cultures aerobically (because bacteria need oxygen for respiration) and in media containing *tryptone* (because this partially hydrolyzed animal protein is especially high in tryptophan) for at least 48 hours.

You can detect indole production with a reagent called ***para*-dimethylaminobenzaldehyde (*p*-DMAB)** prepared with concentrated hydrochloric acid. When the acidified *p*-DMAB reacts with indole, a red-violet dye called *rosindole* is formed (Figure 34-2). The

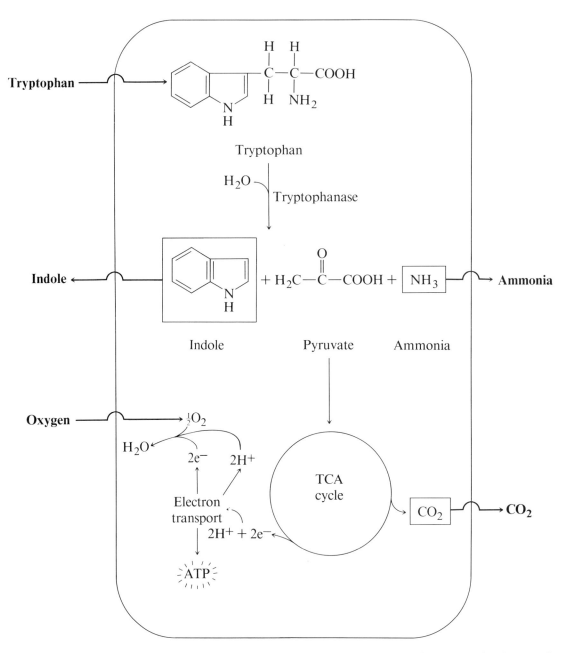

FIGURE 34-1 Formation of indole and other compounds from the amino acid tryptophan. Arrows leading out of the cell indicate the waste products excreted. Tryptophan is transported into the cell and then is cleaved by the enzyme tryptophanase. Indole (and small amounts of ammonia) are excreted, and pyruvate is metabolized further via the TCA cycle. Note that pyruvate catabolism is *not* fermentation; it is respiration, with oxygen serving as the terminal electron acceptor.

problem is that tryptophan also reacts with p-DMAB to produce rosindole dye. Since tryptone broth is rich in tryptophan, how does one use p-DMAB to determine if indole is formed by pure cultures?

In 1928, Kovacs modified the Ehrlich–Boehme procedure so that the test reagent contained acidified p-DMAB and **amyl alcohol.** The amyl alcohol is critical to this test because (1) indole is soluble in amyl alcohol and tryptophan is not and (2) amyl alcohol is not miscible with water, that is, it forms a layer on top after being mixed with an aqueous culture medium.

So, when you vigorously mix Kovacs' reagent with a culture in tryptone broth that has produced indole, the amyl alcohol selectively extracts the indole, and the p-DMAB that is dissolved in the amyl alcohol reacts with indole to produce the red-violet rosindole

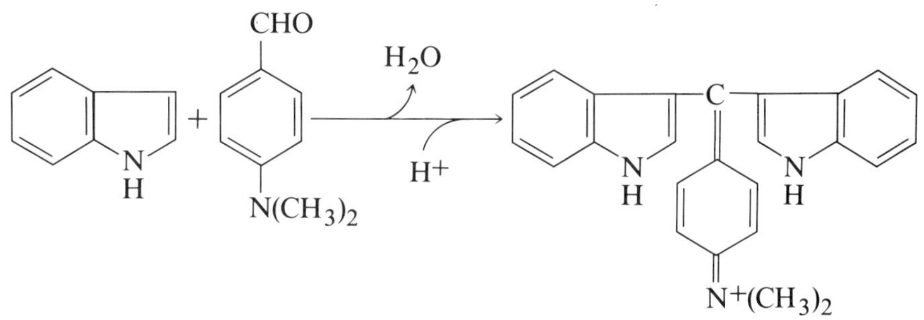

34 INDOLE PRODUCTION FROM TRYPTOPHAN

FIGURE 34-2 Reaction between indole and p-DMAB to form the complex known as rosindole dye. Acidified p-DMAB is dissolved in amyl alcohol to form Kovacs' reagent. The amyl alcohol selectively extracts indole from culture fluids; then an indole-p-DMAB-dye complex is formed, and this dye complex collects in an immiscible, red-violet layer on top of the culture medium.

dye (see Figure 34-2). After standing awhile, the immiscible amyl-alcohol droplets rise to the top of the culture medium, carrying the red indole-*p*-DMAB complex along with them, and the droplets fuse to form an immiscible layer on top of the aqueous broth. Thus, a red-violet layer on top of the culture fluid is a *positive* test for indole formation.

A *negative* test does not show a red-violet (amyl alcohol) layer on top of the aqueous culture fluid. However, a negative test may exhibit some red-violet color in the bottom (culture) layer because of a reaction between the unaltered tryptophan in the culture fluid and *p*-DMAB. Note, however, that a negative test shows little or no color in the top layer because tryptophan is not soluble in amyl alcohol.

LEARNING OBJECTIVES

- Realize that amino acids can be used by microorganisms as energy sources.

- Understand that indole is a product excreted by some bacteria that use the amino acid tryptophan as a source of energy during aerobic respiration, and that indole is *not* a fermentation product.

- Know how to test for indole in culture media so that this test can be used in identifying unknown microorganisms.

- Understand the concept of catabolite repression and know how to test for it.

MATERIALS

Cultures *Escherichia coli*
          Enterobacter aerogenes

Media   1 percent tryptone broth
          1 percent broth + 5 percent glucose

Supplies Kovacs' reagent

PROCEDURE: FIRST LABORATORY PERIOD

1. **Inoculate each culture separately into both tryptone broth and tryptone + glucose broth.** The result should be four inoculated tubes: two types of media contain *E. coli* and two types contain *E. aerogenes.*

2. **Ask your instructor to label several 1 percent tryptone broth tubes for uninoculated (negative) controls** and to incubate these along with your inoculated tubes.

3. **Incubate all four tubes at 30°C for 48 hours.**

PROCEDURE: SECOND LABORATORY PERIOD

1. **Add one-half (0.5) milliliter of Kovacs' reagent to each culture, and mix thoroughly by** thumping the bottom of the tube with your

finger. If possible, use a burette to accurately dispense the caustic Kovacs' reagent.

Note that this assumes that your culture contains 5 ml of broth, which represents a 1:10 proportion of reagent to culture. This same proportion may be used regardless of the amount of culture tested.

Ask your instructor to treat the uninoculated control tubes in the same way.

2. **After mixing, set the tube aside for 5 minutes before recording the result.** This allows time for separation of the immiscible amyl alcohol from the aqueous culture medium. A *positive* result is a dark red-violet color in the *top* (amyl alcohol) layer; this is the indole-*p*-DMAB complex (rosindole dye) dissolved in amyl alcohol. A *negative* result is any color other than dark red-violet (or no color) in the top (amyl alcohol) layer.

3. **Compare tests on your inoculated cultures with those on the uninoculated controls.**

4. **Record your results,** and properly discard your tubes.

INDOLE TEST SUMMARY

Test name	Indole test
Medium used	1-percent tryptone broth
Incubation	30°C for 48 hours
Substrate metabolized	Tryptophan
Product formed	Indole
Reagents used	Kovacs' reagent
Positive reaction	Dark red-violet color in the amyl-alcohol layer on top of the broth.
Negative reaction	Any color other than dark red-violet (such as pink, orange, or amber) in the amyl-alcohol layer on top of the broth. A red color in the broth layer means that tryptophan is still present.
Purpose of test	To determine indole production from microbial catabolism of tryptophan

34 INDOLE PRODUCTION FROM TRYPTOPHAN

NAME _____

DATE _____

SECTION _____

RESULTS

Record the results from each tube in two ways: (1) Record whether positive (+) or negative (−) for indole production, and (2) write your perception of the color found in the Kovacs' layer. Use the words for color given in the Procedure or in the Indole Test Summary.

Culture	Indole production (+ or − and description of color in top layer)	
	Tryptone broth	Tryptone + glucose broth
Escherichia coli		
Enterobacter aerogenes		
Uninoculated		

CONCLUSIONS

Ability of *Escherichia* and *Enterobacter* to form indole on tryptone broth:

Ability of these same cultures to form indole on tryptone broth when glucose is added:

How these results reflect the concept of catabolite repression:

QUESTIONS

Completion

1. The test for indole formation is part of a series of tests used to help identify the bacterial genus called _____. This series of tests is called the _____ test.

2. The reason it is important to identify this genus is that it is an indicator of _____.

3. The name of the compound (substrate) metabolized by some bacteria to produce and excrete indole is _____.

4. A positive indole test is one in which there is a _____ color in the _____ (top or bottom) of the tube.

5. To test for indole production, you use _____ reagent and determine the results after _____ time has elapsed.

6. What purpose do the uninoculated tubes serve?

True – False (correct all false statements)

1. Tryptophan is one type of sugar. _____

2. The production of indole is evidence of tryptophan fermentation. _____

3. Glucose inhibits indole formation because some intermediate from its catabolism represses the synthesis of an enzyme that is needed for tryptophan breakdown. _____

4. For optimum tryptophan catabolism and for indole production and excretion, one should incubate cultures aerobically in the absence of glucose. _____

35

Products Formed in Milk: The Litmus Milk Test

Milk is not only a nutritious food for mammals, it is also an excellent growth medium for many bacteria. Milk contains the water-soluble sugar **lactose,** also called milk sugar. Lactose is a *disaccharide,* that is, a molecule composed of two single-sugar molecules (monosaccharides) covalently bonded. One of these monosaccharides is *glucose;* the other is *galactose.* Many bacteria can transport lactose into the cell as an intact disaccharide, without having to first split the molecule into monosaccharides outside the cell. Once the lactose is inside, enzymes cleave it and use both glucose and galactose as sources of energy for synthesis and growth. Some bacteria metabolically oxidize these sugars via respiration until the oxygen is depleted, and then they ferment them. Other bacteria, such as the **lactic-acid bacteria,** exclusively ferment sugars. The lactic-acid bacteria are commonly found in milk and excrete large quantities of lactic acid when fermenting glucose or galactose. Members of genera called *Streptococcus, Lactobacillus,* and *Leuconostoc* are examples of lactic-acid bacteria. One metabolic pathway these bacteria use to form lactic acid in milk is shown in Figure 31-1.

Acids excreted by lactic-acid bacteria affect **casein,** the major protein in milk (see Exercise 28). Casein is a phosphoprotein that exists as large molecules finely dispersed as a colloid in milk. It is this protein colloid that gives milk its white, turbid appearance. If the microbes excrete sufficient acid during fermentation of the sugars in milk, casein will *denature.* **Acid denaturation** causes a structural change in the protein molecule, from well-organized, tightly folded, large protein molecules (existing as a colloid) to a formless meshwork of large, unfolded protein molecules that form a **curd.** The process of denaturing a soluble protein or a proteinaceous colloid into a semisolid, gel-like mass is called **coagulation.** This semisolid coagulated protein (curd) is the major constituent of most cheeses and other fermented milk products such as yogurt. When curd separates from acid-denatured milk, the remaining clear, colorless or straw-colored fluid is called **whey.**

When inoculated into sterile skim milk, lactic-acid bacteria grow, ferment lactose, and excrete lactic acid. So much acid is produced, and so much curd is formed, that the milk becomes one semisolid gel. When tubes containing these cultures are inverted, the curdled milk remains in the bottom of the tube, and only a small amount of separated whey runs down the side of the tube.

Some bacteria produce large volumes of gas along with acids when they ferment lactose. These gases may physically break apart the curd and force pieces of it far up into the tube. This process, known as **stormy fermentation,** is characteristic of a number of *Clostridium* species.

Like lactose, casein can serve as a source of energy for bacterial growth, especially for gram-positive, spore-forming rods such as species of *Bacillus* and *Clostridium.* These bacteria characteristically form and excrete extracellular enzymes, the **proteases,** that break certain peptide bonds that hold amino acids together to form the protein molecule. Other enzymes, the **peptidases,** further hydrolyze these molecules into smaller polypeptides that can be taken into the cells. (See Exercise 28 for a review of these concepts.)

When these proteolytic enzymes attack casein, the protein changes from large molecules existing as a colloid or denatured gel into much smaller molecules that go into solution. This process of converting an insoluble protein into much smaller water-soluble molecules is called **proteolysis.** Proteolysis eventually converts the milky-white suspension into a relatively clear fluid. When proteolytic bacteria grow in plain skim milk, this clear fluid usually has a brownish or amber color.

Once proteolysis breaks down the proteins outside the cell, the resulting amino acids and small molecular weight polypeptides are taken into the cell and oxidized. The first step in the metabolic oxidation of any amino acid is usually removal of the amine group (Figure 35-1). This enzyme-catalyzed removal is called **hydrolytic deamination,** and the reaction produces ammonia (NH_3), much of which is excreted from the cell. Ammonia excretion makes the milk alkaline, which is detected by adding a pH indicator to the sterile skim milk. Therefore, proteolysis is accompanied by (1) solubilization of the suspended casein (or the denatured curd) and by (2) ammonia excretion, which results from amino acid oxidation inside the cell and causes the milk to become alkaline.

If you use whole milk to grow bacteria, fermentation usually causes the breakdown of fats, and the resulting milk becomes rancid. To avoid this further complication, microbiologists most often use dried skim milk to culture bacteria (because the fat has been removed before drying). Skim milk has only two major substrates that bacteria can attack and use as energy sources: lactose and casein.

Some bacteria ferment lactose and produce great quantities of acids. Others split casein, transport the amino acids into the cell, and ferment them while excreting basic and neutral end products. Still other bacteria simultaneously ferment lactose and the amino acids from casein proteolysis. Since the type of fermentation accomplished by bacteria when growing in skim milk is characteristic of certain bacterial types, you can use this medium to help identify certain bacteria, but the interpretation of results is often difficult.

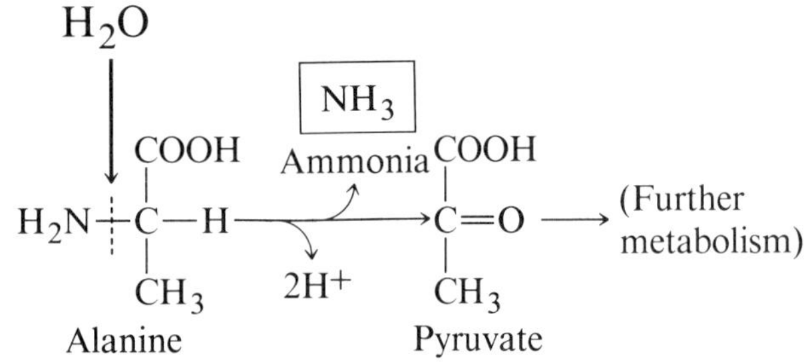

Microbiologists may use **litmus milk** as one of many differential tests for the identification of a microorganism. This medium contains only skim-milk powder and an indicator called litmus.

Litmus is a substance obtained from **lichens** (mainly from the genera *Lecanora* or *Rocella*). These lichens are colorless; but when they are treated with ammonia, lime, and potash or soda, blue/gray colors develop due to certain phenolic compounds, which can be extracted from these lichens. Litmus is only one type of phenol-containing compound obtained in this way. The exact chemical composition of litmus is not known (it may be more than one compound.) It is a feeble dye and so is never used as a biological stain, but it has long been used as a **pH indicator.** At one time (late 1800s), litmus was the best-known acid-base indicator. Litmus is *red* below pH 5, *light-blue/gray* around pH 7, and *blue-violet* above pH 8. The pH of litmus milk, after autoclaving, is about 6.8, so it appears light-blue to gray.

Litmus is also an **oxidation/reduction (O/R) indicator.** (See Exercise 18 for an explanation of O/R potentials.) When the O/R potential of the medium is *lowered,* by removing oxygen, litmus loses its color. For example, this happens during the autoclaving of litmus milk. Heat expels oxygen from the medium, thereby *reducing* litmus to its *colorless (leuco)* form. When the medium begins to cool, however, oxygen is reabsorbed, and the light-blue/gray color comes back.

Another example of litmus reduction occurs when bacteria rapidly consume oxygen during growth. Facultative microorganisms surrounded by oxygen support growth with a respiratory metabolism of the lactose, a metabolic activity that rapidly consumes oxygen. As the oxygen supply becomes depleted, the facultative microbes shift to fermentation and produce sufficient acid to form a curd. The curd slows oxygen diffusion into the medium; so a little oxygen is available only in the top part of the tube. This results in a tube that is filled with curd and that has a *red top* (acid litmus in the presence of oxygen) and a *white bottom* (reduced litmus in the absence of oxygen). Thus, litmus in culture media serves both as a pH indicator and as an O/R indicator.

Note that reduced litmus (leuco form) created in the absence of oxygen does *not* function as a pH indicator; only the oxygenated (colored) forms of litmus are pH indicators, and the type of color depends upon the hydrogen-ion concentration (pH).

Microbiologists have employed litmus for over 70 years as an indicator in media used for culturing bacteria. Although, chemically speaking, it is an inaccu-

FIGURE 35-1 Enzymatic hydrolysis (oxidative deamination) of the amino acid alanine, resulting in the production of ammonia. If ammonia is excreted from the cell, it will eventually turn the medium alkaline, and this can be detected by an acid-base indicator. Enzymatic deamination is the first step in the catabolism of all amino acids. The keto acid product (such as pyruvate) is oxidized further and converted to a usable form of energy and/or intermediates for synthesis.

35 MICROBIAL PRODUCTS FORMED IN MILK

rate indicator for either pH or O/R determinations, it has served microbiologists in a way that no other single indicator has been able to do.

Visual inspection of litmus-milk cultures can indicate four major types of microbial activity:

Changes in pH resulting from excretion of acid or alkaline fermentation products

Changes in O/R potential resulting from oxygen depletion

Curdling of milk (denaturation of casein) resulting from large quantities of excreted acid

Proteolysis (hydrolysis) of casein resulting from the action of excreted extracellular enzymes

Note that one microorganism may accomplish one or more of these types of activity while growing in litmus milk.

It is not a simple task to interpret results from litmus milk-grown cultures, but much information can be gained and applied to the identification of unknown microbes and to the understanding of microbial spoilage of milk. Therefore, it is important that you see and learn how to interpret the four major types of microbial activity on litmus milk.

LEARNING OBJECTIVES

- Realize the kinds of energy sources in milk that can be used by different organisms.

- Be able to recognize proteolysis, acid production, coagulation, and reduction occurring in inoculated litmus milk tubes.

- Realize, after observing the results of this experiment, that this test can be used to help identify different bacteria and to better understand the types of microbial spoilage of milk.

MATERIALS

Cultures *Streptococcus lactis*
          Escherichia coli
          Bacillus cereus
          Proteus mirabilis
Media   Litmus milk tubes
Supplies None

PROCEDURE: FIRST LABORATORY PERIOD

1. **Inoculate each bacterial species into a separate litmus milk tube.**

2. **Incubate at 30°C for one week.** You should examine these cultures after *24 hours, two days,* and *one week* of incubation.

3. **Label each tube** with your *name, date* of inoculation, and *culture name.*

4. **Ask your instructor to label and incubate several uninoculated (negative) control tubes** for the entire class so that you can use these to compare with the appearance of the inoculated tubes after each incubation period.

PROCEDURE: 24-HOUR INCUBATION

1. **Observe (do not shake) each tube, and record your observations** according to the guidelines given in the Interpretation Notes.

INTERPRETATION NOTES

No change Color remains light-blue/gray (same as uninoculated control). No change in texture of fluid.

Acid Change in color of litmus from light-blue/gray to pink/red. Continued incubation may yield a white color (reduced litmus) with pink/red color only in a band at the top of the tube. Often accompanied by formation of curd.

Slightly acid Slight change in color from light-blue/gray to pink/red.

Litmus reduction Change in color of litmus from light-blue/gray to white (total lack of color), caused by the lack of oxygen.

Curd Change in milk from fluid to solid. Clear straw- or amber-colored fluid (whey) often evident at surface of curd. Accompanied by reduced litmus (white at bottom) and acid litmus (pink) band near surface.

Alkaline Change in litmus color from light-blue/gray to darker blue. Frequently precedes proteolysis.

Proteolysis Decrease in turbidity of milk (loss of casein colloid) eventually resulting in a clearer, bluish-brown to amber colored fluid. Milk becomes more watery (less viscous).

Compare each tube with an uninoculated control.

Do not shake these tubes at any time during the 7-day incubation.

Check the culture medium for color change and alterations in turbidity and viscosity. To check viscosity, tip the tube gently to one side to see how runny it appears.

2. **Return the tubes to the incubator** for an additional 24-hour incubation.

PROCEDURE: 2-DAY INCUBATION

1. **Observe each tube, and record your observations.** Compare each tube with an incubated, uninoculated control tube.

2. **Return the tubes to the incubator** for 5 more days.

PROCEDURE: 7-DAY INCUBATION

1. **Observe each tube, and record your observations.** Compare each inoculated tube with an uninoculated control tube.

2. **Properly discard these tubes.**

LITMUS MILK TEST SUMMARY	
Purpose of test	Fermentation of lactose and/or proteolysis of casein
Medium used	Litmus milk
Incubation	30°C for 1, 2, and 7 days
Substrates metabolized	Lactose and/or casein
Products formed	Acid, gas, and/or ammonia
Reagents used	None (litmus already in medium)
Positive reactions	Six possibilities (see Interpretation Notes)
Negative reaction	No change in appearance of medium (compare with uninoculated control)

35 MICROBIAL PRODUCTS FORMED IN MILK

NAME _____

DATE _____

SECTION _____

RESULTS

Record your observations of color, texture, and turbidity, and record the meaning of each observation by using the following abbreviations: A = acid reaction, sA = slightly acid, R = litmus reduction, C = curd formation, P = proteolysis, Alk = alkaline, Neg = no change.

Culture	Record	24 hours	2 days	7 days
Streptococcus lactic	Observation			
	Meaning			
Escherichia coli	Observation			
	Meaning			
Bacillus cereus	Observation			
	Meaning			
Proteus mirabilis	Observation			
	Meaning			
Uninoculated	Observation			
	Meaning			

CONCLUSIONS

QUESTIONS

Completion

1. Another name for lactose is _____. Lactose is composed of two covalently linked _____ molecules whose names are _____ and _____.

2. The lactic-acid bacteria are commonly found in mammalian milk. Name two genera of bacteria that are included in this group: _____ and _____.

3. The main protein in milk is called _____. When it is denatured by acid, it is called _____, which is the major constituent in a dairy product known as _____ _____. When denatured or native milk protein is broken down by extracellular enzymes excreted by bacteria, the process is called _____.

4. The two predominate sources of energy in litmus milk are _____ and _____.

5. Litmus serves two functions in this medium: as a _____ indicator and as a _____ indicator (use words only).

6. Casein in the litmus milk may also change in response to two kinds of microbial activity. The two consequences of these activities are called _____ and _____.

True – False (correct all false statements)

1. Lactose can be metabolically broken down by facultative microorganisms using either respiration or fermentation. _____

2. Proteolysis is often followed by catabolism of amino acids and the subsequent release of ammonia. This is why proteolysis is often accompanied by an alkaline reaction. _____

3. The uninoculated (negative) control in this experiment functions to determine if there is any nonbiological change in the pH indicator during incubation. _____

4. Microbiologists use whole milk to prepare litmus milk. _____

5. Litmus is red at alkaline pH values, light-blue/gray at acid pH values, and colorless in the absence of oxygen. _____

6. Bacterial production of lactic acid is a consequence of respiratory metabolism. _____

36

Test for Cytochrome c (Oxidase) and Catalase Activities

Respiratory catabolism (the oxidative, energy-yielding part of respiratory metabolism) is many times more efficient than fermentative catabolism in converting nutrients into usable energy (ATP). Respiratory catabolism involves enzymes that rapidly break down (oxidize) compounds such as sugars via pathways such as the Embden–Meyerhoff–Parnas (EMP) pathway and the tricarboxylic acid (TCA) cycle. These oxidative pathways transfer large numbers of hydrogens and their accompanying high-energy electrons to NAD^+ and $NADP^+$ and then gives them to cytochromes in the electron transport chain (Figure 36-1). Here the electrons are passed from one cytochrome to another; and with each step they transfer some of their energy to the phosphorylation of ADP to ATP. This process is called **oxidative phosphorylation**.

Respiratory catabolism is characterized by rapid and complete oxidation of organic compounds to carbon dioxide and the subsequent removal of the many electrons that are passed through an electron transport chain to a *terminal electron acceptor*. In other words, the terminal electron acceptor removes energy-spent electrons, which are no longer of any use to the cell. If the terminal electron acceptor is oxygen, the process is called *aerobic respiration* (see Figure 36-1); if the terminal electron acceptor is any other inorganic compound (sulfate, for example), the process is called *anaerobic respiration* (see Exercise 33).

This exercise deals with two catalytically active proteins (enzymes) that are involved in the utilization of oxygen by microbes that carry out aerobic respiration: **cytochrome** *c* and **catalase**.

LEARNING OBJECTIVES

- Understand how cytochrome *c* and catalase function for microbial respiratory metabolism.

- Realize that both tests can be used to help identify unknown microorganisms.

- Know how to perform the oxidase test (for the presence of cytochrome *c*) and the test for catalase.

MATERIALS

Cultures   **cytochrome *c* (oxidase) test**
    Escherichia coli
    Pseudomonas aeruginosa
  catalase test
    Streptococcus lactis
    Staphylococcus epidermidis

Media   None

Supplies   **cytochrome *c* (oxidase) test**
    Filter paper
    Loop made with *platinum wire*
    1 percent aqueous solution of tetramethyl-para-phenylenediamine (must be made fresh each day and covered to exclude light).
    Disposable Petri dishes
  catalase test
    3 percent hydrogen peroxide in dropper bottle
    Microscope slides

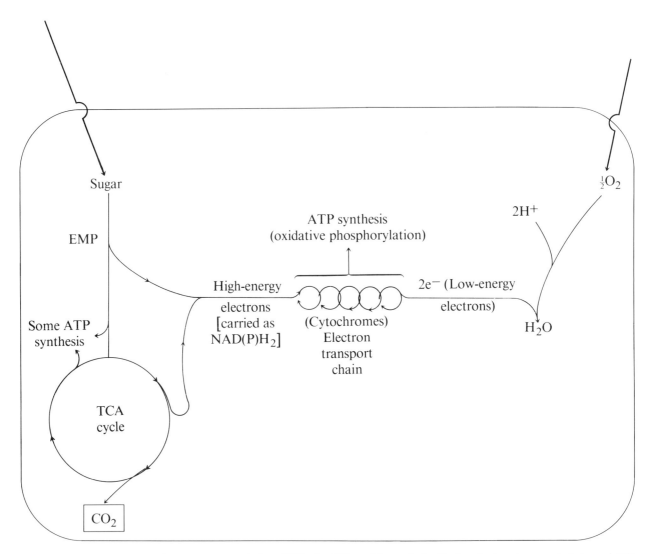

FIGURE 36-1 Overview of respiratory catabolism (EMP plus TCA cycle) coupled with the electron transport chain and ATP production in aerobic respiration. High-energy electrons are removed from the sugar during its catabolic breakdown. These electrons are passed through the electron transport chain to oxygen, and ATP is formed (from ADP and Pi) by a process called oxidative phosphorylation.

A · The cytochrome c (oxidase) test

The oxidase reaction was first observed by P. Ehrlich in 1885 when he noted a blue color at the site of injection of a mixture of alpha-naphthol and dimethyl-*p*-phenylenediamine into animal tissue. F. Rohrmann and W. Spitzer later (1895) reported that this reaction occurs in both animals and plants and that it is caused by an intracellular enzyme that covalently bonds the alpha-naphthol and phenylenediamine (and removes electrons in the process), forming a compound called *indophenol blue;* therefore, they called this enzyme *indophenol oxidase*. In 1924, Otto Warburg described a respiratory enzyme that D. Keilin (1929) demonstrated is the same as indophenol oxidase. Keilin and E. F. Hartree (1938) demonstrated that this enzyme does not react directly with phenylenediamine but with cytochrome *c*; so they renamed it *cytochrome oxidase*. As you may already know, cytochromes are heme-containing catalytic proteins (enzymes) that are tightly bound in the procaryotic cell's plasma membrane.

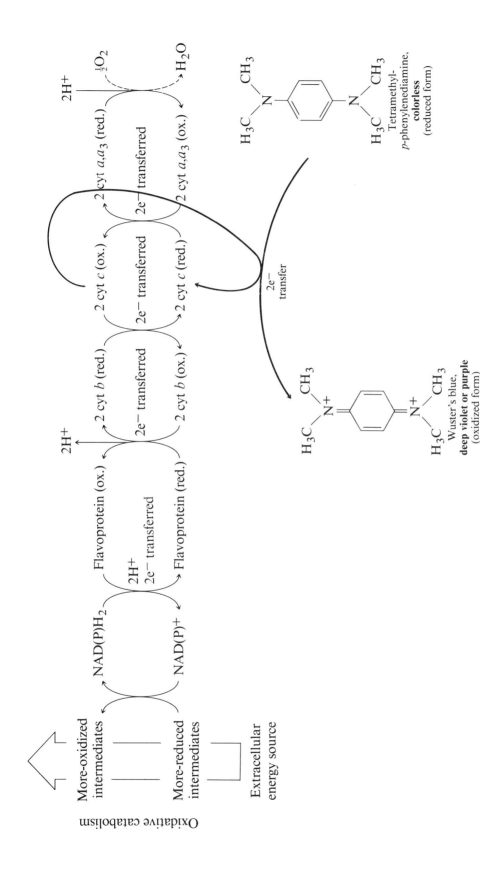

FIGURE 36-2 An aerobic electron-transport chain showing the sequential transfer of electrons from NAD(P)H$_2$ through a flavoprotein and cytochromes b, c, and a,a_3 to oxygen as the final electron acceptor. Cytochrome a,a_3 is the terminal cytochrome in this chain, so it is referred to as the cytochrome oxidase. Note, however, that the oxidase test (application of tetramethyl-p-phenylenediamine) actually tests for the presence of cytochrome c and not the cytochrome oxidase. The designations (ox.) and (red.) refer to the oxidized and reduced forms of the cytochromes, respectively.

It is now known that the oxidase test is actually a test for the presence of cytochrome c. This is somewhat confusing because physiologists and biochemists often refer to the last cytochrome (terminal cytochrome) in the electron transport chain as the *terminal oxidase* or *cytochrome oxidase;* however, the oxidase test does *not* test for the cytochrome oxidase (Figure 36-2). The most sensitive version of this test uses a reagent called tetramethyl-*p*-phenylenediamine. This reagent is an electron donor that can give its electrons to the oxidized form of cytochrome c. The oxidized form of tetramethylphenylenediamine, called *Wuster's blue,* is a dark violet or purple color, whereas the reduced form of this reagent is colorless.

The test for the presence of cytochrome c (the oxidase test) is important in diagnostic bacteriology. For example, the causative agent of gonorrhea *(Neisseria gonorrhoea)* is a gram-negative coccus that contains cytochrome c (oxidase +), which the oxidase test can differentiate from *Moraxella* and *Acinetobacter* species, which are gram-negative cocci or coccobacilli that lack cytochrome c (oxidase −). Another example is that all gram-negative rods known as the enteric bacteria lack cytochrome c (oxidase −), whereas many other gram-negative rods, such as *Pseudomonas* and *Aeromonas* species, contain cytochrome c as part of their electron transport chain (oxidase +).

PROCEDURE

To perform the cytochrome c (oxidase) test (Kovacs' method), carry out the following steps.

1. **Use a ball-point pen to divide a piece of filter paper into three equal sections, and label each section with the name of one organism.**
2. **Place this filter paper inside one-half of a disposable Petri dish, and place a few drops of the reagent** (tetramethyl-*p*-phenylenediamine) **in the center of one section of the filter paper.**
3. **Use a platinum loop (only) to aseptically remove a large quantity of cells from one slant, and rub this culture on the moistened filter paper.**
4. **Check the color of this smear exactly 10 seconds after rubbing the cells on the reagent-moistened filterpaper.** Record a deep violet or purple color developing within 10 seconds as *positive.* Record a light violet or purple color developing within 10 seconds, or a darker color developing after 10 seconds, as *negative.*
5. **Repeat steps 2 through 4 with each of the other two cultures.**
6. **Record your results, and discard the filter paper and Petri dish as contaminated solid waste.**

CYTOCHROME c (OXIDASE) TEST SUMMARY	
Purpose of test	Detect the presence of the plasma-membrane bound, electron transport protein called cytochrome c
Medium used	Culture from any solid medium.
Incubation	Young culture from any solid medium; cytochrome c (oxidase) test requires no incubation.
Substrates metabolized	None (reagent gives electrons to cytochrome c, thereby becoming oxidized)
Products formed	None (after the colorless reagent is reduced, it is called Wuster's blue; see Figure 36-2)
Reagent used	Tetramethyl-*p*-phenylenediamine (colorless)
Positive reaction	Deep violet or purple color within 10 seconds
Negative reaction	No color or a light color within 10 seconds

B · The catalase test

All aqueous environments containing dissolved oxygen (O_2) also contain toxic forms of oxygen, such as the **superoxide radical ($O_2 \cdot$)**. Therefore, if microorganisms survive in aerobic environments, they must be able to detoxify these toxic forms of oxygen. Microbes that grow in environments containing oxygen make an enzyme called **superoxide dismutase** which protonates superoxide to form hydrogen peroxide.

$$2O_2 \xrightarrow{2e^-} 2O_2^- \xrightarrow[\text{Superoxide dismutase}]{4H^+} 2H_2O_2$$

Superoxide radical → Hydrogen peroxide

But, since hydrogen peroxide is also toxic, the microbes produce the enzymes **catalase** to break apart the hydrogen peroxide into water and molecular oxygen.

$$2H_2O_2 \xrightarrow{\text{Catalase}} 2H_2O + O_2$$

Hydrogen peroxide → Water

Thus, catalase is an enzyme produced by and found in essentially all actively growing microbes capable of using oxygen for respiration. In general, bacteria that cannot respire with oxygen do not produce catalase.

The test for catalase in bacteria is simple to perform. You simply place a few drops of 3 percent hydrogen peroxide directly on a colony or on some cells smeared on a dry microscope slide and watch for the vigorous evolution of oxygen bubbles. Catalase exists close to the cell surface and attacks external hydrogen peroxide just as it does that which is formed inside the cell. This test for catalase is best performed on cultures not older than 24 hours because the enzyme seems most active during exponential growth.

The lactic acid bacteria (*Streptococcus*, *Lactobacillus*, and *Leuconostoc* species) carry out fermentative metabolism only and therefore lack catalase. Since certain streptococci are also found in infected throats, clinical laboratories have long used the catalase test to help differentiate these bacteria from other gram-positive cocci. This works well because other common gram-positive cocci, such as *Micrococcus* and *Staphylococcus* species, are catalase positive.

Since strict anaerobic bacteria lack the ability to respire using oxygen, they also lack catalase. Therefore, you can use the catalase test to distinguish between morphologically similar aerobes and oxygen-tolerant anaerobes. For example, both *Clostridium* and *Bacillus* species are gram-positive endospore-forming rods. Some species of clostridia are aerotolerant and are found in environments also containing bacilli. However, species of *Bacillus* are catalase positive, and the clostridia are all catalase negative.

Thus, the test for catalase is an important diagnostic tool for differentiating morphologically similar but metabolically dissimilar microorganisms.

PROCEDURE

To perform the catalase test, carry out the following steps.

1. **Draw a line across the width of a clean microscope slide** with your marking pen. Label one half of the slide *Streptococcus* and the other half *Staphylococcus*.

2. **Use a loop to aseptically remove a large quantity of cells from one slant.** Carefully note the name of this culture.

3. **Apply this loopful of cells to the appropriately marked half of the slide.** Spread the cells over an area no larger than a dime. When you are finished, the cell paste should be very visible to the unaided eye. Do not use water on the slide.

4. **Hold the slide between your fingers, and add a drop of hydrogen peroxide to the cell paste.**

    If you can easily see bubbles being generated in this mixture, call this a *strong positive* reaction.

    If no bubbles are evident, observe the cell paste with a microscope using the 10× objective (only). If you can see bubbles rising from the cell paste only when using the microscope, call this a *weak positive* reaction.

    If you cannot see rising bubbles with your unaided eyes or with the microscope, call this a *negative test*. Remember that this test must not be performed on cultures older than 24 hours because older cultures often give false negative reactions.

5. **Repeat this procedure with the second culture.**
6. **Record your results,** and properly discard your microscope slide.

CATALASE TEST SUMMARY

Purpose of test	Determine presence of the enzyme catalase
Medium used	Culture from any solid medium.
Incubation	Culture must be less than 24 hours old. Catalase test itself requires no incubation.
Substrates metabolized	None (hydrogen peroxide is substrate for enzyme reaction)
Products formed	Oxygen bubbles (and water) from action of catalase on hydrogen peroxide
Reagent used	Hydrogen peroxide
Positive reaction	Oxygen gas evolved (cells appear to be actively bubbling) as detected with the unaided eye or microscope (10× objective).
Negative reaction	No gas bubbles detected either with the unaided eye or microscope (10× objective).

RESULTS A: CYTOCHROME c (OXIDASE) TEST

Microorganism	Test results
Escherichia coli	
Shigella species	
Pseudomonas aeruginosa	

CONCLUSIONS

RESULTS B: CATALASE TEST

Microorganism	Test results
Steptococcus lactis	
Staphylococcus epidermidis	

CONCLUSIONS

36 CYTOCHROME c (OXIDASE) AND CATALASE TEST

NAME _____

DATE _____

SECTION _____

QUESTIONS

Completion

1. Energy obtained from electrons passed down the electron transport chain is used (in part) to form ATP from ADP and inorganic phosphate. This type of ATP formation is called _____ phosphorylation.

2. Respiratory metabolism has two characteristics: rapid and complete _____ of organic compounds to _____, and passage of hydrogen (electrons) to an inorganic terminal electron acceptor.

3. If the terminal electron acceptor is oxygen, this type of metabolism is called _____ respiration. If the terminal electron acceptor is not oxygen, it is called _____ respiration.

4. Catalase is an enzyme produced by most microbes capable of aerobic respiration. The substrate it acts upon is called _____, and the two products formed are _____ and _____.

5. To test for microbial production of catalase, you apply _____ to the cells. A positive test is one in which _____.

6. Why is no negative control used in the catalase-test experiment?

7. The so-called oxidase test does not test for the terminal cytochrome oxidase; instead, it tests for the presence of _____ in the bacteria treated with the reagent.

8. The colorless reagent used in the cytochrome c (oxidase) test passes electrons to the oxidized form of a reagent called _____, and the newly oxidized reagent exhibits a _____ color.

9. Enteric bacteria are cytochrome c (oxidase) _____, whereas *Pseudomonas* species are cytochrome c (oxidase) _____.

True–False (correct all false statements)

1. Respiration is much more efficient than fermentation in converting nutrient energy sources at ATP. _____

2. A positive catalase reaction indicates that the microbe is capable of aerobic respiratory catabolism. _____

3. *Streptococcus* species are catalase negative, but *Staphylococcus* species are catalase positive. _____

4. The reagent used for the cytochrome c (oxidase) test is called tetramethyl-*p*-phenylenediamine, and it is colorless in the reduced state. _____

5. The test for cytochrome c (oxidase) is useful in differentiating pure cultures of enteric bacteria from other types of gram-negative, rod-shaped bacteria. _____

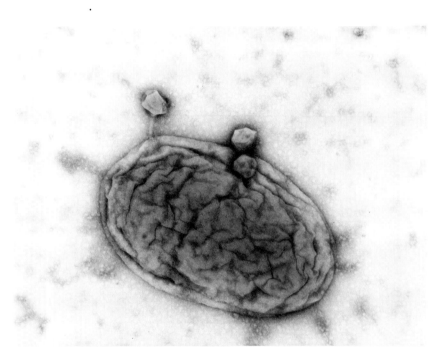

Phage (viruses) attached to *Gluconobacter oxydans* cell.

*So, naturalists observe, a flea
Hath smaller fleas that on him
   prey;
And these have smaller still
   to bite 'em;
And so proceed ad infinitum.*
                        —*Jonathan Swift*

BACTERIAL VIRUSES AND MICROBIAL MUTATIONS

Part IX introduces you to viruses, their lytic and lysogenic pathways in the host cell, and how they are enumerated, and also to the lethal and mutagenic effects of ultraviolet light and certain chemicals on bacterial DNA.

37

Enumeration of Lytic Viruses: The Plaque Assay

A *virus,* by definition, is one type of obligate intracellular parasite, which means that viruses reproduce only inside living cells. The eucaryotic or procaryotic cell type that supports growth of a virus is called the virus's **host.** So far as is known, all cellular types— higher plants and animals, algae, fungi, and bacteria —can be infected by viruses.

There appears to be no single virus capable of attacking more than one cellular type. For example, an animal virus does not infect plant cells, nor does a bacterial virus attack yeast cells. The number of different host types (for example, the number of different strains of bacteria that a virus infects) is called the **host range** of that virus. The host range of most viruses is very limited. For example, not only do bacterial viruses attack only bacterial cells, but one type of bacterial virus usually infects only one strain of a bacterial species. One can accurately say that the *host range specificity* of a virus is very narrow.

Viruses that infect bacteria are called **phages** (pronounced FAAJES), or more specifically, **bacteriophages,** and they are usually found wherever their host cells are found. For example, phages for bacteria that occur in human intestines are also found in the intestines, and phages for bacteria inhabiting the soil are found in the soil.

Because bacteria are so easy to work with in the laboratory, bacterial viruses serve as experimental models for many host/virus interactions.

Bacteriophages exhibit one of two types of interactions with a host (Figure 37-1): the **lytic (reproductive) pathway** or a phenomenon known as **lysogeny.** Which interaction is exhibited depends on the characteristics of both the phage and the bacterium it infects. Let us separately consider these two host–virus interactions.

Examine the *lytic (reproductive) pathway* in Figure 37-1. First, the phage makes contact with the host cell (*adsorption;* step 1). Then the DNA enters the cell (*penetration;* step 2). Note that the protein parts of the phage remain outside the host cell. The genome from the lytic phage carries the information necessary for the synthesis of new viral particles in the infected bacterium. The lytic phage genome immediately shuts down the cell's own synthesis, and directs the cell to make new phage DNA (step 4) and the proteins necessary for coating the DNA (step 5). Steps 4 and 5 are collectively called *replication.* Mature phages are made (*assembly;* step 6). Finally, when all phage parts are assembled into mature, infective phages, the host cells lyse, and release the new phages (step 7). The new bacterial viruses are free to infect other cells and repeat this process.

Note that phage replication can only occur at the expense of young, actively growing bacteria. Phages can not reproduce themselves in nongrowing bacteria.

When lysis occurs in a broth culture, it causes partial or complete clearing of a turbid suspension of growing cells. When lytic phages infect a large industrial bioreactor used to grow bacteria that make a valuable product, the results can be costly, both in time and money. Consequently, industries go to great lengths to develop phage-resistant bacteria.

The number of bacteriophages in medium can be counted by a technique called the **plaque assay,** which is based on the ability of bacterial viruses to infect growing cells and cause some or all of them to lyse,

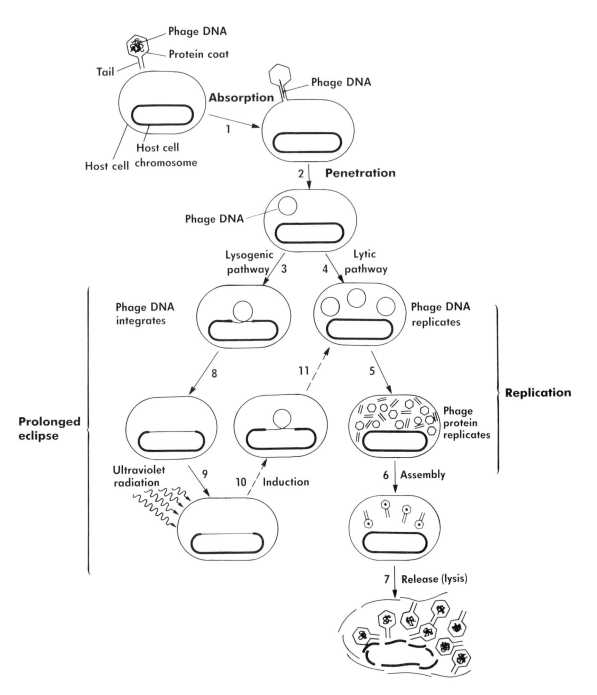

FIGURE 37-1 Schematic diagram of two methods of phage infection of bacterial cells: the lysogenic pathway (steps 1, 2, 3, and 8) and the lytic (reproductive) pathway (steps 1, 2, 3, 4, 5, 6, and 7). (Adapted from M. Ptashne et al., *Scientific American*, November 1982.)

producing even more viruses. When this occurs in a seeded-agar overlay (Figure 37-2), the result is a zone of lysis or clearing called a **plaque**.

Let us now examine how a **seeded-agar overlay** is prepared. First, a tube of semisolid (soft) nutrient agar is autoclaved and held at a temperature just warm enough to keep it from solidifying (42° to 45°C). You may recall that this process is called *tempering*. Bacterial cells are added (seeded) into this liquified agar, phages are mixed with the suspension, and then the entire mixture is poured over a **base layer** of nutrient agar. The seeded overlay hardens on top of the base layer, and the plate is incubated. During incubation, bacteria begin to grow throughout the overlay, thereby creating a uniformly cloudy appearance, sometimes called a baterial **lawn.** On the other hand,

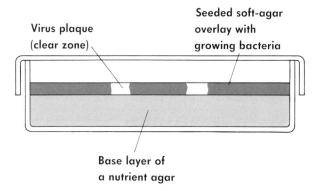

FIGURE 37-2 Side view of a seeded-agar overlay plate used for phage plaque assay. Tubes containing autoclaved and tempered semisolid medium (soft agar) are inoculated (seeded) with actively growing bacteria and bacteriophages. The suspension is poured onto the surface of a solid nutrient agar plate (base layer), allowed to solidify, and incubated to allow growth of the bacteria in the soft-agar overlay (lawn). Where phages are present in the overlay, cell lysis occurs, and a clear zone (plaque) results. the number of plaques represent the number of phages originally added to the soft-agar overlay.

wherever a virus occurs, these cells become infected, lyse, and produce more phages, which infect and lyse more cells. The bacteria are immobile in this soft-agar overlay, but the much smaller virus particles freely diffuse through it. Wherever a single lytic virus exists, its continuing cycle of lysis/reinfection/lysis produces a **plaque,** a zone of clearing in the lawn. The number of plaques in the overlay equals the number of lytic viruses; thus the lytic viruses in a culture are enumerated. Very lytic phages produce very clear plaques, but not all phages are lytic phages.

Lytic animal and plant viruses are counted in a similar manner, except that cells from animal or plant tissue must be used as hosts instead of bacteria. Whenever plant or animal cells are separated from tissue and grown in an artificial medium, the preparation is called **tissue culture** to distinguish it from methods of growing free-living microorganisms. As these animal or plant cells grow, they cling to the inside surface of the culture vessel, and plaques form within the tissue-culture cell layer.

The *lysogenic pathway* involves infection of bacterial cells with viruses called **temperate phages;** these are bacterial viruses that are not totally lytic. Instead, most temperate phages establish a condition known as *lysogeny* when their host cells are infected, and only a few of the temperate phages replicate inside their hosts and lyse these cells.

In the lysogenic pathway (see Figure 37-1), a bacterium is first infected with a temperate phage (adsorption; steps 1 and 2). Note that the first two steps are the same as for the lytic pathway. Once the temperate phage's DNA (chromosome or genome) is inside the bacterial cell (penetration; step 2), it becomes incorporated into the cell's chromosome (step 3); the phage genome is then called the **prophage,** and the infected host is called a **lysogenic** bacterium. The prophage remains in the chromosome and is replicated along with the host cell's chromosome during subsequent cell divisions. Occasionally, something *induces* the prophage to take over a cell's metabolic machinery and stimulate it to make phages (steps 9, 10, 11, 5, 6, and 7). In other words, the prophage can be induced to enter the lytic pathway. Sometimes this occurs spontaneously, but, in the laboratory, we often promote **induction** by subjecting cells to chemical or physical *mutagens,* such as ultraviolet radiation.

Some microbiologists believe that temperate phages are far more numerous than lytic phages in nature, and that lysogeny is the more common behavior of bacterial viruses. Infection of cells by temperate phages may give the new hosts a selective advantage over their noninfected neighbors if the prophages accidentally carry new bacterial genes from a previous host cell. Using temperate phages to transfer new bacterial genes from one cell to another is called **transduction;** it may be very common in the bacterial cell's natural environment.

Temperate phages are more difficult to study than lytic phages in a beginning microbiology laboratory because their effect on a bacterial culture is usually more subtle and more difficult to measure. For example, temperate phages produce plaques that are only slightly less turbid than the surrounding seeded-agar overlay and thus are much more difficult to see than the clear plaques of lytic phages. Thus, lysis is seldom a consequence of the lysogenic pathway (see Figure 37-1); however, at any given time, a culture of lysogenic bacteria will have some cells in which the prophages are spontaneously entering the lytic cycle. But only those few cells will lyse. Therefore, only partial clearing of a seeded-agar overlay (a cloudy plaque) will occur at the site of infection with a temperate phage.

Because viruses are a part of our environment and affect us in so many ways, it is important to understand something of how microbiologists enumerate and work with these submicroscopic, parasitic particles.

LEARNING OBJECTIVES

• Realize that animals, plants, and bacteria all have viruses and that these intracellular parasites are specific to the host type.

- Understand basic aspects of the lytic cycle and that not all viruses are lytic.

- Understand how lytic viruses form plaques and how their numbers are determined by a plaque assay.

MATERIALS

Cultures *Escherichia coli* phage T-1 (the viruses)
Escherichia coli strain B (the host cells)

Media Trypticase-soy agar plates (for basal medium)
Soft-agar tubes (for preparing overlays)
9-ml trypticase-soy broth dilution blanks (for diluting phages)

Supplies 1.0-ml sterile pipettes
Waterbath set at 45°C.

PROCEDURE: FIRST LABORATORY PERIOD

Review the section called "Principles of Serial Dilution" in Exercise 20, and practice these principles in this exercise.

1. **Read this entire procedure before you begin.** This is a complex procedure that requires much coordination in a classroom. We assume that you are working with at least one partner so that, as a group, you can accomplish several things at the same time. Take time to plan together before you begin your work.

2. **Obtain four agar plates, and label them** 10^{-2}, 10^{-3}, 10^{-4}, and *uninoculated control*. (Figure 37-3). Add your *name*, experiment *number*, and *date* inoculated.

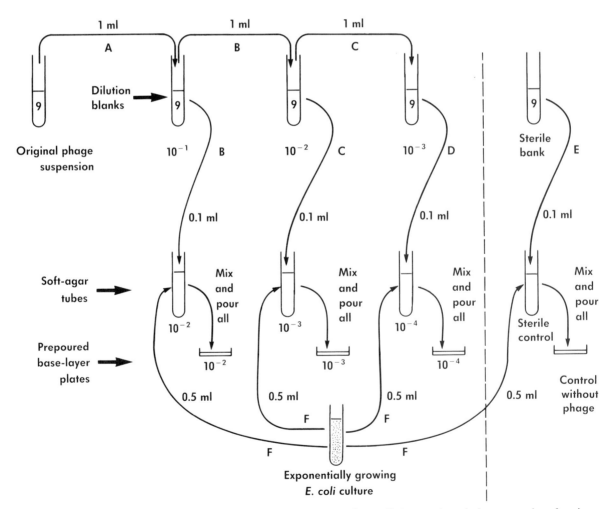

FIGURE 37-3 Serial dilution scheme and inoculations for preparing phage dilutions and seeded-agar overlays for phage enumeration.

3. **Obtain four 9-ml dilution blanks, and label them** (10^{-1}, 10^{-2}, and 10^{-3}) (Figure 37-3). Note that these are for diluting the phage suspension only.

4. **Obtain the phage suspension.** Your final objective is to calculate the number of phages in this (original) phage suspension. Note that the phage suspension is optically clear (not turbid). Unlike bacteria and fungi, virus particles are too small to scatter light; thus their concentration cannot be estimated by determining turbidity.

5. **Label four soft-agar tubes** 10^{-2}, 10^{-3}, 10^{-4}, and *sterile control,* like those shown in Figure 37-3.

6. **Note that the phage dilutions, the addition of host cells and phages to the soft agar, and the spreading of the seeded-soft agar must all be done at about the same time. Plan ahead!**

7. **Add *E. coli* to the soft-agar tubes.** You are supplied with an actively growing *E. coli* broth culture and tubes containing a sterilized preparation of 0.7 percent agar in nutrient broth. This agar preparation is called *soft agar* because it has only about half of the concentration of agar normally used in solid media (1.5 percent). These soft-agar tubes are sterilized (and liquified) in an autoclave and then tempered in a 45°C waterbath.

    Recall that agar becomes liquid at 100°C but does not gel (solidify) until cooled to 38° to 40°C. This characteristic enables you to keep agar in the liquid state at 45°C until you can mix the host cells and phages with it. This temperature is also low enough so that most bacteria and phages are not harmed by short exposures to those temperatures.

    Add 0.5 ml of *Eschericia coli* to all soft-agar tubes, but only after your partner indicates that it is time to do so. You should not expose the cells to these elevated temperatures any longer than necessary.

8. **Begin diluting the phage suspension,** as shown in Figure 37-3. Each capital letter in that figure (A–F) refers to one pipette needed to make aseptic transfers. Note that you need only six pipettes to accomplish these transfers because some pipettes are used more than once. For example, pipette B is used to make two transfers from the 10^{-1} phage dilution. **Do not lay the pipette down between transfers.**

9. **Inoculate phage-dilutions with *E. coli*.** As partner I prepares the 10^{-1} phage dilution, partner II should inoculate only one soft-agar tube (labeled 10^{-2}) with *E. coli* (see Figure 37-3). Partner III, if the distance is great, can deliver the coli-inoculated 10^{-2} soft agar tube to partner I, so that pipette B can be used to transfer phage from the 10^{-1} tube to the coli-inoculated 10^{-2} soft-agar tube immediately after 1 ml of phage suspension is transferred from the 10^{-1} to the 10^{-2} phage dilution tube.

    Partner II should **add *E. coli* cells to one soft-agar tube at a time,** only as needed, so that the cells are not subjected to a elevated temperature longer than necessary.

10. **Thoroughly mix bacteria and phages in the soft-agar tubes immediately after each addition.** Do not shake the tubes, but rotate them vigorously for about 10 seconds by rolling the tubes between the palms of your hands. You need to mix this suspension thoroughly, but do not introduce bubbles into the agar because bubbles in the solidified-agar overlay will look like small, clear plaques.

11. **Pour the seeded-agar overlay** immediately after mixing the host cells and phage suspension for *each* dilution. **It must be poured while the soft-agar suspension is above 40°C,** otherwise it will begin to solidify, making it impossible to evenly cover the surface of the base-layer plate.

    a. Partially raise the plate cover with one hand, and, using good aseptic technique, quickly pour the entire contents of the tube onto the surface of the agar.

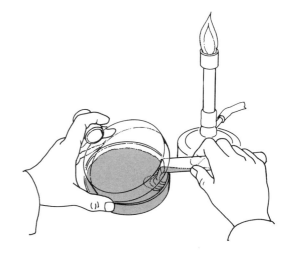

b. Immediately cover the plate, and slide the plate on the bench top with a slow, smooth, circular motion. This swirling action should evenly distribute the soft agar over the surface of the plate without slopping it on or over the edge of the plate. As soon as the base-layer agar is covered, stop this motion to let the soft-agar overlay harden.

c. If an inoculated, tempered, soft-agar overlay tube should happen to solidify before you can pour it on the base-layer plate, discard it as you would any other contaminated tube, and prepare a new tube.

12. **Note that each plate must be prepared separately** in the manner stated above. When you have finished your preparation, you should have (1) three seeded-agar overlay plates, each containing *E. coli* and one of the 10^{-2}, 10^{-3}, or 10^{-4} phage dilutions and (2) one seeded-agar overlay plate containing only *E. coli* (no phage).

13. **Incubate these plates at 30°C for 6 to 10 hours only.** This incubation time is *very* important because plaques will enlarge and fuse together during a longer incubation time, making it impossible to count individual plaques. If the phages are too numerous, the plaques will fuse to form what is called *confluent lysis*, which results from total elimination of the host cells on the plate. **Before you leave the laboratory,** make sure you know who is responsible for refrigerating these plates after 6 to 10 hours incubation.

14. **Properly discard your used materials, and clean all work areas.**

PROCEDURE: SECOND LABORATORY PERIOD

1. **Examine all phage-dilution plates for the presence of plaques.** Compare these plates with the seeded overlay lacking phage (the control plate).

2. **Count the number of plaques on each plate** using the 30-to-300 rule (review Exercise 20).

3. **Calculate the number of phages in the original phage suspension,** and record it as the number of plaque-forming units per milliliter (pfu/ml) of the original suspension.

4. **Write your results on the board** so that the class results can be compared and discussed.

5. **Record your results and the class results; then properly discard your plates.**

"Biggest damn virus I've ever seen!"

37 ENUMERATION OF LYTIC VIRUSES: THE PLAQUE ASSAY

NAME _____

DATE _____

SECTION _____

RESULTS

Place *your* results in the following table.

Dilution plated	Number of plaques per plate	Phages in original suspension (pfu/ml)
10^{-2}		
10^{-3}		
10^{-4}		
Control		

Record the *class* results in the following table.

Student or group number	Calculated phage numbers in original suspension (pfu/ml)
1	
2	
3	
4	
5	
6	
7	
8	
9	
10	
11	
12	
13	
14	
15	

Class average	

CONCLUSIONS

QUESTIONS

Completion

1. A virus is one type of obligate intracellular _____.

2. Viruses that infect bacteria are commonly called _____.

3. The process by which a virus enters a cell and reproduces itself is called the viral _____ or _____ cycle.

4. The zone of lysis or clearing in a seeded-agar overlay is called a _____.

True – False (correct all false statements)

1. The process of holding an agar medium in a liquid state (from 42° to 45°C) is called *tempering*. _____

2. Lysogenic phage can be induced to continue the replicative cycle by treating the lysogenized bacterial culture with a chemical or physical mutagen. _____

3. Not all viruses are lytic. _____

4. Lytic viruses can transfer host cell chromosomal material. _____

38

Using Mutants to Detect Potential Carcinogens: The Ames Test

We now know that some forms of cancer in humans are caused by certain types of chemicals. Therefore, it is important that we have a rapid and inexpensive method for testing the large number of chemicals that we release into our environment. However, to test each chemical for its **carcinogenic** (cancer-causing) potential in laboratory animals often takes from 2 to 3 years and costs more than $100,000 per chemical. Considering the thousands of synthetic chemicals that must be tested and the cost of each test, it is apparent that animal testing is too time-consuming and expensive for routine use.

Because 90 percent of all carcinogens have proved to be mutagens (which suggests that human cancers result from a mutantlike alteration in the cellular DNA), Bruce Ames and colleagues at the University of California developed a test that uses special bacteria that are very sensitive to **mutagenic agents** (chemicals that cause mutations). Scientists are now using the Ames test to screen many chemicals quickly and inexpensively to determine which are mutagenic and, therefore, potentially carcinogenic. Those few chemicals that are mutagenic are then tested more extensively in animals to see if they are also carcinogenic. The Ames test takes only about three days and costs only several hundred dollars per chemical tested.

The Ames mutants

Nonmutated cultures of the bacterium *Salmonella typhimurium* grow in media without amino acids. This is possible because they have pathways for making all of the amino acids needed to make proteins. Each amino acid is made from an intermediate formed during the breakdown (catabolism) of the cell's energy source. Each amino acid has a separate pathway for synthesis, and each pathway begins from one catabolic intermediate. For example, Figure 38-1 shows the pathway for **histidine** synthesis, which begins with catabolic intermediate C and uses numerous enzymes (1–9) to convert C to histidine.

In contrast, growth of the Ames mutants of *Salmonella typhimurium* requires the presence of the amino acid histidine. This means that a mutation has occurred in one of the genes that produces an enzyme used in the pathway for histidine synthesis (schematically shown in Figure 38-1). This mutation is so severe that the gene can no longer code for a functional enzyme. Without this enzyme, the cells no longer convert the catabolic intermediate to histidine. Therefore, the Ames mutants grow only if the growth medium supplies histidine. These mutants are called *histidine-dependent* or *his⁻ mutants* because they depend on the medium to supply them with this amino acid.

The Ames mutant that we use in this exercise has two characteristics that are important for you to understand. First, it is a **deletion mutant,** which has a single nucleotide deleted from the DNA strand that codes for the synthesis of one gene in the mutant's histidine biosynthetic pathway. Second, this Ames mutant lacks a **DNA excision–repair mechanism,** which would normally correct such deletions.

A mutation resulting in a deletion of a single nucleotide from the **sense strand** (+) of the DNA molecule causes a reading-frame shift in such a way that protein synthesis terminates at that point during transcription (Figure 38-2). Protein synthesis stops because the triplet that follows the deletion appears as a nonsense triplet during translation. Thus, a critical enzyme in the pathway for histidine biosynthesis is

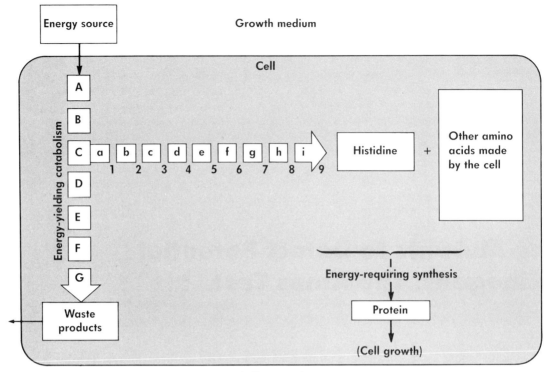

A Prototroph

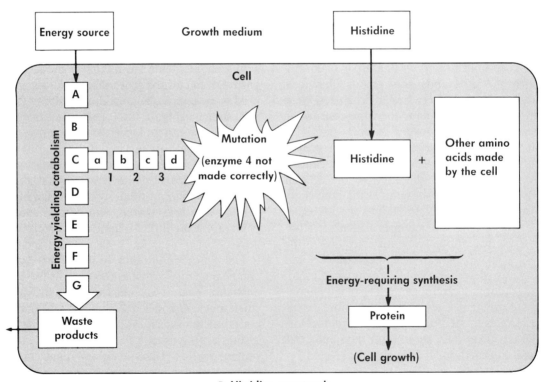

B Histidine auxotroph

FIGURE 38-1 Schematic overview of *Salmonella typhimurium* metabolism and the effect of a nonrepaired deletion mutant on histidine biosynthesis. (A) Metabolism of prototrophic (nonmutated) *S. typhimurium*. Here, the catabolism (breakdown) of the energy source produces both the energy and the precursor (C) needed for histidine synthesis. The formation of each intermediate in the histidine pathway (a–i) is catalyzed by a different enzyme (1–9). Each enzyme is a protein whose synthesis is directed by one structural gene on the histidine operon. Any mutation that results in the lack of a correctly functioning enzyme in this pathway also results in no histidine being made by the cell. (B) Metabolism of an auxotrophic mutant of *S. typhimurium*. Here, the catabolism (breakdown) of the energy source still produces the energy needed for synthesis and the precursor (C) needed for histidine synthesis, but one of the enzymes in the pathway for histidine synthesis is not correctly made. This nonfunctioning enzyme keeps the pathway for histidine synthesis from working correctly, which results in the inability of the mutants to make their own histidine. Consequently, these his^--mutants can grow only if the growth medium supplies the histidine.

38 POTENTIAL CARCINOGENS: THE AMES TEST

> **TECHNICAL NOTE**
>
> We use only one Ames mutant in this exercise. In the actual Ames test, many different mutants of *Salmonella typhimurium* may be used. Dr. Bruce Ames and his colleagues have assembled a series of strains of histidine auxotrophs, each of which is auxotrophic because of a different type of mutation in the histidine operon. All appear to be either frame-shift or base-substitution mutants, and each is thought to test for a different type of reversion (back mutation). To simplify the test, we use only one type of mutant, as described in the text. An excellent summary of a recommended procedure for the Ames test is given by B. C. Carlton and B. J. Brown, "Gene Mutation," in *Manual of Methods for General Bacteriology*, P. Gerhardt et al., Eds., Washington, D.C.: American Society for Microbiology, 1981.

not made, and the mutant requires histidine for growth. Nutritional mutants such as this are called **auxotrophs**. Since the Ames mutants lack the ability to make histidine *(his⁻)*, they are more specifically called **histidine auxotrophs**.

The fact that these histidine auxotrophs lack a mechanism for repairing nucleotide deletions means that these mutants are quite stable. If normal cellular repair were possible, these mutants would quickly revert to their prototrophic (parental) state, and histidine would no longer be required for growth.

All that would be necessary to return these auxotrophs to the prototrophic state by mutation would be for the cell to add the deleted nucleotide at that point on the gene. When that happens, it is called an **insertation mutation, back mutation,** or **revertant** because this second mutation returns the mutant (or reverts it) to the parental or prototrophic state.

Mutagens are chemicals that increase the frequency (rate) of mutation above that which occurs with no apparent cause (the spontaneous rate). Thus, mutagens increase the frequency of back mutation (reversion) of the Ames mutants. The stronger the mutagen, the greater the number of reversions; so the number of revertants reflects the strength of the mutagen. Strong mutagens are also likely to be potent carcinogens.

Selection of revertants

To determine the number of revertants following exposure to a mutagen, we must have a way to differentiate the mutants and the spontaneously formed revertants. The Ames test uses a *chemically defined (synthetic) medium* for this purpose. This is a medium for which we know both the type and the quantity of each chemical compound (nutrient) because we have added precise amounts of known chemical compounds to distilled water in preparing the medium. In one procedure, the Ames mutants are inoculated onto the surface of plates containing the chemically defined medium, and then the cells are immediately exposed to the mutagenic agent. This is followed by an incubation to allow colony formation by the revertants.

The chemically defined medium used for the Ames test contains only trace (growth-limiting) amounts of histidine. When we plate the Ames mutants on this medium, they grow only until they run out of histidine. If the quantity of cells plated is high, growth should appear only as a slight haze of confluent background growth on the plate (a faint lawn). On the other hand, the revertants should form normal-sized colonies because their growth (colony size) is not limited by the amount of histidine present in the medium. Trace amounts of histidine in the medium are preferable because some mutagenic agents act only on replicating DNA. Thus, the medium should have enough histidine to support a few rounds of DNA replication, to allow mutation to be reflected in both (+ and −) strands and to permit a few generations of growth but not enough histidine to permit colony formation.

The Ames test

If you were working for a testing laboratory, you would be required to follow a strict set of standard methods set up by Ames and his colleagues. These standards are defined so that everyone follows the same procedures and reliable results are obtained. The actual Ames test involves many controls, various mutant strains, and various concentrations of the chemical to be tested. The purpose of this exercise is to familiarize you with the theories involved and the general methods used to perform the test, rather than to teach you how to perform it correctly.

In general, scientists in testing laboratories perform the Ames test in the following way. Every chemical to be tested is treated with an induced liver-enzyme preparation; this *activates* the test chemical (if it is one that is normally converted to a mutagen by the liver). An identical quantity of an Ames mutant is spread on a plate that contains a trace (growth-limiting) concentration of histidine. Then, a filter-paper disk is saturated with either distilled water (control) or one of several concentrations of the "activated" test chemi-

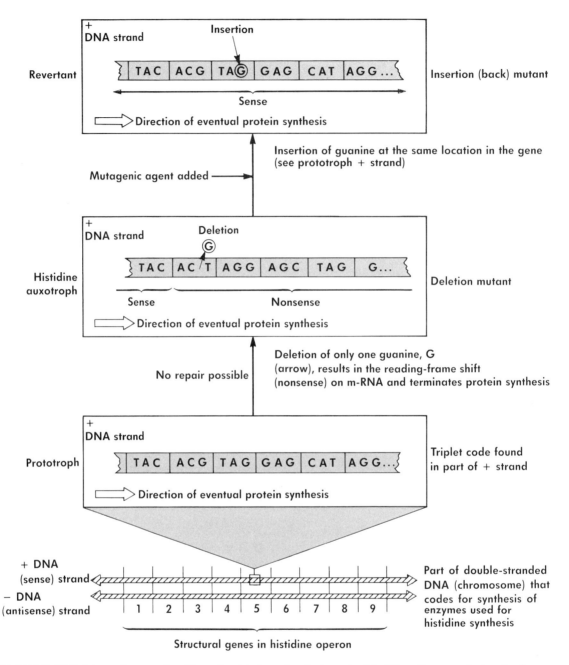

FIGURE 38-2 Schematic example of how the deletion of a single nucleotide (G) on the gene that codes for one enzyme can cause auxotroph formation and how the reinsertion of that nucleotide by mutation can cause the revertant to behave as a prototroph.

cal, after which the disk is placed in the center of a plate. The test chemical diffuses outward from the disk forming a concentration gradient. If the test chemical is a mutagen or is converted to a mutagen by the liver enzymes, then some of the Ames mutants revert (back mutate) and form colonies on the plate. Stronger mutagens cause larger numbers of revertants, and their colonies form further away from the disk. Weaker mutagens cause fewer revertants, and their colonies form closer to the disk. In other words, the reversion frequency depend on both the mutagen's *concentration* and its *potency* as a mutagen. Plates containing disks saturated with distilled water demonstrate the rate of spontaneous mutation; thus they serve as negative controls for the experiment.

Quantitative analyses of the mutagen's strength can be obtained by adding known numbers of cells and a known concentration of the test chemical to the agar

38 POTENTIAL CARCINOGENS: THE AMES TEST

TECHNICAL NOTE

Many chemical substances are not mutagenic (carcinogenic) until after they are consumed by an animal. The chemical changes that make them carcinogenic often take place in the liver after it has been induced to form detoxifying enzymes. These chemical conversions are part of a process that normally allows the liver to detoxify chemicals that would otherwise be harmful. However, some chemicals react with these detoxifying enzymes (oxygenases) to form epoxides that then react strongly with cellular DNA. Thus, the body may convert relatively harmless chemicals into potent mutagens (carcinogens).

For this reason, all test chemicals being examined with the Ames test are routinely treated with an induced, rat-liver enzyme preparation prior to adding the chemical to the plate containing the Ames mutants. Rats are induced to produce detoxifying enzymes in their livers by injecting them with a commercial preparation called Aroclor. The rats are then killed, their livers are removed and homogenized, and this homogenate is processed to make it suitable for the Ames test.

You will be testing a class of chemicals known as the nitrocarcinogens. These chemicals are also relatively harmless when they are consumed. However, induced-liver enzymes are not needed to work with nitrocarcinogens. Instead of being converted to mutagens in the liver, the intestinal bacteria convert the nitrocarcinogens to nitrosamines, which are potent mutagens (carcinogens). Since the Ames mutants are intestinal bacteria, and we use only nitrocarcinogens in this exercise, we delete the liver-homogenate treatment step.

LEARNING OBJECTIVES

- Learn what is meant by the terms *auxotroph, prototroph,* and *revertant* and how each term applies to the Ames test.

- Understand what is meant by a chemically defined medium and why one is needed in the selection procedures used in the Ames test.

- Know the basic steps and principles involved in the Ames test.

- Know why the Ames test is effective in screening for potentially carcinogenic chemicals.

MATERIALS

Cultures	*Salmonella typhimurium,* Ames strain TA98
Media and solutions	Ames minimal-medium plates (for basal medium)
	Ames soft-agar tubes (must be supplemented with Ames trace-histidine solution before overlays are used)
	Ames trace-histidine solution (for the instructor's use in preparing the Ames supplemented soft-agar overlay medium).
	Trypticase-soy agar plates (for instructor's use only)
	Mutagen solutions (nitrocarcinogens; *for the instructor's use only;* chemicals must be placed in screw-capped tubes with contents clearly labeled)
Supplies	0.1-ml pipettes (sterile)
	Waterbath set at 45°C

overlay during its preparation (instead of adding the chemical to a paper disk and letting it diffuse out at unknown concentrations).

The Ames test has been applied to thousands of chemicals such as industrial reagents, cosmetics, food additives, hair dyes, and pesticides. Potent mutagens have been found in each chemical category. A high frequency of reversion, however, does not always mean that the chemical is a *carcinogen.* Ultimate proof of animal carcinogenicity is obtained only from animal testing. On the other hand, only a few of the more than 300 known animal carcinogens have failed to increase the reversion frequency of the Ames mutants.

PROCEDURE: FIRST LABORATORY PERIOD

> **WARNING!**
>
> If students are allowed to perform this exercise, they must not handle vessels containing stock solutions of mutagenic chemicals (potential carcinogens), nor should they be allowed to transfer (pipette) stock solutions of these chemicals. Stock solutions of mutgenic chemicals must be handled by the instructor only! Students who handle overlay agar containing suspected mutagenic chemicals must wear protective gloves and take other precautions to avoid contact with the skin. All vessels that contain suspected mutagenic chemicals should be given special treatment so that the chemical residues can be properly discarded. This includes (but is not limited to) containers used for preparation and storage of stock chemicals, pipettes used for chemical transfers, both spent and unused soft-agar overlay tubes, plates that were discarded before incubation because of improper preparation, and properly prepared plates examined after incubation.
>
> All procedures used for this exercise must receive prior approval from your college or university health and safety office, and all student use of suspected mutagenic agents must follow the strict guidelines established by the United States Environmental Protection Agency. Students should be shown written documentation of approval by the proper college or university officials before beginning this experiment.

Your instructor should aseptically add either distilled water or a mutagen solution and the Ames trace-histidine solution to each melted and tempered tube containing the Ames soft-agar. The contents of each tube should be thoroughly mixed; then each tube should be returned to the waterbath so that the agar remains in a liquid (tempered) state.

1. **Obtain four Ames minimal-medium plates, and label each plate** with your *name,* today's *date,* and the exercise *number.* Label one plate with the word *control* and the remaining three plates with the words *chemical 1, chemical 2,* and *chemical 3,* respectively.

2. **Determine the location of the soft-agar overlay tubes.** Prior to class, your instructor will have heated these tubes to liquify the agar, then added the trace-histidine solution to each tube, and, finally, added either distilled water (control) or one of the test chemicals (1–3) to different tubes. You will need to work with one of each of these four separately labeled tubes. Ask your instructor the names of the chemicals referred to as 1, 2, and 3.

3. **Record the name of each chemical** in the appropriate place in your Result section.

4. **Aseptically add 0.1 ml of *Salmonella typhimurium* culture to the tube of liquified soft-agar** labeled *control.*
   Thoroughly mix this suspension by rotating the tube between the palms of your hands. **Immediately pour the contents of this tube onto the appropriately labeled Ames minimal-medium plate,** and **swirl the plate gently to cover the surface** of the minimal agar base layer.

5. **Repeat step 4** with each of the other labeled soft-agar tubes (1–3) until all four labeled plates are covered with an inoculated overlay.

6. **Ask your instructor to prepare four identical plates using T-soy agar** instead of Ames minimal-medium agar as the basal medium. One set of four plates (three mutagens and one control) will be sufficient for the entire class.
   The purpose of these plates is to demonstrate that a histidine-deficient medium is necessary for detecting a chemically induced increase in mutation frequency in this culture.

7. **Incubate all plates at 30°C for 48 to 72 hours.**

PROCEDURE: SECOND LABORATORY PERIOD

1. **Select the Ames minimal agar plate that contains the greatest number of distinguishable colonies.** If there are too many large colonies on this plate to count accurately (more than 300 per plate), divide the plate into halves or fourths with your marking pen. count the number of large colonies on this plate (or plate fraction), and record this number in the Results section.

2. **Count and record the number of large colonies on each plate** (or the fraction of each plate) for all remaining Ames minimal agar plates and all T-soy agar plates.

INTERPRETATION NOTES

The Ames mutant of *Salmonella typhimurium* used here cannot make histidine. The basal medium used in this test lacks histidine, but the overlay agar contains a small (growth-limiting) amount of this amino acid. Therefore, the numerous histidine auxotrophs mixed with the overlay agar will grow for only a few generations and should appear as a faint (almost invisible) lawn of growth within the overlay.

Strong mutagenic agents should cause the Ames mutants (histidine auxotrophs) to revert to their prototrophic (parental) state so that they no longer require histidine for growth. These revertants should be able to form fully developed (large) colonies on this medium because they no longer require histidine for growth. Therefore, each *large colony* represents one back mutation (revertant) that either occurred spontaneously or was induced by a mutagen.

The number of large colonies on the plate containing distilled water instead of a test chemical represents the number of spontaneous mutants (revertants) occurring with this culture under these conditions. *If a chemical agent induces twice the number of mutants as occurred spontaneously, that chemical should be considered mutagenic* and possibly carcinogenic for humans.

3. **Discard these plates** in the manner described by your instructor. Note that these plates may contain mutagenic and/or carcinogenic chemicals; therefore, you will be asked to use special precautions for their disposal.

38 POTENTIAL CARCINOGENS: THE AMES TEST

NAME _____

DATE _____

SECTION _____

RESULTS

Write either the word *full* or the fractions $\frac{1}{2}$ or $\frac{1}{4}$ in each space for number of colonies to indicate the portion of each plate counted. For example, $\frac{1}{2}$-200 shows that one-half the plate contained 200 colonies. Count either the full plate or the fraction of all plates prepared by you and your instructor. If the number of large colonies exceeds 300 per full plate, 150 per $\frac{1}{2}$ plate, or 75 per $\frac{1}{4}$ plate, record this as *TNTC* (too numerous to count).

Plate label	Name of chemical used	Number of large colonies	
		Ames minimal agar	T-soy agar
Control	None		
#1			
#2			
#3			

CONCLUSIONS

List each chemical that is mutagenic and/or carcinogenic, and state the basis for this interpretation.

QUESTIONS

Completion

1. Using laboratory animals to screen for carcinogens is not practical because of the _____ and the _____ required by this procedure.

2. The mutants developed by Ames and colleagues are strains of a bacterial species called _____ _____ (write out in full).

3. If a mutant requires a nutrient that was not required by the parent strain, this mutant is called a(n) _____. In contrast, the parent strain is then called a(n). _____.

4. The Ames mutant used in this study has a single nucleotide deleted from one gene; so this mutant is called a(n) _____ mutant. This mutant also is unable to _____.

5. The number of spontaneous mutants formed under conditions used in the Ames test is found by using the plate containing _____ instead of the one containing _____. What is the number of spontaneous mutants formed under your experimental conditions?

6. A faint background lawn should appear on each plate. What does this lawn represent?

7. How does the lawn formed by the T-soy agar plates appear different from that formed by the Ames minimal medium plates?

   How should you explain the reason for this difference?

8. Can T-soy agar be used to detect revertants in the Ames test? Explain your answer.

True–False (correct all false statements)

1. Since about 90 percent of all carcinogens are also mutagens, it is reasonable to assume that some mutagens are also carcinogenic. _____

2. The amino acid histidine is not needed by the Ames mutants for growth prior to being subjected to chemical mutagens. _____

3. The selection medium used by the Ames test may be called a chemically refined or synthesis medium. _____

4. Only the original Ames mutants will form full-sized colonies on the selection medium used in the Ames test. _____

5. Histidine auxotrophs will not grow on the medium used by the Ames test. _____

39

Effect of Ultraviolet Radiation on DNA, Cell Viability, and Mutation Frequency

Radiation is the process of emitting radiant energy in the form of waves or particles. Whether radiant energy is useful or destructive to microorganisms depends on its **wavelength** (Figure 39-1). The longer wavelengths, such as radio waves ($> 10^6$ nm), have no detectable effect on microorganisms.

The absorption of **infrared rays** given off by the sun (from 800 to 10^6 nm), produces *heat,* and heat is essential for all forms of life. Some of the shorter infrared waves are captured and used by **photosynthetic bacteria** as an energy source. For example, the purple and green photosynthetic bacteria have bacteriochlorophyll inserted within specialized intracytoplasmic membranes, and these molecules absorb light at about 870 nm.

The **visible** portion of the spectrum, which we call *light,* is composed of rays with wavelengths of about 380 to 760 nm. We see these rays as white light and as separate colors when the visible part of the spectrum is split into smaller parts by a prism. The cyanobacteria (blue-green bacteria) and the red algae contain accessory pigments that capture visible light and transfer its energy to adjacent chlorophyll molecules. For example, a red pigment *(phycoerythrin)* absorbs light strongly at 550 nm, and a blue pigment *(phycocyanin)* absorbs light best at around 630 nm. Therefore, the short infrared wavelengths and visible light provide the main sources of energy for photosynthetic bacteria, cyanobacteria, and eucaryotic algae, and the continued existence of photosynthetic microorganisms is essential to our environment.

As one goes to the shorter wavelengths (< 400 nm), one leaves the helpful types of radiation and encounters only damaging forms of energy. The ultraviolet part of the spectrum contains radiation with wavelengths of about 100 to 400 nm (see Figure 39-1). Below 100 nm are the ionizing forms of radiation, including many X rays, gamma rays, and cosmic rays.

Ionizing radiation (< 100 nm) can be used to sterilize materials that are sensitive to heat and, therefore, cannot be autoclaved or subjected to dry-heat sterilization. The killing effect of ionizing radiation is indirect: instead of acting directly on one part of the cell, it reacts with water to form free radicals such as the hydroxyl radical ($\cdot$OH). A *free radical* is an atom or group of atoms having at least one unpaired electron. Free radicals probably react with, and are equally injurious to, all parts of a cell. A microorganism has many copies of most of the critical molecules such as proteins but only one complete molecule of DNA. Therefore, the most harmful interaction may be between free radicals and the cell's DNA. For example, since each bacterium has only one complete chromosome, it has only one copy of most genes. Inactivation of any one critical gene leads to death, because the bacterium is no longer able to code for the synthesis of a protein needed for survival. Bacteria vary greatly in their resistance to the effects of ionizing radiation. For example, *Pseudomonas* species are quite sensitive, whereas species of *Micrococcus* are often very resistant. Sensitivity to ionizing radiation apparently depends on the cell's ability to neutralize free radicals metabolically. Spores are exceptionally resistant to ionizing radiation, probably because they are desiccated, which greatly reduces the chance that ionizing radiation can interact with water to form free hydroxyl radicals.

Nevertheless, if the dose of ionizing radiation is large enough, it can kill even bacterial spores. For

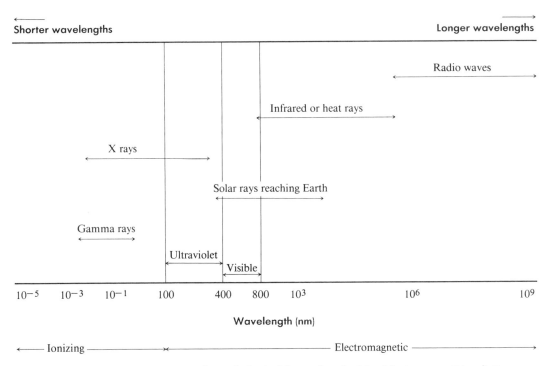

FIGURE 39-1 Energy spectrum that includes ionizing and nonionizing (electromagnetic) radiation.

example, large doses can sterilize materials such as foods. Food sterilization by ionizing radiation has several advantages; for example, this method avoids heat and consequently retains volatile flavors. Foods that are sealed and then sterilized in this way can be stored at room temperature without expensive refrigeration. However, to date, this method of treating foods has not been approved by the U.S. Food and Drug Administration (FDA) because of concern that ionizing radiation may alter substances in food in a way that will have harmful long-term consequences for humans that consume the food. However, the U.S. Armed Forces, which are not under FDA restrictions, have been using irradiated foods for many years. The treatment of foods with ionizing radiation is a controversial subject in the field of food microbiology.

Ultraviolet (UV) radiation (100 to 400 nm) is of special interest because it is used in certain environments to kill microorganisms. For example, hospital operating rooms, clean rooms (used for aseptic transfers), and inoculation rooms or cabinets (used for transfer of virulent pathogens) are often lined with ultraviolet lamps, called *germicidal* lamps because their radiation can kill a variety of microorganisms on the *surfaces* of objects. The bulb that emits UV radiation looks like a clear fluorescent light bulb. When people are not using the areas, these lamps are turned on to kill microbes on walls, floors, ceilings, and bench tops, which helps to keep the work environment relatively sterile. Ultraviolet light is considered to be a germicide rather than a sterilant because it only has the potential for killing vegetative cells and does not kill bacterial spores.

The most effective germicidal region of the UV spectrum occurs from about 240 to 300 nm. The purine and pyrimidine bases in nucleic acids (Figure 39-2) absorb radiation strongly at 260 nm, which is also where the greatest lethal effects occur. Several effects are known, but the most thoroughly studied is the formation of thymine dimers.

Thymine is one of the four **bases** that are building blocks of **deoxyribonucleic acid (DNA)**. Thymine (T) and cytosine (C) are members of a class of chemical compounds called *pyrimidines,* and adenine (A) and guanine (G) are members of a class called *purines*. Each single strand of the double-stranded DNA molecule is made up of alternating sugar and phosphate molecules, with purine or pyrimidine bases attached to the sugars by covalent bonds. The bases on one strand are complementary to the bases on the adjacent strand (A and T are complementary, as are G and C). Weak hydrogen bonds between complementary bases (2 or 3 between each pair) hold the two single strands together in the helical DNA structure.

Two thymine bases are often next to one another on the same strand of a DNA molecule (Figure 39-3). When a UV ray hits such a pair of thymine molecules, they become dissociated from the complementary

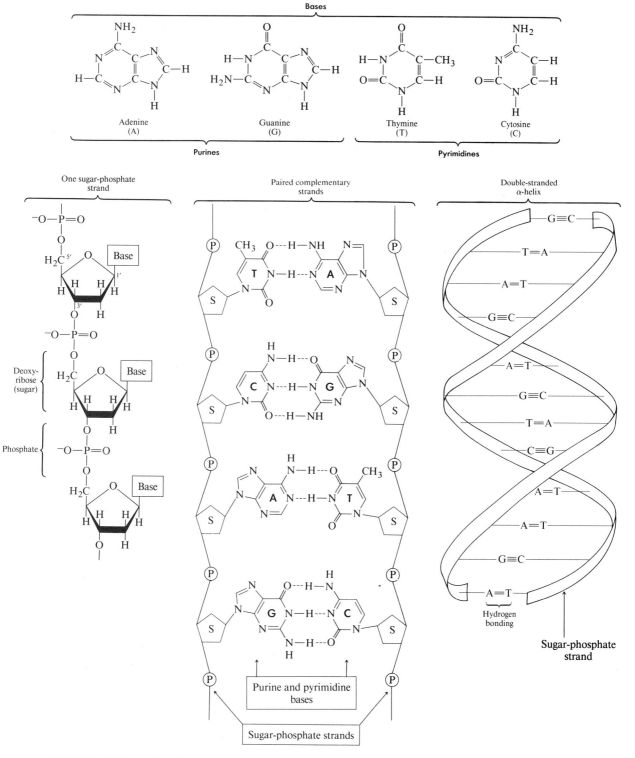

FIGURE 39-2 Relationships among the purine and pyrimidine bases, the sugar-phosphate backbone, and the structure of the double-stranded helical DNA.

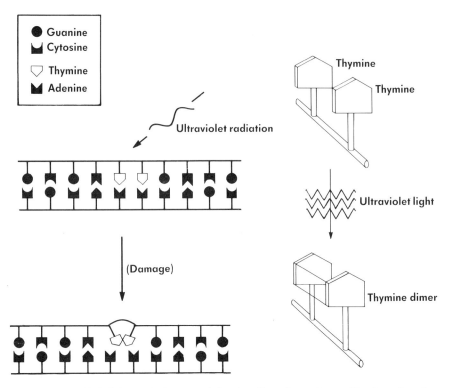

FIGURE 39-3 Schematic diagram of thymine dimer formation and its distorting effect on one strand of the DNA double helix.

bases on the other strand and fuse together to form the structure called a **thymine dimer.** This covalent bonding distorts the shape of the DNA strand; thus, when the strand is later replicated, incorrect bases may be inserted into the new strand of DNA as it is synthesized along the old, distorted DNA strand. This results in a **mutation** (change) in the genetic code. If the altered DNA codes for the cessation of protein synthesis or if it drastically alters synthesis of a critical protein, the cell that contains this mutation will probably die; such an alteration is called a **lethal mutation.** If the protein is not critical or if the alteration in the protein is not severe, the mutant cell may survive in an altered form, and the mutation is called a **nonlethal mutation.**

Most microorganisms have enzymes that can **repair** damage to the DNA strands caused by UV light. Two of these repair mechanisms are photoreactivation and the dark repair mechanism. **Photoreactivation** is catalyzed by a photoreactivation enzyme (PRE) that binds to the dimers. In the presence of visible light, PRE breaks the bonds between the dimer and returns the damaged DNA strand to its original state. Photoreactivation was discovered in bacteria in 1949 by Albert Kelner and has since been observed in fungi, algae, higher plants, and animals, including humans. In **dark repair** *(excision and recombination repair),* enzymes called *nucleases* hydrolyze the phosphodiester bonds at either end of the damaged region of the DNA strand and thereby excise the damaged portion. Then, *lygase* enzymes catalyze the insertion of new nucleotides, complementary to the undamaged strand, into the excised region. The strand thus is restored to its original state and will function properly.

Therefore, UV radiation produces cell death or a nonlethal mutation only when the damage is greater than the cell can repair. A brief exposure to low-intensity UV light may not result in a "hit" on a cell's DNA and thus will have no effect on the cell; or it may cause damage to the DNA that can be repaired; or it may cause damage to the DNA that cannot be repaired, resulting in a lethal or nonlethal mutation. If one increases the intensity or lengthens the exposure time, an increase in the number of unrepaired dimers and a corresponding increase in the numbers of lethal or nonlethal mutations should occur.

One form of UV-induced mutation in human skin cells is a cell that exhibits unrestricted growth, resulting in a cell mass called a **tumor** and a disease known as **skin cancer.** Skin cancer is currently the most common form of cancer among humans. Therefore, you should protect your skin from UV radiation by wearing protective clothing or applying effective chemical sun screens.

39 EFFECTS OF ULTRAVIOLET RADIATION

Since germicidal lamps emit light at around 260 nm, these lamps are very effective in damaging the DNA of cells and should be used only if humans are not present or if their skin and eyes are adequately protected.

LEARNING OBJECTIVES

• Understand how UV light (260-nm radiation) adversely affects microorganisms at a molecular level.

• Realize why time and intensity of exposure are important factors in the design of the following experiments on cell viability and that these two factors are important in the application of any antimicrobial chemical or physical agent.

• Know that UV light can be either mutagenic or lethal for microorganisms, and know how to test for lethality and mutant formation.

A · Effect of UV light on cell viability

This section is designed to show you the *lethal* effects of 260-nm radiation for *Micrococcus luteus,* a bacterium commonly found on dust in the air and known for its resistance to UV light, and for *Escherichia coli,* a bacterium that normally lives in the colons of animals.

MATERIALS

Cultures *Micrococcus luteus* (broth culture)
Escherichia coli (broth culture)

Media Trypticase-soy agar plates

Supplies 1.0-ml pipettes (sterile)
Spreading wheel, glass spreading rod, and 95 percent ethanol
UV lamp (General Electric Model G30TA; 30 watt; 34.5-inch length; placed about 16 inches from the table surface)
4 × 6-inch file cards
Laboratory coats (for those who do not have long sleeves)
Protective gloves (plastic or rubber)
Protective goggles (for those who do not wear glasses)

PROCEDURE: FIRST LABORATORY PERIOD

Steps 1 through 8 should be performed with *Micrococcus luteus* and with *Escherichia coli*. Read all eight steps before you begin your work.

1. **Select three sterile plates, and divide each plate into two equal parts** with a marking pen by drawing a dotted line across the bottom (outside) of each plate. This is to guide your placement of the 4 × 6 card prior to UV exposure.

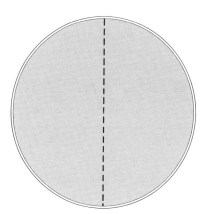

2. **Label each plate** on the bottom with your *name, date* of inoculation, exercise *number,* and the *microorganism* used to inoculate the plate.

3. **Inoculate each of the three plates with 0.1 ml of one of the broth cultures; then aseptically spread the inoculum over the entire surface** with a glass spreading rod, as described in Exercise 20.

4. **Carry all three plates to the UV light source** but do not place them under the light.

> **SAFETY NOTE**
>
> The UV lamp used in this experiment is similar to the sun lamp commonly used in the home, and like the sun lamp, it will burn your skin and eyes if either are exposed too long to the UV rays. *Use some form of shielding* around the lamp to cut down on stray light in your work area. For example, you may place the lamp inside a chemical-fume hood and keep the glass window down while no one is working with the lamp. You should also use various forms of personal shielding, such as eyeglasses (goggles or full-face shield), long sleeves, and gloves.

5. **Put on a laboratory coat** (or roll down your long sleeves), **cover your eyes with glasses** (or other protective wear), and **put on rubber or plastic gloves.**

6. **Remove the plate cover from one plate,** and place this cover right side up on a dust-free surface. **Expose this plate to UV radiation as follows:**
   a. **Place a 4 × 6 card over the top** of this plate so that its edge is aligned with the dotted line drawn on the plate bottom.
   b. **Move this plate so that it is directly under the UV source,** and immediately begin timing the exposure.
   c. **After 10 seconds, remove the plate** from under the UV source, remove the 4 × 6 card, and replace the cover. Label this plate, on the bottom, with *10 sec,* and place the letter *C* on the half of the plate that was covered by the card.

7. **Repeat step 6 with the second plate, but use a 30-second exposure.** Label this plate, on the bottom, with *30 sec,* and mark the half of the plate that was covered by the card (*C*).

8. **Repeat step 6 with the third plate, but use a 60-second exposure.** Label this plate, on the bottom, with *60 sec,* and mark the half of the plate that was covered by the card (*C*).

9. **Each class should also prepare two "control" plates,** as follows:
   a. **Divide** both plates into two parts by drawing a dotted line across the bottom of the plate as before.
   b. **Label** each plate with your class meeting time, the date, and the name *E. coli.*
   c. **Inoculate** each plate with 0.1 ml of the *E. coli* broth culture; then aseptically spread the inoculum over the entire surface with a glass spreading rod.
   d. **Set one control plate aside without additional treatment.** This will serve as the control to show you how many cells you placed on each plate. Label this plate *untreated control.*
   e. **Leave the cover on the second control plate, and expose it to UV radiation for 60 seconds.** This plate will help you deter-

mine if UV radiation is stopped by the clear, plastic, Petri-dish cover. Label this plate *60 sec w/cover on—control.*
10. **Incubate all plates at 30°C for 48 hours.**

PROCEDURE: SECOND LABORATORY PERIOD

Steps 2 through 5 should be performed with plates inoculated with *Micrococcus luteus* and with plates inoculated with *Escherichia coli.*

1. **Determine or estimate the number of colonies on one-half of the control plates** (*untreated* and *60 sec w/cover on*) as follows:
   a. **If all or most of the colonies are well separated,** count the colonies on the entire plate and divide by two. This corresponds to the way in which you will express the data from your experimental plates. Record that number (number of colonies per one-half plate) in the Results section for Section A of this exercise.
   b. **If there are more than 300 colonies on the entire plate** (150 per half plate), try to divide the plate into quarters by carefully drawing lines with your marking pen on the plate bottom; then count the numbers on one quarter of the plate and multiply by two. Record the number as *TNTC* (too numerous to count), and place your estimate (numbers per one-half plate) after TNTC in parentheses.
   c. **If most of the colonies have grown together,** so that counting is impossible, record this as *confluent growth.*
2. **Select the 10-sec experimental plates** (those partially exposed to UV while uncovered), and proceed as follows:
   a. **Count the number of colonies on the exposed half** of the plate. Use the same method to estimate large numbers as described in step 1. Record that number (number per one-half plate) in the Results section.
   b. **Count the number on the card-covered half** of the plate. Use the same method to estimate large numbers as described in step 1. Record that number (number per one-half plate) in the Results section.
3. **Repeat step 2** using the 30-sec experimental plates.
4. **Repeat step 2** using the 60-sec experimental plates.
5. **Properly discard all plates.**

B · Effect of UV light on mutation frequency

This section is designed to show you the *mutagenic* effects of UV radiation. You will be asked to use the Ames mutants (see Exercise 38) to study the ability of a physical (rather than a chemical) agent to increase the frequency of mutation.

You will plate a histidine **auxotroph** of *Salmonella typhimurium* on a medium that lacks histidine. The only cells that will be able to grow on this histidine-deficient medium are the **revertants** (cells that back mutate and regain their ability to make histidine). You will then expose these plates either to several levels of UV radiation or to no radiation to see if the number of mutants (revertants) can be correlated with the amount of UV exposure.

For a control, you will also plate this histidine auxotroph of *Salmonella typhimurium* on a medium that contains histidine and expose the plate to the same levels of UV radiation. This control will enable you to see the total number of cells that survived the UV radiation. Since this plate contains histidine, both nonmutants (auxotrophs) and mutants (revertants) will grow. You can compare the number of colonies on this plate with the number on the histidine-deficient plate to determine what proportion of the survivors were back mutations.

MATERIALS

Cultures   *Salmonella typhimurium* (a diluted broth culture of the Ames strain TA-102)

Media   Minimal medium (agar plate) without histidine (MIN − H)
Minimal medium (agar plate) with histidine (MIN + H)

Supplies   UV light (germicidal)
1.0-ml pipette (sterile)

Glass spreading rod and 95 percent ethanol

Index card (4 × 6-inches)

Laboratory coats (for those who do not have long sleeves)

Protective gloves (plastic or rubber)

Protective goggles (for those who do not wear glasses)

PROCEDURE: FIRST LABORATORY PERIOD

1. **Select one sterile MIN − H plate and one sterile MIN + H plate. Divide each plate** with a marking pen **into five equal parts** by drawing dotted lines on the bottom (outside) of the plate.

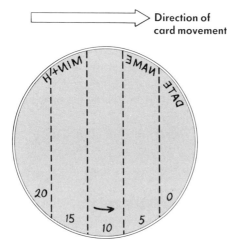

Final appearance of plate after exposure and labeling

2. **Label these plates** (on the bottom along the outer edge) with your *name,* the *date, MIN − H* or *MIN + H* (as appropriate), the total exposure *time,* and the *direction* of card movement.

Because the timing of steps 3 through 8 is critical, you should perform them on *one plate at a time.* Be sure to reread the Safety Note in Section A.

3. **Inoculate both plates (MIN + H and MIN − H) with 0.1 ml of** *Salmonella typhimurium* (Ames strain) culture, and aseptically spread the inoculum over the entire agar surface with a glass spreading rod.

4. **Carry both plates to the UV light source,** but do not place them under the light yet.

5. **Remove the cover of the inoculated MIN + H plate,** and place it (right side up) on a clean table top.

6. **Place a 4 × 6-inch card over the plate** so that the open plate is *completely covered* by the card.

7. **Place the plate under the UV light.**

8. **Manipulate the 4 × 6 card as follows** (use the second hand on your watch or a nearby wall clock):

    a. **At 0 seconds,** slide the 4 × 6 card to the *first* dotted line, so that the first one-fifth of the plate is exposed to the UV light. That is, start timing this procedure when you first move the card. Note that this section of the plate will eventually receive a total of 20 seconds of exposure.

    b. **At 10 seconds,** slide the 4 × 6 card over to the *second* dotted line so that the second one-fifth of the plate is exposed to the UV light. Note that this section of the plate will eventually receive a total of 15 seconds of exposure.

    c. **At 15 seconds,** slide the 4 × 6 card over to the *third* dotted line so that the third one-fifth of the plate is exposed to the UV light. Note that this section of the plate will eventually receive a total of 10 seconds of exposure.

15 seconds

d. **At 20 seconds,** slide the 4 × 6 card over to the *fourth* dotted line so that the fourth one-fifth of the plate is exposed to the UV light. Note that this section of the plate will eventually receive a total of 5 seconds of exposure.

20 seconds

e. **At 25 seconds,** completely cover the plate with the 4 × 6 card. Note that the last one-fifth of the plate is not exposed to UV radiation.

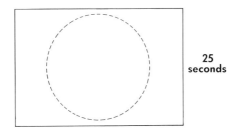

25 seconds

f. Move the plate away from the UV source, remove the 4 × 6 card, and replace the plate cover.
9. **Perform steps 3 through 8 with the inoculated MIN − H plate.**
10. **Incubate both plates at 30°C for 48 hours.**

PROCEDURE: SECOND LABORATORY PERIOD

1. **Select the plate labeled *MIN + H*. Count the number of colonies in each of the five sections.** Record these numbers in the results section. Use the guidelines given in Section A (Procedure: Second Laboratory Period) of this exercise for recording confluent growth or numbers that are too numerous to count.

   Note that the number of colonies on this medium represents *all viable cells* (auxotrophs plus back mutations) that survived in each section of the plate (four different UV exposure times).

2. **Select the plate labeled *MIN − H*. Count the number of colonies in each section of the plate.** Record these numbers in the Results section.

   Note that the number of colonies on this medium represents only the number of *mutants* (revertants) that survived the four levels of exposure to UV radiation. This is because only those that back mutated can grow in the absence of histidine.

3. **Calculate the approximate frequency of mutation** (percentage of back mutation) at each level of exposure, and record these values in the Results section.

   This calculation can be done with some accuracy with this method because each section (on MIN + H and MIN − H plates) for a given exposure time has approximately the same surface area. Therefore, you are comparing the total number of survivors with only the number of mutants for the same exposure time and the same surface area. (Only one variable.)

4. **Properly discard both plates.**

39 EFFECTS OF ULTRAVIOLET RADIATION

NAME _____

DATE _____

SECTION _____

RESULTS A: EFFECT OF UV LIGHT ON CELL VIABILITY

Control Plates

Control plates	*E. coli* colonies (number per one-half plate)
Untreated (0 sec, no UV)	
Cover on (60 sec UV)	

Experimental Plates

Time of treatment (sec)	Your results (number per one-half plate)				Class averages (number per one-half plate)			
	M. luteus		*E. coli*		*M. luteus*		*E. coli*	
	Open	Covered	Open	Covered	Open	Covered	Open	Covered
10								
30								
60								

CONCLUSIONS

The effect of clear plastic on UV radiation:

The effect of thin cardboard on UV radiation:

Comparison of the effect of UV radiation on *M. luteus* versus *E. coli*:

RESULTS B: EFFECT OF UV LIGHT ON MUTATION FREQUENCY

Medium used	What should grow?	UV exposure time (sec)	Number of colonies formed (cfu/area)	Calculated mutation frequency (%)
MIN + H	All survivors (histidine auxotrophs and back mutations)	0		N/A
		5		N/A
		10		N/A
		15		N/A
		20		N/A
MIN − H	Only back mutations (histidine not required)	0		
		5		
		10		
		15		
		20		

Use the following formula to calculate the mutation frequencies, and place the result of each calculation in the right-hand column of the table.

$$\text{Mutation frequency (\%)} = \frac{\text{number of colonies (back mutations) on MIN} - \text{H at single exposure time}}{\text{number of colonies (auxotrophs and back mutations) on MIN} + \text{H plate at same exposure time}} \times 100$$

Show one sample of your calculation below:

CONCLUSIONS

1. Closely examine the MIN − H plate; then answer the following questions:

    a. Do any of the sections exposed to UV radiation have *more* colonies than the section that received no UV exposure? If so, explain these results.

39 EFFECTS OF ULTRAVIOLET RADIATION

NAME _____

DATE _____

SECTION _____

    b. Are there any colonies on the *unexposed* part of the plate? If so, explain these results.

    c. Does the section that received the *most* UV exposure have fewer colonies than any other section of this plate? If so, explain these results.

2. Closely examine the MIN + H plate. Is there evidence here of significant killing correlated with the length of UV exposure?

What is your evidence for this?

After what time period of exposure did the killing become noticeable?

_____ seconds

3. What is the effect of UV radiation on mutation frequency?

QUESTIONS

Completion

1. Most of the germicidal radiation reaching earth from the sun has a wavelength of about _____ nm.

2. Germicidal UV lamps usually emit radiation with a wavelength of about _____ nm. The cellular material most affected by that radiation is the _____.

3. The range of the energy spectrum covered by UV radiation is about _____ to _____ nm.

4. Briefly describe what happens to adjacent thymine bases when hit by ultraviolet radiation.

5. Thymine is a type of molecule called a pyrimidine _____, which makes up part of a macromolecule called _____.

True–False (correct all false statements)

1. If your results from Section A of this experiment turned out as expected, they show that the amount of killing of *E. coli* is inversely proportional to the time of exposure. _____

2. Ionizing radiation (less than 100 nm) is useful for killing microbes on the surfaces of things that are heat labile. _____

3. Radiation in the UV range of the energy spectrum is an example of ionizing radiation. _____

4. In the design of Section A of this experiment, radiation intensity varied from 30 to 90 nm. _____

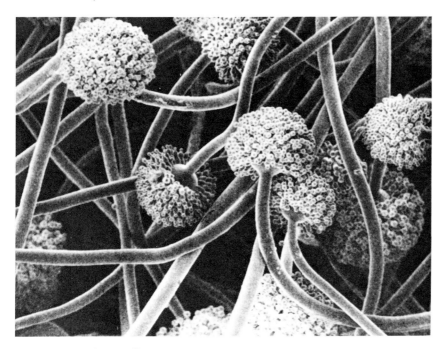

Fruiting bodies of *Aspergillus niger*.

And the little things that our naked eye cannot penetrate into, have in them a greatness not to be seen without astonishment . . . how exquisite, how stupendous must the structures of them be!

—*Cotton Mather, Thanksgiving sermon, 1689*

THE FUNGI

The purpose of Part X is to introduce you to the structural features, methods of reproduction, classification, and significance of the filamentous fungi (molds) and the single-cell fungi (yeasts).

40

Introduction to the Fungi

The **fungi** (singular = *fungus*) are a diverse group of eucaryotic organisms. Some, such as the mushrooms and wood-decay fungi, have tissuelike differentiation and exist in nature as multicellular masses that are large enough to be seen from 75 to 100 feet away with the unaided eye. Other fungi, such as the bread- and fruit-molds, are composed of filamentous cells that are microscopic in size and capable of a free-living existence; yet these microscopic filaments often form large colonylike masses of cells that can easily be seen with the unaided eye on the surface of various foods. Still others, such as the yeasts, exist only as free-living single cells (or small clusters) and tend to live in more fluid environments. Even the largest, most differentiated fungus, however, lives at least one part of its life cycle in a single-celled form called a spore. (Note, however, that fungal spores are not nearly as resistant to environmental stresses as are bacterial endospores.)

Words having the prefix *myc-*, or *myco-*, come from the Greek word *myket*, or *mykes*, meaning *fungus*. Examples are *mycology* (the study of fungi), *mycologist* (one who studies fungi), and *mycosis* (a fungal disease). You will find other examples of this prefix throughout Part X. Note too that the letters *-myc-* may appear together in the middle of a word referring to taxonomic groupings of fungi, such as *Ascomycetes* and *Basidomycetes*. Knowing this should help you recognize words that refer to fungi or fungallike microorganisms.

General characteristics of fungi

The fungi are a group of living organisms that are like plants in that they have definite cell walls, they are usually **nonmotile,** and they reproduce by means of seedlike **spores.** However, fungi are unlike all plants in that they **lack chlorophyll** and are unlike the higher plants in that they lack stems, true roots, and leaves. Fungi are often filamentous and multicellular. All fungi have **eucaryotic** cell structure, and their nuclei can be easily shown. Their masses of cells exhibit little differentiation and almost no division of labor. Fungal filaments elongate by apical growth (Figure 40-1); but most parts are capable of growth, and a small fragment from almost any part of the fungus is able to produce a new growing point and to start a new individual.

Unique characteristics of fungi

One of the first decisions that is made by a diagnostic microbiologist is whether the cells seen under the microscope are fungi, bacteria, algae, or protozoa. The four characteristics that, *when used together,* distinguish fungi from bacteria and all other microorganisms relate to internal cell structure, cell-wall chemistry, method of deriving energy from the environment, and the size of individual cells. These characteristics are summarized in Table 40-1 and explained more fully below. Note that most cannot be readily determined by the beginning student of microbiology.

All fungi have a *eucaryotic cell structure* (as do the protozoa and eucaryotic algae) whereas all bacteria are procaryotes. Determination of internal cell structure requires the use of an electron microscope and procedures for embedding and ultrathin sectioning of cells; such equipment and procedures are not usually available to beginning students.

All fungi (except the slime molds) have *cell walls that contain chitin, cellulose, or hemicellulose* as the structurally rigid component, as do the eucaryotic algae. On the other hand, bacteria contain only peptidoglycan as the structurally rigid component, and protozoa totally lack a structurally rigid wall. How-

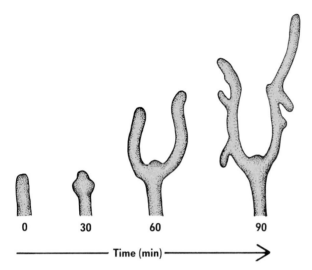

FIGURE 40-1 Successive stages of apical growth (growth from the tip) of a filamentous fungus. Microscopic observations shown at half-hour intervals. (Adapted from C. J. Alexopoulos and C. W. Mims, *Introductory Mycology*, 3rd ed. New York: Wiley, 1979; with permission.)

ever, cell-wall isolation and chemical analysis are procedures not normally accomplished by beginning microbiology students.

All fungi are *strict chemoorganotrophs* (like all protozoa and many bacteria); that is, they require reduced organic compounds as a source of energy and carbon for growth. Therefore, all fungi are *nonphotosynthetic* and grow well in total darkness. This characteristic excludes the eucaryotic algae and the photosynthetic bacteria. If enough time is available, this characteristic can be tested by beginning students.

Almost all fungi have a *cell diameter that is from 5 to 10 micrometers (μm)*. Conversely, diameters of bacteria rarely exceed 2 μm. Cell size can be easily and quickly approximated by beginning students who use the procaryote *Escherichia coli* as a **ruler**. These cells are about 1.0 μm long; so, if you prepare a smear of *E. coli* beside a smear of the unknown microbe, or if *E. coli* cells are mixed on the slide with the unknown, then you can make fairly accurate size estimates. *Remember, the diameter of the typical fungal cell is from 5 to 10 times larger than the length of the average* E. coli *cell*.

Fungal morphology

The "body" of a fungus, called a **thallus** (plural = *thalli*), can vary from a single cell (free-living during its entire life) to a complex organism that is composed of masses of cells and spores.

If the thallus is a single cell under most or all environmental conditions, the fungus is called a **yeast** (Figure 40-2). Most yeasts are much larger than most bacteria. They are commonly egg-shaped, but some are spherical and others elongated. Each species has a characteristic shape, but pure cultures often exhibit many variations in size and shape, depending on cell age and the growth environment. Yeasts can reproduce either (1) asexually by transverse, binary fission or by forming **buds**, or (2) sexually or by conjugating with another cell and forming spores.

If the thallus is composed of very long filaments under most or all environmental conditions, then the fungus is commonly called a **mold** or a *filamentous fungus* (Figure 40-3). The long filaments are usually called **mycelia** (singular = *mycelium*), but they may also be called **hyphae** (singular = *hypha*). The mycelia are often *branched*. If a branch terminates in a group of spores, this filament is called a *reproductive mycelium* to differentiate it from the branched and seemingly endless mat of interwoven *vegetative mycelia* from which it arose (Figure 40-3). The filamentous fungi (molds) can reproduce either asexually, by forming spores at the end of reproductive mycelia, or sexually when adjacent cells conjugate to form spores.

TABLE 40-1 Comparison of fungi with bacteria

Characteristic	Fungi	Bacteria
Cell size[a]	Usually about 5–10 μm in diameter	Seldom greater than 2 μm in diameter
Cell structure	Eucaryotic	Procaryotic
Structurally rigid component of the cell wall	Chitin, cellulose, or hemicellulose	Peptidoglycan
Metabolism	Strictly chemoorganotrophic	May be chemoorganotrophic, chemolithotrophic, photoorganotrophic, or photolithotrophic

[a] Cell size is the only characteristic that can be readily determined in most classroom laboratories.

40 INTRODUCTION TO THE FUNGI

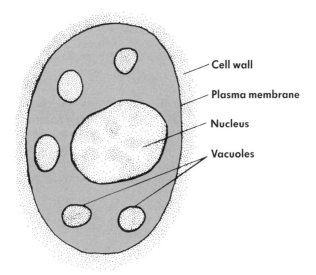

FIGURE 40-2 Typical single-cell fungus (yeast) showing cell wall, plasma membrane, and internal structures that are often seen with simple stains and the brightfield microscope.

Some fungi (called **dimorphic**), can exist in either a single-celled (yeast) form or a filamentous form. For example, some fungi that are pathogens of humans and other animals grow in a yeastlike form in the animal but in a filamentous form in the soil or on artificial media in the laboratory. However, some plant pathogens produce filaments in the host but grow as single cells on artificial media. Identification of these pathogens often requires the demonstration of dimorphism.

Classification of the fungi

Biologists order (classify) all organisms into groups **(taxa)** according to their relatedness. The largest groups are the biological kingdoms, such as the animal and plant kingdoms. The way in which one distinguishes between higher (more highly developed or complex) organisms, such as trees and elephants, is well established and leads to almost no arguments. We can examine distinct general characteristics for the kingdoms, such as the way in which energy is generated and the chemical composition of tissues. We can even clearly indicate species designations by examining fossil records or by determining whether two individuals can both contribute compatible cells that will fuse and form a new individual (carry out sexual reproduction).

On the other hand, the correct way to classify microorganisms into distinct taxa (such as kingdoms, divisions, or classes) is not at all clear because (1) microbes do not fit well into the criteria established for either animals or plants (indeed, some microbes have characteristics of both animals and plants in the same organism), and few have unique characteristics; (2) the chemical composition and metabolism of many microbes is not thoroughly understood; (3) there is no fossil record for most microorganisms; and (4) people who study either eucaryotic or procaryotic microorganisms do not communicate with one another about classification. Therefore, you will find that classification of microorganisms, such as the fungi, into large taxa is largely a matter of personal opinion and often reflects the background and prejudices of the microbiologist.

Recent developments suggest that there is hope for definitive classification of the higher taxa of microorganisms. Molecular techniques that compare the DNAs and RNAs of two organisms may one day help to show the degree of relatedness between them. Until sufficient molecular data are available, however, we must adhere to the present system that deals only with observed phenotypic characteristics.

Fungi viewed as part of the plant kingdom Historically speaking, the fungi were first considered to be plants that lacked chlorophyll, and they were studied by persons having classical training in botany. Indeed, later studies demonstrated that most fungal cells have rigid cell walls that contain chitin, cellulose, or hemicellulose; therefore, it seems reasonable to place fungi in the plant kingdom (Table 40-2). In this view, the plant kingdom is divided into five divisions. The fungi

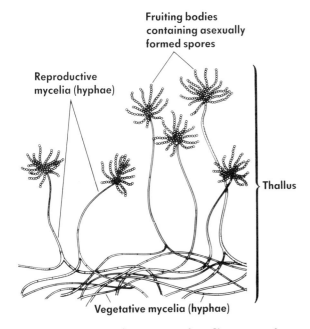

FIGURE 40-3 Typical structure of a filamentous fungus (mold) showing reproductive mycelia and fruiting bodies as observed with low magnification using a brightfield or phase-contrast microscope.

NOTES ON THE HISTORY OF MICROBIAL CLASSIFICATION

Prior to the nineteenth century, biologists tried to classify all living things into either the plant kingdom or the animal kingdom. However, during the 1800s a German evolutionist, Ernst Haeckel, proposed that there should be a third biological kingdom that would include all microorganisms, and he suggested that this kingdom be called *Protista*. For an organism to be considered part of this kingdom, Haeckel proposed that the organism must be unicellular in at least one stage of its life history and that it must not develop into specialized tissues under any conditions. Haeckel's revolutionary proposal received little attention by eighteenth- and early nineteenth-century biologists. However, his kingdom *Protista* has recently been revived by some authors of general biology textbooks. Many of today's biology students are trained to think of all microorganisms as belonging to the kingdom *Protista*.

Although most biologists acknowledge that microorganisms should be classified separately from plants and animals, those who have extensively studied microorganisms feel that it is a vast oversimplification to place all microbes into a single kingdom. To classify all microorganisms together, regardless of whether they are bacteria, protozoa, algae, or fungi, is to ignore the large body of recent molecular evidence that demonstrates a past, separate, divergent evolution among the microorganisms.

After Haeckel's pioneering suggestion, the next significant proposal for microbial classification came during the middle 1900s from Roger Stanier and his colleagues at the University of California at Berkeley. Their ideas paralleled the development of the electron microscope and the accumulating knowledge of the internal structure of microbial cells. They proposed that all microorganisms be divided into two kingdoms according to the structure of their nuclear material.

Stanier and colleagues proposed that one kingdom be called *Procaryotae* (prenucleated cells), because their evolution is thought to precede the development of cells with a more complex, membrane-bound nucleus. They proposed that the other kingdom be called *Eucaryotae* (cells with a membrane-bound or "true" nucleus). Their proposal was based not only on the structure of the cell's nucleus but also on many other internal structural features. Since their ideas were first proposed, the list of differential characteristics between procaryotes and eucaryotes has grown as microbiologists have continued to examine both the structure and the molecular properties of these cells. These findings have greatly strengthened the proposal that microorganisms need to be placed in separate kingdoms because of their fundamental dissimilarities.

Very recently, Carl Woese at the University of Illinois and other molecular geneticists proposed that the procaryotes themselves be divided into two groups. One group, called the *Archaeobacteria*, is distinguished from the other procaryotes by three properties: (1) the sequence of bases on their ribosomal RNA, (2) their cell-wall structure, and (3) the chemical composition of their membrane lipids. The *Archaeobacteria* are found only in very special and extreme habitats. They include the thermoacidophilic bacteria (which grow only in very hot, acid pools like those in Yellowstone National Park), the extreme halophilic bacteria (found only in nearly saturated salt pools), and the methane-producing bacteria (found only in extremely anaerobic environments such as marsh muds).

Therefore, an overwhelming body of evidence suggests that microorganisms should be divided into three kingdoms: the *Eucaryotae*, the *Procaryotae*, and the *Archaeobacteria*. Most of this laboratory manual deals with the *Procaryotae*. In Part X, however, we briefly explore one type of eucaryotic microorganism, the fungi.

---

are in division *Thallophyta*, along with the algae and the lichens. Note that the fungi are then subdivided into five classes that include the slime molds *(Myxomycetes)*.

Fungi as a separate kingdom (Myceteae) Fungi are unlike other plants in that they totally lack photosynthetic pigments. For this reason, and others, some prominent mycologists feel that fungal taxa should be separate from plant taxa. The classification scheme given in Table 40-3 is a somewhat simplified version of the scheme proposed by C. J. Alexopoulos and C. W. Mims (1979). They envision the fungi as a separate kingdom, called the *Myceteae*, and subdivide this kingdom into three divisions.

The first two divisions contain the slime molds and the aquatic fungi. The slime molds are included because their fruiting bodies appear similar to those of the other fungi, and because mycologists usually study these organisms. In many other ways, they are quite unlike the other fungi. The third division, the *Amastigomycota*, contains all of the terrestrial fungi

TABLE 40-2 The classical view of fungi as part of the plant kingdom

Kingdom *Plants*

  Division *Spermatophyta* (seed plants that are well differentiated into roots, stems, leaves, and flowers)

  Division *Pteridophyta* (green plants having roots, stems, and leaves but no flowers; includes ferns and club mosses)

  Division *Bryophyta* (green plants somewhat differentiated into stems and leaves; includes liverworts and mosses)

  Division *Thallophyta* (simple forms are unicellular; the plant body of higher forms is called a thallus and is not differentiated into stems and leaves)

    Subdivision Algae (organisms containing chlorophyll; diatoms and red, brown, and green algae)

    Subdivision Fungi (organisms without chlorophyll)

      Class *Myxomycetes* (slime molds)

      Class *Phycomycetes* (aquatic fungi)

      Class *Ascomycetes* (terrestrial fungi)

      Class *Basidiomyetes* (terrestrial fungi)

      Class *Deuteromycetes* (terrestrial fungi)

    Subdivision Lichens (composite organisms; fungi and algae living together symbiotically)

  Division *Protophyta* (primitive plants: unicellular, multiplication by simple fission; includes bacteria, richettsia, and viruses)

NOTE: In this view, the fungi are part of the Division *Thallophyta* (simple plant forms) along with the algae and lichens. Bacteria, rickettsia, and viruses are considered primitive plants *(Protophtya)*.

(terrestrial yeasts, molds, and the more highly developed fungi such as the mushrooms); these are the fungi that we consider in Exercises 41 and 42.

Separating fungi into classes Mycologists often disagree on where to place a particular fungus, especially in groups known as classes (see Tables 40-2 and 40-3), because they often do not know enough about the development, structure, physiology, or genetics of a fungus to properly classify it. Mycologists also have difficulties because phenotypic characteristics can change drastically with a shift in environment. In other words, data on the structure, life cycles, and physiology of many fungi are fragmentary. Thus, fungal classification schemes are often based, in part, on opinion rather than on confirmable data.

Fungal classification is primarily based on the characteristics of the *sexually formed spores* and the associated fruiting bodies. The process of sexually forming spores is considered part of the *perfect* life cycle of the fungus. This is a major problem because some fungi (the so-called *imperfect* ones) have never been observed to form spores sexually, or they have done so only on rare occasions that are difficult to confirm in other laboratories. Therefore, these imperfectly described fungi must be classified on other bases, such as the morphology of their mycelia and asexually formed spores; and they are confined to the class *Deuteromycetes*. If or when sexual stages are observed, the fungi will be reclassified into one of the remaining three classes of terrestrial fungi.

Nevertheless, mycologists must sort the fungi into groups (taxa) and name these groups, so that they can communicate with one another about their findings concerning a certain fungus and its relationship to other fungi and other living organisms. The current way in which fungi are separated into classes according to Alexopoulos and Mims (1979) is shown in Table 40-4. This classification scheme will be used throughout Part X (Exercises 41 and 42). Note that the name of each fungal class ends with -*mycetes*. When considering the *Ascomycetes*, you should realize that it is a class of fungi made up of many genera that all share the characteristics given in Table 40-4.

The Future of fungal classification One of the most exciting areas of research in fungal classification deals with a relatively modern technique known as **nucleic-acid hybridization.** This is a method whereby one type of nucleic acid (NA) is extracted from a fungus, separated (if necessary) into single strands (ssNA), and then compared with the ssNA obtained from another type of fungus. After mixing the single strands together and encouraging the formation of new double strands *(hybrids)*, one can determine the genetic relatedness of the two fungal types. If the two fungi are genetically identical, both ssNAs completely recouple *(hybridize)*; the two fungal types are then said to be 100 percent homologous.

Nucleic-acid hybridization techniques have been used very successfully in studying the genetic relatedness of procaryotic microbes. Thus, the future application of these techniques to the study of fungal classification should eventually eliminate the dependence on observed phenotypic characteristics and on the opinion of experts. Instead, known genetic relationships can be shown and a classification scheme developed that is based on genotype rather than on a fragmented phenotype.

Fungal physiology and growth

The **temperature** at which most fungi *grow* best is between 20° and 30°C. Just as with bacteria, the fungi can withstand extremely low temperatures, which

TABLE 40-3 Classification of the fungi as a separate kingdom

Superkingdom *Eukaryota* (eucaryotic macro- and microorganisms)

  Kingdom *Myceteae* (fungi)

    Division *Gymnomycota* (slime molds)
    Organisms that ingest particulate nutrients (like bacteria) and that lack cell walls in the vegetative stage; includes the following genera: *Dictyostelium, Polysphondelium, Physarum,* and *Didimium*

    Division *Mastigomycota* (aquatic fungi)
    Aquatic fungi producing motile cells; includes the following genera: *Allomyces, Rhizidiomyces, Hyphochrytrium, Hyphochrytrium,* and *Saprolegnia*

    Division *Amastigomycota* (terrestrial fungi)
    Terrestrial fungi having absorptive nutrition; except for the yeasts, most produce a well-developed mycelium consisting of septate or nonseptate hyphae; motile cells are completely absent; sexual reproduction, where known, culminates with the production of zygospores, ascospores, or basidiospores; asexual reproduction is by budding, fragmentation of hyphae, or sporangiospores of conidia formation; includes the yeasts, molds, mildews, cup fungi, rusts, smuts, bracket or shelf fungi, puffballs, and mushrooms

      Class *Zygomycetes*

      Class *Ascomycetes*

      Class *Basidiomycetes*

      Class *Deuteromycetes*

NOTE: According to the view of C. J. Alexopoulos and C. W. Mims, *Introductory Mycology*, 3rd ed., New York: Wiley, 1979, in which the fungi are separated into three divisions, and the terrestrial fungi are subdivided into four major classes according to the type or presence of sexual reproduction.

TABLE 40-4 Classification and description of the major classes of terrestrial fungi (*Amastigomycota*)

Division *Amastigomycota* (terrestrial fungi)

  Class *Zygomycetes*
  (Gr. *zygos* = yoke) Sexual reproduction by gametangial fusion; the resulting zygote is transformed into a thick-walled spore called a zygospore, which is formed within a zygosporangium; asexual reproduction by means of sporangiospores produced within a sporangium; most species exhibit nonseptate hyphae; includes the following genera: *Rhizopus, Phycomyces,* and *Mucor*

  Class *Ascomycetes*
  (Gr. *askos* = sac) Sexually produces an ascus, a saclike cell usually containing a definite number of ascospores; asexual reproduction by fission, fragmentation, chlamydospores, or conidia formation; mycelium, if formed, usually composed of septate hyphae; includes the higher mushrooms, such as morels and truffles, the yeasts, such as the genus *Saccharomyces,* and the molds, such as the following genera: *Neurospora, Chaetomium, Endothia,* and *Ceratosystis*

  Class *Basidiomycetes*
  (Gr. *basidion* = small base) Sexual reproduction culminates in the production of basidiospores on the outside of a specialized spore-producing structure called the basidium; asexual reproduction by budding, fragmentation of the mycelium, or production of conidia, arthrospores, or oidia; mycelium consists of septate hyphae that penetrate the substratum and absorb nutrients; includes some higher mushrooms, the puffballs, rusts, smuts, jelly fungi, and bracket fungi

  Class *Deuteromycetes*
  (Gr. *deutero* = second class, or imperfect) Sexual reproduction absent; asexual reproduction by means of conidiospore formation; mycelium, if formed, typically contain septate hyphae; includes many molds and mildews and some yeasts (the helpful genus *Penicillium* belongs here); also found are many plant pathogens and most human fungal pathogens, such as the casual agents of ringworm and athlete's foot, and the following genera: *Alternaria, Aspergillus, Candidia,* and *Trichophyton*

NOTE: According to C. J. Alexopoulos and C. W. Mims, *Introductory Mycology*, 3rd ed. New York: Wiley, 1979.

makes possible the long-term storage of fungal cultures in liquid nitrogen ($-196°C$).

In contrast to bacteria, fungi prefer to grow in a medium with an acid **pH**—about 6 is optimum for the growth of most fungi.

The mycelium (or hypha) of a filamentous fungus usually emerges from a germinating spore as a tubular cell called a **germ tube.** As the germ tube grows, it forms a mycelium that elongates and forms branches as it grows. An intact hypha grows only at its tip; therefore, this is the region where most synthesis takes place in the developing fungus. Because of frequent branching of the hyphae, the colony (thallus) grows equally in all directions. This is most readily observed with *liquid* laboratory media, where the developing thallus forms a spherically shaped, fluffy-appearing structure similar to a loose ball of cotton or a dandelion flower that has gone to seed.

When grown on a *solid* laboratory media, the colony (thallus) often appears to be dry, powdery, or fuzzy. The fungal colony is often capable of almost indefinite growth. Given the proper environment, a single colony can completely cover the solid medium inside a petri dish after 4 to 5 days. With longer incubation, the mycelium can actually lift the cover off the plate!

Alexopoulos and Mims (1979) state that, in nature, fungal colonies have been known to continue growing for more than 400 years, and it is possible that some mycelia are thousands of years old.

Unlike the bacteria, the metabolism and **nutrition** of the fungi fit into only one category: chemoorganotrophic (see Table 40-1). This means that the fungi are capable of using only reduced organic compounds as a source of carbon and energy. Conversely, the fungi are incapable of using cabon dioxide as a sole carbon source, and they can use neither light nor reduced inorganic compounds as an energy source.

The fungi have diverse ways of obtaining energy for growth. Some are **parasites,** which infect and get their food from living tissue. Many fungi form a **symbiotic** or **mutualistic** relationship with plants, such as the lichens and the mycorrhizal fungi. The **lichens** are structures that contain both algae and fungi living together for their mutual benefit. The **mycorrhizae** are fungi that live closely associated with plant roots to the mutual benefit of both plant and fungus.

Most known fungi are **saprophytes** (Gr. *sapros* = rotten). This means that fungi are capable of growing on nonliving organic materials, such as artificial or synthetic laboratory media. For the growth of most fungi in the laboratory, you need only provide a carbohydrate (such as glucose or maltose), an organic or an inorganic source of nitrogen and phosphorous, and certain minerals.

Many saprophytic fungi can also use more complex carbohydrates (polysaccharides) as a source of metabolic carbon and energy. These compounds are frequently too large to be transported into the cell; so the fungus excretes **extracellular enzymes** that break down the polysaccharide into smaller units for transport (see the exercises in Part V). This is one reason why fungi grow on a wide variety of materials. For example, if given a little moisture, the common green mold *Penicillium* and the common black mold *Aspergillus* can grow on anything from cheese to shoe leather.

When grown on synthetic media containing excess carbohydrates, many fungi form intracellular storage **vacuoles** that contain glycogen or lipids. These compounds are stored by the fungus for later use as carbon and energy sources when the environment becomes depleted of such materials. The vacuoles may be examined with the brightfield light microscope after they are stained. For example, *iodine* is used to stain starch and glycogen, and *Sudan black* stains vacuoles that contain lipids.

Laboratory cultivation of fungi

Molds and yeasts are studied by the same cultural methods that are used for bacteria. Either solid media or small amounts of liquid media (with a lot of surface exposed for oxygen absorption) are used because almost all fungi either require oxygen (molds) or grow better in the presence of oxygen (yeasts).

Fungi are often characterized by slow growth; that is, most fungi grow more slowly than most bacteria. This relatively slow growth can be a problem, especially when you want to isolate fungi from environmental sources. To prevent bacteria from outcompeting the fungi for nutrients, you can use a selective medium that contains a relatively high sugar concentration (about 4 percent) and is adjusted to an acidic pH (5 to 6).

One of the best-known selective media that meets these criteria was developed by Sabouraud and later modified by substituting glucose for maltose. **Sabouraud's modified medium** contains 4 percent glucose (carbon and energy source), 1 percent peptone (source of organic nitrogen and minerals), water, and agar (if desired). This medium is adjusted to pH 5.0 to 5.6 before it is autoclaved. Sabouraud's medium is widely used for the isolation and subsequent cultivation of molds and yeasts.

Although an intact hypha grows only at the tip, all parts of the mycelium are capable of growth. The transfer of even a small, broken fragment of a hypha is sufficient to cause a new colony to form, which can usually be done with the same loop or needle used to transfer bacteria. The colonies of some filamentous fungi, however, are very hard and difficult to penetrate with a bacteriological loop. For this reason, mycologists sometimes use a stiffer needle with a flattened tip that is sharp enough to cut into these hard colonies.

Asexual reproduction of the fungi

Most fungi reproduce both sexually and asexually. Alexopoulos and Mims (1979) believe that asexual reproduction is the more important in nature because (1) it results in the production of many more individuals, and (2) the asexual cycle is usually repeated several times a year whereas the sexual stage of many fungi occurs only once a year.

Asexual reproduction, sometimes called *somatic* or *vegetative* reproduction, does not involve the union of nuclei, sex cells, or sex organs. In the broadest sense, there are four methods of asexual reproduction: fragmentation, transverse fission, budding, and asexual spore formation.

Fragmentation Some fungi naturally fragment their hyphae into components known as **arthrospores** (Gr. *arthron* = joint + *spora* = seed). Arthrospores are illustrated in Figure 40-4a. If the cells develop thick walls before they separate from each other, they are called **chlamydospores** (Gr. *chlamys* = a mantle or stone arch surrounding a fireplace + *spora* = seed). Chlamydospore formation is shown in Figure 40-4b. Fragmentation of the mycelium may also occur accidently. When it occurs in nature under conditions that favor growth, each bit of the torn mycelium may form a new thallus. When microbiologists artificially fragment the thallus with an inoculation loop and transfer the fragments to a sterile medium, a new culture will start.

Transverse fission Fungi can also reproduce asexually by transverse, binary fission of one cell into two daughter cells, as shown in Figure 40-5a. This is characteristic of the division of a number of yeasts as well as of the septate hyphae (those with cross walls) of some filamentous fungi.

Budding Many biology students are aware that an asexual reproduction called budding occurs in most yeasts, as illustrated in Figure 40-5b. However, many do not realize that budding also occurs in many other types of fungi at certain phases of their life cycles or under certain growth conditions.

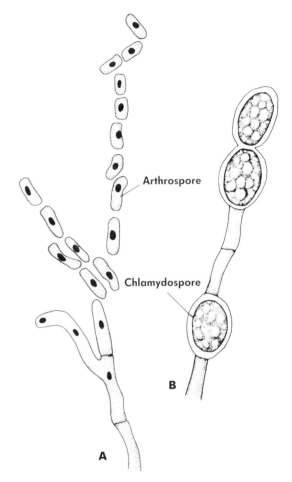

FIGURE 40-4 Asexual reproduction of the filamentous fungi by "natural" fragmentation. (A) In one type, hyphae are fragmented into arthrospores. (B) In another type, hyphae form chlamydospores prior to fragmentation. Although the resulting structures are called "spores," they should not be confused with the type of fungal spores produced as a result of sexual or asexual reproduction. (Adapted from C. J. Alexopoulos and C. W. Mims, *Introductory Mycology*, 3rd ed. New York: Wiley, 1979; with permission.)

Budding is the formation of an outgrowth from the parent cell that is accompanied by nuclear division in the parent cell. One nucleus migrates into the bud, and the other stays with the parent cell. The bud increases in size and then breaks off from the parent cell to form a new individual.

Asexual spore formation In all the fungi, the most common form of asexual reproduction is the formation of spores. This type of spore formation is called asexual because all genes in each spore come from a single parent. Spore formation is considered reproduction because many individuals (spores) come from a single parent cell (the reproductive mycelium). An almost endless variety in the size, number, shape, and arrangement of asexually

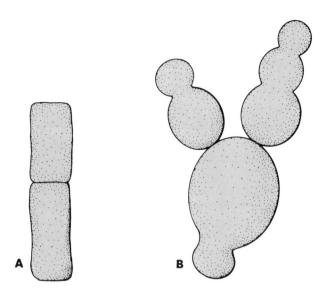

FIGURE 40-5 Two examples of asexual reproduction: (A) transverse cell division (binary fission) and (B) budding.

formed spores occurs among the fungi. In general, however, asexually formed spores either form in saclike structures at the end of reproductive (aerial) mycelium or develop, without a sac, at the tip and sides of the reproductive mycelium.

If the asexually formed spores are formed inside sacs at the tip of the reproductive mycelium, these saclike structures are called **sporangia** (singular = *sporangium;* Gr. *spora + angeion* = seed vessel), and the spores are called **sporangiospores** (Figure 40-6). Sometimes the sporangiospores are motile because of the presence of one or two flagella.

If the asexually formed spores are formed at the tops or sides of the reproductive mycelia (aerial hyphae) and are not enclosed in a saclike structure, the spores are called **conidia** (singular = *conidium;* Gr. *konis* = dust). The fungi form many different types of conidia, one of which is shown in Figure 40-7.

Once the sporangiospore is released from its sac or the conidia are freed from the reproductive mycelium, each spore can **germinate** and form a new individual.

Resistance of the fungal spore

Sporulation by the fungi serves two purposes: reproduction (one or two cells produce many spores) and survival (resistance to environmental stresses).

All of the filamentous fungi (molds) and some of the single-celled fungi (yeasts) form spores, either sex-

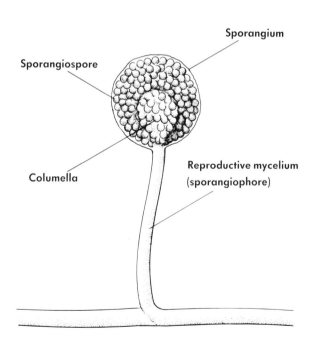

FIGURE 40-6 Cross-sectional view of the first of two possible arrangements of asexually formed spores produced by the filamentous fungi. In this arrangement, the spores are formed inside a sac (sporangium) at the end of the reproductive mycelium. This arrangement of asexually formed spores inside a sporangium is characteristic of the class Zygomycetes, which contains genera such as *Rhizopus* and *Mucor*. (See Table 40-4.)

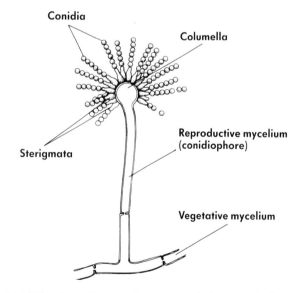

FIGURE 40-7 Cross-sectional view of the second of two possible arrangements of asexually formed spores produced by the three-dimensional fungi (the three-dimensional appearance looks more like that shown in Figure 40-3 or that of a dandelion flower that has gone to seed). In this arrangement, the spores (conidia) are *not* formed inside a sac; instead, they are formed at the top or side of the reproductive mycelium. Although this is an illustration of only the mature fruiting body the genus *Aspergillus*, it does show the septated mycelium and non-sac-enclosed spores, which are general structure features of both the class Ascomycetes and the class Basidomycetes. (See Table 40-4.)

ually or asexually. The fungal spore is not nearly as resistant as the bacterial endospore; nevertheless, the fungal spore is more resistant to environmental stresses than the vegetative cell of the same species. The fungal spore's resistance to the penetration of chemicals is largely attributed to its thickened wall. Resistance to elevated temperatures and drying is probably due to the already desiccated state of the spore. A wide variation in heat resistance exists among fungal spores; but, in general, most are resistant to a 5- to 15-degree elevation beyond the temperature at which the vegetative cell of the same species will survive. On the other hand, most fungal spores are killed by holding them for 5 to 30 minutes at 60°C. Essentially all fungal spores are killed by pasteurization conditions.

Sexual reproduction of the fungi

Alexopoulos and Mims (1979) define sexual reproduction in the fungi, as in all other living organisms, as "the union of two compatible nuclei." This union occurs in three distinct stages: plasmogamy, karyogamy, and meiosis.

- **Plasmogamy** (Gr. *plasma* = a molded object + *gamos* = marriage) is the union of two protoplasts to bring the nuclei close together in the same cell. During this *protoplast fusion,* the two cell membranes fuse, and all constituents of the two cells mix. This event is followed, either immediately or after some time, by karyogamy.
- **Karyogamy** (Gr. *karyon* = nucleus + *gamos* = marriage) is the fusion of the two nuclei. This time, the two nuclear membranes in this cell fuse, so that the nuclear contents mix and the nucleus becomes *diploid.* Karyogamy is followed by meiosis.
- **Meiosis** (Gr. *meiosis* = reduction) is the splitting of the diploid nucleus that was formed during karyogamy, which reduces the nucleus to the *haploid* state. This completes the sexual reproductive cycle.

Sexual reproduction among fungi is often dependent on the formation of sexual organs. Some fungi produce male and female sex organs on each thallus; such species are called **hermaphroditic** fungi. Other species form distinct male or female thalli and are called **dioecious** fungi. At least five distinct methods exist by which the fungi use their sexual organs to accomplish reproduction. Descriptions of these methods are beyond the scope of this introduction, but the methods are a source of fascinating study for students of mycology.

Habitat and significance

Fungi are found in diverse environments. Some, the *aquatic* fungi, live primarily in fresh water; and a few, the *marine* fungi, live in the oceans. However, most fungi are *terrestrial* and live in soil or on dead plant matter; these play a principal role in the complete degradation (mineralization) of plant and animal tissue. For example, you may have seen plant-decay fungi growing on dead trees as you walked through the woods. The decay of dead plant and animal tissue is necessary to the maintenance of our environment because this process returns essential chemicals to the soil to serve as nutrients for new living things.

On the other hand, the **decay** produced by fungi in the artificial environments created by humans can be a problem. For example, particularly in the warm and humid parts of the United States, fungi are often found growing on leather goods, books, painted surfaces, and on shower curtains and bathroom tiles. We often refer to these decay fungi as **mildew.** They do not usually cause much damage if one promptly removes them by washing, kills them with solutions containing phenolic compounds, destroys them with solutions containing household bleach (sodium hypochlorite), or prevents their growth by greatly reducing the humidity. When not controlled, however, decay fungi can cause severe loss of timber, textiles, food, and other materials stored for and used by humans, especially in tropical climates. Whether the problem of decay fungi is a mere nuisance or inflicts severe economic loss depends upon three things: how humid the climate is, whether people know how to kill or control fungal growth, and whether people can afford these preventative measures. The study of how to prevent the growth and action of decay fungi on food, clothing, equipment, and other human materials is a challenging area of current research.

Spores from molds (and some yeasts) are commonly found as contaminants by microbiologists who use poor aseptic technique in the laboratory, because mold and yeast spores resist desiccation and because they are often present on dust particles in the air or on the laboratory bench.

Some human diseases are caused by fungi. These are collectively called **mycoses** and can be diseases of the skin, such as ringworm and athlete's foot, or of the deep tissue (systemic or subcutaneous mycoses), such as candidiasis, histoplasmosis, and coccidioidomycosis. The fungal skin diseases cause discomfort and an unpleasant appearance, but they are rarely life threatening and can be easily and effectively treated with antifungal powders and creams. The deep-tissue mycoses are much more serious, but they seem to be more rare and often occur as a secondary infection in a person who is weakened by another disease or exposed to chemotherapeutic drugs (such as antibiotics or hormones). Antifungal antibiotics are usually very effective in controlling systematic mycoses. For these reasons, human mycoses do not have a large economic impact in most parts of the United States.

Of much greater significance are the **plant-pathogenic fungi** because most economically significant diseases of plant crops are caused by fungi. Other fungi can grow on stored grains (when the moisture content is too high), and some of these fungi excrete substances called **mycotoxins** or **aflatoxins.** These fungal excretory products are extremely toxic or carcinogenic (or both) to humans and other animals that consume the chemically contaminated grain.

Not all fungi are bad, however. Some have been put to use by humans. For example, single-celled fungi (yeasts) are used to leaven bread and to prepare fermented beverages such as beer, wine, and hard cider. Some filamentous fungi (molds) are added to certain cheeses (such as Roquefort, blue, and Camembert) to produce characteristic flavors during the cheese-ripening process. Other similar filamentous fungi produce **antibiotics** that are used as chemotherapeutic drugs to kill or inhibit the growth of harmful bacteria, thereby controlling infections. Another area of current research is to genetically modify these eucaryotic cells in ways that will allow the cells to produce more of these desirable products.

40 INTRODUCTION TO THE FUNGI

NAME _____

DATE _____

SECTION _____

QUESTIONS

Completion

1. Unlike all other types of higher plants, fungi lack stems, roots, and leaves; and unlike all plants, fungi lack _____.

2. Of all the characteristics that differentiate fungi and bacteria, the only one that beginning microbiology students can easily use is _____.

3. The entire body of a fungus is called a(n) _____. With filamentous fungi, this body is made up of many filaments, which are called _____.

4. According to Alexopoulos and Mims (1979), the terrestrial fungi *(Amastigomycota)* should be thought of as divided into _____ classes, and these classes are primarily distinguished by the characteristics of their _____.

5. Dimorphic fungi are so called because they can exist in either the _____ or the _____ form.

6. Selective media for fungal isolation often contain a relatively high concentration of _____ and have a pH adjusted to a value of about _____ to _____. A selective medium that is commonly used for that purpose is called _____ medium.

7. Fungi make use of four general types of asexual reproduction. These are _____, _____, _____, and _____.

8. Conidia are spores that are formed asexually, not surrounded by a sac, formed by filamentous fungi, and placed in the class called _____.

9. Sporangiospores are surrounded by a saclike structure that is called the _____; these asexually formed spores are formed by filamentous fungi that are in the class called _____.

10. Plant pathogenic fungi may adversely affect the health of humans because some produce human toxins called _____ or _____.

11. Some filamentous fungi produce chemicals that are used as chemotherapeutic drugs that are effective against other types of microorganisms; these drugs are called _____.

True – False (correct all false statements)

1. All fungi have a procaryotic cell structure. _____

2. One can use the bacterium *Escherichia coli* as a rough ruler for making length estimates with the light microscope, by assuming the average *E. coli* cell to be about 1.0 μm long. In contrast, an average fungal cell diameter is about 5 to 10 times that size. _____

3. Fungi prefer to grow in a medium that has an acid pH, and all fungi either require oxygen or grow better in the presence of oxygen. _____

4. Natural fragmentation of mycelia produces structures, which are not truly spores, called sporangiospores and conida. _____

5. Budding is commonly found as a sexual reproduction mechanism in the yeasts although it is not limited to this fungal type. _____

6. Sexual reproduction is not possible among the fungi. _____

41

Morphology and Reproduction of the Molds

If you examine moldy bread or fruit or the black patches that develop on your shower curtain with a hand lens, you will see tiny, white, threadlike structures interspersed with darkly colored dots. The white threads are vegetative mycelia, and the dark dots are pigmented fruiting bodies at the end of reproductive mycelia. These structural features are common to the microscopic, filamentous fungi commonly called the molds and mildews.

Many of the filamentous fungi are helpful, even indispensable, to human life. For example, some produce antibiotics, which have revolutionized the treatment of bacterial infections. Other fungi excrete commercially valuable acids or help form certain human foods. Still others help decompose dead animal and plant tissue or synthetic materials in such a way that the decomposition products are available as nutrients to living plants and animals.

However, some filamentous fungi are pathogenic for plants, producing diseases such as blights, rots, rusts, and black spots. Others produce fungal diseases (called micoses) of humans or other animals, which affect the hair, skin, or deeper tissues such as the lungs. Some fungi that are pathogenic for plants also produce potent toxins that poison and even kill the animals that eat the infected plants. If healthy plant material harvested for human consumption is too moist when placed in storage or is stored under damp conditions, decay fungi may grow and decompose much of the produce. (It is estimated that more than 50 percent of all food harvested worldwide for human consumption is lost to microbial decay or other types of microbial spoilage.) Still other decomposition fungi damage fabrics, leather, rubber, wood, paper products, or synthetic materials (especially in humid climates).

This exercise introduces the terrestrial, microscopic, filamentous fungi. The views on fungal classification are taken from Alexopolous and Mims (1979). These authors list four classes of terrestrial fungi, three of which contain examples of the microscopic, filamentous fungi: the *Zygomycetes,* the *Ascomycetes,* and the *Deuteromycetes.* (See Table 40-4.)

Filamentous fungi in the class Zygomycetes

Significance Most of these fungi (Table 41-1) are **saprophytic** (rather than parasitic), and many are capable of synthesizing industrial products. For example, the common bread mold *Rhizopus stolonifer (R. nigricans)* has been used for the manufacture of fumaric acid and for some steps in cortisone manufacture. Another mold, *Rhizopus oryzae,* produces large quantities of ethanol. These and other species of *Rhizopus* can also produce lactic acid. Other genera produce citric, oxalic, and succinic acids and other chemicals for industry. A species of the genus *Actinomucor* is grown on soybeans to produce a food *(sufu)* eaten in Oriental countries; another Oriental food *(tempeh)* is made with various species of *Rhizopus.*

Structure and growth of vegetative mycelia In most species of zygomycetes, the vegetative mycelia have no separating cross walls; that is, they are *without septa (coenocytic),* as shown in Figure 41-1. Note that several reproductive mycelia may be formed close to one another on a vegetative mycelium. The vegetative mycelia of some species, especially in the genus *Rhizopus,* form **rhizoids** (Gr. *rhiza* = roots), which are rootlike structures that project from the vegetative mycelia and that form especially where the mycelia touch a solid surface. Rhizoids are difficult to see

TABLE 41-1 Characteristics of molds in the class *Zygomycetes*

Mycelium (hypha)	No cross walls (called nonseptate or coenocytic)
Asexual reproduction	Production of spores, called sporangiospores, formed within a membranous sac (sporangium) found at the end of a reproductive mycelium
Sexual reproduction	By gametangial fusion. Results in a thickened wall, called a zygosporangium, that surrounds one diploid zygospore. (Gr. *zygos* = yoke)
Selected genera	*Rhizopus* (black, bread mold) *Phycomyces* (phototrophic) *Mucor*
Pathogenic species	Very few

NOTE: This table describes only the microscopic, filamentous fungi in the class *Zygomycetes*, which is one of four classes of terrestrial fungi according to C. J. Alexopoulos and C. W. Mims, *Introductory Mycology*, 3rd ed., New York: Wiley, 1979. The class *Zygomycetes* does not contain forms commonly called yeasts but does have some filamentous fungi with dimorphic tendencies.

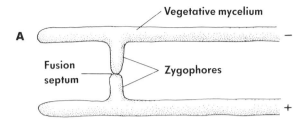
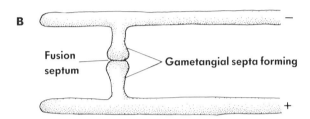
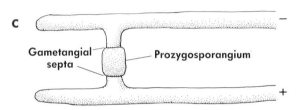
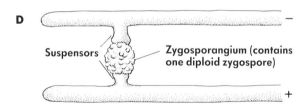

FIGURE 41-2 Sexual reproduction by filamentous fungi in the class *Zygomycetes*. Steps A–D represent four time periods during the development of a zygosporangium. The plus (+) and minus (−) refer to adjacent cells that have different genotypes.

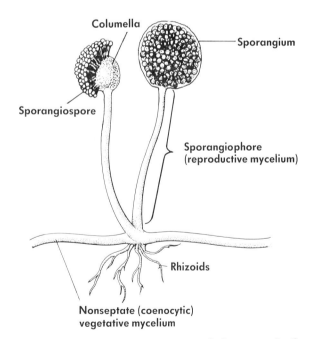

FIGURE 41-1 Asexual reproduction of *Rhizopus stolonifer* (*R. nigricans*), a member of the class Zygomycetes. The fruiting body on the right is shown in cross section; that on the left is a partially dissected fruiting body. Structural features include nonseptate vegetative mycelia, reproductive mycelia, and asexually produced sporangiospores contained within a sporangium. Rhizoids, root-like structures that anchor the mycelia to solid growth medium, are not typical of all zygomycetes.

when they are growing down into nutrient agar unless they are located at the agar–coverglass interface of a slide culture.

Initial growth is usually white and fuzzy, reflecting the development of vegetative mycelia, which may spread over the entire surface of a Petri plate. Continued incubation results in a thickening of the thallus, pigmentation of its top surface (exposed to the air), and wrinkling and discoloration of its underside. Pigmentation of the top surface is usually correlated with the formation of reproductive mycelia, sporangia, and sporangiospores.

FIGURE 41-3 Mature zygosporangium (ZS) with adjacent suspensors (S) as seen with scanning electron microscopy. (From K. L. O'Donnell J. T. Ellis, and C. W. Hesseltine, USDA, *Canadian Journal of Botany* 55, 662–675, 1977, with permission.)

Asexual reproduction This type of reproduction is characterized by the formation of reproductive mycelia, called **sporangiophores** (see Figure 41-1). At the distal end of the sporangiophore is a swelling called the **columella**, which is surrounded by a membrane (sporangium) that encloses the asexually formed spores (sporangiospores).

Sexual reproduction This type of reproduction is characterized by **gametangial fusion**, in which two adjacent vegetative mycelia form projections (specialized hyphae) called **zygophores** that grow toward each other (Figure 41-2a). Zygophores are typically formed near the tips of the actively growing vegetative hyphae. Compatible zygophores are attracted to one another and bond together to form a fusion septum.

After fusion, the tip of each zygophore swells to form a **progametangium**. A **gametangial septum** then forms near the end of each zygophore (Figure 41-2b). When fully formed, the septum seals off a cell at the end of the zygophore called a **terminal gametangium**, which contains one set of chromosomes from the parent mycelium. What remains of the original zygophore is now called the **suspensor cell**.

Next, the fusion septum dissolves, the protoplasts of the two gametangia mix (**plasmogamy**), and the nuclei of the two cells fuse (**karyogamy**) to produce a diploid nucleus. After fusion of the two gametangia, the newly formed diploid cell, called the **prozygosporangium** (Figure 41-2c), enlarges, develops a thick, multilayered wall that often has a rough appearance. Once matured, this thick wall is called the **zygosporangium**, and a single diploid ($2n$) **zygospore** develops inside the zygosporangium (Figures 41-2d and 41-3). In the genus *Mucor*, meiosis occurs about one week after the mature zygospore is formed and produces four haploid ($1n$) cells inside the zygosporangium. In the genus *Rhizopus*, the diploid zygospore undergoes meiosis just prior to or during zygospore germination.

At the time of germination, the zygosporangium cracks open, and a tubelike structure grows from one of the haploid cells inside. This becomes a reproduc-

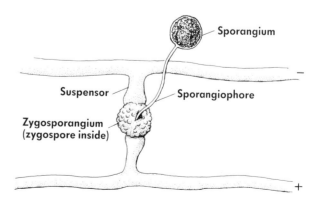

FIGURE 41-4 A zygospore germinating from within a zygosporangium. The sporangiophore extends outward through the cracked zygosporargium and culminates in asexually formed spores within a sporangium.

tive mycelium (sporangiophore), which emerges through the cracked zygosporangium. Usually, a sporangium forms on the distal end of this sporangiophore (Figure 41-4).

Although this magnificent display of differentiation and sexual reproduction has been observed in all laboratory-grown zygomycetes, zygospore formation is rarely seen in nature. Thus, some mycologists question the importance of sexual reproduction for the zygomycetes.

Filamentous fungi in the class *Ascomycetes*

Significance Many of the ascomycetes are found in the soil and greatly contribute to degradation of dead plant and animal tissue (Table 41-2). These degraders are essential for maintaining our ecosystem, but when they invade our homes and factories, they can cause problems. Of special economic importance are the cellulolytic ascomycetes, such as *Chaetomium* species, which destroy vast quantities of document papers and cotton fabrics, especially in the more humid climates of the tropics.

Perhaps the most damaging of the ascomycetes are the plant pathogenic strains that destroy crop plants as well as timber and ornamental trees. These microbes cause the plant diseases frequently referred to as *scabs, rots,* and *blights*. Examples of crop diseases are apple scab, ear rot of corn, brown rot of stone fruits, and foot rot of cereal grains. Two examples of tree or timber diseases are especially notable: the ascomycete *Endothia parasitica* causes chestnut blight, which has almost completely destroyed all mature American chestnut trees; and the ascomycete called *Ceratosystis ulmi* causes Dutch elm disease, which is well on the way to causing extinction of the American elm tree.

Very few of the true ascomycetes cause diseases of domestic animals and humans. One of the exceptions, the **ergot** fungis, *Claviceps purpurea,* is especially interesting because it is a pathogen to both plants and animals. This ascomycete invades and destroys the fruit of rye and other grasses. It also produces sporelike structures containing **alkaloids** that are deadly to humans and other animals when consumed in large quantity. Yet these same alkaloids, when administered correctly and in very small quantities, are important drugs for the treatment of human ailments such as migraine.

Structure of vegetative mycelia The vegetative mycelia of the ascomycetes contain **cross walls,** also called septa (singular = septum); see Figure 41-5. Septum formation begins at the outside and grows toward the center, like the iris diaphragm of a camera. In most cases, a small hole or **pore** is left near the center of the septum, but often this pore appears plugged with membranous material. It is difficult to think of these mycelia as many individual cells when the mycelia form incomplete septa. Because of this problem, some mycologists refer to the ascomycete

TABLE 41-2 Characteristics of molds in the class *Ascomycetes*

Mycelium (hypha)	Contains incomplete cross walls (called septa)
Asexual reproduction	Production of spores, called conidia, formed at the end of a reproductive mycelium; conidia not surrounded by a membranous sac
Sexual reproduction	By ascus formation (a saclike cell) containing a definite number of ascospores (usually eight)
Selected genera	*Neurospora* (red, bread mold) *Chaetomium* (cellulolitic) *Endothia* species (plant pathogens) *Ceratosystis* species (plant pathogens)
Pathogenic species	Few human pathogens but many plant pathogens

NOTE: This table describes only the microscopic, filamentous fungi in the class *Ascomycetes,* which is one of four classes of terrestrial fungi according to C. J. Alexopoulos and C. W. Mims, *Introductory Mycology,* 3rd ed., New York: Wiley, 1979. In addition to the microscopic filamentous fungi, this class contains ascus-forming yeasts, powdery mildews, cup fungi, morels, and truffles.

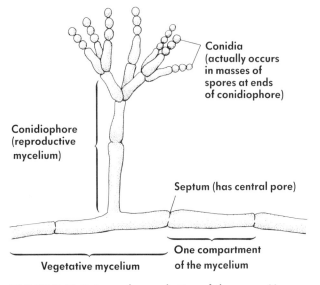

FIGURE 41-5 Asexual reproduction of the genus *Neurospora*, a member of the class *Ascomycetes*. Characteristic features include septate mycelia and asexually formed conidia. Note that other fungi, such as the genera *Penicillium* and *Aspergillus*, may produce conidia and septate mycelium but appear to lack sexual formation of asci; therefore, they are placed in the class *Deuteromycetes*.

mycelium as composed of *compartments* rather than individual cells.

The ascomycete mycelia are often profusely branched. Compartments may be either uninucleate or multinucleate. In some cases, cell nuclei migrate from one compartment to another. Whether nuclei are carried through the septum pores by protoplasmic streaming or whether they have motility of their own is not yet known.

Asexual reproduction The majority of the filamentous ascomycetes appear to reproduce asexually by forming **conidia** (see Figure 41-5). Conidia can arise either directly from the vegetative mycelium or at the end of a reproductive mycelium called a conidiophore. Conidiophores range from short branches of vegetative mycelium to long and intricately branched structures.

The common red bread mold, *Neurospora sitophila,* is an ascomycete and has the following characteristics. The conidia are oval and pink and form in masses at the ends of branched conidiophores (see Figure 41-5). The mycelia can also be pigmented, depending on the composition of the growth medium. When these fungi invade the microbiological laboratory, they play havoc with pure-culture technique because (1) they produce enormous quantities of conidia that are easily dispersed, (2) the conidia are quite

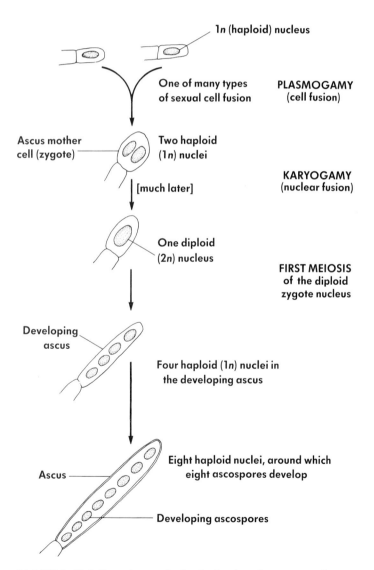

FIGURE 41-6 Sexual reproduction in the class *Ascomycetes* showing plasmogamy, karyogamy, and meiosis. The sexual method by which plasmogamy (cell fusion) takes place varies widely among genera within this class.

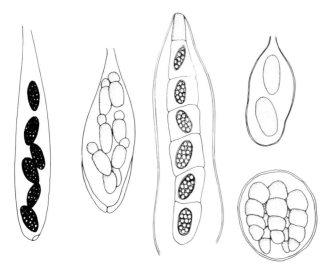

FIGURE 41-7 Various types of sexually formed asci containing ascospores that are found in the class *Ascomycetes*. (Adapted from C. J. Alexopoulos and C. W. Mims, *Introductory Mycology*, 3rd ed., New York: Wiley, 1979; with permission.)

resistant to environmental stress and germinate readily, and (3) the colonies (thalli) grow rapidly.

Sexual reproduction The ascomycetes employ a great many methods to fuse adjacent compatible cells (plasmogamy). Once cell fusion takes place, the resulting binucleated cell is called the **ascus mother cell** (Figure 41-6), which will become the **ascus** (plural = *asci*). Eventually, nuclear fusion (karyogamy) takes place in the ascus mother cell, resulting in one diploid nucleus. Most commonly, meiosis takes place twice to produce eight haploid nuclei, each of which develops into an **ascospore** within the ascus.

Most ascomycetes produce elongated asci shaped like a clubs or cylinders, and most contain eight ascospores (Figure 41-7). Ascospores vary greatly in size, shape, and color; thus their characteristics are often used to identify the genus and species of an ascomycete.

Filamentous fungi in the class *Deuteromycetes*

Overview These fungi apparently lack a sexual phase (Table 41-3). Because of this they are commonly referred to as the "imperfect fungi" or, more technically, ***fungi imperfecti.*** The deuteromycetes have **septate** mycelia and, so far as anyone has observed, reproduce only asexually by forming conidia at the ends of their reproductive mycelia. In other words, the mycelia and fruiting bodies of the deuteromycetes are identical to those of the ascomycetes. Therefore, mycologists consider *the deuteromycetes to be ascomycetes whose sexual stages have never been discovered or do not exist.*

The similarity between the two classes can be confusing to the beginning student because mycologists often use examples from the class *Deuteromycetes* when describing the structural features of the class *Ascomycetes*. For example, genera such as *Penicillium* and *Aspergillus* are used as examples of ascomycetes, even though sexual means of reproduction is unknown among species of these genera. To avoid later confusion for students wanting to continue with mycology, this presentation will consider such genera to be members of the class *Deuteromycetes*. However, remember that the structural features of genera within the class *Deuteromycetes* are identical to those in the class *Ascomycetes*. For a discussion of this controversial issue, see P. J. VanDemark and B. L. Batzing, *The Microbes,* Benjamin/Cummings, Menlo Park, 1987.

Significance The fungal class *Deuteromycetes* consists of more than 15,000 species. Some of these, such as those in the genus *Penicillium* and the genus *Cephalosporium,* produce antibiotics that have been of extreme importance to human medicine and domestic animal production. Others, such as species in the genus *Aspergillus,* produce commercial chemicals. Some of these fungi infect other fungi or trap and consume nematodes.

TABLE 41-3 Characteristics of molds in the class *Deuteromycetes*

Mycelium (hypha)	Contains incomplete cross walls (called septa)
Asexual reproduction	Production of spores, called conidia, formed at the end of a reproductive mycelium; conidia not surrounded by a membraneous sac
Sexual reproduction	None observed
Selected genera	*Aspergillus* *Alternaria* *Cephalosporium* *Cladosporium* *Emmonsiella (Histoplasma)* *Geotrichum* *Microsporium* *Penicillium* *Sporotrichum* *Trichophyton*
Pathogenic species	Many are pathogenic for humans, other animals, and plants.

NOTE: This table describes only the microscopic, filamentous fungi in the class *Deuteromycetes,* which is one of four classes of terrestrial fungi according to C. J. Alexopoulos and C. W. Mims, *Introductory Mycology*, 3rd ed., New York: Wiley, 1979. These authors consider the *Deuteromycetes* to be ascomycetes with unobserved sexual reproductive forms. In addition to the microscopic filamentous fungi, this class also contains some yeasts for which sexual means of reproduction have not been observed.

Some deuteromycetes are human pathogens that produce fungal diseases, collectively called mycoses (singular = *mycosis*). If the fungus infects the *keratinized areas* of the body (such as the skin, hair, or nails), the fungus is called a **dermatophyte,** and the resulting diseases are **dermatomycoses;** examples are ringworm and athlete's foot. These diseases are usually self-limiting and nonfatal. Frequently, they are caused by species from three genera: *Microsporium, Epidermophyton,* and *Trichophyton.* Oral use of a fungistatic agent called *griseofulvin* is often effective in combating dermatophyte infections.

If the fungus infects the *deeper tissues* (such as the lungs, inner eye, or bone), the disease is called a **deep** or **systemic mycosis.** Although the systemic mycoses are frequently caused by yeasts or dimorphic fungi, some filamentous fungi are also causative agents. Regardless of the species of fungus, the disease is usually named after the genus that causes that disease. For example, *aspergillosis* is a deep-tissue infection caused by one of about seven species from the genus *Aspergillus.* It is a very serious disease, especially when lung tissue is infected. Another example is a disease known as *histoplasmosis,* which is caused by a filamentous fungus with no known sexual stages; it used to be called *Histoplasma capsulatum* but is now known as *Emmonsiella capsulata.* This fungus infects lung and internal-eye tissue and is fatal in some cases. Because it can be picked up from bird and bat droppings, people who are around birds or bats frequently show positive skin tests for this fungus. Antifungal antibiotics such as *nystatin* and *amphotericin*-B, available since the late 1950s, are quite effective in treating deep-seated fungal infections.

Out of thousands of known fungi in this class, only about 60 species are commonly pathogenic for humans. If the conditions are right, however, fungi that are normally harmless can cause human infection. Fungal infections by these **opportunistic** microorganisms seem to occur mostly in people whose immune systems are compromised by another disease or who are on antibiotic, steroid, or hormone therapy, which disrupts the normal microbial flora or alters the chemistry of the body.

Two types of deuteromycetes are of particular note because of their ubiquity and their economic importance: the genus *Aspergillus* and the genus *Penicillium. Aspergillus* contains about 200 known species. Conidia of the *Aspergillus niger* fungi, commonly called the black molds, pervade the air and upper layers of soil over our entire planet. These molds use a great variety of substances for food, requiring only a little organic matter and moisture to grow. Some species produce various mycotoxins when growing on nuts and grain stored under moist conditions. The most important of these mycotoxins are the **aflatoxins,** some of which are potent carcinogens. Other species grow on leather and fabrics and produce an unsightly appearance and a musty odor. On the other hand, members of this genus produce many helpful chemicals. For example, microbiologists use *Aspergillus niger* to make citric acid, which is the basis for an enormous soft-drink industry. Citric acid used to be isolated from lemon juice until it was found that it was less expensive to have this fungus make it from sucrose solutions. As a group, the aspergilli form reproductive mycelia and conidia in such abundance that their color—black, brown, yellow, or green—is predominant in the thallus. The exact color depends on both the species and the medium composition, especially the inorganic trace elements.

Species in the genus *Penicillium,* which are as common in the air and soil as the aspergilli, are often called the **green** and **blue molds.** They are frequent pathogens of citrus fruits and can be found on decomposing apples or citrus fruits during long-term storage. The penicillia are also important degraders of leather and fabrics and, like the aspergilli, are capable of commercially producing organic acids, such as citric, fumaric, gluconic, gallic, and oxalic acid. Perhaps the most important industrial use of the penicillia, however, is in the manufacture of cheese and antibiotics. In some ripened cheeses, the fungus produces both the color and the flavor; the flecks of color in the cheese are pieces of reproductive mycelia and conidia that have grown during the ripening process, and the flavor is the consequence of aromatic organic compounds excreted by the growing fungus during ripening. For example, *Penicillium roqueforti* is used to produce Roquefort cheese, and *Penicillium camemberti* to produce Camembert cheese. Danish blue cheese is also ripened with *Penicillium* species. Several species of penicillium produce the antibiotic penicillin, although high-yielding mutants from *Penicillium chrysogenum* are most commonly used. *Penicillium griseofulvum* produces griseofulvin, the most effective antifungal antibiotic used for fungal skin diseases (dermatomycoses).

Structure of vegetative mycelia The filamentous deuteromycetes typically produce incomplete **septate** and branched mycelia. The compartments are usually multinucleate. In all respects, the mycelia are identical to the ascomycetes.

Asexual reproduction The filamentous deuteromycetes, like the ascomycetes, asexually reproduce by forming conidia at the ends of reproductive mycelia (conidiophores). The entire fruiting body may be a ball-like mass of conidia that looks like a paint brush or pom-pom. A microscope shows only the outside layers of conidia. To see the internal arrangement of conidia and the cells connecting the conidia to the reproductive mycelium, one must tease apart the fruiting body. Such exposed arrangements of the conidia of *Aspergillus* and *Penicillium* species are shown in Figures 41-8 and 41-9.

Sexual reproduction As mentioned previously, fungi are placed in the Class *Deuteromycetes* because there are no known sexually reproductive stages.

Introduction to this exercise

You will now study two harmless fungi whose mycelia and asexually formed fruiting bodies have different structural features: (1) *Rhizopus nigricans* (the black, bread mold), which exhibits structural features that are characteristic of filamentous fungi in the class *Zygomycetes,* and (2) *Penicillium notatum,* which is placed in the class *Deuteromycetes* because no one has seen its mode of sexual reproduction, although the structure of its mycelia and asexual fruiting bodies are characteristic of those found in the class *Ascomycetes.*

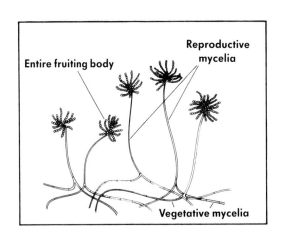

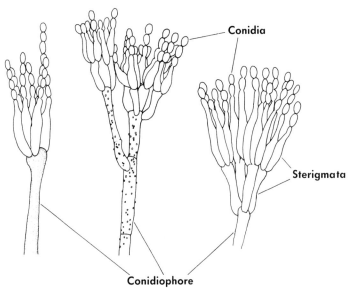

FIGURE 41-8 Structure of three types of asexually formed fruiting bodies found in the genus *Penicillium* (placed in the class *Deuteromycetes*). These illustrate three partially dissected fruiting bodies. The entire fruiting body microscopically often looks like a bottle brush (see inset), and part must often be dissected away to see the number and arrangement of sterigmata and the branching of the conidiophore. (All but inset adapted from C. J. Alexopoulos and C. W. Mims, *Introductory Mycology*, 3rd ed., New York: Wiley 1979; with permission.)

41 MORPHOLOGY AND REPRODUCTION OF THE MOLDS

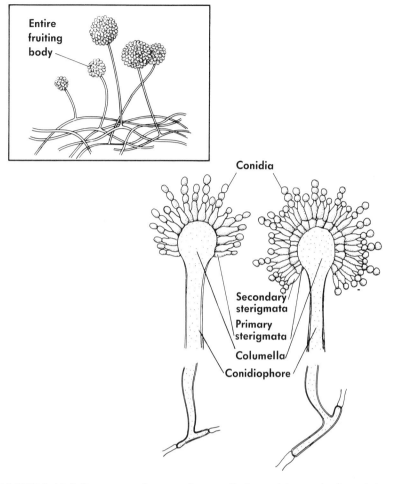

FIGURE 41-9 Structure and types of asexually formed fruiting bodies of the genus *Aspergillus* (placed in the class *Deuteromycetes*). The entire fruiting body usually appears as a sphere of packed conidia (see inset). Only when part is dissected away, can one determine the number and arrangement of sterigmata and conidia. (All but inset adopted from C. J. Alexopoulos and C. W. Mims, *Introductory Mycology*, 3rd ed., New York: Wiley, 1979; with permission.)

LEARNING OBJECTIVES

• Learn to prepare a slide culture and a moist chamber for growing the filamentous fungi.

• Learn to use certain structural features to distinguish between species belonging in the class *Zygomycetes* and those belonging to the class *Ascomycetes (Deuteromycetes)*.

• Learn the significance of the fungi to humans.

MATERIALS

Cultures *Rhizopus nigricans* (a zygomycete)
 Penicillium notatum (a deuteromycete)

Media Sabouraud's agar (melted and tempered for slide-culture preparation)
 Sabouraud's agar plates

Supplies 1.0-ml pipettes (sterile) for slide-culture preparation
 Microscope slides and cover slips
 95 percent ethanol and forceps
 Vaspar (Vaseline and paraffin mixture)
 Small brush

PROCEDURE: FIRST LABORATORY PERIOD

Preparing moist chambers

1. Place a filter-paper disk inside each of two **glass** Petri dishes, **and place a glass micro-**

scope slide on top of each filter-paper disk. Sterilize both assemblies in the autoclave.

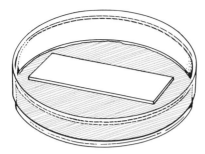

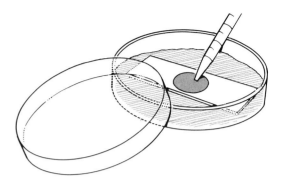

2. **Prepare two slide cultures,** as described below.

3. **Aseptically add 2 ml of sterile distilled water to the filter-paper disk** inside each Petri dish. This apparatus is now called a **moist chamber.** The water provides a humid atmosphere for growth of the filamentous fungi during incubation of the slide culture. The aseptic technique inside the chamber keeps contaminating organisms from growing on the moist filter paper during extended incubation.

4. **Aseptically add sterile water** to the filter-paper disks **every day** that the slide cultures are incubated. You must **keep this paper moist at all times during the incubation period.**

Preparing slide cultures

1. **Prepare two sterile moist chambers with slides,** as described in step 1 of *Preparing moist chambers.*

2. Remove one dish lid, keep it inverted, and rest it on one edge of the bottom of the Petri dish.

3. Without removing the microscope slide, aseptically place one drop of sterile, melted, and tempered agar at the center of the slide. Repeat this procedure with the other chamber.

4. **Replace Petri-dish lids,** and allow the agar drops to solidify.

5. Once again, set one lid aside. **Sterilize an inoculation loop, and use it to cut the agar slab in two. Use the sterile loop to remove and discard the top half of the agar slab.** Repeat this procedure with the other chamber.

   Keep the lid on the Petri dishes when you are not working in these sterile chambers.

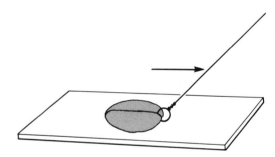

6. Sterilize the loop again, and **inoculate the cut edge of one agar slab with a pure culture of** *Rhizopus nigricans.*

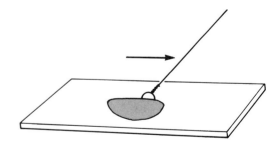

7. Use forceps to dip a coverslip in 95 percent ethanol; then carefully **pass the coverslip through your burner flame** to ignite the ethanol. **Be careful not to drip burning ethanol onto any combustible surface.**

8. After 1 to 2 seconds of cooling, place this sterile coverslip on top of one inoculated agar slab. Repeat this procedure with the inoculated slab in the other chamber.

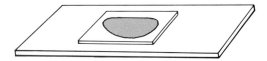

9. Use a small brush to apply melted Vaspar *only* around sides 1, 2, and 3 of each coverslip. Leave side 4 of each open to the air.

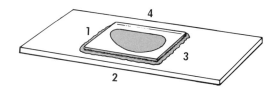

10. Repeat this procedure to inoculate *Penicillium notatum* onto the second slide culture.

11. **Incubate both slide cultures in the moist chambers at 30°C for one week. Add sterile water each day** during this incubation. Take care not to disturb the slide culture. You do not want to detach spores so that they escape from the slide culture and contaminate the moist chamber.

    Because of the need for daily moistening of the chamber, these cultures may be provided for you as a demonstration.

Preparing agar plates

1. Inoculate two agar plates, one with *Penicillium notatum* **and the other with** *Rhizopus nigricans,* **after labeling,** as follows.
   a. Gently touch a sterile inoculating loop to the pigmented surface of a well-sporulated agar culture of one fungus.
   b. Aseptically transfer these spores to the center of one sterile plate. Rub the spores on the surface of the plate so that you cover an area that is no greater than 5 mm in diameter.
   c. Repeat this procedure with the other fungus on a separate, well-labeled, plate.

2. **Tape the cover onto each plate** to prevent its accidental removal.

3. Incubate these plates for two consecutive weeks at 30°C.

PROCEDURE: AFTER ONE WEEK (SLIDE CULTURE)

1. **Examine the cut edge of one slide culture with the high-dry objective** of the light microscope; **then examine the fungus with the oil-immersion objective.** Especially note the following for each type of fungus:
   a. The structure of the reproductive mycelium and the fruiting body with its asexually formed spores.
   b. The septate or nonseptate structure of the vegetative mycelia.

2. **Sketch these features for each fungus** in the Results section.

PROCEDURE: AFTER TWO WEEKS (PLATE CULTURE)

Leave the plates taped during all observations.

1. **Observe the agar-plate cultures, and record the following observations** in your Results section:
   a. The color and textural appearance of the top of each culture as viewed with the unaided eye.
   b. The color and general appearance of the bottom of each culture as viewed with the unaided eye.

Note that both the top and bottom views of these colonies (thalli) are characteristic of the fungal culture.

2. **Use a dissecting microscope to examine each fungal colony (thallus)** for the structure of reproductive mycelia, asexual spore arrangement, and vegetative mycelia structure. Concentrate particularly on the edge of the colony for these observations.

What do you conclude about the suitability and ease of preparation of the slide culture versus the agar plate for *microscopic* characterization of the filamentous fungi?

NAME _____

DATE _____

SECTION _____

RESULTS

Slide Culture of Rhizopus nigricans

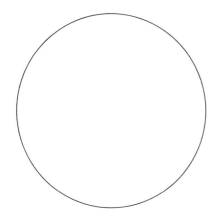

Structure of the reproductive
mycelium and fruiting body
Magnification: _____×

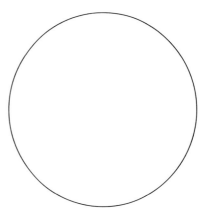

Structure of the vegetative mycelia
Magnification: _____×

Plate Culture of Rhizopus nigricans

Color and appearance of colony viewed from *top* of plate:

Color and appearance of colony viewed from *bottom* of plate:

Word observations from viewing with a dissecting microscope:

Slide Culture of Penicillium notatum

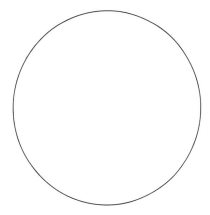

Structure of the reproductive
mycelium and fruiting body
Magnification: _____ ×

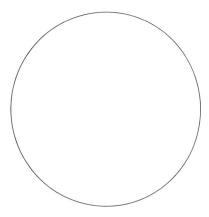

Structure of the vegetative mycelia
Magnification: _____ ×

Plate Culture of Penicillium notatum

Color and appearance of colony viewed from *top* of plate:

Color and appearance of colony viewed from *bottom* of plate:

Word observations from viewing with a dissecting microscope:

CONCLUSIONS

41 MORPHOLOGY AND REPRODUCTION OF THE MOLDS

NAME _____

DATE _____

SECTION _____

QUESTIONS

Completion

1. Diseases of humans and other animals caused by fungi are called _____.

2. Two beneficial products that are produced by the genus *Rhizopus* (a zygomycete) are _____ and _____.

3. The following structural features are characteristic of filamentous fungi in the class *Zygomycetes*.
   a. Describe vegetative mycelia (septate or nonseptate?)

   b. Describe the appearance of asexually produced spores.

   c. Describe the appearance of sexually produced spores.

   d. Name one genus in this class.

4. The following structural features are characteristic of filamentous fungi in the class *Ascomycetes*.
   a. Describe vegetative mycelia (septate or nonseptate?)

   b. Describe the appearance of asexually produced spores.

   c. Describe the appearance of sexually produced spores.

   d. Name one genus in this class.

   e. What general type of fungal pathogens are commonly found in this class?

5. Filamentous fungi in the class *Deuteromycetes* have structural characteristics that are identical to those in the class *Ascomycetes* except that there is a lack of evidence for _____.

6. Two antibiotics commonly used to fight systemic mycoses or dermatomycoses are _____ and _____.

7. Three products manufactured by using filamentous fungi in the genus *Penicillium* are _____, _____, and _____.

8. One harmful substance produced and excreted by some filamentous fungi in the genus *Aspergillus* is _____, and one useful product made by other members of this same genus is _____.

9. _____ is a commonly used solid medium for cultivation of the filamentous fungi.

True – False (correct all false statements)

1. Chestnut blight and Dutch elm disease are two fungal diseases caused by fungi in the class *Deuteromycetes*. _____

2. Most fungal diseases of humans are caused by genera found in the class *Deuteromycetes*. _____

3. Filamentous fungi in the class *Ascomycetes* have septate vegetative mycelia, yet the mycelia are not thought of as being composed of separate cells. _____

4. Most fungal diseases of humans are caused by genera found in the class *Deuteromycetes*. _____

5. Human diseases commonly called ringworm and athlete's foot are generally considered to be systemic mycoses. _____

42

Morphology and Reproduction of the Yeasts

Yeasts are fungi and have characteristics in common with all other fungi. For example, they lack chlorophyll; have the eucaryotic cell structure; have chitin, cellulose, or hemicellulose in their cell walls; are strict chemoorganotrophs; and are from two to ten times larger than the average bacterial (procaryotic) cell.

Yeasts also have unique characteristics. The word *yeast* refers to single-celled fungi that occur in clusters or chains. This definition creates a problem in the description of yeasts because some filamentous fungi exist as individual cells during only one part of their life cycle, whereas other fungi exist as yeasts in one environment but form true filaments in other environments. These morphologically varying fungi are called **dimorphic**.

For the sake of simplicity, let us use the word *yeast* to refer to fungi that are known to exist only as single cells (or chains or clusters of single cells) and that reproduce asexually by budding or transverse fission. Some yeasts that fit this description are also capable of sexual reproduction.

It should be emphasized that the word *yeast* has no taxonomic significance; that is, mycologists do not place yeasts in taxa that are separate from the filamentous fungi but include them with the other terrestrial fungi in the division *Amastigomycota* (see Table 40-3). As with all other fungi, classification depends upon whether sexual reproduction has been observed and, if so, upon its mechanism.

Morphology and asexual reproduction

When yeasts grow on media that contain excess carbohydrates, they form large, membrane-bound, intracellular **vacuoles** in which they store starch or lipid for later use as carbon and energy sources when the environment becomes depleted. These internal structures can be examined with the brightfield light microscope after they are stained (**iodine** stains vacuoles containing starch; **Sudan black** stains those containing lipid). In this exercise, you will use iodine to stain starch-containing vacuoles inside yeast that have been grown on a rich medium. The yeast used for this study is a species called *Saccharomyces cerevisiae,* the species commonly added to bread dough; it produces carbon dioxide and leavens (raises) the bread before baking. This same yeast is also used for making ethanol in the fermentation of fruit juices and grain extracts.

There is a large variation in size among the yeasts: Their lengths range from 2 to 50 μm and their widths from 1 to 10 μm. There also is a great variation in shape. Many are ovoid (oval-shaped) or cylindrical, some are spherical, a few are apiculate (lemon-shaped), and a few others are flask-shaped. You will soon see that the shape of a cell reflects its method of division.

You should recall from Exercise 40 that **asexual** reproduction (also called **somatic** or **vegetative** reproduction) does not involve the union of two cells *(plasmogamy)* with its eventual nuclear fusion *(karyogamy)*. Asexual reproduction in yeasts is accomplished by the division that accompanies growth. Three types of division are apparent among the yeasts: budding, bud fission, and septation (cross-wall formation).

Budding The most common type of division is **budding** (Figure 42-1). A bud is an outgrowth on the parent cell that accompanies the division of the nucleus. One nucleus migrates into the bud, and the other nucleus stays with the parent cell. The bud increases in size and is eventually walled off from the parent cell (by one of several mechanisms) to form a new individual. Budding that occurs at or near the end of most

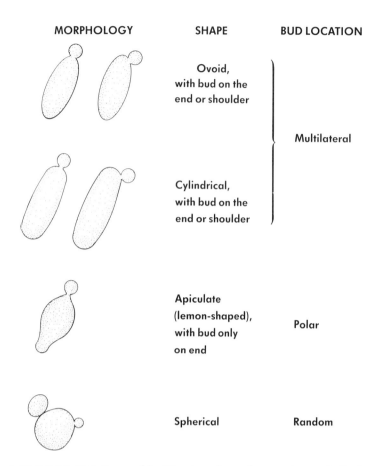

FIGURE 42-1 Types of budding (one form of asexual reproduction) found among the yeasts. Most ovoid or cylindrical yeasts that bud do so both at the ends and near the ends of the cell (multilateral budding), whereas apiculate yeasts form buds only at the ends (polar budding). Spherical yeasts may form buds at any location on the cell as long as a bud has not previously formed there.

ovoid or *cylindrical* yeasts is called **multilateral** budding. Since ovoid or cylindrical yeasts are the most common forms, multilateral budding is the most common mechanism of division (asexual reproduction) among the yeasts.

A yeast bud usually separates from the parent cell before the next bud is fully formed. When separation occurs *after* the next bud is fully formed, the result is a *chain* of fully separated cells called a **pseudomycelium,** or false mycelium (Figure 42-2). Thus, the pseudomycelia of yeasts consist of individual (completely separated) cells. (In contrast, the mycelia formed by the "septated" filamentous fungi do not contain fully formed crosswalls.) In the succeeding paragraphs, you will see that pseudomycelia are also formed by yeasts that divide by other mechanisms.

If the yeast cell has an *apiculate* (lemon) shape, the buds form only at the ends of the cell; this is called **polar budding** (see Figure 42-1) and is characteristic of a few genera of yeasts such as *Hanseniaspora* and *Kloeckera.* These yeasts are typically found during the early stages of the natural fermentation of fruits and other materials having a high sugar content. The apiculate yeasts usually start life as a newly separated, spherical-to-oval-shaped bud (Figure 42-3). Buds de-

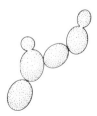

FIGURE 42-2 A primitive pseudomycelium formed by an ovoid yeast when progeny cells (buds) remain attached instead of separating from the parent cell before the next bud on the parent is fully formed. Other types of pseudomycelia in yeasts are shown in Figures 42-5B and 42-5C. Bud separation among the ovoid yeasts is more common than pseudomycelium formation.

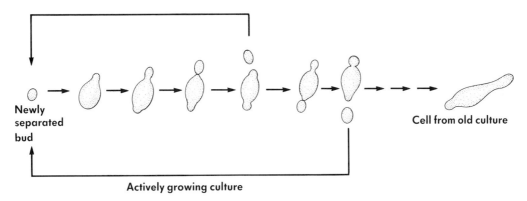

FIGURE 42-3 Ontogeny (developmental history) of an apiculate yeast exhibiting polar budding. (Adapted from Phaff et al., 1966).

velop only at opposite ends of this new cell, which gives rise to the cell's typical lemon shape. Cells taken from old cultures often have an irregular elongated shape, possibly as a result of repeated bipolar budding.

In spherical yeasts such as *Debaryomyces* species, no axes are present; so buds randomly arise at any place on the surface of the cell (see Figure 42-1).

The number of buds produced by yeast cells varies according to the type of yeast. Studies show that cells from various strains of the ovoid-shaped *Saccharomyces cerevisiae* produce from 9 to 43 buds (24 average) during their lifetimes. The budding process in most of the common yeasts leaves a **bud scar** on the cell wall where no other buds arise; consequently, a cell can produce only a limited number of buds. After the production of 3 or 4 buds (and bud scars), the cell begins to assume an altered shape. These surface irregularities can be seen even with a light microscope. Only in apiculate (lemon-shaped) yeasts, in which buds form only at a cell's poles, are new buds formed on top of bud scars. The stacked bud scars on the apiculate yeasts give a collar-like appearance to the cell surface that contributes to the cell's lemon shape.

Under ideal conditions, most yeasts are able to complete the budding process in 1 to 2 hours, but the time required for budding increases with the number of buds produced by a cell (up to about 6 hours per division just prior to cell death).

Bud fission A few yeast genera reproduce by a process called **bud fission** (Figure 42-4). This process is a form of division that appears to be intermediate between budding and a division process called *septation*. Bud fission has been observed in genera called *Nadsonia* and in *Pityrosporum*. Bud fission is similar to budding except that more wall material separates the bud and parent cell. In budding, the bud appears to be pinched off the parent cell. In bud fission, a constriction exists between parent and progeny during division, and a definite cross wall is formed between them. Yeast cells that are dividing by bud fission often are shaped like flasks or bottles.

Septation Some cylindrical genera such as *Schizosaccharomyces* and *Endomyces* divide by **septation** (see Figure 42-4). In this type of division, the cell elongates, and a cell wall and plasma membrane (septum) form approximately in the center and across the short (transverse) axis of the cell. Note that septation is not accompanied by constriction of the cell wall where the septum is located. Once septation is complete, the septum usually thickens, which usually separates the parent and progeny cells.

If cylindrical cells do *not* come apart after septation, then the long strands of cells are called pseudomycelia. Members of the genus *Trichosporon* usually grow in this way. When this type of yeast is grown on solid media, the cells at the end of the pseudomycelium often break off to form individual cells called **arthrospores,** which often appear disjointed, like cars of a derailed train (Figure 42-5A). Arthrospores are not

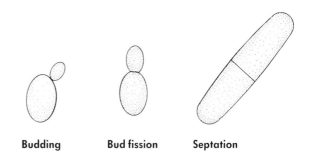

Budding   Bud fission   Septation

FIGURE 42-4 The three mechanisms of division (asexual reproduction) found among the yeasts. Budding appears to be the most common type of division. Bud fission appears to be a mechanism that is intermediate between budding and septation. Septation occurs across the short axis of the cell without invagination (constriction) of the cell wall.

resistant forms of cells like the conidiospores formed by molds or the endospores formed by bacteria.

It is unfortunate that mycologists often use the term *spore* to refer to any single cell that is capable of reproduction. Another example of this usage is the term *blastospore,* which refers to buds formed on pseudomycelia by the blastomycetous yeasts (see Classification and Taxonomy section below for an explanation of this type of yeast). Thus, the microbiology student must take care to determine whether a mycologist is referring to a resistant spore or to a single cell produced by a pseudomycelium.

The chain of cells that results when budding cells do not detach from parent cells is also called the **pseudomycelium.** This behavior is characteristic of the elongated cells of the genus *Endomycopsis* and of some species of the genus *Candida.* The buds of these cells elongate but remain attached, forming structures called *primitive pseudomycelia,* such as those illustrated in Figure 42-5B. A pseudomycelium is called *primitive* if the main chain has branches that are composed of only a few cells and if there is little or no difference between the appearance of the cells of the branch and those of the main chain.

Septation and budding Some yeasts divide by both septation and budding. For example, members of the genus *Candida* usually grow as budding individual cells, but some species can be induced to form pseudomycelia like those illustrated in Figure 42-5C. In this case, the cylindrical cells form chains by septation, and some of these chained cells form buds that form buds, resulting in structures called *well-developed pseudomycelia.* (Some mycologists call them true mycelia; but unlike mold mycelia, these cells are separate from each other.) The pseudomycelium is called well-developed or fully formed if cells in the chain appear markedly different from those in the branches. In this type of differentiation, the buds around the chains of cylindrical cells are called **blastospores** (see Figure 42-5C). Again, these are not resistant spores but are differentiated individual cells. Blastospores may remain spherical or oval, or they may elongate into cylindrical cells and produce branched pseudomycelia.

Yeast identification by means of asexual reproduction The *ability* to form pseudomycelia is a valuable characteristic to use in yeast identification; however, the *type* of pseudomycelium formed has little taxonomic value because various types of pseudomycelia can be formed by one species when growth conditions are altered (compare Figures 42-5B and 42-5C).

The sizes and shapes of yeast cells, the way they asexually divide (and either cling together or separate

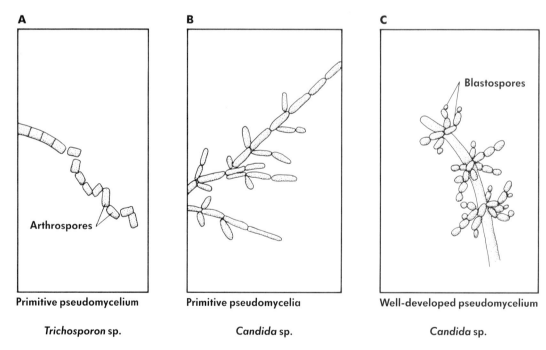

A	B	C
Primitive pseudomycelium	Primitive pseudomycelia	Well-developed pseudomycelium
Trichosporon sp.	*Candida* sp.	*Candida* sp.

FIGURE 42-5 Three examples of pseudomycelia (false fungal mycelia) formed by yeasts. (A) Long chains of cylindrical yeasts with separation into arthrospores (single cells) at one end of the pseudomycelium. (B) Primitive pseudomycelia showing chains of cells with similarly shaped cells branching from the chain. (C) Well-developed pseudomycelia having a chain of cylindrical cells dividing by septation, with many buds (blastospores) formed at septation sites. (Adapted from Phaff et al., 1966).

42 MORPHOLOGY AND REPRODUCTION OF THE YEASTS

after division), and the way they sexually reproduce all help to identify a pure yeast culture. However, each of these parameters can vary with **culture age** and with the **growth environment**. Therefore, it is important to standardize the conditions of growth and culture maturation so that the conditions you use match those used by others who have morphologically characterized the yeasts.

In this exercise, you are asked to observe only one form of asexual reproduction accomplished by yeast (budding in smears of *Saccharomyces cerevisiae* that have been stained with either crystal violet or iodine). However, you may also be able to observe septation (fission) by *Schizosaccharomyces* cells after you stain for ascospores, if there are still some actively dividing cells in this culture.

Sexual reproduction

If one defines yeasts as fungi that only exist as individual cells or as clusters or chains of individual cells, then sexual reproduction is confined mostly to yeasts in the class *Ascomycetes*. As the name suggests, these yeasts produce an **ascus** (saclike cell, plural *asci*) in which **ascospores** develop.

You may recall from Exercise 40 that true fungal sporulation (both asexual and sexual) serves two pur-

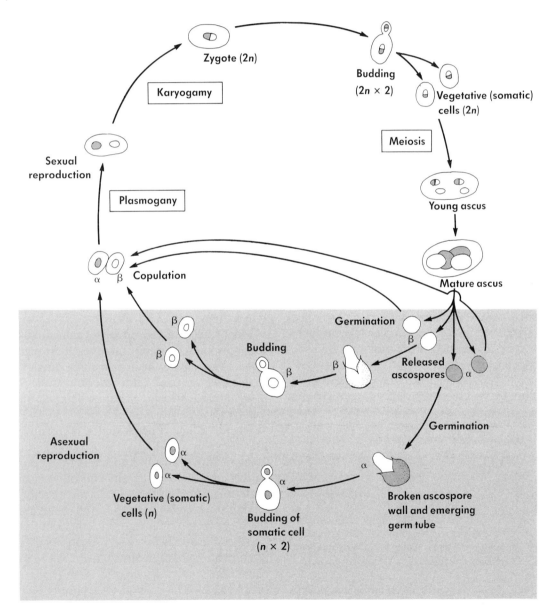

FIGURE 42-6 The life cycle of one type of ascosporegenous yeast (*Saccharomyces cerevisiae*) that illustrates both sexual and asexual forms of reproduction.

poses: reproduction (in which one or two cells produce many spores) and survival (because the true fungal spore, as opposed to arthrospores and basidiospores is more resistant to environmental stresses than is its corresponding vegetative cell).

Sexual reproduction among the *ascomycetous* (ascospore-forming) yeasts takes place either between two copulating somatic (vegetative) cells or between two copulating ascospores (Figure 42-6). After copulation, there is plasmogamy, karyogamy, and then the formation of a zygote. Eventually (after multiple divisions of the zygote), each zygote is capable of producing an ascus. The number of ascospores formed in the ascus depends on the number of meiotic divisions that have taken place and the subsequent development of the nuclei. Most ascomycetous yeasts produce either four or eight ascospores per ascus. The ascus wall eventually ruptures, releasing the mature ascospores into the environment.

The appearance of mature ascospores varies according to the type of yeast that produces them (Figure 42-7). For example, *Debaryomyces, Saccharomyces, Schizosaccharomyces,* and *Saccharomycodes* ascospores are ovoid. All species of *Pichia* and some species of *Hansenula* form hat-shaped ascospores, whereas other species of *Hansenula* form hemispherical or saturn-shaped ascospores.

The yeast ascospore is more resistant to environmental stresses than is the yeast vegetative (somatic) cell. The thickened cell wall of the ascospore is probably what gives it resistance to the penetration of chemicals. The ascospore is also resistant to temperatures that are 5 to 15 degrees higher than those tolerated by the vegetative cell. Nevertheless, ascospores are killed by pasteurization; so this technique is commonly used to treat beverages made from fruits and grains to prevent spoilage by yeasts.

When conditions are correct for germination, a break occurs in the ascospore wall through which a new somatic cell emerges (see Figure 42-6). The emerging cell, called a **germ tube,** often looks like a sausage emerging from an egg. Once the somatic (vegetative) cell is free from the ascus, it may grow and divide asexually, thus completing the life cycle, or it may sexually reproduce with another vegetative cell.

In this exercise, you will microscopically examine a species of *Schizosaccharomyces* after staining it with malachite green to determine the presence, number, and appearance of ascospores. The ability to recognize ascospores in mature yeast cultures is important for identifying yeasts.

Classification and taxonomy

From Exercise 40, you should remember that mycologists do not classify yeasts separately from the filamentous fungi (molds). Therefore, all terrestrial yeasts are placed in the division *Amastigomycota* along with all other terrestrial fungi (see Exercise 40; Table 40-3). According to Alexopoulos and Mims, the division *Amastigomycota* contains the following four classes: *Zygomycetes, Ascomycetes, Basidiomycetes,* and *Deuteromycetes.* If we define the yeasts as fungi which are found only as individual cells or as clusters or chains of individual cells, then we find that yeasts are placed in only the last three of the classes. (A few of the fungi in the class *Zygomycetes* are dimorphic, but none fit the definition for yeasts used here.) At present, there are only a few yeasts placed in the class *Basidiomycetes,* and these yeasts appear to be poorly defined. Therefore, almost all yeasts are placed into one of two classes: the *Ascomycetes* or the *Deuteromycetes.*

Most yeasts that reproduce sexually do so by forming ascospores; therefore, these yeasts are placed in the class *Ascomycetes,* the asci-forming fungi. All yeasts in this class are placed into one family, the *Saccharomycetaceae.* The characteristics, habitats, and genera that typify these yeasts are shown in Table 42-1. These yeasts reproduce either sexually, by forming asci, or asexually, by budding or septation. Perhaps their greatest significance to humans is their contribution to the initial decay of fruits and other sugary plant materials. Some of these yeasts are also used by the baking and brewing industries, and others are used for the manufacture of a vitamin called riboflavin.

All of the yeasts for which sexual forms of reproduction have not been observed are placed in the subclass *Blastomycetidae* in the class *Deuteromycetes* (Table 42-2). These yeasts are commonly called the **asporogenous** or **imperfect** yeasts, and they appear to reproduce only by budding. A few genera (*Bullera* and *Sporobolomyces*) are placed in the order *Sporobolo-*

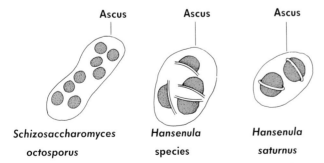

FIGURE 42-7 Three common shapes of ascospores found in asci formed by the ascosporogenous yeasts. This illustrates the shapes encountered, not the number found in the asci. (Adapted from C. J. Alexopoulous and C. W. Mims, *Introductory Mycology,* 3rd ed., New York: Wiley, 1979; with permission).

TABLE 42-1 Characteristics of the ascospore-forming yeasts

Class *Ascomycetes* (the ascus-forming fungi)[a]

Subclass *Hemiascomycetidae*[b]
Fungi in which mycelia are primitive, scant, or lacking and sexually formed asci are developed either directly from a single-celled zygote or at the end of an ascophore; cellular forms that are commonly called yeasts are placed in the order *Endomycetales*.

Order *Endomycetales*[c]
Fungi in which the sexually (or parthenogenetically) formed zygote either directly forms an ascus or produces an erect, septate ascophore that bears the ascus. Contains six families that are differentiated by whether mycelia are formed, by whether asci are formed free or born at the end of an ascophore, and by the number of ascospores formed in the asci. Only the family *Saccharomycetaceae* contain cellular forms commonly called the yeasts.

Family *Saccharomycetaceae*
(Gr. *saccharon* = sugar) Have a unicellular thallus that asexually reproduces by forming buds, septa, or both; sexual reproduction results in ascospores contained in a free ascus originating either from a zygote or parthenogenetically from a single vegetative (somatic) cell; sometimes asexually dividing cells form chains (pseudomycelia). Genera inhabit plant exudates, such as tree slime and flower nectar; also occur abundantly on the surfaces of healthy or decaying fruits and grow on their sugary exudates; some are symbiotic with insects. The budding yeasts are especially important in baking and brewing industries; best known species is *Saccharomyces cereviseae* (baker's and brewer's yeast). Yeasts in this family are also used in cacao-bean fermentation (part of chocolate manufacture), as supplements for human foods, and in riboflavin manufacture; they are also important in the spoilage of foods, and some are destructive plant or animal pathogens.

Representative genera (method of asexual reproduction)
  Saccharomyces (budding)
  Saccharomycodes (bipolar budding)
  Schizosaccharomycetes (septation)
  Kluyveromyces (septation)
  Debaryomyces (septation)
  Hansenula (septation)
  Pichia (septation)

SOURCE: From C. J. Alexopoulos and C. W. Mims, *Introductory Mycology*, 3rd ed., New York: Wiley, 1979, with permission of the publishers.
[a] This class of fungi also contains the ascus-forming filamentous fungi, powdery mildews, cup fungi, morels, and truffles (see exercise 42; Table 42-1).
[b] The Subclass *Hemiascomycetidae* contains three orders: *Protomycetales*, *Endomycetales*, and *Taphomycetales*. All ascomycetous yeasts, as defined in this exercise, are in the order *Endomycetales*.
[c] The Order *Endomycetales* contains six families: *Ascoideaceae*, *Dipodascaceae*, *Endomycetaceae*, *Saccharomycetaceae*, *Cephaloascaceae*, and *Spermophthoraceae*. Only one family, *Saccharonmycetaceae*, contains yeasts as defined in this exercise.

mycetales, and they appear to be most important in the decay of dead plant tissue. However, most of the asporogenous yeasts are placed in the order *Cryptococcales,* and genera in this order are important primarily because some species are opportunistically pathogenic for humans.

Physiology

In contrast to most filamentous fungi, which are strictly aerobic, most yeasts are facultatively anaerobic, which means that yeasts grow either in the presence or absence of oxygen (but grow more luxuriantly in its presence). This occurs because yeasts carry out a respiratory catabolism with oxygen and a fermentative catabolism without oxygen; as you know, more energy is obtained from respiratory catabolism. Use is often made of this characteristic in the fermentation industry; the inoculated yeasts are first incubated aerobically to rapidly increase cell numbers and then are placed in an anaerobic environment to shift them into fermentative catabolism, by which the desired product is made.

The optimum temperature for metabolism and growth of most yeasts is between 20° and 30°C, just as it is with all fungi. You should note, however, that cold temperatures slow or stop metabolism and preserve yeasts just as with all other microorganisms. It is common practice to store pure yeast cultures in liquid nitrogen ($-196°C$), where they remain viable indefinitely.

As with all fungi, the yeasts grow best in media that are slightly acid. A pH of about 6 is optimum for most yeasts; however, some types of organic acids (such as acetic acid) inhibit or even kill yeasts.

TABLE 42-2 Characteristics of yeasts in the class *Deuteromycetes*

Class *Deuteromycetes* (the imperfect fungi)
Fungi for which sexual forms of reproduction have not bee n observed

  Subclass *Blastomycetidae*
  Asporogenous (imperfect) yeasts; true mycelia, if present, are not well developed; either form or do not form pseudomycelia; this subclass contains only two orders: *Sporobolomycetales* and *Cryptococcales;* most yeasts in the class *Deuteromycetes* are placed in the order *Cryptococcales*

    Order *Sporobolomycetales*
    Thought to be asexual stages of fungi that would otherwise be placed in the class *Basidiomycetes;* all are common inhabitants of leaf surfaces; this order contains four genera, but only two genera (*Bullera* and *Sporobolomyces*) are yeasts as defined in his exercise; these yeasts form buds (called balistospores) that may be forcibly released from the parent cell; sometimes called mirror yeasts, because the ballistospores that are released will accumulate on the lids of inverted culture plates and form mirror immages of the colonies from which the buds were discharged; the released buds are among the most numerous types of fungal "spores" normally present in the air, especially during cool, humid weather (at these times, there may be as many as 10^6 ballistospores per cubic meter of air just from species within the genus *Sporobolomyces*)

      All yeast genera
        Bullera

        Sporobolomyces

    Order *Cryptococcales*
    Thought to be asexual stages of either ascomycetous or basidiomycetous yeasts. This order contains about a dozen genera. All genera reproduce by budding; none form ballistospores. A few genera produce either primative or well-developed pseudomycelia, and a few genera form arthospores. Some species are pathogenic for humans; other species may be used as a food or as a vitamin source. All species may be easily cultivated on organic waste materials.

      Some yeast genera (with characteristic features)
        Candida
        (Heterogeneous characteristics; normally grow as single cells, but may be induced to form either primative or well-developed pseudomycelia in the laboratory; commonly isolated from soil, water, plants, or animals; considered part of the normal flora of the mucous membranes of warm-blooded animals; *C. albicans* is opportunistically pathogenic for humans (especially those who are immunologically compromised) that may cause a disease (generally called candidiasis or moniliasis) of the mucous membranes of the mouth (specifically called thrust), vagina, and alimentary tract; *C. utilis* (formerly called *Torula utils*) is cultivated on waste materials from the paper and pulp industry and used as a nutritious cattle-feed supplement

        Cryptococcus
        Commonly isolated from soil or from plants and animals; *C. neoformans* is opportunistically pathogenic for humans and may cause chronic infections of lung or nervous tissue with symptoms like meningitis or lymphoma

        Pityrosporum
        P. orbiculare causes a human skin disease

        Rhodotorula
        Forms pink, red, or orange colonies; little economic importance; common laboratory contaminants

        Torulopsis
        Exist almost exclusively in the single-cell phase; otherwise similar in morphology and habitat to *Candida* species

        Trichosporon
        T. cutaneum (sometimes called *T. beigelii*) causes white piedra, a disease of the hairy portions of the human body

SOURCE: From C. J. Alexopoulos and C. W. Mims, *Introductory Mycology,* 3rd edition, John Wiley & Sons, New York, 1979.

The inhibition of yeasts by acetic acid probably occurs commonly in nature. The yeasts found naturally on fruit surfaces frequently begin the decay process as the fruit ages and dies; the yeasts then grow aerobically on the sugars and quickly use up the oxygen just below the surface. Next, under these anaerobic conditions, the yeasts shift to fermentation of the sugars and produce ethanol. The ethanol is used as an energy source by the acetic acid bacteria that are also naturally found on the surface of the fruit. The acetic-acid bacteria are strictly aerobic, so they grow only on the fruit surface where oxygen is available; and they oxidize (catabolize) ethanol and excrete acetic acid. The acetic acid diffuses into the fruit, where it slows and then stops yeast growth and eventually kills the yeasts, leaving the acetic acid bacteria to proliferate. Of course, these bacteria are depleting their source of ethanol, but they are able to compete effectively with other microorganisms for other nutrients. Excretion of acetic acid not only inhibits yeast but also lowers the pH of the plant tissue to the point at which growth of other bacteria and fungi are inhibited. Eventually, the acetic acid bacteria deplete another critical nutrient, their growth stops, and then microbes that can grow on acetic acid take over as the predominant microorganism in the decay process. Similar growth cycles occur with other microbes, and this scenario continues until all organic nutrients are consumed. This is an example of **microbial succession,** in which various microbes predominate in cyclic succession during the process of tissue decomposition.

The fact that yeasts are nutritionally categorized as chemoorganotrophic (see Table 40-1) means that they are capable of using only reduced organic compounds as sources of carbon and energy. In other words, yeasts cannot use light or reduced inorganic compounds as energy sources, nor can they use carbon dioxide as a sole carbon source.

Laboratory cultivation

Yeasts are grown and enumerated by the same methods used for bacteria. Either solid media (incubated aerobically) or liquid media (incubated aerobically by shaking) is used for yeast growth because yeast grows better in the presence of oxygen. If you want to study the fermentation process as well as growth, you should incubate liquid cultures aerobically first to promote rapid growth and then anaerobically so that cells will shift to a fermentative metabolism.

The majority of yeasts are **saprophites** (Gr. *sapros* = rotten); this means that they are capable of growing on nonliving organic materials, such as artificial, or synthetic, laboratory media. To grow most yeasts in the laboratory, you need only to provide carbohydrates (such as glucose or maltose), organic or inorganic sources of nitrogen and phosphorous, and other minerals.

Pure cultures of yeast are often grown on malt extract agar, on Wickerham's maintenance agar, or on glucose/yeast-extract agar. The use of malt extract for yeast cultivation stems historically from the close association of the brewing industry and the study of yeast. Malt extract is the concentrated liquid formed during one step of beer manufacturing. It is either spray-dried or concentrated to a heavy syrup. For yeast cultivation in the laboratory, the malt extract powder or syrup is dissolved and diluted so that the concentration is about 5 percent soluble solids. Wickerham's maintenance medium is often recommended for growing and storing yeasts over a long period; it contains malt extract, yeast extract, peptone, and glucose. Glucose/yeast-extract medium contains 5 percent glucose and 0.5 percent yeast extract (a powder prepared from the water-soluble fraction of broken yeast cells).

The **colony morphology** of a yeast closely parallels the type of yeast; therefore, just as with bacteria, a pure yeast culture should yield a single colony type. Colony morphology is used by the experienced mycologist, along with cell morphology, as an aid in yeast identification. Particular attention is given to variations in color, topography, ability to reflect light, probed texture, diameter, and the appearance of the colony periphery. However, *colony morphology varies with the cultivation medium!* For example, a pure culture streaked on trypticase-soy agar appears different from the same culture streaked on Sabouraud's agar. Even the simple substitution of gelatin for agar as a solidifying agent has a profound impact on colony morphology (Figure 42-8). But, when the medium is a familiar one and the growth conditions standardized, an experienced mycologist can often guess the genus by observing the colony morphology.

Isolation of yeasts from natural materials often requires the use of *enrichment* or *selective media* (see Exercise 21), because other microorganisms may be present in greater numbers and easily outgrow the yeasts when cultured in the laboratory. Several strategies can be used for selective yeast growth, depending upon what types of microbes predominate.

If bacteria predominate and you use a liquid medium for enrichment, you can lower the pH to 3.7 by adding 1 *N* HCl prior to autoclaving. If you prefer an agar medium, you should add a predetermined amount of 1 *N* HCl after autoclaving because agar breaks down during autoclaving at this pH. Only a

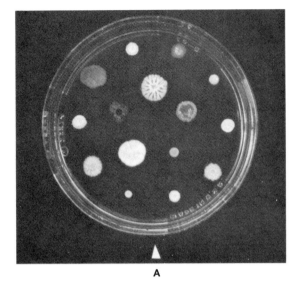

A

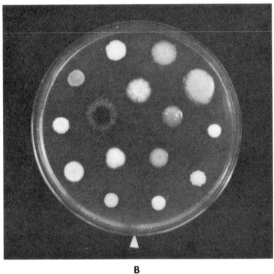

B

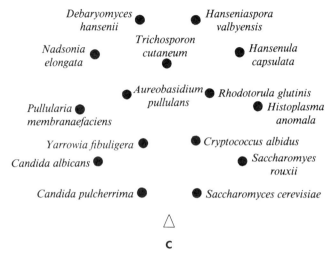

C

FIGURE 42-8 Colonies of 15 selected yeast cultures grown for 8 days at 18°C in Petri dishes containing (A) 5 percent malt gelatin or (B) 5 percent malt agar. The chart (C) identifies the cultures by position. (From H. J. Phaff, M. W. Miller, and E. M. Mrak, 1978, *The Life of Yeasts*, 2nd edition, Harvard University Press, Cambridge, Mass.; with permission.)

few bacteria that are normally found with yeasts can grow at pH 3.7 (notably the acetic acid bacteria), and you can often suppress the growth of these bacteria by adding certain antibiotics. From Exercise 40, you may remember that **Sabouraud's medium** (with glucose) is recommended for cultivating fungi; it contains a high sugar concentration (4 percent) and selects against the growth of most bacteria because of its low pH (from 5.0 to 5.6). But you can see now that other media have an even lower pH and thus are even more selective; these are often preferred for selective yeast growth from samples that contain large numbers of diverse bacterial types. You will use Sabouraud's medium in this exercise, not because of its selective properties (you will be using pure yeast cultures) but because it is a good, general-purpose medium for yeast cultivation and is commercially prepared and readily available.

One can sometimes use broad-spectrum antibiotics in place of severe pH adjustments to control bacterial growth in cultures inoculated with habitat materials. If this antibiotic strategy is to be effective, however, one must choose an antibiotic that inhibits procaryotic but not eucaryotic cells.

If molds predominate in the natural sample, the problem of selectively growing yeasts is more difficult because molds and yeasts are both eucaryotic cell types and have similar metabolic properties and growth requirements. The most troublesome molds are those, such as species of *Mucor* and *Rhizopus,* that quickly spread over an agar surface and any yeast colonies that are present. To handle this problem on solid media, most microbial ecologists add antifungal agents to the medium at a concentration that is inhibitory to most molds but not to the yeasts present. The most commonly used agents are *propionic acid* (most effective at low pH) or a dye called *rose bengal* (used in very low concentrations). Both of these agents seem to work best in nutritionally dilute media. Alternatively, one can incubate the natural inoculum in a shaken liquid medium, in which the mold cells grow as suspended balls of entangled mycelia that do not sporulate, and the yeasts grow as individual cells. After incubation of the mixed culture, the yeasts can be separated from the molds by passing the culture through sterile glass wool, which retains only the molds. The yeasts from the filtrate are cultivated in the usual way.

Scope and purposes of this exercise

The yeasts are fascinating eucaryotic, single-celled microorganisms that reproduce both asexually and sexually; therefore, their study is included in most introductory microbiology courses at the university level. However, the cytology and reproductive strategies of yeasts are so diverse that no introductory course can do more than begin your acquaintance with these creatures. If you become interested in further study, you should take an introductory course in mycology (the study of all fungi); but first, you should determine that the yeasts are given adequate coverage (many courses deal almost exclusively with the filamentous fungi).

The purposes of this exercise are to introduce you to the size of a typical yeast cell (so that you will recognize it as a yeast when you see it again), to let you examine one type of internal organelle in these eucaryotic cells (a starch vacuole), to show you one method of asexual reproduction accomplished by the yeasts (budding), and to let you examine the results of sexual reproduction among the yeasts (ascospore formation). Good hunting!

LEARNING OBJECTIVES

- Become familiar with the relative sizes and typical shapes of yeast cells so that you will know how to recognize yeast cells when you examine them with various types of light microscopes.

- Learn to recognize typical intracellular storage inclusions in yeasts and how to identify the inclusion type with specific stains.

- Know how to microscopically recognize the consequence of reproductive methods (asexual and sexual) in yeast cultures.

MATERIALS

Cultures	*Saccharomyces cerevisiae* *Schizosaccharomyces octosporus* *Escherichia coli*
Media	None
Supplies	Crystal violet Diluted Gram's iodine (1:1 with water) Malachite green Paper toweling (cut square and slightly smaller than the width of a microscope slide)
Demonstration	Stained smear containing a mixture of yeast and *E. coli* cells for size comparisons

PROCEDURES

Before you begin these procedures, review the following concepts and techniques:

Wet-mount, Exercise 6

Phase-contrast microscope use, Exercise 6

Simple staining, Exercise 5

The *E. coli* ruler, Exercise 40

Then observe the demonstration slide containing a mixed smear of yeast and *E. coli* cells. Use the mixed smear to observe the relative size of yeast and bacterial cells.

Asexual reproduction of yeasts by budding

1. **Examine the size, shape, and evidence of budding** of *Saccharomyces cerevisiae* using one or both of the following methods:
   a. **Prepare a wet mount** of unstained *Saccharomyces cerevisiae,* and examine it with a phase-contrast microscope. Use either the high-dry or the oil-immersion objective.
   b. **Prepare a simple stain** of *Saccharomyces cerevisiae* using crystal violet, and examine it with a brightfield microscope. Use either the high-dry or the oil-immersion objective.
2. **Record the relative size, shape, and evidence of budding** in the Results section. Draw ex-

amples rather than the entire microscope field. Be particularly careful in drawing the position of the bud on the mother cell and the exact shape of the mother cell.

Storage inclusions in yeast cells

1. **Add one drop of diluted iodine** to the center of a clean microscope slide.

2. **Aseptically remove one loopful of** *Saccharomyces cerevisiae* **culture, and thoroughly mix this with the drop of iodine.**

3. **Place a coverslip on top of the iodine-treated culture** to prepare a wet mount.

4. **Examine** this preparation with brightfield microscope using the oil-immersion objective.

INTERPRETATION NOTE

Both animal and plant cells form reserve polysaccharides under suitable nutritional conditions. The reserve polysaccharide formed by animal cells is called *glycogen,* and that formed by plants cells is called starch. You may recall from Exercise 27 that there are two forms of starch, and each type reacts differently with iodine. The first type, *amylose,* is a straight-chain polymer composed of glucose molecules; it reacts with iodine to form a deep blue color. The second type, *amylopectin,* is a branched-chain polymer composed of glucose molecules; it reacts with iodine to form a red to brown color. Plant cells may produce starch granules that contain a mixture of amylose and amylopectin; therefore, the color in the presence of iodine is determined by the percent of each type of starch in the storage inclusion.

---

5. **Record the size, shape, arrangement, and color** of structures that appear inside these cells. Note also whether the cell itself is stained by the iodine.

Ascospore formation (sexual reproduction) in yeast

1. **Prepare a thin smear** of *Schizosaccharomyces octosporus,* and **heat fix this smear.**

2. **Stain these cells with malachite green** using the following procedure:
   a. **Place the slide on a staining rack.**
   b. **Cut a piece of paper toweling** that will cover the smear and be slightly narrower than the width of the slide.
   c. **Flood the smear** with malachite green, and place a cut piece of paper toweling over the smear so that it completely covers the smear and soaks up the stain.

   The toweling should now be saturated with stain. If it is not, add more stain directly to the toweling.

   Remember that malachite green is difficult to remove from skin and clothing, so be careful in handling this strong dye!

   d. **Stain for 5 minutes without heating.** Check periodically to make sure the toweling remains saturated during this time.

   This procedure is similar to that used to stain bacterial endospores, except that here you do not apply heat from a bunsen burner.

   e. **Carefully remove the toweling** with a forceps while you hold a clean paper towel under the stain-saturated toweling. Discard both the saturated toweling and the drip-catcher towel as directed by your instructor.
   f. **Very thoroughly (but gently) rinse** the slide with water until all excess stain is removed; then continue rinsing for an additional 30 seconds.
   g. **Counterstain** with basic fuchsin for 30 seconds.
   h. **Wash the slide** gently with water until all excess stain is removed; then blot and air dry.

3. **Examine the cells** using the brightfield microscope and an oil-immersion objective.

4. **Record** the size, shape, number, and general appearance of the cells and any internal structures seen. Label each structure drawn.

   You may wish to exaggerate size in your drawings to illustrate structural detail more accurately; if so, be sure to state that your drawings are not to scale.

   Several references state that *Schizosaccharomyces octosporus* characteristically forms *eight* ascospores in each ascus after sexual reproduction. Do you find this to be so? (Note the translated meaning of the Latin word *octosporus,* the specific epithet of the species name.)

NAME _____

DATE _____

SECTION _____

RESULTS: ASEXUAL REPRODUCTION OF YEASTS BY BUDDING

Label all illustrated structures.

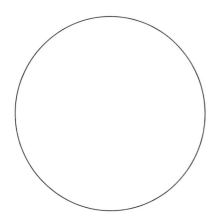

Saccharomyces cerevisiae
Unstained wet mount
(phase-contrast microscope)
Magnification: _____ ×

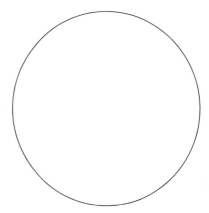

Saccharomyces cerevisiae
Simple stain (crystal violet)
(brightfield microscope)
Magnification: _____ ×

CONCLUSIONS

What proportion of cells contain buds? About _____ percent of the cells.

How many buds does the average cell have? About _____ per cell.

Using *E. coli* as a ruler, about how wide and how long is the average *S. cerevisiae* cell? About _____ μm wide by

_____ μm long.

RESULTS: STORAGE INCLUSIONS IN YEAST CELLS

Label and list the color of all illustrated structures.

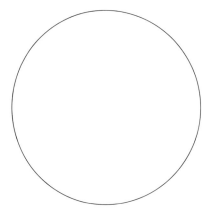

Saccharomyces cerevisiae
Gram's iodine : water (1 : 1)
(brightfield microscope)
Magnification: _____ ×

CONCLUSIONS

What proportion of cells contain storage inclusions? About _____ percent of the cells.

How many storage inclusions does the average cell have? About _____ per cell.

Does the average inclusion appear to contain more amylose- or more amylopectin-type of starch? Give the reasoning upon which you base this answer.

NAME _____

DATE _____

SECTION _____

RESULTS: ASCOSPORE FORMATION (SEXUAL REPRODUCTION) IN YEAST

Label and list the color of all illustrated structures.

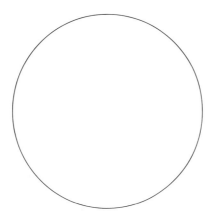

Schizosaccharomyces octosporus
Malachite green stain (no heat)
(brightfield microscope)
Magnification: _____ ×

CONCLUSIONS

What proportion of cells contain ascospores? About _____ percent of the cells.

How many ascospores are there per cell (for those that have them)? About _____ per cell.

What do your procedures and results suggest about the resistance of yeast ascospores as compared with bacterial endospores? (Also give the reasoning upon which your answer is based.)

QUESTIONS

Completion

1. There are three types of asexual reproduction common among the yeasts. These are _____, _____, and _____.

2. When yeast cells do not readily separate after division but form chains of individual cells, these chains are called a _____.

3. Most yeast grow best when the medium pH is _____ (one word) and the temperature is between about _____° and _____°C.

4. Most yeasts are placed into one of _____ classes of fungi. If true fungal spores are sexually formed, the yeasts are placed in the class _____. If no sexual forms have been observed, the yeasts are placed in the class _____.

5. One compound that is frequently added to media for the cultivation of yeasts that allows it to be selective against the growth of molds is _____.

6. _____ is the name of a medium that is commonly used to cultivate yeasts.

7. A typical yeast cell is from _____ to _____ times larger than the average length of the *E. coli* cell.

8. Sometimes, yeasts form internal accumulations of materials such as _____, and these accumulations are called _____.

True-False (correct all false statements)

1. Spores formed inside yeast cells are the result of sexual reproductive mechanisms, and these spores are usually called conidia. _____

2. In the natural environment, yeasts are often found associated with enteric bacteria, and both appear to participate in the natural degradation of fruits and other sugary plant materials. _____

3. There are no known yeasts that are opportunistically pathogenic for humans. _____

4. The term *yeast* has no taxonomic significance; that is, the yeasts are not separately classified from the molds. _____

5. Yeasts are incapable of sexual reproduction. _____

6. There are two types of structures formed by yeasts that are very resistant to environmental stresses (such as drying and heat); these are called arthrospores and blastospores. _____

7. Both the cellular and the colony appearance of yeasts are very stable regardless of the environment used to cultivate them. _____

8. Some yeasts are eucaryotic and others are procaryotic. _____

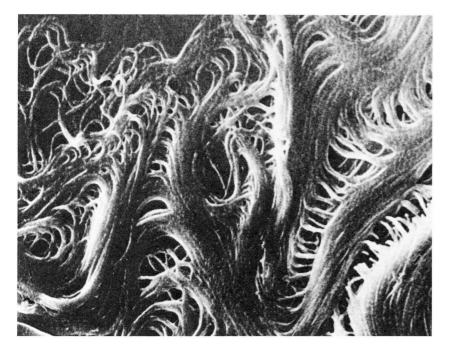

A mighty creature is the germ,
Though smaller than the
  pachyderm.
His customary dwelling place
Is deep within the human race.
His childish pride he often pleases
By giving people strange diseases.
Do you, dear reader, feel infirm?
You probably contain a germ.
  —*Ogden Nash*

The edge of a *Bacillus anthracis* colony.

MEDICAL MICROBIOLOGY AND IMMUNOLOGY

Part XI is a brief introduction to the positive and negative ways that microorganisms affect human health; it does not pretend to be an extensive introduction to diagnostic microbiology. Many of the tests that are used for pathogen identification are located in other parts of this manual because they apply to microorganisms other than human pathogens. For example: all the procedures in Part VIII (except Exercise 35) are used to identify pathogenic bacteria; the structural characteristics given in Part X are used to identify pathogenic fungi; and the principles given for selective and differential media (Exercise 21) and enrichment techniques (Exercise 22) are used extensively in medical microbiology. This arrangement was chosen to emphasize microbial characteristics rather than the application of these tests to pathogen identification.

43

Evaluating the Effectiveness of Common Antiseptics and Disinfectants

Antiseptics, disinfectants, and antibiotics can be referred to as **chemical antimicrobial agents,** in contrast to physical antimicrobial agents such as heat and ionizing or nonionizing radiation (see Exercise 39).

Microbiologists usually distinguish between antiseptics and disinfectants in the following way. **Antiseptics** are preparations of chemicals that are meant to be applied to skin or other living tissue. Examples are chemical preparations designed to treat wounds or infected throats. *Alcohol,* which is probably the most widely used antiseptic, denatures proteins, extracts lipid from membranes, and dehydrates cells, all of which probably contribute to the effectiveness of alcohol as an antiseptic. *Iodine,* a good antiseptic with bacteriocidal and sporicidal properties, is often mixed with other chemicals, such as alcohol or detergents, and used to pretreat the incision area prior to surgery. Iodine is a strong oxidizing agent whose effect on microbes is not understood; it probably interferes with the structure and active site of many enzymes. *Heavy metals* are also used in antiseptic preparations; for example, laws require that silver nitrate be applied to the eyes of newborn humans, primarily to prevent the transmission of gonococci from the mother to the infant. Heavy metals bind to cellular enzymes and inactivate them. *Detergents,* especially the quaternary ammonium salts, are frequently used alone or in combination with other agents in antiseptic preparations. Detergents denature enzymes and destroy membranes.

Disinfectants are preparations of chemicals (usually liquids) that are intended for application to the surfaces of nonliving materials. The chemical used to wipe down your laboratory work area probably contains a strong disinfectant. Many household cleaning agents such as ammonia and bleach (hypochlorite) are also disinfectants. Most household disinfectants are strong oxidizing agents.

If you examine the labels of antiseptic and disinfectant preparations, you will find the above-mentioned chemicals and many others of interest.

Two important rules for you to remember concerning the use of antiseptics and disinfectants are (1) always use the most concentrated form that will cause the least damage to the tissue or inanimate surface, and (2) if long-term use is necesssary, apply either a combination of agents or frequently change to another effective agent. The continued application of a single agent (especially at low concentrations) will select for microbial mutants that are resistant to that agent. That risk is minimized when you use higher concentrations and is almost eliminated when you use a combination of effective agents.

One simple way to evaluate the relative effectiveness of antiseptics and disinfectants is to use the **zone-of-inhibition method.** With this method, you apply the chemical to a freshly inoculated plate, incubate the culture, and then look for a zone of inhibition. The presence of a *clear zone* (a lack of growth) surrounding the chemical shows either that the cells have been killed or that their growth has been inhibited (but you cannot tell which). In other words, a zone of inhibition does *not* discriminate between **bacteriostatic** and **bacteriocidal** chemicals.

There are two popular ways to inoculate the plate with the test microorganism. The simpler is to use a sterile cotton swab to heavily inoculate the plate surface. Another is to inoculate a volume of tempered nutrient agar, pour it into a sterile Petri dish, and then allow it to harden; this method distributes microorganisms throughout the agar instead of just on the surface.

After you inoculate the plate, you can apply the test chemicals in a number of ways. For example: you can cut into the inoculated agar with a sterile cork-bore, remove the agar plug, and replace it with a solution of the test chemical; you can place the test chemicals inside sterile metal or ceramic rings set on the surface of the inoculated agar; or you may saturate sterile, filter-paper disks with the test chemical, drain the disks, and place them on the surface of an inoculated plate. With each application technique, the liquid is absorbed into the surrounding agar where it makes contact with the test organism.

When heavily inoculated plates are incubated, microbial growth should be **confluent** on the surface (if streaked) or throughout the agar (if seeded); in other words, growth should be continuous instead of forming separate colonies, thus making the agar appear *very* opaque or cloudy. In places where the microbe has been killed or where its growth has been inhibited, there will be a clear area called the *zone of inhibition.* The size of the clear (inhibition) zone depends not only on the effectiveness of the antiseptic or disinfectant but also on its ability to diffuse into the medium. For example, even a very effective chemical that does not readily diffuse out from its source will not produce a very large zone of inhibition. Therefore, zone size alone should not be used to determine the effectiveness of an antiseptic or disinfectant.

In this exercise, you will use the inhibition-zone method to presumptively examine the antimicrobial activity of various antiseptics and disinfectants on three bacteria commonly found on or in humans.

LEARNING OBJECTIVES

• Understand the common usage of the terms *antiseptic* and *disinfectant.*

• Know how to employ the inhibition-zone method to test the relative effectiveness of an antiseptic or disinfectant.

• Understand how a zone of inhibition is formed.

• Know the significance of using each of the three microbes employed in this exercise.

MATERIALS

Cultures *Streptococcus saliverius* (broth)
          Escherichia coli (broth)
          Staphylococcus aureus (broth)

Media   Trypticase-soy agar (tempered in 100-ml bottles)

Supplies Various antiseptics and disinfectants for selection by student
          Sterile filter-paper disks
          Forceps and 95 percent ethanol
          Sterile Petri dishes
          Metric ruler (second period only)

PROCEDURE: FIRST LABORATORY PERIOD

1. **Select three sterile Petri dishes; label each** with your *name, date* of inoculation, and *experiment number.* **Label each with the name of a different bacterium** (see the materials list).

2. **Take these labeled sterile dishes to a central location for adding inoculated (seeded) agar.**

Ask your instructor to demonstrate the inoculation of tempered agar. Then (if the instructor agrees), one student can seed enough agar for the entire class. The **first step** is to calculate the approximate amount of agar you need to supply the class, using about 25 ml of seeded agar per plate.

The **second step** is to aseptically transfer one tube of one culture into one 100-ml volume of tempered T-soy agar. Mix this seeded agar thoroughly, using about ten gentle inversions, (so that bubbles do not mix into the agar). It is important that these cells be thoroughly mixed before plates are poured. **Repeat this procedure** for each culture.

3. **To prepare each plate, aseptically add tempered seeded agar to the plate until one-half to two-thirds of the bottom of the plate is covered;** then **cover the plate and swirl it until the agar spreads out to completely cover the bottom of the plate.**

    When finished, you should have three plates, and each should be seeded with a different bacterium.

    Wait until the agar has completely hardened before you use the plate. (Remember that molten agar is transparent, but it becomes translucent (opaque) when it hardens.

4. **Divide each hardened plate into six equal parts.** Use your felt-tip marking pen to draw three lines across the bottom of the plate so that they go through the center, as you might cut a pie.

5. **Choose six chemicals from the group of antiseptics and disinfectants.** Record their names in the Results section of this exercise. Also record a letter or number code for each chemical. **Use the same six chemicals for all three plates.**

   Note the type of microorganisms used in this experiment. All are bacteria. One is a gram-negative rod commonly found in human fecal material *(E. coli)*, one is a gram-positive sphere (coccus) commonly found on the skin of humans *(S. aureus)*, and the other is a gram-positive sphere (coccus) commonly found in the oral cavity of humans *(S. saliverius)*. You might want to select your antiseptics or disinfectants with these habitats in mind.

6. **Use your code to label each plate** so that you will know which antimicrobial chemical was applied to that section of the plate.

7. **Apply the antiseptics and disinfectants** in the following way:
   a. Flame-sterilize the forceps as follows: Soak the tips in 95 percent ethanol for a few minutes, and then remove them from the ethanol and pass the forceps tips through a bunsen-burner flame just long enough to ignite the ethanol. **Hold the points down so the burning ethanol does not run onto your hand.** The technique here is identical to that for sterilizing the spreading rod used for preparing spread plates.
   b. Aseptically remove one filter-paper disk, and touch it to the liquid antiseptic or disinfectant.
   c. Remove excess liquid by touching the wet disk several times to the inside of the antiseptic or disinfectant container.
   d. Gently lay the wet disk (in the appropriate position) on the surface of the seeded agar. Take care not to drip chemical on other parts of the plate.

   When finished, you should have three plates, each of which is seeded with a different microbe and all of which have the same combination of six chemicals added to paper disks.

8. **Incubate right-side-up at 30°C** until the next laboratory period.

PROCEDURE: SECOND LABORATORY PERIOD

1. **Examine your plates for zones of inhibition.**
2. **Measure each inhibition zone** with a metric rule. Measure from one edge of the zone to the other edge (across the paper disk).

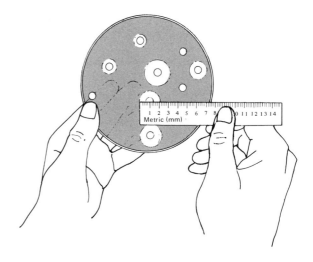

3. **Record your results** in the Results section. Your instructor may ask you to share your results with the class by writing them on the chalkboard.
4. **Properly discard your plates.**

43 EVALUATING ANTISEPTICS AND DISINFECTANTS

NAME _____

DATE _____

SECTION _____

YOUR OWN RESULTS

Bacterium used	Chemical used	Code on plate	Inhibition zone size (mm)
Streptococcus			
Escherichia			
Staphylococcus			

CONCLUSIONS

INTERESTING CLASS RESULTS

Bacterium used	Chemical used	Code on plate	Inhibition zone size (mm)
Streptococcus			
Escherichia			
Staphylococcus			

CONCLUSIONS

43 EVALUATING ANTISEPTICS AND DISINFECTANTS

NAME _____

DATE _____

SECTION _____

QUESTIONS

Completion

1. The term *antiseptic* is commonly used to refer to antimicrobial chemicals that are applied to _____, whereas the term *disinfectant* is used for chemicals that are applied to _____.

2. Often the size of the inhibition zone is used to determine the relative effectiveness of two antimicrobial chemicals on a test organism. A small zone does not necessarily indicate a lack of effectiveness; alternatively, it may mean that _____.

3. Agar used to test the effectiveness of antiseptics and disinfectants may be inoculated in two ways: _____ or _____.

4. The name of the test organism used in this exercise that is most commonly found on the skin of humans is _____, whereas the organism that is an indicator of fecal contamination is _____.

True-False (correct all false statements)

1. A clear zone surrounding a disk saturated with an antimicrobial chemical after incubation shows that the inoculated bacteria have all been killed. _____

2. To effectively test antiseptics or disinfectants, agar is inoculated with the test organism and incubated (to allow for growth of these bacteria), and then the test chemicals are applied. _____

3. Zones of inhibition are caused by lysis of bacterial cells adjacent to the filter-paper disks. _____

44

Antibiotic Evaluation by The Kirby-Bauer Method

Any chemical or physical agent that inhibits the growth of a microorganism or kills it is called an **antimicrobic** or an **antimicrobial agent.** There are many types. For example, in Exercise 43, we learned that antiseptics and disinfectants are chemical antimicrobial agents that are used to treat the surfaces of living and nonliving things.

In the present exercise, we examine another type of antimicrobial agents known as **chemotherapeutic drugs,** which are chemical compounds that are *taken internally* to ease the symptoms of a disease or to speed the patient's recovery. Chemotherapeutic drugs may be as simple as aspirin or as complex as antibiotics. A chemotherapeutic drug is called an **antibiotic** if it meets three criteria: it is a chemical substance produced by a microorganism; it stops the growth of or kills other microorganisms; and it is effective in very small doses.

Antibiotic production by microorganisms is common, especially by microorganisms normally found in the soil. Most of the microbes that produce medically valuable antibiotics fall into three groups: (1) fungi, especially those in the genus *Penicillium,* which produce antibiotics such as penicillin and griseofulvin; (2) bacteria in the genus *Bacillus,* which produce antibiotics such as bacitracin and polymyxin; and (3) bacteria in the genus *Streptomyces,* which produce antibiotics such as chloramphenicol, erythromycin, streptomycin, and tetracycline. Of these three types of microorganisms, the *Streptomyces* produce the largest number of medically important antibiotics.

The genus *Streptomyces* is one member of a group of unusual bacteria called the **actinomycetes,** which are similar to the filamentous fungi in many ways (see Exercise 40). For example, their cells are filamentous and often branched, and older colonies form an aerial mycelium with long chains of spores (conidia) at their ends. Unlike the fungi, however, species in the genus *Streptomyces* have a procaryotic cell structure, and they are smaller in diameter (usually about 0.5 to 1.0 micrometer) than the fungi. The streptomycete conidia are more resistant to environmental stresses than their corresponding vegetative cells, but streptomycete conidia are not nearly as resistant as the endospores formed by species of *Bacillus* and *Clostridium.* The streptomycete aerial filaments and conidia are often pigmented, thus giving a characteristic color to mature colonies. The color, dusty appearance, and compact nature of streptomycete colonies make them fairly easy to recognize on an agar plate.

Although they are effective against bacteria and fungi, *antibiotics* (as defined here) *are not effective against viruses.* Thus, the principal use of antibiotics is to help the body fight bacterial and/or fungal infections. The course of an infection is often likened to a race between the pathogen's ability to grow in the host tissue and the tissue's ability to capture and destroy the invading pathogen. Antibiotics are given to weaken or kill some of the invading pathogens; hopefully, the body's tissues can then destroy the rest.

When a physician or veterinarian sees a patient with an infection, a good guess can usually be made as to the probable infectious agent. Based on this guess, one or more appropriate broad-spectrum antibiotics are prescribed. If the patient shows signs of recovery from the infection within 24 to 48 hours, no further investigation is made. However, if the infection appears **chronic** (marked by long duration or frequent recurrence), then the pathogen will need to be isolated and its **susceptibility** to various antibiotics determined.

It is the job of the clinical microbiologist to isolate the pathogen, to determine its probable identity, and to determine its susceptibility to various antibiotics. Physicians or veterinarians need to know as soon as possible which antimicrobial agent will be most effective as a chemotherapeutic drug so that they can begin effective treatment.

After isolating the pathogen, the microbiologist heavily streaks the plates and then uses **antibiotic sensitivity disks** to determine the antimicrobial activity for each antibiotic. Disks are placed on the agar surface after inoculation and *before incubation.* Each disk contains a known concentration of a different antibiotic. During incubation, the antibiotic diffuses from the disk into the agar, creating a concentration gradient such that the greatest concentration at the edge of the disk. Cells on the heavily streaked plate grow confluently to form a thick lawn of microbial growth except where the microorganisms are either killed or stopped from growing by the antibiotic. If this happens, a clear zone in which there is no microbial growth surrounds the disk (see Exercise 43). This is called a **zone of inhibition,** even though killing may also have occurred in this zone.

The effectiveness of an antibiotic is preliminarily determined by the size of the zone of inhibition, but zone size varies according to how easily the antibiotic diffuses through the agar, the type of medium used, and many other factors. Therefore, one must standardize such things as medium type and inoculum size, so that one can relate inhibition-zone size to whether an isolate is *resistant, sensitive,* or *intermediate* in its response to a particular antibiotic (Table 44-1).

The **Kirby-Bauer method** is the standard recommended by the U.S. Food and Drug Administration and the Subcommittee on Antimicrobial Susceptibility Testing of the National Committee for Clinical Laboratory Standards. We follow the Kirby-Bauer Method, for the most part, in this exercise. This method is highly standardized to avoid varying results due to alterations in procedure. For example, only **Mueller-Hinton agar** is used, and it is prepared under exacting conditions such that its chemical composition is always the same, and the pH is always adjusted to 7.2 to 7.4. The medium should always be poured in Petri dishes to a uniform thickness of 4 mm (25 ml in a 100-mm diameter plate). Plates must be thoroughly inoculated, using a cotton swab, from a broth culture having a turbidity that matches a defined standard. Excess culture must be aseptically rolled off the swab prior to inoculation. Only disks containing certain concentrations of antibiotics are used, and the disks must be pressed firmly on the agar surface to ensure contact. Then plates are incubated for 16 to 18 hours before inhibition zones are measured, and the words used to describe the effectiveness of each antibiotic are rigidly defined (see Table 44-1).

Let us assume that the Kirby-Bauer method is used to standardize the conditions under which many antibiotics will be tested on a microbial isolate. Given this, then the size of the zone of inhibition is related to two things: (1) the minimum concentration that kills or inhibits that microbe and (2) the ability of the antibiotic to diffuse through the agar. If all antibiotics diffused through agar in the same way, one could say that all zones of a particular size would indicate that a microbial isolate was sensitive to that antibiotic. However, the diffusion characteristics of different antibiotics are not the same. Thus the size of zone that indicates sensitivity to an antibiotic varies greatly. For example, a 10-mm zone surrounding a Vancomycin disk indicates that the microbe is *resistant* to Vancomycin, whereas a 10-mm zone around a Colistin disk indicates that the microbe is *sensitive* to Colistin (Table 44-1). These differences reflect the different concentration gradients formed by these two antibiotics during diffusion through the agar and under the standard conditions of the Kirby-Bauer method. Recommendations on how to interpret the size of the zone of inhibition for various antibiotics, such as those given in Table 44-1, are available to the clinical microbiologist.

In this exercise, you will use four different bacteria to test the effectiveness of a number of antibiotics: (1) a gram-negative rod commonly found in fecal material *(E. coli),* (2) a gram-negative rod commonly found in hospital-acquired (nosocomial) infections *(P. aeruginosa),* (3) a gram-positive sphere (coccus) commonly found on the skin of humans *(S. aureus),* and (4) a gram-positive sphere (coccus) commonly found in the oral cavity of humans *(S. salivarius).*

LEARNING OBJECTIVES

• Understand what constitutes an antibiotic and learn something about the types of microbes that form antibiotics.

• Understand the meaning of chemotherapeutic drugs and how antibiotics differ from other types of chemotherapeutic drugs.

• Know why the Kirby-Bauer method is called a *standard* method of analysis and how to employ this

TABLE 44-1 Significance of inhibition-zone size in evaluating the effectiveness of antibiotics with the Kirby-Bauer method

Chemotherapeutic agent[a]	Disk potency[b]	Inhibition zone diameter (mm)		
		Resistant (R)	Intermediate (I)	Sensitive (S)
Amikacin	10 μg	<12	12–13	>13
Ampicillin				
For Gm(−) bacteria	10 μg	<12	12–13	>13
Staphylococcus and penicillin-G susceptibles	10 μg	<21	21–28	>28
Bacitracin	10 U	<9	9–12	>12
Carbenicillin				
For *Proteus* species and *Escherichia coli*	50 μg	<18	18–22	>22
For *Pseudomonas aeruginosa*	50 μg	<13	13–14	>14
Cephalothin				
For cephaloglycin only	30 μg	<15		>15
For all other cephalosporins	30 μg	<15	15–17	>17
Chloramphenicol	30 μg	<13	13–17	>17
Clindamycin	2 μg	<15	15–16	>16
Colistin	10 μg	<9	9–10	>10
Erythromycin	15 μg	<14	14–17	>17
Gentamicin (for *P. aeruginosa*)	10 μg	<13		>13
Kanamycin	30 μg	<14	14–17	>17
Lincomycin (Clindamycin)	2 μg	<17	17–20	>20
Methicillin (penicillinase-resistant penicillin class)	5 μg	<10	10–13	>13
Nafcillin	1 μg	<11	11–12	>12
Nalidixic acid	30 μg	<14	14–18	>18
Neomycin	30 μg	<13	13–16	>16
Nitrofurantoin	300 μg	<13	13–16	>16
Novobiocin	30 μg	<18	18–21	>21
Oleandomycin	15 μg	<21		>21
Oxolinic acid	2 μg	<11		>11
Penicillin G				
For staphylococci	10 U	<21	21–28	>28
For other organisms	10 U	<12	12–21	>21
Polymyxin B	300 U	<9	9–11	>11
Rifampin (for *Neisseria meningitidis* only)	5 μg	<25		>25
Streptomycin	10 μg	<12	12–14	>14
Tetracycline	30 μg	<15	15–18	>18
Tobramycin	10 μg	<12	12–13	>13
Triple sulfonamides*	250 μg	<13	13–16	>16
Vancomycin	30 μg	<10	10–11	>11

[a] All chemotherapeutic agents listed are antibiotics except those whose names are followed by an asterisk (*).
[b] Disk potency refers to the concentration of the agent in the disk in either micrograms (μg) or Units (U).

method to test a microorganism for its antibiotic sensitivity.

• Understand how a zone of inhibition is formed by an antibiotic.

MATERIALS

Cultures   *Streptococcus salivarius* (broth)
           Staphylococcus aureus (broth)
           Escherichia coli (broth)
           Pseudomonas aeruginosa (broth)
Media   Mueller-Hinton agar plates
Supplies   Various antibiotic disks and application tools
           95 percent ethanol and forceps
           Sterile cotton swabs
           Metric rulers (second laboratory period only)

PROCEDURE: FIRST LABORATORY PERIOD

1. **Select four plates of Mueller-Hinton agar, and label each plate** with your *name, date* of inoculation, *experiment number,* and the *name of the bacterium* used to inoculate the plate (see the materials list). Write with small letters around the outside rim on the bottom of each plate.
2. **Inoculate each agar plate with a different organism,** as follows:
   a. Thoroughly mix the broth culture so that the cells are evenly suspended.
   b. Aseptically remove one sterile cotton swab from its container, and immerse the cotton tip in the turbid broth culture. Lift the cotton tip above the broth, and roll the tip against the inside of the tube so as to squeeze the excess fluid from the cotton.

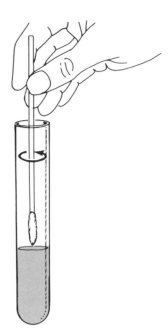

   c. Use this culture-moistened cotton tip to inoculate the entire surface of one agar plate. Continuously streak the agar surface in *one direction;* then repeat this streaking *two more times,* each in a different direction. Finish inoculating the entire plate surface by rubbing the cotton tip *around the outside edge* of the agar surface.

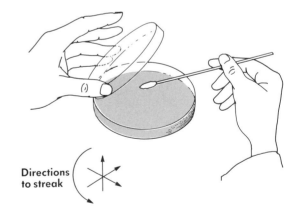

Directions to streak

   d. Discard this cotton swab as directed by your laboratory instructor.
   e. When finished, you should have four plates, each inoculated with a different bacterium.
   f. Wait about 5 minutes after inoculation for the culture fluid to soak into the agar surface before applying the antibiotic disks.

3. **Apply the sterile disks containing antibiotics to each plate** so that each plate contains the same types of antibiotic disks.
   a. *If a mechanical dispenser is used,* follow the instructions given to you by your laboratory instructor.

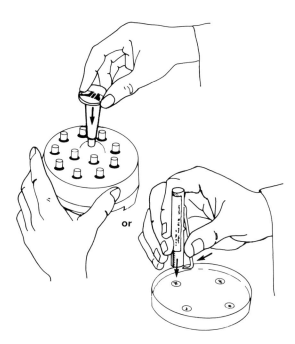

b. *If forceps are used,* sterilize them in the following way. Soak the tips in 95 percent ethanol for a few minutes, and then remove them from the ethanol and pass the forcep tips through a bunsen-burner flame. This is done only to ignite the ethanol. **Hold the points down so the burning ethanol does not run onto your hand.** The technique here is identical to that for sterilizing the spreading rod used for preparing spread plates. Aseptically remove one antibiotic sensitivity disk at a time, and place it on the surface of the inoculated plate. Resterilize forceps only if they touch something other than sterile disks. Arrange the disks on the plate, using a predesignated pattern suggested by your instructor.

c. **Use sterile forceps to press down on each disk,** regardless of the method used to transfer disks. To assure that it makes thorough contact with the inoculated medium.

d. You should have four plates when finished, each seeded with a different microbe and all containing the same combination and pattern of antibiotic disks.

4. **Record the antibiotics used and the corresponding codes** printed on each disk in the

TABLE 44-2 Identification codes for antimicrobial disks from Baltimore Biological Laboratories (BBL)

Code[a]	Antimicrobial agent[b]	Type of microbe[c]	Code[a]	Antimicrobial agent[b]	Type of microbe[c]
AM-10	Ampicillin	F	N-30	Neomycin	S
AN-10	Amikacin	S•	NA-30	Naldixic acid	*
B-10	Bacitracin	B	Nb-30	Novobiocin	S
C-30	Chloramphenicol	S	NF-1	Nafcillin	F
CB-50	Carbenicillin	F	NN-10	Tobramycin	S
CB-100	Carbenicillin	F	OA-2	Oxolinic acid	*
CC-2	Clindamycin	S	OL-15	Oleandomycin	S
CF-30	Cephalothin	F	P-10	Penicillin-G	F
CL-10	Colistin	B	PB-300	Polymyxin-B	B
DP-5	Methicillin	F	RA-5	Rifampin	S
E-15	Erythromycin	S	S-10	Streptomycin	S
FM-300	Nitrofurantoin	*	SSS-25	Triple sulfonamides	*
GM-10	Gentamycin	S	Te-30	Tetracycline	S
K-30	Kanamycin	S	Va-30	Vancomycin	S
L-2	Lincomycin	S			

[a] The numbers following each combination of letters refer to the quantity in micrograms (μg) or Units (U) for the chemical on that disk (see the column labeled "Disk potency" in Table 44-1).
[b] All antimicrobial agents listed are used as chemotherapeutic agents. All are also antibiotics except where the type of microbe is indicated with an asterisk (*).
[c] Microbial type that produces the antibiotic is abbreviated as follows: S, a species of the bacterial genus *Streptomyces*; B, a species of the bacterial genus *Bacillus*; and F, a species of filamentous fungi. An asterisk (*) indicates that this agent is not an antibiotic as defined in this exercise. A bullet (•) means that the organism first makes the antibiotic and then the antibiotic is chemically modified; the chemically modified antibiotic is called a *semisynthetic* antibiotic.

Results section of this exercise. See Table 44-2 for an interpretation of the code printed on each BBL disk.

5. **Incubate the inoculated plates right-side-up at 30°C for 18 hours; then refrigerate them until the next laboratory period.**

PROCEDURE: SECOND LABORATORY PERIOD

1. **Measure each inhibition zone** with a metric ruler to the nearest whole millimeter. Measure across the *diameter* of the inhibition zone, as determined with the unaided eye. All commercially produced antibiotic-sensitivity disks have a standard diameter; therefore, you should measure the zone of inhibition across the entire diameter of the zone (from one outer edge to the opposite outer edge of this zone).

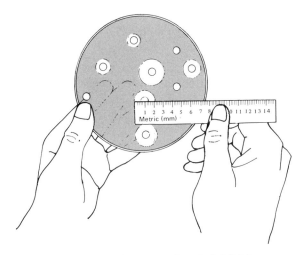

2. **Record the diameter of each inhibition zone** in the Results section. Compare your measurements with those obtained by others in the class for the same antibiotics. This will help you determine whether you are measuring the inhibition zones in the same way as your fellow students. You may be surprised at the variation in measurements due only to the perceptions of the people doing the measuring.

3. **Record the response of all four organisms to each antibiotic** in the Results section [such as resistant (R), intermediate (I), or sensitive (S)], according to the standards given in Table 44-1.

4. **Properly discard all plates.**

NAME _____

DATE _____

SECTION _____

RESULTS

Bacterium used	Antibiotic used	Code on disk	Inhibition zone size (mm)	Culture response from Table 44-1 (R, I, or S)
Streptococcus salivarius				
Staphylococcus aureus				
Escherichia coli				
Pseudomonas aeruginosa				

QUESTIONS

Completion

1. Any chemical or physical agent that inhibits the growth of a microbe or kills it is called a(n) _____.

2. Chemical compounds that are taken internally to ease the symptoms of a disease or to speed the patient's recovery are called _____.

3. For drugs to be called antibiotics, they must meet three criteria: (1) _____, (2) _____, and (3) _____.

4. In your microbiology laboratory, the easiest way for you to tell that an isolate is a streptomycete rather than a fungus is to _____.

5. The way that an antibiotic diffuses into the agar under and around an antibiotic-sensitivity disk creates a concentration _____ such that the greatest concentration is _____ the least concentration is _____.

6. The thick, confluent growth of bacterial cells on a plate prepared for antibiotic-sensitivity testing is called a(n) _____, and the clear area surrounding an antibiotic sensitivity disk is called a(n) _____.

7. The effectiveness of an antibiotic on a microorganism is usually stated by one of three terms. These are _____, _____, or _____.

8. The standard agar diffusion method used to determine the effectiveness of antibiotics is called the _____ method.

9. In general, the more effective an antibiotic is in adversely affecting a microbe the _____ will be the inhibition-zone diameter.

True–False (correct all false statements)

1. All commercially produced antibiotic-sensitivity disks have a standard diameter. _____

2. Bacteria of the genus *Streptomyces* and the genus *Penicillium* often produce antibodies. _____

3. The most common natural habitat for antibiotic-producing microbes is the soil. _____

4. A number of antibiotics are effective in inhibiting growth or killing viruses. _____

5. All cells within a zone of inhibition are either killed or inhibited from growing. _____

45

Coagulase Production by Pathogenic Staphylococci

Staphylococci, common inhabitants of the skin and mucous membranes of all mammals, are gram-positive, nonmotile, non-spore-forming, aerobic, and facultatively anaerobic bacteria that are not usually encapsulated. In healthy persons, these normal inhabitants do not cause problems; but when the opportunity arises, they can cause infections. Thus, the staphylococci are called **opportunistic pathogens.** Even as such, the staphylococci are responsible for over 80 percent of the pus-forming (suppurative) diseases of the skin (such as pimples and boils). In addition, some types of pneumonia and infections of nerve tissue and bone are known to be caused by staphylococci. The staphylococci also cause a high percentage of all cases of food poisoning because they excrete exotoxins when growing in certain types of food.

Two distinct species of staphylococci are most often found associated with humans:

Staphylococcus epidermidis is nonpathogenic, rarely pigmented, nonhemolytic, and **coagulase-negative.**

Staphylococcus aureus is opportunistically pathogenic, usually pigmented (golden-yellow colonies), hemolytic, and **coagulase-positive.**

It is because of this correlation between a positive coagulase reaction and potential pathogenicity that the **coagulase test** has become an important diagnostic tool.

If a microbiologist examines a pustulating wound and finds both *S. aureus* and *S. epidermidis,* one usually assumes that the *S. aureus* is the causative agent of the infection and that the *S. epidermidis* is a secondary contaminant from the skin around the wound.

The coagulase test uses a substance called blood plasma. Blood has both formed and soluble components (Figure 45-1). The *formed* components are larger (nonsoluble) entities such as red and white cells and platelets. The *soluble* components are large protein molecules such as albumin, globulins, fibrinogen, and complement. When blood is prevented from coagulating and the formed components are removed by centrifugation, the remaining liquid is called **blood plasma.** Plasma contains all of the soluble components of the blood. The liquid that remains after clotting is called **serum;** it contains all of the soluble components of plasma except for those proteins involved in clotting.

Let us examine how you can obtain plasma from whole blood. First, you immediately treat freshly withdrawn blood with an anticoagulant, such as sodium citrate or heparin or ethylenediaminetetracetic acid (EDTA), to keep the blood from forming a natural (or fibrin) clot. Next, you centrifuge this anticoagulant-treated whole blood to remove all of the formed components. The soluble proteins remaining in the supernatant fluid are collectively called *blood plasma.*

If you had not added the anticoagulant, the blood would have formed a natural (or fibrin) clot. Let us examine how a fibrin clot is formed because an understanding of this process will help you understand how coagulase works. The best-characterized part of natural clotting is the conversion of fibrinogen into fibrin by a proteolytic enzyme called *thrombin.* **Fibrinogen** is a soluble blood protein whose overall structure looks like that illustrated in Figure 45-2. In circulating blood, fibrinogen probably exists as sepa-

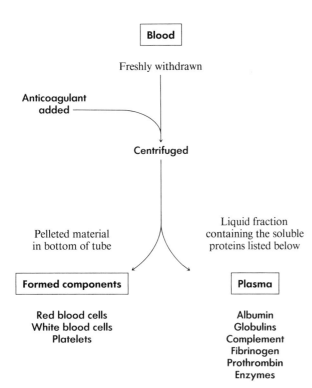

FIGURE 45-1 Major components of blood from vertebrate animals and how plasma is obtained.

searchers believe that it is an enzymelike protein (note the *-ase* ending on the word). It has antigenic activity; but unlike many enzymes, it is a heat-stable protein. Genes that code for coagulase production occur on **plasmids** (extrachromosomal genetic material). Certain staphylococci appear to have two forms of coagulase: one is bound to the cell wall, and the other is excreted by the cell as an extracellular protein.

Coagulase is believed to protect opportunistically pathogenic staphylococci from **phagocytosis** (engulfment by phagocytic white blood cells) because it clots the host's blood around the coagulase-positive cells, effectively walling them off from the host's phagocytes. (Phagocytic activity is one way an animal fights infection by invading microorganisms.) Thus, coagulase-positive bacteria may be better equipped than coagulase-negative cells to survive and grow (produce an infection) in animal tissue. Microbiologists say that coagulase-positive bacteria are more **virulent** than coagulase-negative bacteria. Coagulase is one of

rate molecules because each molecule contains many negatively charged amino acids that give the fibrinogen molecule a large overall negative charge. Therefore, fibrinogen molecules repel one another in circulating blood. Microbiologists presently believe that to form the clot a proteolytic enzyme called **thrombin** attacks the fibrinogen proteins and selectively removes the negatively charged amino acids (Figure 45-3). The release of negatively charged amino acids gives these proteins a less-negative surface charge, which leads to their specific aggregation (fibrin clot). But this clot is not very stable. To make the clot more stable, an enzyme in the blood (called transamidase) forms covalent peptide bonds (cross-links) between amino acids on the adjacent fibrinogen molecules.

By now you are probably asking where the thrombin comes from. Circulating blood contains an inactive soluble protein called **prothrombin** (see Figures 45-2 and 45-3). When something abnormal happens to circulating blood (such as spilling into a wounded area), an enzyme is activated that breaks the prothrombin molecule (see cleavage site in Figure 45-2), thereby releasing the active thrombin to attack fibrinogen and begin the clotting process (see Figure 45-3).

Now let us examine what **coagulase** is and how it works. The chemical composition of the substance(s) called coagulase is not well understood. Some re-

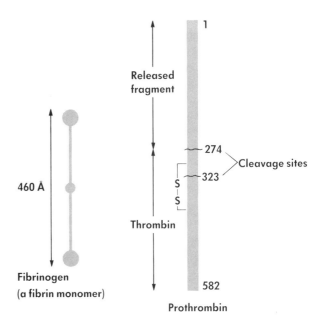

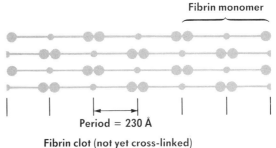

FIGURE 45-2 Graphic representation of fibrinogen, prothrombin, and a non-cross-linked fibrin clot.

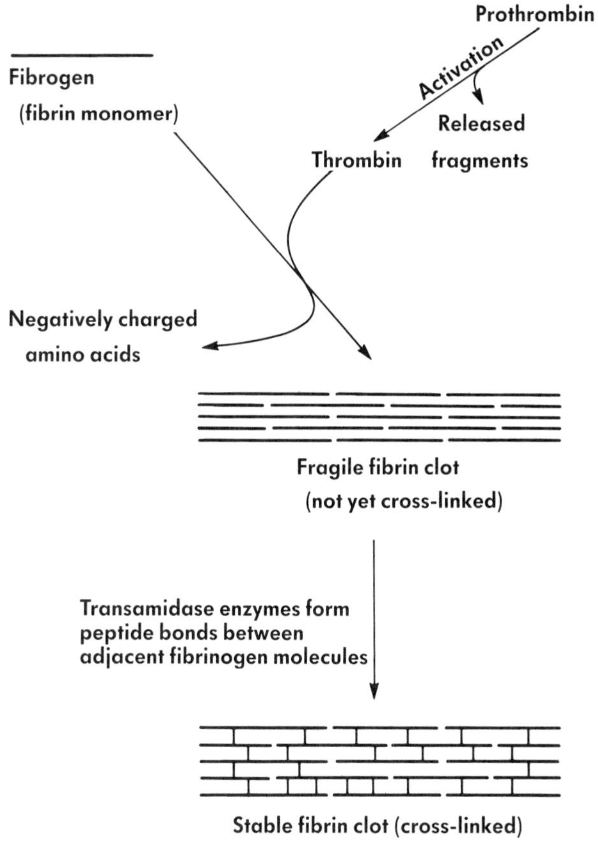

FIGURE 45-3 Schematic representation of reactions involved in forming a stable fibrin clot.

the bacterial **virulence factors,** defined as substances produced by bacteria that help them survive and grow (infect) animal or plant tissue. Other examples of virulence factors produced by *S. aureus* are *hemolysins, fibrinolysins,* and *enterotoxins,* all of which are extracellular proteins, probably coded for on plasmids.

Microbiologists do not completely understand how coagulase interacts with blood components to form a clot around the staphylococcal cell. At present, it appears that coagulase requires some component found in animal blood plasma, possibly prothrombin or a modified prothrombin molecule (rabbit and horse plasma seem to have a particularly high concentration of this substance). In any event, coagulase seems to combine with this plasma component; and the complex that is formed appears to act on fibrinogen, modifying it so that a fibrin clot is formed.

One of the primary goals of examining staphylococci in the clinical laboratory is to differentiate *S. aureus* and *S. epidermidis.* Of all the characteristics that are used for distinguishing between these two bacteria, the production of coagulase is the most convenient and reliable test. Note, however, that a coagulase test should not be used as the sole indicator of pathogenicity since coagulase-negative pathogenic staphylococci are occasionally isolated from infected tissues.

Note too that false positive tests may result from mixed cultures. For example, if a few coagulase-positive bacteria contaminate a coagulase-negative bacterial culture, they may produce enough coagulase to give a positive reaction. Also, when citrate is used as an anticoagulant, a false positive can result if the bacteria use up the citrate because removing the anticoagulant allows the plasma to clot. This latter problem may occur with certain gram-negative bacteria, such as *Pseudomonas* species. For this reason, many microbiologists prefer to use the inorganic compound EDTA as an anticoagulant because it is not used by bacteria.

LEARNING OBJECTIVES

• Understand the function of coagulase in protecting coagulase-positive bacteria.

• Realize that this test is used to help identify the pathogenic staphylococci and that it is used primarily to distinguish between *S. aureus* and the other species of staphylococci.

• Know how to perform this test.

MATERIALS

Cultures	*Staphylococcus aureus*
	Staphylococcus epidermidis
Media	None
Supplies	Small test tubes (coagulase tubes)
	1.0-ml pipettes (sterile)
	Commercially prepared coagulase plasma (EDTA-treated rabbit or horse plasma)

PROCEDURE: FIRST LABORATORY PERIOD

1. Label two small test tubes (coagulase tubes) with your *name, date,* and experiment *number.*

2. Aseptically transfer **0.5 ml of coagulase plasma** to each of these two tubes.

3. Aseptically transfer a *large* loopful of cells from one of the slant cultures into one tube of coagulase plasma. Transfer as many cells as

you can get on your loop. The resulting suspension should be *very* turbid.

TECHNIQUE NOTE

Cultures should be 18 to 24 hours old. An inoculum can be (1) as many cells from a slant as you can get on an inoculating loop, (2) 0.1 ml of a very turbid broth culture, or (3) one colony from a plate, transferred with your loop.

a. **Stir this mixture** with your loop until the cells are evenly suspended.
b. **Label this tube** with the name of the bacterium it contains.
c. **Cover this tube** tightly with Parafilm.

4. **Repeat step 3** with the other culture and the other tube of coagulase.
5. **Incubate both tubes at 30°C** until your next laboratory period.

TECHNIQUE NOTE

The standard clinical laboratory procedure is to incubate coagulase-test mixtures at 37°C and examine them for clotting at 30-minute intervals for 4 hours. If no clot is observed, the mixture is examined again at 6 and at 24 hours. If lower temperatures are used, longer incubation times must be used. Typical introductory microbiology laboratory sessions, however, do not last for 4 hours; so we have modified this technique accordingly.

PROCEDURE: SECOND LABORATORY PERIOD

1. **Examine both tubes for positive or negative coagulase reactions.** *Do not shake these tubes* because it is possible to break up a clot formed by a coagulase-positive culture. Instead, gently tip the tube to a nearly horizontal position, and compare the appearance of the tube contents with Figure 45-4.

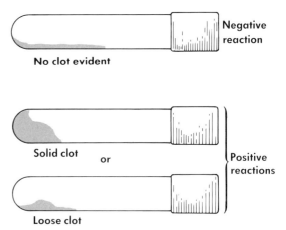

FIGURE 45-4 Illustration of negative- and positive-coagulase reactions after inoculating 0.5 ml of commercially prepared coagulase plasma with an 18- to 24-hour culture and incubating this mixture for 24 to 48 hours at 30°C.

INTERPRETATION NOTES

A *positive test* is any degree of clotting, ranging from a loose clot suspended in plasma to a solid clot of the entire mixture in the bottom of the tube. To check for the presence of clotting, tip the tube on its side. A loose clot will be evident as a raised, jellylike mass in the plasma as it runs down the side of the tube (Figure 45-4). A solid clot will hardly move as the tube is tipped on its side: The entire mixture will be a jellylike mass at the bottom of the tube (Figure 45-4). Note, however, that the clot is never so solid that the tube can be completely inverted without disturbing the clotted plasma. Most coagulase-positive strains of staphylococci produce a clot within the first 8 hours at 30°C (or 4 hours at 37°C). Some strains produce a clot within 3 hours at 30°C (or 1 hour at 37°C).

A *negative test* exhibits no evidence of a clot after 48 hours of incubation at 30°C (or 24 hours at 37°C). The plasma remains as fluid as it was when first inoculated.

2. Record your results and properly discard these tubes.

COAGULASE TEST SUMMARY	
Purpose of test	Determine bacterial production of coagulase.
Reagent used	Commercially prepared coagulase (EDTA-treated rabbit or horse) plasma (0.5 ml added to a small tube).
Cells used	An 18- to 24-hour culture of gram-positive cocci from any solid (large loopful) or liquid medium (0.1 ml) or one colony mixed with the 0.5 ml of coagulase plasma.
Incubation	Coagulase plasma plus cell mixture requires 4 to 48 hours of incubation in a capped tube at 30°C.
Positive reaction	Any evidence of clot formation up to 48 hours of incubation at 30°C.
Negative reaction	No evidence of clot formation after 48 hours of incubation at 30°C.

NAME _____

DATE _____

SECTION _____

RESULTS

Microorganism	Incubation time (hours) at _____ °C	Coagulase test (+ or −)
Staphylococcus aureus		
Staphylococcus epidermidis		

CONCLUSIONS

QUESTIONS

Completion

1. The coagulase test is primarily used to differentiate opportunistically pathogenic and nonpathogenic species in the genus _____.

2. A species of coagulase-positive, gram-negative, pigmented cocci commonly found on the skin of humans is called _____.

3. When blood is treated to prevent coagulation and then is centrifuged to remove the cells and platelets, the remaining straw-colored fluid is called _____.

4. The genes that code for bacterial coagulase production are located on a genetic structure called a _____.

5. Coagulase is believed to function by forming a thin clot around the bacterial cell that makes the cell resistant to _____. Thus, coagulase is considered to be one of the _____ factors, and coagulase-positive bacteria are said to be more _____ than coagulase-negative bacteria.

6. One way to avoid false-positive coagulase reactions is to make sure that your culture is pure. This is because a coagulase-negative isolate may be _____.

True–False (correct all false statements)

1. The staphylococci are commonly found to cause skin infections and are also known to cause nerve tissue, bone, and lung infections. _____

2. Some coagulase molecules, when produced by bacterial cells, remain bound to the cell wall; others are excreted into the extracellular environment. _____

3. The inorganic compound that is abbreviated EDTA is preferred as an anticoagulant over citrate because false-positive reactions may result if the bacteria transport and use all of the citrate. _____

4. In contrast to the procedure you used, clinical laboratories usually check coagulase preparations every 30 minutes for a total period of 4 hours incubation at 37°C. _____

46

Hemolysis of Red Blood Cells

Several pathogenic or opportunistically pathogenic bacteria produce proteins that can interact with and damage animal cell membranes. This damage eventually leads to **lysis** of cells in the infected tissue, a process in which cells burst open and die. Proteins, produced by bacteria that cause animal cells to lyse are called **hemolysins** (Greek *haima,* blood) because their effects are detected most easily when they lyse **erythrocytes** (red blood cells), and release hemoglobin. Most hemolysins can also damage or kill platelets, macrophages, lymphocytes, neutrophils, and other types of cells. Consequently, some authors now prefer to call these proteins **cytolytic toxins** instead of hemolysins.

Many (possibly all) hemolysins are **extracellular proteins** (made in the bacterial cell and excreted into the surrounding medium) and therefore are examples of the *exotoxins* (toxic proteins excreted by bacteria) that enhance the *virulence* (the likelihood of being able to produce disease) of many potentially pathogenic microorganisms. Hemolysins do not possess many other traits in common, however. Some have enzymatic properties, whereas others do not. They vary in sensitivity to oxygen, heat, and acid; in their immunological traits; in the types of animals in which they lyse cells; and in the optimum temperatures at which they cause lysis. Hemolysins also vary with respect to the mechanisms by which they interact with and damage animal cell membranes.

The hemolysins most significant to humans are made by species in three bacterial genera: ***Clostridium, Staphylococcus,*** and ***Streptococcus*** (Table 46-1). The primary clostridial hemolysin is **lecithinase** (also called *alpha-toxin* or *phospholipase C*) produced by *Clostridium perfringens* and several other closely related species. *Clostridium perfringens* causes one form of clostridial food poisoning and is the leading cause of gangrene in wound infections. Clostridial lecithinase is actually a phospholipid-degrading enzyme that splits lecithin molecules in the animal cell membrane into phosphorylcholine and diglyceride (see Exercise 29), thereby disrupting the membrane enough to bring about cell lysis. This enzyme can destroy red blood cells, platelets, and several types of white blood cells. It also weakens the plasma membrane of muscle cells and plays a direct role in the gangrene process.

The primary hemolysin-producing species in the genus ***Staphylococcus*** is *Staphylococcus aureus.* This bacterium is a common resident of the human nasal passage and throat, where it usually resides without causing any detectable harm to the body. However, *S. aureus* is also an opportunistic pathogen that can cause disease when the resistance mechanisms of the body are weakened or when a particular strain of the bacterium happens to possess a combination of virulence factors (for example, it might be able to produce several potent exotoxins including hemolysins) that render it able to invade and damage the body. *Staphylococcus aureus* usually causes **abscesses** (localized pus-containing infections) in skin tissue; but sometimes it invades more deeply into the body and infects the bones and joints, the kidneys, the central nervous system, the lungs, and/or the heart. Some strains of *S. aureus* produce as many as four hemolysins, called **alpha-, beta-, gamma-,** and **delta-hemolysin** (see Table 46-1), and almost all strains produce alpha- and beta-hemolysin. The *S. aureus* hemolysins vary with respect to the types of cells they lyse and the mechanism by which they lyse them. Their precise role in the disease process is not yet understood in most cases, but they (along with other virulence factors) help the bacteria invade the body and overcome some of its defense mechanisms.

TABLE 46-1 Properties of selected bacterial hemolysins

Hemolysin	Types of cells or organelles affected (lysed)	Mechanism of action	Role in the disease process
Clostridial hemolysins			
Lecithinase (alpha-toxin, phospholipase-C)	Erythrocytes, platelets, and most leucocytes	Enzymatic; splits membrane lecithin into phosphorylcholine and diglyceride	Directly involved in the gangrene process
Staphylococcal hemolysins[a]			
Alpha-hemolysin	Erythrocytes, platelets, most leucocytes, and fibroblasts	Unknown; possibly detergent-like action on membrane	Not completely defined, but affects central nervous system and cardiovascular system in animals
Beta-hemolysin	Erythrocytes, fibroblasts, leucocytes, and macrophages	Enzymatic; hydrolyzes membrane sphingomyelin	Unknown
Gamma-hemolysin	Erythrocytes, leucocytes, and lymphoblasts	Unknown, but probably not enzymatic	Unknown
Delta-hemolysin	Erythrocytes, lysosomes, and mitochondria	Detergent-like action on membrane	Probably involved in intestinal disturbances
Streptococcal hemolysins			
Streptolysin S (SLS)	Erythrocytes, leucocytes, platelets, lysosomes, and tumor cells	Damages membrane sterols; alters membrane permeability	Damages parenchymatous organs and kidneys, may induce arthritis
Streptolysin O (SLO)	Erythrocytes, leucocytes, mesenchymal cells, platelets, tumor cells, and lysosomes	Damages membrane sterols; alters membrane permeability	Damages heart tissue, constricts coronary arteries, and stimulates release of acetylcholine

[a] The name of each hemolysin refers to the *order* in which they were isolated; *it does not refer to the type of hemolysis* exhibited on a blood-agar plate.

Several species of the genus ***Streptococcus,*** especially some of the more important pathogenic species (see below), produce two major hemolysins: **streptolysin S (SLS)** and **streptolysin O (SLO)**. These hemolysins, which lyse a variety of blood cells (see Table 46-1), alter membrane permeability or bring about lysis by damaging sterols in the infected tissue cell membrane. Streptolysin O is probably more important than SLS in the disease process because it affects heart tissue, causing systolic contractions and constriction of coronary arteries. The two hemolysins also differ in sensitivity to oxygen; SLO is inactivated by oxygen, whereas SLS is not.

One detects hemolytic activity by plating the bacteria on **blood agar.** Blood agar is a complex, all-purpose medium (usually trypticase-soy agar or heart infusion-tryptose agar) to which 5 percent sterile, defibrinated sheep or rabbit blood has been added. The blood provides nutrients that would not otherwise be present in all-purpose media and that are required by certain types of **fastidious** microorganisms (microorganisms that have unusually complex nutrient requirements, a common trait among pathogenic species that are used to living in the body). Blood agar is often referred to as an **enriched all-purpose medium** because it supports the growth of fastidious organisms. However, *enriched* media such as blood agar should not be confused with *enrichment* media (see Exercise 22) because they work differently (enriched media are not very selective) and have different applications.

Blood agar functions as a **differential medium** (see Exercise 21) because it discriminates between microorganisms that affect red blood cells in one or more of the following ways.

Beta-hemolysis involves the complete destruction (lysis) of red blood cells, usually brought about by one or more of the bacterial hemolysins described. Beta-hemolysis makes a distinct clear zone (a colorless, transparent area) in the agar around beta-hemolytic bacterial colonies.

Alpha-hemolysis involves only partial lysis of red blood cells, accompanied by the chemical reduction of hemoglobin (which is red) to methemoglobin (which is green). Consequently, alpha-hemolysis makes a green or greenish-brown zone (in which there may be a limited amount of clearing) around alpha-hemolytic bacterial colonies.

Gamma-hemolysis is not hemolysis at all. This is a deceptive term used to describe situations in which there is no change in the agar around a bacterial colony. Organisms that produce no reaction on blood agar are best referred to as *nonhemolytic* rather than gamma-hemolytic.

The primary use of blood agar in clinical laboratories is to distinguish species within the genus *Streptococcus* because there is a good correlation between hemolytic activities and pathogenicity in this genus. (Clinical laboratories seldom use blood agar to study other bacterial genera because hemolytic reactions are a much less reliable indicator of pathogenicity in other genera.)

Several frequently encountered *Streptococcus* species, their reactions on blood agar, and their roles in human disease are listed in Table 46-2. The streptococcus that clinical laboratories are looking for when they plate a throat swab on blood agar (see Exercise 48) is *Streptococcus pyogenes*. This organism is best known as the cause of streptococcal pharyngitis ("strep throat"), which sometimes develops into scarlet fever and can later lead to rheumatic heart disease. Less frequently, *S. pyogenes* is involved in wound infections, infections of the skin (pyoderma), and kidney disease. Another clinically important *Streptococcus* species that produces beta-hemolysis on blood agar is *S. agalactiae,* which causes neonatal septicemia (infection of the circulatory system in newborn infants) and a form of meningitis.

Most of the alpha-hemolytic and nonhemolytic (gamma-hemolytic) streptococci (see Table 46-2) are either saprophytic (strains seldom associated with the human body) or opportunistic pathogens (strains that normally inhabit various parts of the human body and produce disease only under specific circumstances). The most dangerous of the opportunistic species in the genus *Streptococcus* is *S. pneumoniae,* which causes about 70 percent of all human pneumonia. Other important opportunistic species include *S. sanguis,* a resident of the human mouth that sometimes produces gingival (gum) abscesses or other diseases; *S. mutans,* a resident of the mouth that is involved in the tooth decay process; and *S. faecalis,* a resident of the intestine that sometimes causes urinary-tract infections or endocarditis (inflammation of the lining of the heart and its valves).

Although blood agar is most frequently used to differentiate streptococci, it does have other applications. Being an enriched medium, it is highly suitable for maintaining cultures of several types of fastidious

TABLE 46-2 Pathogenicity and hemolytic activity of important *Streptococcus* species

Species	Hemolysis reaction(s) on blood agar	Diseases caused by this species
S. agalactiae	Beta[a]	Neonatal septicemia and meningitis
S. equi	Beta	Pharyngitis, pyoderma, and endocarditis
S. faecalis	Alpha, beta, or NH[b]	Endocarditis, urinary-tract infections, and neonatal septicemia and meningitis
S. lactis	Alpha or NH	Not pathogenic; found in milk and other dairy products
S. mitis	Alpha	Gingival (gum) abscesses and endocarditis
S. mutans	Alpha or NH	Gingival abscesses, endocarditis, and dental caries
S. pneumoniae	Alpha	Bacterial pneumonia
S. pyogenes	Beta	Pharyngitis (strep throat), scarlet fever, rheumatic fever, pyoderma, kidney disease, and wound infections
S. salivarius	NH	Endocarditis
S. sanguis	Alpha	Gingival abscesses and endocarditis

[a] Most strains
[b] NH = nonhemolytic (gamma-hemolysis)

microorganisms, including some significant human pathogens. Blood agar can also be made selective (see Exercise 21) by the addition of antibiotics. In this form, it is especially useful for detecting and/or isolating antibiotic-resistant pathogenic microorganisms. Rapid detection and isolation of antibiotic resistant microorganisms are important because these microbes are rapidly becoming a serious problem in the treatment of human infections.

This exercise gives you an opportunity to observe the three responses that are produced by microorganisms on blood agar. You will inoculate blood agar with pure cultures to produce either alpha- or gamma-hemolytic (nonhemolytic) reactions. Sealed demonstration plates, on which *Streptococcus pyogenes* has grown, are provided to illustrate beta-hemolysis. (You will not handle *S. pyogenes* because it is pathogenic and potentially dangerous.) You will also inoculate blood agar with material swabbed from your teeth, so that you can see whether there are any hemolytic microorganisms residing there.

LEARNING OBJECTIVES

• Learn about hemolysins and understand their significance in human disease.

• Learn about the major hemolysin-producing groups of microorganisms.

• Learn about blood agar and its various applications.

• Understand the significance of blood agar with respect to the differentiation of streptococci in clinical microbiology laboratories.

• Learn how to interpret hemolysis reactions on blood agar in the laboratory.

MATERIALS

Cultures *Streptococcus sanguis,* alpha-hemolytic strain (broth)
Streptococcus salivarius, nonhemolytic strain (broth)
Streptococcus pyogenes, beta-hemolytic strain, blood-agar streak (sealed demonstration plate)

Media Blood-agar plates

Supplies Tube of saline (sterile)
Cotton swab (sterile)
Glass microscope slides
Gram-staining reagents

PROCEDURE: FIRST LABORATORY PERIOD

1. **Label both blood-agar plates** with your *name,* the exercise *number,* and the *date.*

2. **Divide the bottom of one of the blood-agar plates** into three sections with your marking pen, and label one section of this plate *S. sanguis,* the second section *S. salivarius,* and the third section *control.*

3. **Inoculate two sections of the first plate with the appropriate organisms.** Transfer only a small amount of culture to the plate and streak it throughout the section.

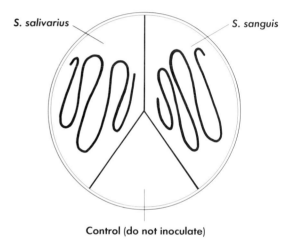

4. **Inoculate the second blood-agar plate with material swabbed from your teeth** according to the following method.
   a. Label the bottom of the plate *tooth swab.*
   b. Moisten a sterile cotton swab in sterile saline and rub it vigorously over your teeth. Rub it over the front and back surfaces of your teeth in several different locations.
   c. Inoculate the blood-agar plate by gently rubbing the cotton swab across a small area near the edge of the plate.
   d. Use a sterile inoculating loop to streak the inoculum across the rest of the agar surface. Streak for isolation (see Exercise 14).
   e. Discard the contaminated cotton swab as directed by your instructor.

5. **Incubate both blood-agar plates in the inverted position at 37°C for 24 to 48 hours.**

PROCEDURE: SECOND LABORATORY PERIOD

1. **Examine the blood-agar plate that was inoculated with** *S. sanguis* **and** *S. salivarius* **and the sealed demonstration plate** showing beta-hemolysis. **Record** your observations in the Results section.

2. **Examine the blood-agar plate that you inoculated with material from your teeth.** Note any alpha- or beta-hemolytic colonies that are present on this plate. Record your observations in the Results section.

3. **Prepare Gram stains of representative hemolytic colonies on the plate** inoculated with material from your teeth. If your plate does not contain any hemolytic colonies, obtain a plate that does from one of the other students.

4. **Examine the Gram stains** with the oil-immersion (100×) objective. Record your observations in the Results section.

5. **Properly discard all plates.**

NAME _____

DATE _____

SECTION _____

RESULTS

1. Identify and describe the hemolytic reaction that was produced by each of the following organisms:
   a. *S. sanguis*

   b. *S. salivarius*

   c. *S. pyogenes* (demonstration plate)

2. Did you obtain any hemolytic colonies on the plate that you inoculated with material from your teeth? If so, describe each distinct type of hemolytic colony, and indicate the type of hemolysis (alpha or beta) that resulted.

3. In the following table, describe each of the Gram stains that you prepared from hemolytic colonies on tooth-swab blood-agar plates.

Type of hemolysis (alpha or beta)	Gram reaction (+ or −)	Cell morphology and arrangement (shape, presence of chains, etc.)

CONCLUSIONS

Do any of the hemolytic organisms that you isolated from your teeth and Gram stained appear to be streptococci? If so, what species do you think they may be? What reasoning have you used?

XI MEDICAL MICROBIOLOGY AND IMMUNOLOGY

NAME _____

DATE _____

SECTION _____

QUESTIONS

Completion

1. Bacterial hemolysins cause _____ of animal cells, during which the cells burst open and die.

2. The hemolysin produced by *Clostridium perfringens* is a phospholipase and is involved in the _____ process that sometimes accompanies wound infections.

3. Blood agar is sometimes classified as a(n) _____ medium because it contains extra nutrients that can support the growth of fastidious organisms.

4. On blood agar, the formation of a green or greenish-brown zone in the agar around a bacterial colony is called _____.

5. Blood agar can be made selective by the addition of _____.

True–False (correct all false statements)

1. Some authors feel that bacterial hemolysins should be called *cytolytic toxins* because they affect only erythrocytes. _____

2. *Staphylococcus aureus* is the primary hemolysin-producing species in the genus *Staphylococcus*. _____

3. Streptococcal hemolysins (SLS and SLO) lyse animal cells by enzymatically attacking the spingomyelin molecules in their cytoplasmic membranes. _____

4. Blood agar is both a differential medium and an enrichment medium. _____

5. Gamma-hemolysis on blood agar is not really hemolysis at all. _____

Multiple Choice (circle all correct answers)

1. Most bacterial hemolysins are examples of
   a. endotoxins.
   b. exotoxins.
   c. extracellular proteins.
   d. virulence factors.
   e. phospholipases.

2. Which of the following bacterial genera are known to produce significant hemolysins?
   a. *Bacillus*
   b. *Streptococcus*
   c. *Salmonella*
   d. *Escherichia*
   e. *Clostridium*

3. Which of the following bacteria produce beta-hemolysis reactions on blood agar?
   a. *Staphylococcus aureus*
   b. *Streptococcus salivarius*
   c. *Streptococcus sanguis*
   d. *Streptococcus agalactiae*
   e. *Streptococcus pyogenes*

4. Which of the following bacteria produce alpha-hemolysis reactions on blood agar?
   a. *Staphylococcus aureus*
   b. *Streptococcus pyogenes*
   c. *Streptococcus mitis*
   d. *Streptococcus pneumoniae*
   e. *Streptococcus salivarius*

47

Bacteria on Human Skin

Normal resident microorganisms

Each of us harbors billions of microorganisms and carries them around all the time, usually without harm. They are sometimes called our *normal microflora,* a term from the days when bacteria were thought to be microscopic plants. A more descriptive term is **resident** microorganisms. Not only do these microbes usually cause us no harm, some of them may contribute to our well being.

Our resident microbes live on the moisture and chemicals that we excrete or accumulate in our daily activities. These microorganisms are either nonpathogenic or are prevented from infecting us by the various mechanical, chemical, and microbial defenses of our bodies. Consider the following examples:

Mechanical defenses

Our bodies are covered by several layers of skin that keep microbes outside where they can cause no harm.

Some parts of our bodies contain cells with cilia, which physically sweep microbes outward, preventing them from entering our body.

Mucous membranes, such as those that line the respiratory, alimentary, and uriogenital tracts, and the conjunctiva, are coated with mucous that physically traps microbes.

Sneezes and coughs forcibly remove microbes from our respiratory tracts and oral cavities.

Chemical defenses

Various of our external tissues excrete salt, acidic compounds, or enzymes that chemically kill or inhibit the growth of microorganisms.

Microbial defenses

We normally have microorganisms on our skin and in our mouths and intestines (for example) that create environments that prevent the establishment of other, more harmful, organisms.

In addition to these defense mechanisms, good nutrition, adequate rest, and low stress levels also contribute to the prevention of infection, in ways that are not clearly understood.

On the other hand, many of our resident microbes are opportunistic pathogens, which are normally harmless but may cause an infection if given an opportunity. For example, if the skin is broken, a resident microbe may grow in the wound and damage that tissue. If this happens to a person suffering from malnutrition or lack of adequate rest, or to a person undergoing immunosuppressive chemotherapy or infected with a virus that causes immunosuppression, then the usually benign microbe may cause a dangerous infection. Theoretically, it is possible for *any* microorganism that is capable of growth in a tissue to infect that tissue *if* the body's defensive mechanisms are disrupted. (This is a good reason why you should practice good aseptic technique in the laboratory even when you are not working with human pathogens.)

Anatomy and defenses of the skin

Uninjured skin is an impenetrable barrier to the passage of microbes. Most nonresident microorganisms fail even to colonize the skin surface, let alone grow enough to damage the skin; nor can they penetrate it to damage underlying tissues. But if the outer (epidermal) layer of the skin is broken, otherwise harmless microbes may grow. These **cutaneous infections** are usually restricted to the site of entry and are usually more irritating than harmful. However, if the epider-

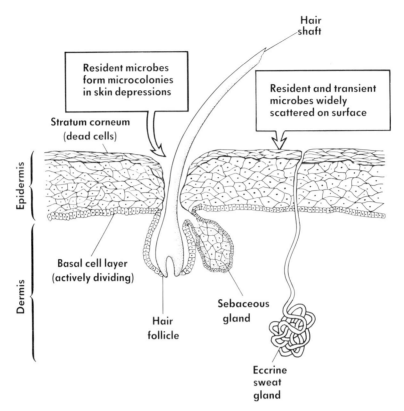

FIGURE 47-1 Cross-sectional anatomy of human skin showing the location and frequency of resident microorganisms.

mal and dermal layers of the skin are broken, the microorganisms have access to deeper tissues and may enter the bloodstream to be distributed to other parts of the body. That is called **parenteral inoculation,** and may occur through cuts, puncture wounds, severe burns, or bites of infected animals or arthropods, such as mosquitoes.

To understand how the skin protects an animal's body from microbial invasion, consider the simplified cross-sectional anatomy of skin diagrammed in Figure 47-1. The skin is composed of two parts: **epidermis,** the outer, epithelial layer, and **dermis,** the inner, sensitive, vascular, mesodermic layer.

The epidermis has an outer layer called the **stratum corneum,** which is composed of several layers of overlapping, flattened, dead cells that are rich in a water-insoluble protein called **keratin.** At the inner level of the epidermis is the **basal cell layer,** the cells of which grow rapidly, continually forcing the older, progeny cells toward the surface. These cells produce keratin during this migration process; and as they approach the outer surface, they flatten and die, becoming part of the stratum corneum. The skin maintains a constant thickness because the dead cells of the stratum corneum are constantly shed. Some physiologists estimate that our skin sloughs dead epithelial cells at the rate of about one million every 40 minutes, and resident microorganisms are shed along with them.

The dermis, which lies immediately under the epidermis, consists mainly of connective and muscle tissue. It contains nerve endings, blood and lymphatic vessels (which supply the epidermis with nutrients), and protective cells (which are capable of inflammatory response and wound repair). Also embedded in the dermis are **hair follicles** (where the keratin-rich hair strands are produced), **sebaceous glands** (which secrete an oily substance called **sebum** into the hair follicle), and **sweat glands** (which secrete a fluid composed mainly of water, with low concentrations of sodium chloride, nutrients, and nitrogenous wastes).

As long as the skin remains intact and relatively dry and retains a normal rate of epithelial-cell shedding, most nonresident microbes cannot permeate, colonize, or even survive on the skin. The two most important factors in preventing extensive microbial growth on the skin are both *mechanical* mechanisms:

the impenetrability of the stratum corneum and the rapid shedding of dead epithelial cells and attached microorganisms.

Chemical factors play a lesser, though important, role. For example, skin has a normal pH of between 3 and 5 (somewhat higher in moist regions). This low pH is thought to be due to lactic and other acids that are excreted by resident microbes, such as the staphylococci, as they metabolize (ferment) nutrients found in skin secretions. Low pH discourages the growth of many other microorganisms. Another example is the lack of moisture in and on the stratum corneum, which seems to inhibit growth of many gram-negative microbes, although gram-positive bacteria seem to be less inhibited by a dry environment. And yet another example is that sweat and sebum contain moisture and chemicals that inhibit the growth of some microorganisms. Sebum contains lipids and fatty acids that are mildly fungicidal and make the skin resistant to some fungal infections. Also, the concentrations of sodium chloride and lysozyme in sweat is mildly bacteriostatic, but their presence probably does not outweigh the tendency of sweat to promote microbial growth by providing moisture. Altogether, these inhibitory chemical factors seem to be less important than mechanical factors in protecting the body from microorganisms.

Normal microflora versus transient microorganisms of the skin

No region of the skin is sterile, but in moist areas such as the armpit, there may be one million bacteria per square centimeter (about 100 times the population of the dryer human forearm). Our most common resident microorganisms are listed in Table 47-1. Of those listed, the two types almost universal are the staphylococci and the proprionobacteria. The staphylococci are gram-positive, nonmotile, non-sporeforming, aerobic, and facultatively anaerobic bacteria that are usually not encapsulated (see Exercise 45). The propionobacteria are gram-positive rods of irregular shape (straight to slightly curved rods that exhibit swellings, club shapes, or other deviations from a uniform rod shape). Both of these bacteria are considered opportunistically pathogenic. These fairly stable populations of resident microorganisms that inhabit our skins live widely scattered on the surface of the stratum corneum and also in **microcolonies** of 1000 to 10,000 cells in the depressions where hairs emerge from their follicles (see Figure 47-1).

Transient microorganisms, which we pick up from our environment, such as *Escherichia coli* and *Bacteroides fragilis* from fecal contamination and *Bacillus subtilis* and *Enterobacter aerogenes* from soil contamination, usually fail to become permanent residents of the skin for several reasons. Perhaps the most important reason is that the established residents are more effective in competing for nutrients. Transient microbes usually come from other environments and are poorly adapted to the skin surface. Studies show that many transients die within 2 to 3 hours after being deposited on the skin, and most disappear within 24 hours after inoculation. Many may die because of lack of sufficient nutrients, but most are probably shed with the dead stratum corneum cells.

Washing is the most important way to remove transient microorganisms from the skin. Such microbes, usually found scattered on the smooth surfaces of the stratum corneum, are exposed to mechanical abrasions and removal during washing. However, one must do a lot of scrubbing to remove even transient microorganisms.

Reduction of normal skin defense mechanisms

Many situations reduce the skin's effectiveness as a barrier to microbial infections. For example, if the skin is not allowed to dry, the normally impervious stratum corneum softens and is no longer an effective barrier. People such as homemakers and bartenders, who frequently immerse their hands in water, often have problems with inflamed or infected skin. High

TABLE 47-1 Microorganisms that are common long-term residents on human skin

Name	Type	Microscopic description
Staphylococcus epidermidis	Bacterium	Gram-positive coccus; clusters
Staphylococcus aureus	Bacterium	Gram-positive coccus; clusters
Propionobacterium acnes	Bacterium	Gram-positive pleomorphic rods
Corynebacterium species	Bacterium	Gram-positive pleomorphic rods
Pityrosporum species	Yeast	Small yeasts; often budding
Streptococcus pyogenes	Bacterium	Gram-positive coccus in chains

humidity and excessive sweating keep moisture on the stratum corneum and allow microbes, both residents and transients, to grow; thus, people in tropical climates often have problems with skin infections. Waterproof materials, such as a baby's plastic diaper cover, also create a local high-moisture environment that promotes microbial growth. Overweight and aged people have folds of skin that trap moisture and provide favorable environments for bacterial growth. Tight clothing and friction caused by movement can cause skin abrasions and force microbes into them.

The administration of some drugs to combat other problems, such as the corticosteroids, decreases the skin's rate of epithelial-cell shedding, giving pathogens a longer time to become established. If a skin infection is already present when corticosteroids are used, the infection usually becomes worse. People who suffer from malnutrition, diabetes, cancer, diseases of or defects in the immune system, or who are undergoing immunosuppressive treatments, are also more prone to skin infections.

Traumatic injuries that produce punctures or breaks in the skin, or abrasions or burns that remove skin, are often responsible for dangerous infections. Such wounds expose the underlying tissues to opportunistic pathogens, and they may also cut blood vessels, thus impairing normal tissue circulation. This, in turn, reduces the oxygen supply to the tissue, creating an anaerobic environment that is favorable for growing the anaerobic bacteria that are the most frequent and most dangerous causes of wound infections. Improper surgical techniques, insertion of nonsterile hypodermic needles (or sterile needles through untreated skin), penetration of the skin by the proboscus of an insect, or a deep bite from an insect that is carrying a pathogen are other ways that the natural barrier of the skin can be circumvented by microorganisms.

The primary purpose of this exercise is to show you some of your own natural resident microorganisms and, possibly, to let you see the effect of one or more skin treatments on the numbers or types of microbes present on your skin.

LEARNING OBJECTIVES

- Introduce the concepts of normal microbial residents and opportunistic pathogens.

- Understand where microorganisms are normally found on the skin and the normal defensive mechanisms of the skin.

- Demonstrate the types of microorganisms that are usually found on healthy skin.

MATERIALS

Cultures	You are carrying them with you.
Media	Trypticase-soy agar plates
Supplies	Cotton swabs (sterile)
	Tubes of sterile diluent
Demonstration	Proper way to swab a surface for collecting microorganisms

PROCEDURE: FIRST LABORATORY PERIOD

1. **Divide the agar plate in half** by placing a mark across the bottom with your marking pen, and label this plate with your *name* and today's *date.*

2. **Aseptically remove one swab from its dispenser and moisten it with sterile diluent.** Roll the wet swab on the inside of the tube of diluent to squeeze out excess diluent from the cotton tip.

3. **Rub the swab on your skin,** covering an area of about four square inches. (Please be discreet!).

4. **Use the same part of the swab to inoculate one-half of the plate.** Streak this half of the plate so that the entire surface has been touched by the same part of the swab that touched your skin.

5. **Label** this half of the plate so that you will remember the skin area sampled.

6. **Repeat steps 2 and 3 to obtain microbes from another part of your body** (maintaining discretion), and transfer these microorganisms to the second half of your T-soy agar plate.

7. **Label** this half of the plate so that you will remember the skin area sampled.

8. **Incubate the plate at 30°C** until the next laboratory period.

ALTERNATIVE PROCEDURES

Your laboratory instructor may suggest that you try one or more of the following alternative procedures.

Swab the palm of one hand, and inoculate one-half of the plate. Then thoroughly scrub your hands with plain or antiseptic soap and water, and let them air dry or dry them with a *sterile* towel. Use a second sterile swab to sample the same hand in the same area as before. Purpose: to estimate the effectiveness of washing in removing microorganisms.

Thoroughly scrub your hands with soap and water, and let them air dry or dry them with a *sterile* towel. Swab the palm of your right hand, and inoculate one-half of the plate. Go around the room, shaking hands with all who have the time and inclination. Now, swab the same area of the right hand with a second sterile swab, and streak the second half of your plate. Purpose: to determine whether microorganisms are transferred by touching.

Your instructor may suggest other alternatives or may ask you to devise your own experiment to answer questions that you might have about skin microflora. If you are given permission to devise your own experiment, have it approved by your instructor before beginning.

PROCEDURE: SECOND LABORATORY PERIOD

1. **Describe each colony type and determine the numbers of each.** Record these descriptions in the Results section for each half of the plate. Do all of the colonies seem to be of the same type? Would you expect to find molds and yeasts after this length of incubation?

INTERPRETATION NOTES

If you have already carried out Exercise 46, you have learned that two species of staphylococci are most often found on human skin. *Staphylococcus epidermidis* is a gram-positive coccus, rarely pigmented, and considered nonpathogenic. *Staphylococcus aureus* is also a gram-positive coccus, but it is usually pigmented (golden-yellow colonies after 48 hours) and considered opportunistically pathogenic. Look also for propionobacteria and the transient microbes described at the beginning of this exercise.

---

2. **Prepare Gram stains for each predominant colony type on the plate.** Record these results in the Results section. Do the gram-staining characteristics seem to fit the generalizations given at the beginning of this exercise?

NAME _____

DATE _____

SECTION _____

RESULTS

Separately examine each half of the Petri plate. Then enumerate and describe each colony type and determine the gram morphology of the cells from each colony.

First Half of Plate

Skin area used:

Treatment (if any):

Colony	Number on plate	Colony description	Gram morphology
A			
B			
C			
D			
E			

Second half of plate:

Skin area used:

Treatment (if any):

Colony	Number on plate	Colony description	Gram morphology
A			
B			
C			
D			
E			

CONCLUSIONS

47 BACTERIA ON HUMAN SKIN

NAME _____

DATE _____

SECTION _____

QUESTIONS

Completion

1. Human skin is composed of an outer region called the _____ and an inner region called the _____. The outermost layer of the outer region is called the _____.

2. The two most important mechanical ways in which the skin resists microbial invasion are _____ and _____.

3. Possibly the two most common types of bacteria found on the skin are the _____ and the _____.

4. Three (different) conditions that may lead to a higher incidence of skin infections are _____, _____, and _____.

True–False (correct all false statements)

1. The stratum corneum is the rapidly dividing layer of cells deep within the dermis. _____

2. Hand washing is effective in controlling microbial numbers on the skin because of the bacteriocidal activity of hand soaps. _____

3. The highest concentrations of resident microbes on the skin are found in the sebaceous glands. _____

4. *Staphylococcus aureus* is a common transient type of bacterium on the skin. _____

48

Bacteria in the Human Throat

Anatomy of the upper respiratory tract

The respiratory tract (Figure 48-1) can be thought of as divided into the upper respiratory tract (URT) and the lower respiratory tract (LRT). The upper respiratory tract includes everything above the **glottis,** which is the opening of the **trachea.** The glottis is covered by the **epiglottis** when we swallow, so that food goes into the **esophagus** and stomach instead of into the **trachea** and lungs. Note that the eyes and middle ears are physically connected to the URT by the **nasolacrimal ducts** and **auditory canals.** These passages are considered to be accessory upper respiratory structures.

The word *throat* is the lay term for the *rear of the oral cavity,* which includes the opening to the **oropharynx,** the tonsils (or areas that once contained the tonsils), and the rear wall of the oropharynx.

Microbiology of the nasopharynx and oropharynx

Microorganisms in the URT are either normal residents or transients. Most residents are part of the normal microflora, which are either nonpathogenic or opportunistically pathogenic. *Nonpathogens* do not cause tissue infection, regardless of conditions. *Opportunistic pathogens* can cause an infection if given the opportunity. A few people have resident microbes that are *virulent pathogens.* Such people are called **carriers,** because they carry with them a virulent microbe that is pathogenic to others, while they are not infected by it. The resident microbes carried by most people, however, are not virulent pathogens. A list of resident microbes commonly found in the nasopharynx and oropharynx are given in the right-hand column of Table 48-1.

Large numbers of transient microorganisms are brought into our throats with the air we breathe and the food we eat. (Note in Figure 48-1 that the oral cavity and oropharynx are shared by both the alimentary canal and the upper respiratory tract.) Airborne microbes can enter the URT with inhaled dust particles or aerosol droplets. Foodborne microbes that are part of the natural microflora of the food not removed during food preparation also serve as a source of transient microbes. Alternatively, the microbes can come from your own hands or those of the person who prepared the food, or from contaminated food utensils, or they may develop during improper storage of the food after preparation. Regardless of how transient microorganisms enter the throat, most are nonpathogenic, and most never survive because of the defenses provided by the body's tissues.

Nutrient sources for throat microbes

The food that we consume to nourish us also nourishes the resident and transient microorganisms in the URT. This food is probably the most important source of nutrients for growth of throat microbes, unless damage is done to the throat tissue. Discharge from the nasopharynx into the throat also supplies nutrients to these microbes.

Saliva, however, is not an especially good source of nutrients. Saliva contains about 0.5 percent dissolved solids, and about half of these are inorganic. The predominant organic constituents of saliva are proteins, such as salivary enzymes, microproteins, and serum proteins. Two enzymes found in saliva actually show antibacterial activity: **lysozyme,** which destroys cell-wall peptidoglycan, and **lactoperoxidase,** which generates oxygen radicals that are poisonous to many microorganisms. Therefore, the presence of saliva is

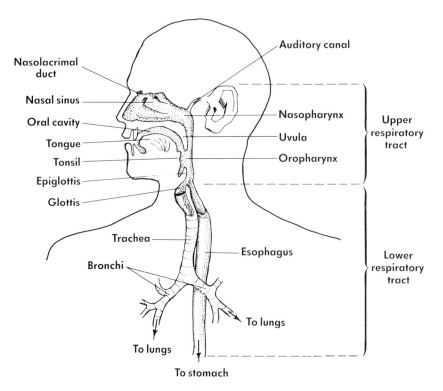

FIGURE 48-1 Cross-sectional anatomy of the upper respiratory tract.

often more of a hindrance than a help to throat microbes. The exact chemical composition of saliva can vary with changes in physiology and stress levels; so it is difficult to predict the effect of saliva on the growth of throat microbes.

Defenses of the upper respiratory tract

The average person takes in about 11,500 liters of air per day, a volume of air that contains about 10,000 microorganisms. Once these microbes contact the mucous membranes that line the URT, they find a warm, moist environment that is bathed in nutrients. Fortunately, there are mechanical, chemical, and microbial defenses in the URT tissues.

Examples of *mechanical defense* are sneezing to discharge particles (and microbes) from the URT; inhaling air past nasal hairs to create turbulence, which increases the chance that microbes will be trapped in the mucus that lines the nasal passages; and sweeping by the ciliated mucosa that line our nasal passages to remove trapped microbes from the URT.

Examples of *chemical defense* are the production of **interferon** by host cells, which protects them from viral infection, and the release of lysozyme and lactoperoxidase.

TABLE 48-1 Microorganisms common to the human oropharynx (throat) and nasopharynx

Common transient pathogens	Common nonpathogenic or opportunistic residents
Beta-hemolytic streptococci (*Streptococcus pneumoniae*)	Alpha- and nonhemolytic streptococci
Staphylococcus aureus	Other staphylococci
Neisseria meningitidis	Other neisseria
Corynebacterium diphtheriae	Other corynebacteria
Haemophilus influenzae	Other haemophilus
Candida albicans	Actinomycetes
Viruses	Bacteroides

Most resident microorganisms found in the URT are nonpathogenic and contribute to the defense of the URT by successfully outcompeting transient pathogens for essential nutrients. This is known as the *microbial defense* of the URT. Maintaining the normal balance of these microorganisms is important to the health of URT tissues. Anything that disrupts this delicate balance may allow a pathogenic microbe to outcompete the normal microflora and begin an infection.

Some resident microbes, as well as some transients, are opportunistic pathogens which can cause infections if predisposing factors compromise resistance. Many factors may **predispose** a person to develop an infection caused by an opportunistic pathogen. Predisposing factors are usually situations that cause damage to URT tissues or upset the balance among the normal microflora that coat these tissues. For example, tobacco smoking, decreased humidity, viral infections, hay fever, and general anesthesia can damage ciliated mucosa and/or interfere with phagocytosis (mechanical defenses). One of the most common causes of upset to the balance of normal microbial residents is antibiotic therapy for an infection somewhere else in the body. Antibiotics encourage the antibiotic-resistant opportunistic pathogens that are among the residents to outcompete all other microorganisms. Other predisposing factors are extreme age, hormonal imbalances (such as corticosteroid excess or deficiency), chronic stress, diabetes, immunosuppressive therapy (used after organ transplant), anticancer chemotherapy, immunological deficiencies, and severe malnutrition.

Not all diseases of the URT are caused by predisposing factors and opportunistic pathogens. Healthy individuals occasionally get mild to life-threatening diseases if the inoculum is large enough or if the pathogen is sufficiently virulent. Both the causative agent of pneumonic plague *(Yersinia pestis)* and the measles virus are virulent pathogens that can overwhelm a healthy person's normal defenses.

This exercise will introduce you to some of your own nonpathogenic and opportunistic throat microbes and show you the types and the variety normally present.

LEARNING OBJECTIVES

- Introduce the concepts of normal microbial residents and of opportunistic and virulent pathogens of the human throat.

- Understand some of the normal defensive mechanisms that exist in and around the throat.

- Demonstrate the type of microorganisms that are normally found in a healthy person's throat.

MATERIALS

Cultures	You are carrying them with you.
Media	Blood (sheep) agar plates
Supplies	Cotton swabs (sterile)
	Tubes of sterile diluent
Demonstration	Method for swabbing the throat

PROCEDURE: FIRST LABORATORY PERIOD

1. **Have your partner swab your throat, by twirling the swab tip, and then ask your partner to hand you the swab.**

TECHNIQUE NOTES

Throat specimens are taken in the clinical laboratory to aid the diagnosis of upper-respiratory tract diseases. They are usually taken on a sterile, flexible swab tipped with a nonabsorbent, cottonlike material. One should avoid drying the specimen by immediate inoculation onto a suitable culture medium or by placing the swab in a tube of liquid transport medium and taking it directly to the laboratory for transfer to a culture medium.

Take care to avoid contact with the teeth, gums, cheeks, and tongue because these areas contain large numbers of microbes that are normally found in the mouth. You want a throat specimen!

Avoid touching the uvula. Such touching stimulates the **gag reflex,** and you may create a mess.

Throat specimens should be taken at the oropharynx opening and the rear wall of the oropharynx. If your partner has tonsils, you might start by sampling one of these. If no tonsils are present, sample the area that previously contained a tonsil. Continue sampling in a clockwise fashion until the swab contains material from both tonsils (or tonsil areas) and the extreme rear of the oral cavity (rear wall of the oropharynx).

To sample each area, twirl the swab somewhat as you stroke the area gently. When you have finished, the entire tip of the swab should contain sample. **Be gentle!** Do unto others because they *will* do unto you!

2. **Inoculate the blood agar plate** with the swab containing your throat specimen. Streak the plate surface so that it is entirely covered; then turn the plate 90 degrees and streak the plate again at right angles to the first streak.

   As you streak the plate, roll the cotton tip of the swab somewhat, so that the entire tip is used to streak the plate.

3. **Discard the swab with other contaminated material** according to your instructor's directions.

4. **Label the plate** with your *name*, today's *date*, the *name* of the *medium* used, and the body *area* sampled.

5. **Incubate the plate at 30°C for 48 hours.** Refrigerate the plate after 48 hours of incubation if the time of your next laboratory period is more than 48 hours away.

PROCEDURE: SECOND LABORATORY PERIOD

1. **Examine each colony type, and determine the approximate number of each type present.** Look at colony morphology and the blood agar immediately surrounding that colony. **Record** these observations in the Results section.

INTERPRETATION NOTES

Blood-agar plates are used so that types of hemolysis (lysis of the red blood cells) can be detected. In *beta-hemolysis,* the developing microbial colony excretes a substance that lyses the red blood cells (RBCs), leaving a clear (colorless) zone around the colony. Some streptococci are characteristically beta-hemolytic. Other streptococci, in *alpha-hemolysis,* partially lyse the RBCs and also reduce the hemoglobin (the compound that gives the red color to an animal's blood) to methemoglobin (which is green). Still other types of microorganisms, called *nonhemolytic* neither lyse RBCs nor reduce hemoglobin; so there is no zone surrounding their colonies.

---

2. **Prepare Gram stains for each predominant colony type present on the plate.** Record your observations in the Results section.

   Do the Gram-staining characteristics seem to fit the colony descriptions and generalizations given in Table 48-2?

3. **Compare your plate with your partner's plate.**

INTERPRETATION NOTES

You incubated your plates under aerobic conditions; so you will isolate no strict anaerobic microbes from in the oral cavity. If you Gram stain cells taken from an area of the plate containing confluent growth, however, you may see gram-negative rod-shaped cells of *aerotolerant anaerobes.* Refer to Table 48-2 for the morphology and gram-staining characteristics of the more common microorganisms found in the nose and throat.

TABLE 48-2 Bacteria commonly found in the human nose and throat

Name	Oxygen tolerance	Gram characteristics and morphology
Staphylococci	Facultative anaerobic	Gram-positive cocci in clusters
Streptococci	Facultative anaerobic	Gram-positive cocci in chains
Branhamella catarrhalis		Gram-negative cocci in pairs
Corynebacterium species (diphtheroids)	Facultative anaerobic	Gram-positive rods with varying shapes (pleomorphic)
Haemophilus species	Facultative anaerobic	Gram-negative rods that are pleomorphic
Actinomyces species	Strictly aerobic	Gram-positive rods that are filamentous
Bacteroides species	Strictly anaerobic	Gram-negative rods that are pleomorphic

NAME _____

DATE _____

SECTION _____

RESULTS

Examine and describe each colony type and determine the gram morphology of cells from each colony type.

Colony type	Number on plate	Colony description	Hemolysis type	Gram morphology
A				
B				
C				
D				
E				

CONCLUSIONS

QUESTIONS

Completion

1. The upper respiratory tract refers to everything above the _____.

2. The term *throat* refers to the rear of the oral cavity, which includes _____ and _____.

3. Many resident microbes in our throats are considered to be opportunistically pathogenic; this means that these microbes _____.

4. Two examples of transient pathogens are _____ and _____.

5. Our throat tissue may be thought of as having three types of defenses against the invasion by microbial pathogens. For each type listed below, give one example.
   a. Mechanical:
   b. Chemical:
   c. Microbial:

6. Two examples of predisposing factors are _____ and _____.

7. The oropharynx is best described as _____.

8. The gag reflex is brought about by touching the _____.

9. Alpha-hemolysis of red-blood cells is characterized by the _____ of the red-blood cells, which makes the areas surrounding colonies appear _____.

10. Two examples of facultatively anaerobic, gram-positive cocci that you may have found on your plates are _____ and _____.

True-False (correct all false statements)

1. Transient microbes may be obtained from our food and water, from the air around us, from our unclean hands when we place our fingers on our lips, and from those we kiss. _____

2. The primary source of nutrients used to sustain throat microbes is our own saliva. _____

3. *Neisseria meningitidis* is but one example of a bacterium that is also a virulent pathogen. _____

4. Throat cultures are best obtained by swabbing only the posterior gums, back of the tongue, and tonsils. _____

5. Alpha-hemolysis is characterized by partial lysis of red-blood cells, accompanied by reduction of the released hemoglobin to yield a brown color. _____

49

Slide-Agglutination Test for Serotyping Pathogens

An important part of the broad subject of **immunity** is the reaction between antibodies and antigens. When a microbe invades an animal, the animal's immune system recognizes the microbe as foreign and responds by producing **antibodies,** which are proteins that specifically react with this foreign material. Foreign substances that stimulate antibody production are called **antigens.**

Antibodies are soluble proteins found in both plasma and serum (see Exercise 45). *Plasma* is the fluid portion of blood that remains after blood (with an anticoagulant added) is centrifuged. If freshly drawn blood (having no anticoagulant) is placed in a test tube and allowed to stand, it separates into two fractions: the semisolid *clot* and a clear, straw-colored fluid called *serum.* As blood clots, prothrombin is activated and becomes thrombin; thrombin reacts with fibrinogen to form a fibrin network, the clot; and this fibrin network traps the formed parts of the blood (such as the cells and platelets). Therefore, serum is the same as plasma, except that it lacks the proteins that participate in clot formation (such as prothrombin and fibrinogen).

Consider an animal that previously had an infection caused by a *Salmonella* species. The plasma or serum from this animal's blood, which contains antibodies against these bacteria, is called **immune.** If immune serum is mixed with a pure culture of *Salmonella,* the antibodies in the serum bind the cells together into an aggregate. If a visible aggregate is formed, microbiologists call this process an **agglutination reaction** (Figure 49-1). When whole cells are the sources of antigens for an antigen-antibody reaction, this reaction is called an *agglutination.* If the antigen is soluble (such as extracellular toxin excreted by a microbe), the antigen-antibody reaction is called a **precipitation reaction.**

Agglutination reactions can be used to help identify microorganisms. For example, assume that you isolate a suspected pathogen in pure culture from a food recently consumed by a human patient who is bedfast and showing symptoms of salmonellosis (a disease caused by certain *Salmonella* species). You suspect that this is a particularly virulent strain of *Salmonella,* but you need a method of positive identification. You could obtain specific immune antisera (from an animal that was made immune to this strain of *Salmonella*) and see if this serum would agglutinate with the pure culture you isolated from the food source. This method of analysis is called **serotyping** because it uses a specific antiserum and allows you to type (specifically identify) a microbial culture. It is a powerful tool that is used in most modern clinical laboratories to identify disease-causing microbes.

In this exercise, the antigen source is whole bacterial cells which you will add to either immune or nonimmune serum. The agglutination reaction that you observe on the slide is probably identical to the reaction in your blood when circulating antibodies encounter an invading organism to which you are immune. This reaction is extremely important to the health of all animals (including you).

LEARNING OBJECTIVES

- Understand how agglutination is part of the overall concept of immunity and the immune response.

- Know what is meant by an agglutination or a pre-

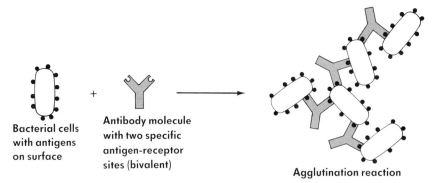

FIGURE 49-1 The agglutination reaction that occurs when colloidally suspended particles (such as bacterial cells) containing specific surface antigens (•) react with bivalent antibodies containing specific binding sites. Since each antibody molecule can react with two adjacent cells, a large lattice is formed. When this lattice becomes visible to the unaided eye, an agglutination reaction has occurred.

cipitation reaction and how microbiologists differentiate between these two immune reactions.

• Know how to perform an agglutination test and how to recognize a positive reaction.

MATERIALS

Cultures	*Salmonella enteritidis* (slant culture)
	Escherichia coli (slant culture)
Media	None
Supplies	Factor 9 (Group D) antiserum for *S. enteritidis*
	Saline (sterile)
	Nonimmune serum
	Wooden applicator sticks
	1.0-ml pipettes (sterile)
	Clean microscope slide
	Wax pencil

PROCEDURE

1. **Obtain a clean microscope slide and a wax pencil. Warm the slide** with a bunsen burner until the wax pencil melts when touched to the slide.
2. **Melt the wax from the pencil to form three closed circles** that are about 15 mm in diameter and well separated. Consider the ring side of the slide to be the *top*.
3. **Label this slide as follows:** Use your felt-tip marking pen to label each wax circle as I-S (immune, *Salmonella*), N-S (nonimmune, *Salmonella*), or I-E (immune, *Escherichia*), on the top of the slide, just outside the wax rings.

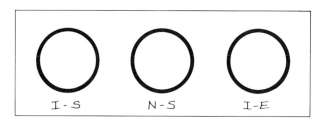

Your instructor will prepare two separate cell suspensions by adding 0.5 to 1.0 ml of sterile saline (0.9 percent NaCl in distilled water) to a 24-hour-old slant of *S. enteritidis* and to a 24-hour-old slant of *E. coli*. This suspension should be *very* dense!

4. Ask your instructor to use a sterile pipette and **add one drop of immune serum inside each of the two rings labeled with an** *I*.
5. Ask your instructor to use a sterile pipette and **add one drop of nonimmune serum inside the ring labeled with an** *N*.
6. Ask your instructor to use a sterile pipette and **add one drop of *Salmonella* suspension inside each of the two rings labeled with an** *S*.

TECHNIQUE NOTE

Care must be taken so that no cross contamination occurs when making these additions. For example, if the pipette containing the cell suspension is allowed to touch the immune serum, all subsequent mixtures will contain antibody and may give confusing results. It is best, if possible, to place the drops of cells and serum next to one another and then mix them together. In this way, pipettes will not transfer contaminating substances to other reaction mixtures.

7. Ask your instructor to use a sterile pipette and **add one drop of *Escherichia* suspension to the center of the ring labeled with an *E*.**

8. **Separately mix the cells with serum in each ring with an unused end of an applicator stick.** Use one end of a wooden applicator stick for one mixture, the other end to stir the second mixture, and one end of a second stick to stir the third mixture. In this way, you will not cross-contaminate these three mixtures. Also, take care not to splatter any material from one ring into another.

9. **Incubate the slide at room temperature for 2 to 3 minutes** while holding the slide horizontally and gently rocking it back and forth. Positive reactions should occur within this time. If you wait too much longer before examining the slide, drying of the reagents may give the appearance of a positive agglutination reaction.

    In a *positive agglutination* reaction, there is visible clumping of the cells. This means that the serum contains antibody that is specific for those cellular antigens.

    In a *negative agglutination* reaction, there is no visible clumping of the cells within 1 or 2 minutes of incubation. This means either that the serum does not contain any antibody specific for those cellular antigens *or* that the serum does not contain sufficient antibody to form an agglutination reaction.

NAME _____

DATE _____

SECTION _____

RESULTS

Slide label	Components	Incubation time (minutes)	Observations
I-S	*Salmonella* cells plus *Salmonella*-immune antiserum		
N-S	*Salmonella* cells plus nonimmune serum		
I-E	*Escherichia* cells plus *Salmonella*-immune serum		

CONCLUSIONS

QUESTIONS

Completion

1. From a chemical standpoint, antibodies are _____ that are made by the animal's body in response to the introduction of a foreign substance called a(n) _____.

2. The fluid blood fractions that contain antibodies are called either _____ or _____.

3. Two types of antigen-antibody reactions can be seen with the unaided eye when an immune blood fraction is mixed with an antigen. If the antigen is part of a colloidally suspended particle (such as a bacterial cell), the reaction is called a(n) _____. If the antigen is soluble, the reaction is called a _____ reaction.

4. Development of one of these antigen-antibody reactions in the laboratory is evident by the formation of a visible _____ in the experimental mixture.

True – False (correct all false statements)

1. Agglutination might be used to determine whether you are immune to infection by a certain pathogen. _____

2. Serotyping refers to the ability of a known type of bacterial pathogen to react specifically with (and therefore identify) the soluble blood fraction from an immune individual. _____

3. The immune response refers to the ability of an animal's body to produce antibiotics in response to the introduction of a foreign material such as bacterial cells. _____

XII

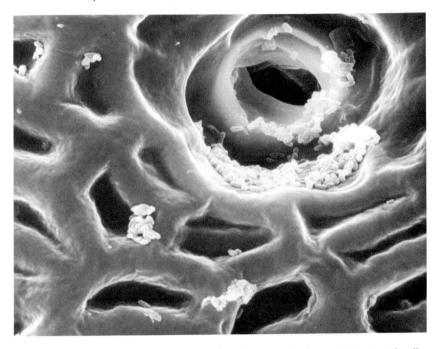

Bacteria associated with a stomate on the surface of a cucumber fermented by *Lactobacillus plantarum*.

In his article on the development of the chemistry of solutions . . . Joel Hildebrand delineates the difference between a true scientist and a mere practitioner. The one has schooled himself to understand; the other has devoted his academic life to learning all the recipes.

—W. W. Knoble, 1965

FOOD MICROBIOLOGY

Part XII will introduce you to two aspects of food microbiology: (1) how microorganisms are used to help prepare and preserve human foods and (2) natural microbial residents on food and how handling and processing influences the number and types of microbes found on human foods.

50

Milk Preservation — Yogurt

Yogurt is but one of many types of fermented milk consumed around the world. All are prepared by culturing certain **lactic acid bacteria** found in milk, but the fermented product differs in name and taste depending on the type of milk used and the type of lactic acid bacteria that accomplish the fermentation. For example, *kefir* is traditionally made in the Balkans by pouring goat, sheep, or cow's milk into a bag made of goat skin and hanging it from a tree branch.

Commercial yogurt in North America is made from bovine (cow) milk and is prepared under controlled conditions in two steps, each at a temperature that favors the metabolism and growth of different types of lactic acid bacteria.

In the *first step,* the inoculated milk is held at 46° to 48°C (115° to 120°F), which encourages the metabolism and growth of a **thermophilic** bacterium called *Streptococcus thermophilus.* While these thermophiles are growing, they rapidly ferment a soluble sugar called lactose (milk sugar). This is a disaccharide (two covalently linked sugars) composed of glucose and galactose which are connected by a glucosidic bond (Figure 50-1). Lactose is transported into the cell and the glucosidic bond is broken by the enzyme *beta-galactosidase.* After the lactose is broken into galactose and glucose, both sugars are converted to glucose-6-phosphate. Then, *S. thermophilus* oxidizes glucose-6-phosphate to lactic acid using the Embden-Meyerhoff-Parnas (EMP) pathway (Figure 50-2).

Streptococcus thermophilus is called a **homolactic acid** bacterium, as are all of the streptococci, because they ferment sugars, mostly to lactic acid, which they excrete. All homolactic acid bacteria appear to ferment sugars using the EMP pathway, and most of the carbon atoms in glucose end up in the excreted lactic acid molecules; therefore, the EMP pathway is called the **major pathway** for the homolactic acid bacteria. This excreted acid increases the hydrogen-ion concentration, thus lowering the pH.

Freshly drawn bovine milk has a pH of about 6.6. At this pH, **casein** (a milk protein) exists as a colloidal suspension of a calcium salt called *calcium caseinate* which gives milk its white color and turbid appearance. When streptococci begin to excrete lactic acid, some of it reacts with the calcium caseinate to form calcium lactate and free (soluble) casein. As the pH approaches 4.6, casein begins to denature (coagulate), forming a smooth semisolid **curd.** Such an acid curd is typical of milk fermentation by the lactic acid bacteria (see Exercise 35).

This first step in yogurt production is continued only until the coagulated milk has the proper texture. If this step continues for too long, the product tastes too sour for most people.

In the *second step,* the temperature is lowered by refrigeration. As the temperature drops, the growth and homolactic-fermenting activity of the thermophile *(S. thermophilus)* slows, and a **mesophilic** bacterium called *Lactobacillus bulgaricus* begins to ferment the remaining lactose in the milk. These lactobacilli are usually more acid resistant than other lactic acid bacteria and therefore are often responsible for the final stages of lactic acid fermentations. *Lactobacillus bulgaricus* also carries out a homolactic fermentation, using the EMP pathway to ferment lactose and convert it primarily to lactic acid. However, *L. bulgaricus* also uses other **minor pathways** to produce volatile organic compounds (Figure 50-3). These minor products give yogurt its characteristic flavor and aroma.

This second step lasts until the temperature falls below the point at which the active metabolism of *L. bulgaricus* stops; consequently, the excretion of both lactic acid and the minor fermentation products also

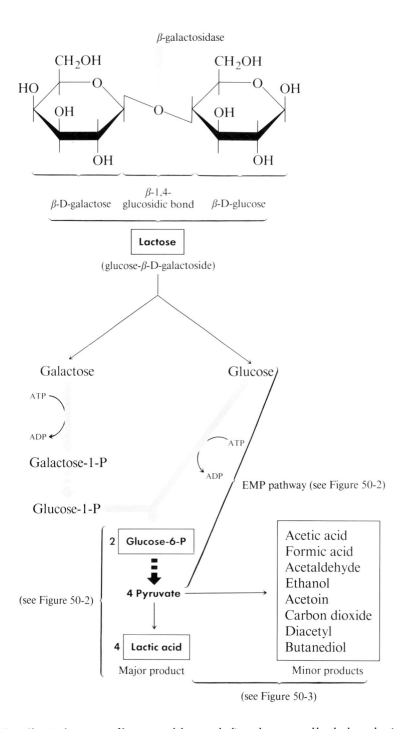

FIGURE 50-1 Chemical structure of lactose and the metabolic pathways used by the homolactic acid bacteria (such as streptococci and lactobacilli) to ferment lactose. Lactose, a disaccharide, is transported into the cell and cleaved by the enzyme beta-galactosidase into galactose and glucose, two sugars which are then converted to glucose-6-phosphate. This molecule is then fermented primarily to lactic acid by the Embden-Meyerhoff-Parnas (EMP) pathway, which provides energy for the lactic acid bacteria to grow. Wide arrows show the *major* fermentation route, the EMP pathway (see Figure 50-2). The thin arrows indicate minor pathways used by some homolactics (see Figure 50-3).

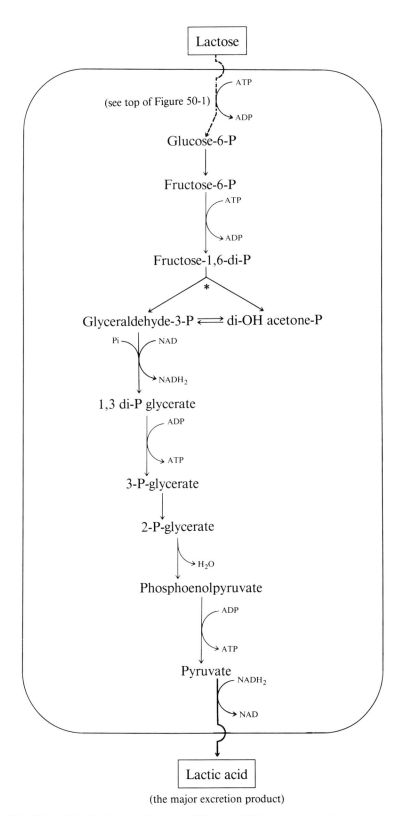

FIGURE 50-2 The Embden-Meyerhoff-Parnas (EMP) pathway used by the homolactic acid bacteria (such as streptococci and lactobacilli) for fermenting lactose to lactic acid. Most of the lactose used by these cells is converted to lactic acid. The EMP pathway provides the energy (ATP) required for growth. Note that the number of each molecule formed is not shown. The dotted arrow between lactose and glucose-6-phosphate denotes a series of reactions (shown in Figure 50-1). The enzyme called fructose diphosphoaldolase (∗) is characteristic of the EMP pathway, and its deletion can be used to help differentiate between homolactic-acid and heterolactic-acid bacteria.

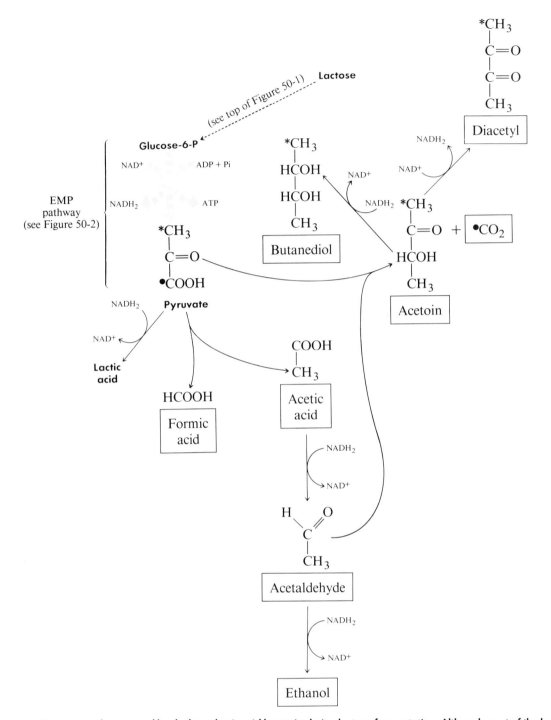

FIGURE 50-3 Minor pathways used by the homolactic acid bacteria during lactose fermentation. Although most of the lactose is converted to lactic acid by the EMP pathway (see wide arrows in Figure 50-2), some pyruvate formed by the EMP pathway is converted to the minor products shown in boxes, and these compounds are excreted from the cell. Note that one pyruvate molecule cannot be converted to all of the products shown. These pathways have different functions, and they may be used at different times according to the cell's needs. The methyl and carboxyl carbons of pyruvate are separately marked so that you can see how pyruvate and acetaldehyde form acetoin and carbon dioxide.

50 MILK PRESERVATION—YOGURT

stops. Refrigeration is not really necessary; it just prolongs the shelf life of the product. As long as the yogurt remains refrigerated, there should be no further excretion of fermentation products. However, the acid alone will prevent growth of other types of microorganisms and, consequently, spoilage of the milk.

In technologically developed regions, mechanical refrigeration increases the shelf life of fresh milk; but before the days of ice boxes and refrigerators, humans fermented milk to increase the time that it would remain fit for human consumption. They used acid fermentations because the lowered pH prevented the metabolism and growth of food spoilage and pathogenic microorganisms. The fermentation of milk is among the oldest known methods of food preservation.

After yogurt is produced by modern techniques, it is refrigerated to keep the pH from going so low that the yogurt is too tart for the American palate. Actually, yogurt will remain unspoiled for a long time at room temperature, but the live bacteria continue to make lactic acid and other fermentation products. Consequently, the texture and taste change.

This exercise shows you how easy it is to make yogurt and demonstrates one method of using microbes for food preservation. Perhaps you would also like to try this at home.

LEARNING OBJECTIVES

- Become aware of acid fermentation as a method of food preservation.
- Learn the types of microorganisms used in the preparation of yogurt and what each contributes to the fermentation.

MATERIALS

Cultures Commercial yogurt preparation, without preservatives or other additives

Media Commercially prepared whole milk; or, for low-fat yogurt, dry (noninstant) skim-milk powder *and* fluid skim milk

Supplies Containers that hold at least 500 ml (about 1 pint)
Aluminum foil (for covering containers)
Incubator set at 46° to 48°C (115° to 120°F)
Crushed pineapple (or other fruit) for flavoring after the yogurt is made
Plastic spoons
Paper cups
Hotplate
pH paper
Gram-staining reagents

PROCEDURE: FIRST LABORATORY PERIOD

You can prepare yogurt with either whole milk or skim milk. Step 2 should be used only if your instructor asks you to use skim milk.

1. Preparation of *whole or skim milk*
   a. **Scald about 500 ml of whole or skim milk** by heating it just to boiling (do not let it boil). This kills almost all of the unwanted bacteria (acid production by the lactic acid bacteria will inhibit the rest).
   b. **Cool the milk to 46° to 48°C (115° to 120°F).** Use a thermometer to check this! It is very important that the milk temperature be no higher than this prior to inoculation because, if it is, most (if not all) of the mesophilic lactobacilli in the inoculum will die.

The procedure in step 2 uses skim milk instead of whole milk so that the yogurt will contain less fat. This procedure also uses dried milk so that it can be rehydrated with less water than normal to increase the protein concentration. *Noninstant* dry skim milk is used because its flavor is often preferred to that of many of the instant dry skim milks. Noninstant dry skim milk can usually be purchased at health-food shops. The following procedure produces a thoroughly blended mixture.

2. Preparation of *dry skim milk*
   a. **Add one cup of noninstant dry skim milk to two cups (400 ml) of tap water,** and thoroughly **blend** with an electric blender.
   b. To this mixture **add one more cup of noninstant dry skim milk and two more cups of tap water.** Thoroughly **blend** this mixture with the electric blender.
   c. To the mixture formed in b, **add one additional cup of noninstant dry skim milk and two additional cups of tap water.** Thoroughly **blend** this mixture with the electric blender.

   You should now have three cups of noninstant dry skim milk thoroughly mixed with six cups of tap water.

d. To the mixture formed in step c, stir in **three cups of liquid skim milk.** This will provide a total of 12 cups (2400 ml) of skim milk mixture.

You are now ready to inoculate the milk regardless of which milk preparation you use.

3. *Inoculation* of milk
   a. **Add one heaping tablespoon of commercially prepared yogurt** to each 500 ml of milk. Let us refer to the commercially prepared yogurt as the **starter culture.**
   b. **Thoroughly mix** the starter culture into the milk. Thorough mixing is very important because the starter culture is a gel and is difficult to mix with the nonviscous fluid (the milk).

   To facilitate mixing, you may wish to place some of the starter culture in a beaker, and add about twice that volume of the milk preparation; then thoroughly mix these together. This produces a more fluid preparation that is easier to mix into the rest of the milk.

4. *Incubation* and *storage*
   a. **Fill containers** to within ¾ inch of the top with the inoculated milk mixture.
   b. **Cover containers with aluminum foil.** This is just a dust cover; no need to seal containers with the foil.
   c. **Incubate this mixture at 46° to 48°C (115° to 120°F) for 3 to 6 hours.** It is important to check the progress of this fermentation periodically. You are now providing conditions suitable for the thermophile *(Streptococcus thermophilus)* to ferment lactose (milk sugar).

TECHNIQUE NOTES

Too long an incubation allows too much acid production; so the final product may be more tart than your taste buds desire. The yogurt is ready to move to the refrigerator as soon as you can tilt the container slightly and see the partially gelled mixture sag. If it flows freely, it needs more incubation to allow for more acid production and more coagulation. If a clear fluid forms on top accompanied by a broken curd (seen through the side of the container), it has incubated too long and will be too sour for most people's taste.

---

d. **Remove the container from the incubator, and place it in a refrigerator.** It will take about 2 hours for the temperature to go from about 48° to about 10°C. During this time, the mesophile *(Lactobacillus bulgaricus)* will begin to grow and ferment the lactose (milk sugar), primarily adding flavor-producing compounds to the yogurt. This mesophile will also produce some lactic acid, further coagulating the milk. This means that the yogurt will become firmer; so do not be misled into thinking that the texture formed at the end of the 48°C incubation is that of the final product.
   e. **Store the new yogurt in the refrigerator until the next laboratory period.** After the temperature reaches refrigeration temperature (about 4°C), little or no growth of either culture will occur; and there should be little or no fermentation. (Remember that cold temperatures slow metabolic activity.)

5. *Examination of commercially prepared yogurt*
   a. **Gram stain this culture.** Use a sterilized loop and proceed as usual. You should be able to see two distinctly different types of bacteria in the commercial yogurt. You will also see a lot of other stained material (background) from the milk.
   b. **Determine the pH of the yogurt.** Sterilize an inoculating loop, and use it to transfer a loopful of yogurt to a piece of pH paper. Examine the opposite side to determine the color of the moistened paper.
   c. **Record your results.**

PROCEDURE: SECOND LABORATORY PERIOD

1. **Examine the physical appearance and texture of your yogurt.**
2. **Determine the pH of your yogurt.** Sterilize an inoculating loop, and use it to transfer a loopful of yogurt (or some of the clear fluid on top of the gel) to a piece of pH paper. Examine the opposite side of the paper to determine the color.
3. **Taste the yogurt.** How does it compare with the commercial preparation? After you taste the plain yogurt, try it with some fruit.

50 MILK PRESERVATION—YOGURT

NAME _____

DATE _____

SECTION _____

RESULTS

1. *Starter culture* (commercially prepared yogurt)

   Appearance of Gram-stained preparation:

   Number of colony types seen:

   Gram reaction and shape of microbes seen:

   Approximate hydrogen ion concentration (pH):

2. *Your culture*

   Appearance and texture:

   Approximate hydrogen ion concentration (pH):

   Amount of tartness:

   Description of flavor:

CONCLUSIONS

QUESTIONS

Completion

1. The types of bacteria that ferment milk are collectively called the _____ bacteria.

2. The two species of bacteria used to ferment lactose are called _____ and _____. Considering their temperature requirements for optimum metabolism and growth, the first species is called a _____, and the second species is considered a _____. The first contributes primarily _____ to the final product, and the second is most useful for making _____.

3. Milk protein is called _____, and it exists in freshly drawn milk as a colloidally suspended salt called _____. During yogurt formation, the bacteria produce acid that drops the pH from about 6.6 to pH _____, which causes the protein to _____ and form a _____.

4. The first stage in yogurt formation occurs at a temperature of about _____ °C.

5. Although yogurt is now produced for its flavor and nutritional value, before the days of mechanical refrigeration, it was made primarily as a way to _____.

True – False (correct all false statements)

1. The name of the trisaccharide that is fermented by lactic acid bacteria in yogurt formation is milktose. It is made by linking two molecules of glucose with one molecule of galactose. _____

2. Sugar fermentation that accompanies yogurt production is accomplished by a catabolic scheme called the Embden-Meyerhoff-Parnas pathway that occurs inside the lactic acid bacteria. _____

3. During the fermentation of milk sugar, glucose is first converted to galactose, and then it is oxidized to pyruvate by the EMP pathway. _____

51

Solid Food Preservation — Sauerkraut

Sauerkraut has probably been made in Europe for at least 4000 years. Originally, the fermentation (natural pickling) of cabbage was a means of preserving (preventing from spoiling) this vegetable, which has served as a major portion of the human diet in many parts of the world. Today, mechanical refrigeration retards the spoilage of cabbage, but sauerkraut is still made because many people enjoy its flavor.

Cabbage fermentation results from the metabolism of **lactic acid bacteria** that exist naturally on healthy cabbage plants, along with many other types of bacteria. These microbes are normally confined to the plant surface; but if a leaf is cut, the microbes gain access to tissue fluid that contains 3 to 6 percent sugar and other water-soluble nutrients, which the microbes can use for metabolism and growth.

Unlike yogurt, sauerkraut results from a **controlled natural fermentation,** which means that humans create an environment that favors the growth and metabolism of only the lactic acid bacteria — that is, an environment that provides selective conditions for **enrichment** of lactic acid bacteria. This control is accomplished in two ways: by adding sodium chloride (table salt) to the leaves, and by creating an anaerobic environment.

Yogurt production is also accomplished by lactic acid bacteria. But in that case, milk is artificially (and heavily) inoculated with two types of lactic acid bacteria; so fermentation does not depend upon naturally occurring microbes.

To make sauerkraut, you shred cabbage leaves, distribute salt uniformly among them, and then pack them tightly in a solid container to promote an anaerobic environment. (Smaller pieces of cabbage are easier to pack so that pockets of air are not trapped to retain an aerobic environment.) Enough noniodized salt (NaCl) is added to give a concentration of 2 to 3 percent by weight. The salt extracts nutrient-containing fluid from the leaf tissue by osmosis, suppresses the growth of undesirable microorganisms, favors the growth of the salt-tolerant lactic acid bacteria, and enhances the flavor of the product. This nutritious, salty, plant-tissue extract is called *brine.* A weighted, solid disk is placed on the shredded cabbage to keep it below the surface of the brine and to help create and maintain an **anaerobic** environment.

The temperature should be kept between 15° and 24°C (60° and 75°F) to provide favorable conditions for the growth of bacteria and the production of acid and flavorful organic fermentation products by the bacteria.

Sauerkraut production is an example of **microbial succession** because the metabolism and growth of several types of microorganisms rise and fall in succession before the final product is formed. This succession usually occurs in three stages.

Early stage

During the first day of cabbage fermentation, some naturally occurring, salt-tolerant microorganisms respire, using up the dissolved oxygen, and the system becomes anaerobic. After the oxygen is removed, the facultatively anaerobic microbes ferment the plant sugars in the salt-extracted juice and begin to produce and excrete various acids, thus lowering the pH. The two most commonly isolated types of microorganisms during the early stage are *Enterobacter cloacae* and *Erwinia herbicola,* both of which are classified in the family *Enterobacteriaceae;* thus both are facultatively anaerobic, gram-negative, straight rods.

Intermediate stage

By the second or third day of fermentation, *Leuconostoc* species are the predominant microbes in the culture. These microorganisms are gram-positive, coccus-shaped, strictly fermentative, salt- and acid-tolerant, lactic acid bacteria. From a metabolic standpoint, these lactic acid bacteria are different from the homolactic acid bacteria that ferment milk (see Exercise 50) in that all are **heterolactic** fermentors. Unlike the homolactic acid bacteria that produce mostly lactic acid from sugar fermentation by using the Embden-Meyerhoff-Parnas pathway, the heterolactic acid bacteria produce a mixture of almost equal amounts of lactic acid, carbon dioxide, and ethanol, using the **pentose phosphoketolase (PPK) pathway,** which is named after the enzyme that is characteristic of these metabolic reactions (Figure 51-1). Heterolactics also produce and excrete small amounts of other compounds such as acetic acid and glycerol. The heterolactic species most often identified during the intermediate stage of the fermentation is *Leuconostic mesenteroides,* which is not as acid tolerant as the lactobacilli, although this species still produces an acid concentration of about 0.6 to 0.8 percent during this fermentation stage.

Final stage

After 4 to 6 days of fermentation, *Lactobacillus* species predominate. These microorganisms are gram-positive, rod-shaped, strictly fermentative, salt-tolerant, lactic acid bacteria that are tolerant to higher acid concentrations than most other lactic acid bacteria. *Lactobacillus* species carry out homolactic sugar fermentation, and therefore, oxidize most of the remaining sugar through the Embden-Meyerhoff-Parnas pathway to lactic acid (see Exercise 50, Figure 50-2). These bacteria produce final acid concentrations ranging from 1.5 to 2.0 percent. However, the lactobacilli also convert a small amount of the sugar to aromatic minor products that are of major importance because they contribute flavor to the product (see Exercise 50, Figure 50-3).

This fermentation process (all three stages) requires about 3 to 4 weeks incubation at 15° to 24°C.

In this exercise, you will prepare sauerkraut and observe the microbiological and chemical changes that take place during this controlled fermentation.

LEARNING OBJECTIVES

- Become aware of acid fermentation as a method of food preservation.

- Observe the succession of microorganisms during a mixed-culture fermentation and correlate this with the presence of NaCl and the acidification of the environment.

- Apply the concepts of enrichment culture conditions (addition of salt, anaerobic conditions, and eventually acid production) for selective growth of the lactic acid bacteria.

- Observe how some microbes alter their own environment so that they out-compete other microbes for available nutrients.

MATERIALS

Cultures None

Media T-soy plates
EMB plates

Supplies *First Laboratory Period:*
Cabbage
Vegetable shredder or grater
Aluminum foil (to catch the shredded cabbage)
Plastic spoons
Noniodized salt (NaCl)
Scale or balance and weighing paper
Clean tamper
Solid vessel (such as ceramic crock or large beaker)
Solid disk to keep shredded cabbage below surface of fluid; should be only a little smaller than the inner diameter of the vessel; may be something like a saucer or a large watch glass
Weight (to keep the solid disk below the fluid surface)
Second Laboratory Period:
pH meter
Gram-staining reagents

PROCEDURE: FIRST LABORATORY PERIOD

1. **Prepare cabbage** by removing the outer leaves of fresh, firm, compact heads. Set these outer leaves aside for later use. Grate or chop the remaining leaves. Fineness is a matter of choice, but strips about 4 mm wide (⅛ inch) or less are recommended. Discard the core.

2. **Weigh the cabbage;** then add 2.3 percent (by weight) noniodized NaCl.

3. **Thoroughly mix** the salt with the cabbage. This is very important!

4. **Pack the salted cabbage** in the vessel by suc-

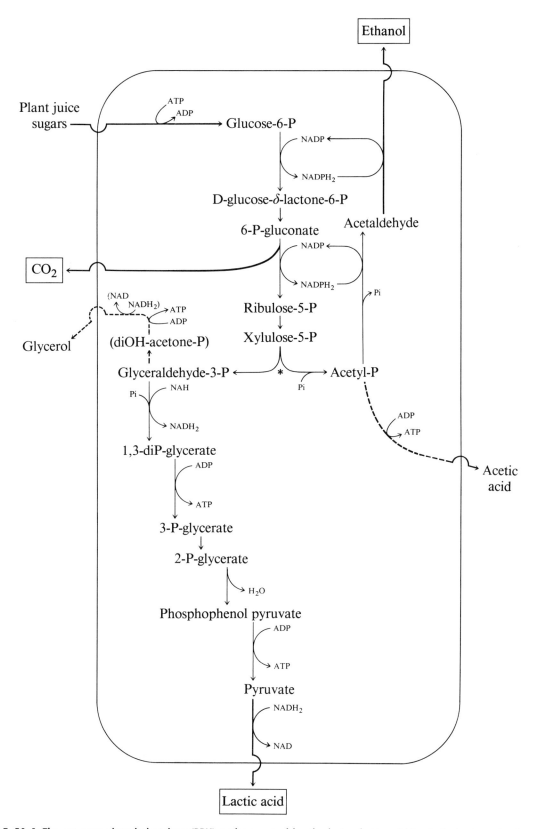

FIGURE 51-1 The pentose phosphoketolase (PPK) pathway used by the heterolactic acid bacteria (such as the *Leuconostoc* species) for fermenting plant sugars to mostly lactic acid, ethanol, and carbon dioxide (solid arrows). The heterolactics can also convert a small amount of the sugar to glycerol and acetic acid, using some of the enzymes of the PPK pathway (dashed arrows). An enzyme called pentose phosphoketolase (*) is characteristic of this pathway and can be used to help differentiate heterolactic and homolactic acid bacteria. The PPK pathway provides the energy (ATP) required for growth of the heterolactics.

cessively adding and tamping (with a large tamper) one-inch layers of shredded cabbage. This tamping begins the juice extraction process and forces out pockets of air trapped in the shredded cabbage.

5. **Continue packing until the vessel is about three-quarters full.** The brine should completely cover the cabbage. If there is not enough brine produced during tamping, add just enough water to cover the top of the packed cabbage.

6. **Place one or more of the outer leaves on top** to completely cover the packed cabbage. Make sure that these whole leaves are also covered with brine because, otherwise, many types of mold will grow on them. Although the molds are not harmful, the leaves will appear very unappetizing. **Refrigerate extra leaves for 24-hour replacements.**

7. **Place the solid disk on top of the whole leaves, and add weight.** The weight should be sufficient to keep the shredded cabbage compressed and the whole leaves submerged in the brine. **This condition should be checked periodically** to insure that anaerobic conditions are maintained.

8. **Incubate for 24 hours at room temperature.**

Laboratory instructors (or interested students) should set up demonstrations for the entire class during the fermentation process that include the following:

Gram stains of brine samples made according to the schedule and procedures shown below. These microscope demonstrations should be arranged in chronological order and appropriately labeled.

T-soy and EMB plates containing streaked brine samples according to the procedures and schedule shown below. Plates should be chronologically arranged and clearly labeled.

All the accumulated Gram-stained slides and plates should be displayed at the appropriate laboratory period for comparison. Students should note the changes in this mixed culture as the fermentation progresses.

PROCEDURE: AFTER 1 DAY

1. **Gram stain a sample of brine.** Save this slide for later observation and comparisons with later slides.
2. **Streak a sample of brine onto one EMB and one T-soy plate.** Incubate plates for 48 hours at 30°C. After incubation, seal each plate with tape and store it in the refrigerator for comparisons with later plates.
3. **Remove the solid disk and weight, replace the outer leaves, adjust the brine level** if necessary.
4. **Replace the solid disk and weight, and continue the room-temperature incubation.**

PROCEDURE: AFTER 3 DAYS, 7 DAYS, AND 3 TO 4 WEEKS

After each time interval, repeat the four steps of the above procedure ("after 1 day").

PROCEDURE: AFTER 1 MONTH

Ask your instructor to collect all previously made Gram-stained slides, EMB plates, and T-soy plates and to arrange them chronologically for your observation. **Record all of your observations.**

Discard the upper layer (about 10 to 20 mm), and smell the remaining fermented cabbage. If it has the pungent aroma typical of sauerkraut, taste it, and record your reaction. (Heating the sauerkraut with or without chunks of meat should provide a delicious snack.)

51 SOLID FOOD PRESERVATION—SAUERKRAUT

NAME _____

DATE _____

SECTION _____

RESULTS

For each time period, record the pH of the brine, describe the approximate number of cellular types (by gram morphology), and describe the colony types present on each medium.

DAY ONE
pH:
Gram stained preparation:
T-soy plate:
EMB plate:
DAY THREE
pH:
Gram stained preparation:
T-soy plate:
EMB plate:
DAY SEVEN
pH:
Gram stained preparation:
T-soy plate:
EMB plate:
3–4 WEEKS (actual number of days: _____)
Gram stained preparation:
T-soy plate:
EMB plate:
Description of the taste:

CONCLUSIONS

QUESTIONS

Completion

1. Before the advent of mechanical refrigeration, cabbage fermentation was primarily used as a form of food _____ .

2. The two chemical conditions provided by you that make possible a selective environment and the enrichment of the lactic acid bacteria during cabbage fermentation are _____ and _____ .

3. _____ is the type of chemical compound in plant-tissue fluids that is used as a source of energy for growth and is converted to lactic acid during sauerkraut formation.

4. Selective conditions that are provided during sauerkraut production create an environment that favors the growth and metabolism of microbes that can function in the absence of _____ .

5. The salty fluid formed by extracting plant juices with sodium chloride is called _____ .

6. The first type of lactic acid bacterium that predominates during cabbage fermentation is the genus _____ , and the lactic acid bacterium that takes over at a later time is the genus _____ . The first genus has a _____ gram reaction and a _____ cell shape, whereas the second genus has a _____ gram reaction and a _____ cell shape.

7. Acid fermentations help preserve foods because _____ .

True–False (correct all false statements)

1. Sauerkraut preparation may be appropriately considered a type of pickling. _____

2. Lactic acid bacteria are the only microorganisms normally found on the surface of plants. _____

3. In addition to forming lactic acid, heterolactic acid bacteria also produce other volatile organic compounds during sugar catabolism. _____

4. Sauerkraut formation occurs largely under anaerobic conditions. _____

52

Numbers of Bacteria on Solid Foods

The mere presence of microorganisms in food does not mean that it is unfit for human consumption (see Exercises 50 and 51). However, foods can be the means by which some **pathogenic microorganisms** are transmitted to humans. If raw food is contaminated with pathogens and is not properly washed, or if the food is contaminated by improper handling during its preparation, or if it is stored in a way that allows pathogens to grow, then the pathogen will be transmitted to the human who consumes the food.

We can prevent food from becoming an agent of transmission of disease-causing microbes by following three simple rules:

1. *Clean food well* to avoid contamination with microbes that are normally present on food materials before preparation. For example, wash freshly harvested food to remove soil-borne organisms, and wash animal carcasses to remove fecal-borne microbes.

2. *Clean your hands and the utensils* used in preparing food so that they will not transfer undesirable microbes to the food.

3. *Store food properly* after preparation and prior to consumption so as to prevent the growth of any undesirable microbes that remain in the food. In technologically advanced countries, refrigerators and freezers are the primary means of food storage. In their absence, food can be canned or pickled (or fermented in other ways) to prevent growth of undesirable microorganisms.

If you follow these simple precautions, you will probably never have problems with food infections or food intoxications.

The detection of microbial pathogens in food is a specialized branch of food microbiology in which various types of selective media are used to encourage the growth of one type of pathogen while inhibiting the growth of all other microbes. Solid differential media that make one type of pathogen look different from others are also used. The food microbiologist isolates pure cultures of a suspected pathogen and performs diagnostic tests to precisely identify it.

Another category of undesirable microorganisms in food is the **food spoilage microorganisms,** which are often not pathogenic, but whose growth result in undesirable textures, aromas, or flavors. Food spoilage causes a tremendous loss of the food produced for human consumption each year. The microbes that cause food spoilage are often those that naturally exist as part of the living plant or animal. In general, food is kept palatable by following the three rules given above.

The numbers of microorganisms on the surface or within a foodstuff is a good indication of how that food has been handled and stored. The presence of large numbers does not mean that pathogens are present or that the food is spoiled, but it does suggest that good sanitation was not followed or that the food was stored in a manner that encouraged the growth of microorganisms. The number of microorganisms is an indicator of the general **microbial quality** of a food. If the microbial numbers are high, one can assume that a food's quality is low and that it may also contain some potential pathogens. But, if the numbers are low, its quality is probably good, and there is a much smaller chance that pathogens are present.

The manner in which plant or animal tissue is prepared for storage greatly affects the number of microorganisms present in the foodstuff. For example, some vegetables are washed, blanched, and quick-frozen. *Washing* should remove most of the soil-borne microbes. *Blanching* denatures the enzymes of plant tissue that cause post-harvest tissue breakdown and kills many of the microorganisms that naturally occur on the surface of all plant tissue. *Quick-freezing* prevents microbial growth after blanching, and keeping the tissue frozen eliminates further microbial growth during storage. Thus, the microbial quality of frozen vegetables is often high. On the other hand, animal tissue is often stored for a considerable length of time, handled by many people during its preparation, and chopped into small pieces by an instrument that has been kept for a long time at room temperature. Making hamburger, sausage patties, and crab cakes are examples of processes that can result in high microbial numbers (low microbial quality). Other meat products require such extensive processing (lunch meats, for example) that chemicals (such as nitrites) are added to kill microbes or prevent their growth. Since microbial numbers vary widely with the type of tissue and the extent of processing, microbial numbers are used mainly to compare the quality of two samples of the same type of food.

In this exercise, you will determine the numbers of microorganisms present in several types of commercially prepared foods.

LEARNING OBJECTIVES

• Understand that all food contains some microorganisms and that, in this way, food is no different from any other part of our environment.

• Understand that the microbiological quality of a food is greatly affected by the way it is handled and stored.

• Understand that pathogenic microorganisms can be transmitted by food intended for human consumption.

• Understand that microorganisms can also cause undesirable changes in texture, odor, and flavor and that this undesirable activity is one type of food spoilage.

• Learn the three rules for keeping food palatable and safe for human consumption.

MATERIALS

Cultures None

Media Standard plate-count agar (melted and tempered)

Supplies 9-ml dilution blanks
1.1-ml pipettes (sterile)
Spreading wheels, spreading rods, and 95 percent ethanol
Solid foods mixed with sterile distilled water (1 gm/9 ml) and treated with a sterilized food blender. Blended samples placed in sterile screw-capped tubes. (Suggestion: Compare solid-food types such as fresh, frozen, and canned beans, or cooked hamburger and raw hamburger.)

PROCEDURE: FIRST LABORATORY PERIOD

Note that food samples have already been diluted ten times because 9 ml of sterile water was added for every gram of food prior to blending and placing samples in screw-capped tubes.

Your instructor will probably assign food samples, so that all food types will be tested for microbial numbers (your requests may be considered).

1. **Construct a dilution scheme** using the flow chart shown in Figure 52-1, but do this only after you have read the following procedure.
2. **Have your instructor check and initial your scheme before you begin your work.** This will give you practice in designing dilution schemes, but it will also allow your instructor to control the number of dilution blanks and sterile pipettes used. When preparing your dilution scheme, consider the following items.
    a. Your goal is to determine how many microorganisms are present per gram of food.
    b. Use the spread-plate technique, as described in Exercise 20.
    c. Transfer only 0.1 ml to each plate.
    d. Prepare one plate each from the following dilutions: 10^{-3}, 10^{-4}, 10^{-5}, and 10^{-6}.
    e. Use only 9.0-ml dilution blanks and 1.1-ml pipettes. The 1.1-ml pipettes are used because they have a wide opening at the delivery tip and will not become clogged as easily with chunks of food. If this is the first time you have used 1.1-ml pipettes, you will need to plan carefully.
    f. As you complete the dilution scheme (see Figure 52-1), *label all tubes and plates with the correct dilution. Show how each pipette will be used* by marking each

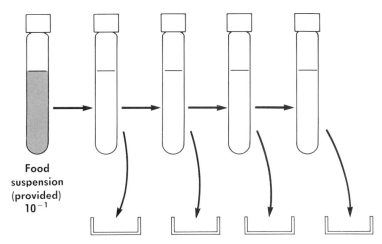

FIGURE 52-1 The form for constructing the dilution scheme for diluting and plating the solid-food preparation used in this exercise. When correctly filled out, this form should show (1) the volume of each dilution blank, (2) the dilution of the original food (represented by each dilution blank and each plate), and (3) each separate pipette used (designated by capital letter). When completed, this scheme should be labeled similarly to that shown in Figure 20-1.

transfer arrow with a capital letter to represent each pipette, and *show the volume transferred* at each step. Your final dilution scheme should be labeled in a manner similar to that shown in Figure 20-1.

3. **Incubate your plates at 30°C for 48 hours only.** If your next laboratory is more than 48 hours away, make sure that you know who is responsible for removing the plates before you leave the laboratory.

PROCEDURE: SECOND LABORATORY PERIOD

1. **Count the colonies on each plate using the 30-to-300 rule** (see Exercise 20).

2. **Calculate the number of viable microorganisms per gram of the food** you used in this experiment. Remember that the foodstuff was diluted 1:10 with sterile distilled water before it was given to you during the first period of this exercise.

3. **Record your numbers and calculations** in the Results section.

4. **Place the name of the food examined and your viable-number calculations on the chalkboard** to share your results.

5. **Record the results from the rest of the class** in the appropriate table in your Results section.

6. **Discard your plates in the appropriate manner.**

NAME _____

DATE _____

SECTION _____

RESULTS

Place **your results** in the following table. For numbers greatly above 300/plate, record as "too numerous to count" (TNTC). Count all other plates. Show your calculations and your calculated number in the right column.

Your results with _____ (name of food)

Dilutions examined	Total number of colonies per plate (show any estimates with an asterisk,*)	Calculated viable number of microbes in the original foodstuff (CFU/gm)
10^{-3}		
10^{-4}		
10^{-5}		
10^{-6}		

Place the **class results** in the following table. Arrange data so that all using the same foodstuff are listed together.

Accumulated class results

Food used	Dilutions Examined	Total number of colonies per plate (show any estimates with an asterisk, *)	Calculated viable number of microbes in the food (CFU/gm)
	10^{-3}		
	10^{-4}		
	10^{-5}		
	10^{-6}		
	10^{-3}		
	10^{-4}		
	10^{-5}		
	10^{-6}		
	10^{-3}		
	10^{-4}		
	10^{-5}		
	10^{-6}		
	10^{-3}		
	10^{-4}		
	10^{-5}		
	10^{-6}		
	10^{-3}		
	10^{-4}		
	10^{-5}		
	10^{-6}		

NAME _____

DATE _____

SECTION _____

CONCLUSIONS

Your results:

Class results:

QUESTIONS ABOUT YOUR DATA

Questions 1 through 4 ask for an opinion. First answer either yes or no on the line provided; then refer to your table of class data to help explain, in writing, the reasoning behind your answer.

1. Do any of these results show the value of washing and blanching food in improving its microbial quality? _____

2. Do any of the results illustrate the effects of temperature during preparation on the microbial quality of the food? _____

3. Do any of these results directly show the presence of pathogens in the foodstuff? _____

4. Do any of these results suggest that these foods are unsafe to eat? _____

5. Healthy animal tissue is sterile prior to slaughter. How then can you explain the results obtained with processed meats?

QUESTIONS

Completion

1. Never enter a culture tube without first _____ your inoculating loop.

2. The three rules to follow for preventing foods from being a vehicle for transmission of human pathogens are _____, _____, and _____.

3. A good indication that a nonfermented food has been properly handled and/or stored is given by the _____.

4. A medium used by microbiologists to plate food samples to make specific types of pathogenic microbes appear different from other types on agar plates is called a _____ medium.

5. The process known as blanching uses heat to accomplish two things: to _____ and to _____.

6. The solid medium used to plate dilutions of food in this exercise is called _____.

True - False (correct all false statements)

1. The presence of microorganisms in or on food means that it is unfit for human consumption. _____

2. Some foods can serve as vehicles by which human pathogens are transmitted. _____

3. Special media used to encourage the growth of specific pathogens in or on food, if they occur, are called selective media. _____

4. Food spoilage microorganisms are those whose growth or metabolic products cause an undesirable change in texture, aroma, or flavor in the food. _____

5. Most food spoilage microorganisms are human pathogens. _____

XIII

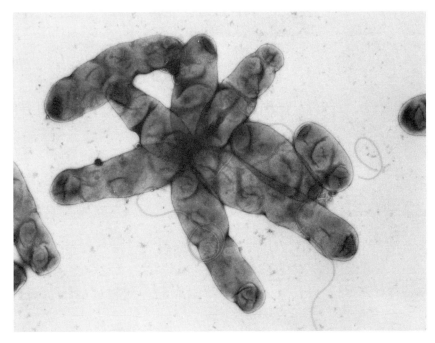

Roseate form of helically sculptured *Seliberia* sp. The organism was isolated from an activated sludge reactor during biodegradation of discarded cutting fluids.

The universe would be incomplete without man; but it would also be incomplete without the smallest transmicroscopic creature that dwells beyond our conceitful eyes and knowledge.

—*John Muir*

ENVIRONMENTAL MICROBIOLOGY

Microorganisms are ubiquitous, occurring everywhere in our environment. Part XIII will introduce you to the microscopic residents of water and soil. The purpose of the following exercises is to show you how the microbiological quality of water is examined and to impress you with the many essential microbial activities that occur within soil.

53

Detecting Coliform Bacteria in Water

We consider fresh-water streams, lakes, and ground water to be **polluted** when some condition makes the water unsafe for human play or consumption or upsets the natural balance of plants and animals living in or near that water. We usually think of two forms of pollution: toxic or unsightly chemicals and pathogenic microorganisms. Probably the largest single source of potentially pathogenic microorganisms in water is animal feces (including human), which contain many billions of bacteria per gram, some pathogenic. For example, some **enteric diseases** (diseases that affect the gastrointestinal tract) are spread from infected to healthy persons through fecally contaminated water supplies. Typhoid fever, cholera, and certain form of dysentery are examples of enteric diseases caused by bacteria. Although they are not considered to be enteric, certain viruses, such as poliomyelitis and hepatitis, also exit the body in feces and can be spread by polluted drinking water. The amoeba that causes dysentery can also be spread in this way.

How can you tell if a water sample contains any of these pathogens? To test for each one separately would be extremely time-consuming and costly; so this simple rule is followed: if water contains *any* microorganisms common to animal intestines, it should not be consumed by humans, because it may also contain pathogens.

Water testing for microbiological safety rests on the ability of microbiologists to detect **coliform bacteria.** The word *coliform* refers to any bacterium that is like *E. coli;* that is, it has the following key characteristics. (1) It is a small gram-negative rod; (2) it does not contain spores; (3) it ferments lactose in the presence of bile, with acid and gas production; and (4) it produces a green metallic sheen on eosin-methylene-blue (EMB) agar. If an isolated bacterium meets these criteria, it is quite probably *E. coli.*

Escherichia coli, which is found in large numbers in the feces of all animals, lives longer in water than intestinal pathogens do. Therefore, microbiologists believe that if no *E. coli* are present, there should be no intestinal pathogens present. That is why the test for coliforms has become universal for determining whether water contains fecal material. Coliform tests are made almost daily in municipal water works, where drinking water is prepared for human consumption, and in waste-water treatment (sewage) plants, where water must be tested for its safety before it leaves the facility. It is almost impossible to overemphasize the importance of these facilities in maintaining human health.

One major problem in testing for fecal coliforms is caused by a bacterium called *Enterobacter aerogenes.* This microbe too is found in fecal material, but it is also found in the soil and on the surfaces of plants; therefore, its presence does not guarantee fecal pollution. However, *Enterobacter aerogenes* has many characteristics in common with *Escherichia coli,* and microbiologists must be sure that they can distinguish between them. One way is with the **IMViC tests** (see Exercises 30, 32, and 34), but these require the isolation of pure cultures and are time-consuming. As an alternative, microbiologists have developed a series of tests that provide faster and easier detection of coliforms in water; these tests can be done with confidence without having to isolate and identify pure cultures. The series includes a *presumptive,* a *confirmed,* and a *completed* test.

The presumptive test

Some of the water to be tested is placed in **lauryl-tryptose broth,** a *selective medium* (see Exercise 21) that enables intestinal bacteria to grow while inhibiting the growth of most other bacteria. In other words, lauryl-tryptose broth selects for the growth of intestinal bacteria.

The selective chemical in lauryl-tryptose broth is a detergent (or surfactant) called **sodium lauryl sulfate (SLS).** Most bacteria are either inhibited from growing or killed by surfactants. However, bacteria that are able to pass through the stomach and enter and grow in the intestines must be able to withstand the surfactants in bile, which is released into the first part of the small intestine (the duodenum). **Bile** is a chemically complex fluid that is secreted by the liver and deposited and concentrated in the gallbladder. Bile contains surfactants that lower the surface tension of (emulsify) the contents of the small intestine.

Lauryl-tryptose broth also contains lactose to support growth (recall that coliforms ferment lactose to acid and gas). It contains no pH indicator, but tubes prepared for the presumptive test do contain an inverted Durham tube to test for gas production from lactose. Therefore, this test distinguishes between those surfactant-resistant microorganisms that produce gas while growing on lactose and those that do not.

To accomplish the presumptive test, add samples of water to lauryl-tryptose broth, incubate the mixture at 35°C, and check for gas production at 24 and 48 hours. Gas production at any time up to 48 hours constitutes a *positive presumptive test.* Only samples that show a positive presumptive test are tested further. The absence of gas formation at the end of the 48-hours incubation constitutes a *negative* presumptive test.

The confirmed test

Culture fluid from positive lauryl-tryptose broth tubes (presumptive test) is transferred to **brilliant green lactose bile (BGLB) broth.** This broth is also a *selective medium,* which allows intestinal bacteria to grow and inhibits the growth of most other microorganisms. It contains a dye called brilliant green and bile extracted from bovine gall bladders. **Brilliant green** is a dye (not a pH indicator) that inhibits most gram-positive but not most gram-negative bacteria. Bile is a chemically complex, surfactant-containing fluid that kills or inhibits most microorganisms that do not survive or grow in animal intestines.

The BGLB broth also contains peptone and lactose to support growth of microorganisms that are not inhibited by bile and brilliant green. Since coliforms produce gas while fermenting lactose, the BGLB broth tubes contain an inverted Durham tube as a gas trap. Note that BGLB broth contains no pH indicator, so it does not detect the acid released during lactose fermentation.

To accomplish the confirmed test, a loopful of culture from positive lauryl-tryptose broth fermentation tubes (presumptive test) is transferred to brilliant green lactose bile broth fermentation tubes. The BGLB broth tubes are incubated at 35°C and checked at 48 hours for gas production. Any amount of gas in the inverted Durham tube at any time within 48 hours constitutes a *positive* confirmed test. Samples that show a positive confirmed test are tested further. The absence of gas formation at the end of the 48-hour incubation period constitutes a *negative* confirmed test.

The completed test

Culture fluid from the positive brilliant green lactose bile broth tubes (confirmed test) is streaked onto plates of **eosin methylene-blue (EMB) agar,** a *selective medium* whose dyes (eosin and methylene blue; Figure 53-1) usually inhibit the growth of gram-positive bacteria. The EMB agar is also a *differential medium* because it differentiates those bacteria that ferment lactose with the production of much acid from those that do not or those that produce only a little acid. The green, metallic sheen that coliforms produce on EMB agar is caused by the excretion of so much acid during lactose fermentation that the dyes precipitate in and on the surface of the coliform colonies. The colonies appear very dark (deep purple or almost black), and the crystals of dye that are precipitated on the surfaces of the colonies give them the metallic sheen (see Figure 53-1). Often, so much acid is excreted into the surrounding agar that the medium itself has this sheen.

To accomplish the completed test, samples from tubes showing a positive confirmed test are streaked onto EMB plates, which are incubated at 35°C for 24 hours. Cells from very dark colonies (with or without a green metallic sheen) are transferred to a lauryl-tryptose broth fermentation tube and to a nutrient-agar slant. The agar slant and broth-fermentation tubes are incubated at 35°C for 48 hours. The formation of gas in the fermentation tube and the demonstration of gram-negative non-spore-forming rod-shaped bacteria from the agar culture constitute a *satisfactory completed test.*

A summary of these tests and the positive results for coliform bacteria is given in Table 53-1.

In this exercise, you will perform only the presumptive and part of the completed tests. You will test several bacteria (including *Escherichia coli*) for their ability to produce gas from lactose fermentation, see typical coliform and noncoliform colonies on EMB agar, and test a water sample for coliforms.

For further information on how microbiologists

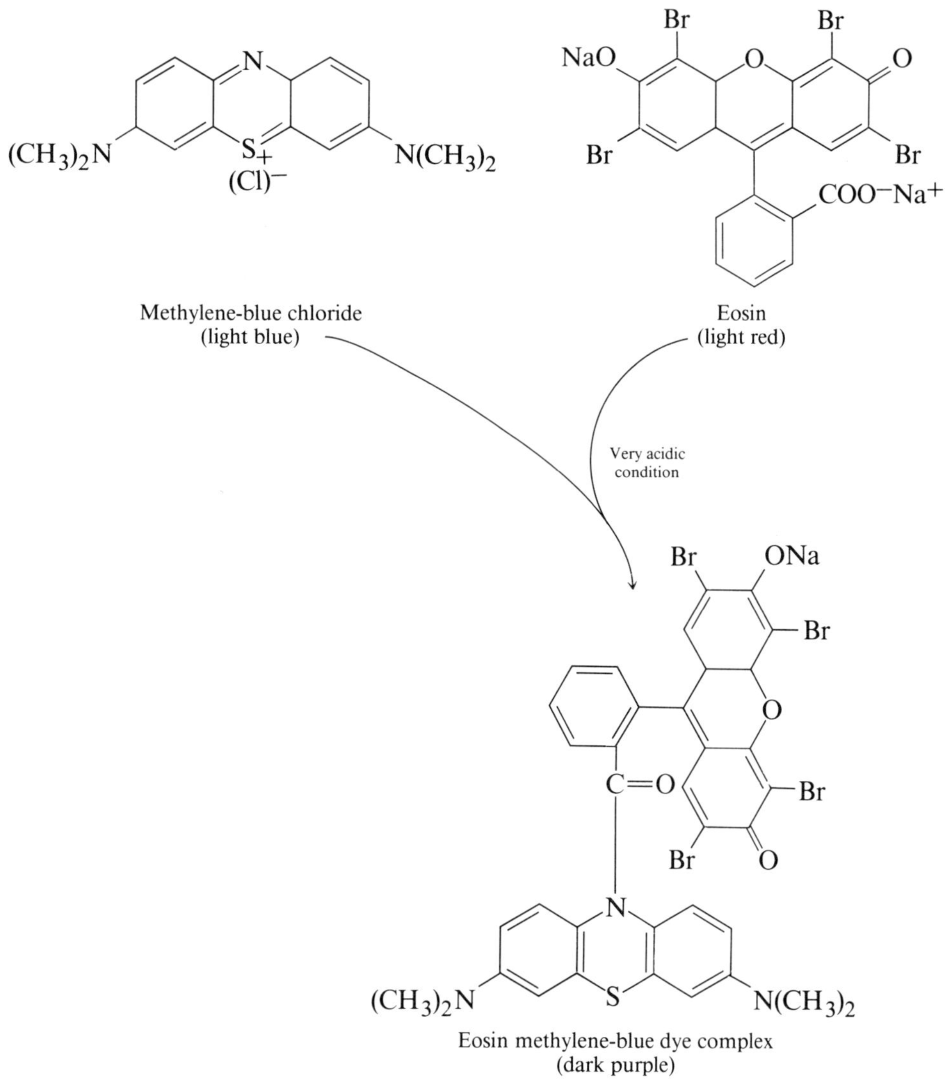

FIGURE 53-1 Molecular structure of the dyes used in eosin methylene-blue (EMB) agar, a violet-colored selective and differential medium used for coliform isolation and detection. As coliform colonies develop, cells in the center of the colony run out of oxygen and begin to ferment lactose and excrete large quantities of acid. The acid causes the eosin (a light red dye) to covalently bond to methylene-blue chloride (a light blue dye), and the resulting dye complex (a dark purple color) forms in the colony. If enough acid is excreted, the dye complex precipitates on the surface of the colony and the surrounding medium and forms a green metallic sheen.

TABLE 53-1 Summary of standard tests used for detecting the presence of coliforms in water and waste-water samples

Test name	Media	Positive test results (reaction of *Escherichia coli*)
Presumptive	Lauryl-tryptose broth	Gas collected in inverted Durham tube at 35°C incubation by 48 hours
Confirmed	Brilliant green lactose bile broth	Gas collected in inverted Durham tube at 35°C incubation by 48 hours
Completed	Eosine methylene-blue agar	Completely dark colonies (often with metallic sheen) at 35°C incubation by 24 hours. Cells from dark colonies are transferred to lauryl-tryptose broth and a nutrient agar slant.
	Lauryl-tryptose broth	Gas collected in inverted Durham tube at 35°C by 48 hours
	Nutrient-agar slant	Cells are gram negative non-spore-forming rods

use the presumptive, confirmed, and completed tests to examine domestic drinking water and waste-water discharges for purity and public safety, see the most recent edition of *Standard Methods for the Examination of Water and Waste-water,* prepared and published jointly by the American Public Health Association, the American Water Works Association, and the Water Pollution Control Federation (M.C. Rand, A.E. Greenberg, and M.J. Taras, editors).

LEARNING OBJECTIVES

- Understand that contamination of fresh water by fecal microorganisms is only one of many types of pollution.
- Understand why *Escherichia coli* is used as an indicator of fecal pollution in aquatic environments.
- Learn the four characteristics of a coliform and how you test for these in the presumptive, confirmed, and completed tests.
- Learn the type and order of tests used for testing water for the presence of coliforms.
- Review the definition of a selective medium (Exercise 21), and learn why EMB agar, lauryl-tryptose, and brilliant green lactose bile broths are selective.
- Review the definition of a differential medium (Exercise 21), and learn why EMB agar fits it.

MATERIALS

Cultures	*Escherichia coli* *Enterobacter aerogenes* Water from a local stream or pond
Media	Single-strength lauryl-tryptose (LT) broth Double-strength lauryl-tryptose (LT) broth Eosine methylene blue (EMB) plates
Supplies	10-ml pipettes (sterile) 1.0-ml pipettes (sterile)

PROCEDURE: FIRST LABORATORY PERIOD

Preparing the presumptive test

1. **Aseptically inoculate four tubes of lauryl-tryptose (LT) broth** as follows:
   a. Place 10 ml of stream or pond water into a tube containing 10 ml of double-strength LT broth. Note that adding this sample dilutes the double-strength broth exactly by half; thus, the broth becomes single-strength, and the presence of at least one coliform in 10 ml can be tested.
   b. Place 1 ml of stream or pond water into a tube containing 10 ml of single-strength LT broth. This will test for the presence of one or more coliforms per milliliter of water sample.
   c. Place one loopful of *Escherichia coli* into another tube of single-strength LT broth, taking care not to mix air into the Durham tube during inoculation.
   d. Place one loopful of *Enterobacter aerogenes* into yet another sterile tube of single-strength LT broth, taking care not to mix air into the Durham tube during inoculation.
2. **Incubate all four tubes at 30°C for 48 hours only.** If it is longer than 48 hours until your next laboratory period, make sure you know (before you leave the laboratory) who is responsible for removing the plates from this incubator and placing them in the refrigerator.

PROCEDURE: SECOND LABORATORY PERIOD

Examining the presumptive test

1. **Examine each tube for the presence of gas in the Durham tube.** Do not shake these tubes because you are likely to mix air into the broth, causing it to become trapped in the Durham tube giving a false-positive result. Formation of any amount of gas within 48 hours constitutes a positive test. The absence of gas formation by the end of 48 hours of incubation constitutes a negative test.

INTERPRETATION NOTES

All strains of *Escherichia coli* grow well in lauryl-tryptose (LT) broth, and all fecal strains produce gas during fermentation of the lactose in this medium. *Enterobacter aerogenes* also grows well in LT broth, but strains differ greatly in their ability to produce gas during lactose fermentation. Thus, you should remember that growth and gas production during lactose fermentation does not mean that the sample has been fecally contaminated. Gas production is a *positive presumptive* test, but more tests must be done to determine if coliforms are present.

2. **Record** your results in the Results section, and then use these tubes as a source of cells for inoculating EMB plates for the completed test.

Preparing the completed test

If you were using standard methods for these tests of water samples for fecal coliforms, you would transfer culture from those tubes showing positive presumptive tests to tubes of brilliant green lactose bile broth to perform the confirmed test. In this experiment, however, we ask you to transfer cells from the presumptive test cultures directly to one of the completed-test media and to transfer cultures even from tubes that show negative results in the presumptive test. The reason is to enable you to examine both positive and negative results of the completed test. Perform the following procedures only after examining and recording the results from the presumptive test.

1. **Label three EMB agar plates** with today's *date,* your *name,* and the *number* of this exercise.
2. **Shake each tube vigorously before you transfer samples** of culture to EMB plates.
3. **Select the tube inoculated with the smallest volume of stream or pond water that shows a positive presumptive test, and streak one loopful onto one sterile eosine methylene-blue (EMB) agar plate.** Label this plate so that you will later know the sample used as an inoculum.
4. **Thoroughly mix and then streak one loopful of *Escherichia coli* culture from the presumptive-test tube onto one EMB agar plate.** Label the place with the name of the inoculum.
5. **Thoroughly mix and then streak one loopful of *Enterobacter aerogenes* culture from the presumptive-test tube onto one EMB agar plate. Label this plate.**
6. **Incubate all three plates at 30°C for 48 hours or until dark purple colonies (with or without a metallic sheen) are evident** prior to 48 hours. If the dark colonies and metallic sheens are evident before the end of 48 hours incubation, the plates may be refrigerated until the end of the 48 hours when they will be examined. If it is longer than 48 hours until your next laboratory period, make sure that someone will remove these plates from the incubator and refrigerate them until your next laboratory period. Make sure you know who is responsible for this before you leave the laboratory.

PROCEDURE: THIRD LABORATORY PERIOD

Examining the completed test

1. **Observe the appearance of colonies and the medium surrounding these colonies on all three plates.** A *positive test* shows completely dark (purple to almost black) colonies that often (but not always) have a purple-green metallic sheen on their surfaces (metallic sheen may also be seen on the medium surface surrounding these colonies). A *negative test* is a plate lacking totally dark colonies.
2. **Record** your results.

TECHNICAL NOTE

According to the *Standard Method of Water Analysis,* cells from dark colonies having green metallic sheen are transferred to lauryl-tryptose broth to make sure that they are typical coliform cells that produce gas on lactose; they also can be transferred to a nutrient agar slant so that they can be Gram stained to make sure that they are short, non-spore-forming, gram-negative rods. The EMB plates and the subsequent inoculation of LT broth and nutrient agar slants are all part of the *completed test.* In this exercise you are asked to skip the second LT broth and Gram-stain preparation because you have already seen what gas formation in LT broth looks like, and by now you have looked at many Gram stains of short, gram-negative rods.

INTERPRETATION NOTES

EMB broth contains eosine (a pink to red dye) and methylene-blue (a weak blue dye) in solution. In the presence of acid, they bond together to form a very dark purple dye complex (see Figure 53-1). When fecal coliforms grow deep inside a colony, they ferment lactose (the sugar in EMB broth) and produce large quantities of acid, which is sufficient to form the eosine-methylene-blue dye complex. Consequently, this dark dye complex accumulates inside and around the colony. When enough of the eosine-methylene-blue dye complex concentrates in any area, light is reflected off the surface, giving a green metallic sheen. All fecal coliforms produce enough acid during lactose fermentation to produce the eosine-methylene-blue dye complex, and enough of this dye complex accumulates to form completely dark colonies which often have a green metallic sheen.

In contrast, *Enterobacter aerogenes* does not produce much acid while fermenting lactose. Consequently, these bacteria do not produce completely dark colonies, nor do they accumulate enough of the eosine-methylene-blue complex to create the green metallic sheen. Typical *Enterobacter aerogenes* colonies are large, pinkish in color (perhaps with dark centers), and mucoid in texture.

53 DETECTING COLIFORM BACTERIA IN WATER

NAME _____

DATE _____

SECTION _____

RESULTS

Sample	Presumptive test (lauryl-tryptose broth)			Partial completed test (EMB agar)	
	Volume of inoculum	Gas produced?	Positive or negative test?	Colony appearance	Positive or negative colonies?
Stream or pond water (_____) source	1 ml				
	10 ml				
Escherichia coli	One loopful				
Enterobacter aerogenes	One loopful				

CONCLUSIONS

QUESTIONS

Completion

1. Enteric diseases can be caused by certain large classes of microorganisms such as _____ and _____.

2. Animal feces contain many billions of bacteria per gram. When this quantity is translated into an exponential number, it might be one such as _____.

3. The four characteristics of *Escherichia coli* that are used to conclude the presence of fecal coliforms in aquatic samples are (1) _____, (2) _____, (3) _____, and (4) _____.

4. The standard methods for water and waste-water analysis consist of three tests. The first is called the _____ test, and it uses a medium called _____. There are two constituents that make this medum selective; these are _____ and _____. Positive results from this test come from the following observation(s):

5. The second test used in the standard methods for water and waste-water analysis is called the _____ test, and it uses a medium called _____ (do not abbreviate). This is considered a selective medium because it allows the microbiologist to select _____ bacteria from those that _____. Positive results from this test produce the following observation(s):

6. Positive results on the third standard methods test is given by observing _____ from a medium called _____ and observing _____ from a medium called _____. The purpose of doing this test is to _____.

7. Both LT and BGLB broths are considered selective, because LT broth contains _____ and _____, and BGLB broth contains both _____ and _____.

True–False (correct all false statements)

1. Diseases other than enteric diseases can be transmitted by fecally contaminated water. _____

2. *Enterobacter aerogenes* is a coliform but not a fecal coliform. _____

3. Microbiologists believe that if *E. coli* is not present in an aquatic sample, then the sample is not fecally contaminated and does not contain intestinal pathogens. _____

4. Each of the media used in the three tests for standard analysis of water contain the same type of fermentable carbon and energy source for bacteria, and this is called lactase. _____

5. No strains of *Enterobacter aerogenes* produce gas while growing in BGLB broth. _____

6. The green metallic sheen formed by fecal coliforms growing on EMB plates results from the large amount of acid formed and excreted when these bacteria grow in the presence of glucose. _____

54

Introduction to the Nitrogen Cycle

Microorganisms in the soil are an important part of the cycle of life. Soil microorganisms are essential for the decay of dead plant and animal tissue, the transformation of this decayed tissue into inorganic materials, and the conversion of the inorganic materials into forms that can be used for plant growth. The complete degradation of dead tissue into CO_2, N_2, inorganic sulfur and phosphorous, and other materials is called **mineralization.** After tissue has been completely degraded, other microorganisms can use the inorganic compounds, converting them into forms that can be incorporated (assimilated) into new living tissue. This continuing process of death, decay, mineralization, and reassimilation of minerals into new living tissue is called a *cycle.*

The **nitrogen cycle** has attracted considerable attention from agronomists and soil microbiologists because nitrogen, in the form of ammonia and nitrate, is often the limiting nutrient in soils for plant growth. Therefore, microorganisms that provide these nitrogen-containing compounds to unsupplemented soil markedly increase plant productivity. (Of course, productivity can also be increased by adding synthetic nitrates, but these are often too expensive for all except those who live in the most highly developed countries.) The production of nitrates is only part of the importance of the nitrogen cycle.

Let us examine the nitrogen cycle shown in Figure 54-1. Begin in the center. **Molecular nitrogen** (dinitrogen, N_2) is a gas that constitutes about 80 percent of the earth's atmosphere, but it is chemically inert and cannot be used by plants or animals unless it is combined with hydrogen. The biological process of reducing inert N_2 with hydrogen to form ammonia (NH_3), and the use of that ammonia to form new, nitrogen-containing cell polymers, is called **nitrogen reduction.** This process is examined in Exercise 55. The vast supply of dinitrogen in the atmosphere and the relative scarcity of fixed nitrogen on the earth's surface suggest that nitrogen fixation (reduction) is the rate-limiting step in the nitrogen cycle. Dinitrogen is fixed (reduced) by ultraviolet light, electrical equipment, lightning, volcanic eruptions, and the internal-combustion engine, but these processes account for less than 1 percent of the total nitrogen fixation that occurs on the earth. Industrial manufacture of nitrogen fertilizers recently accounts for about 5 percent of this total. The remainder occurs by the action of bacteria in the upper few inches of the earth's soil. It is believed that procaryotic microorganisms (bacteria) are the only living things on earth capable of nitrogen fixation. The quantity of dinitrogen fixed by procaryotes is estimated to be about 200,000,000 metric tons per year.

Once N_2 is reduced (fixed) by bacteria, the resulting **ammonium ion** can be assimilated by other procaryotes, the eucaryotic algae, and plants (see Figure 54-1). These higher organisms incorporate NH_3 in their amino acids, purines, and pyrimidines; therefore, microbially fixed nitrogen eventually ends up in the proteins and nucleic acids or algal and plant cells, which then serve as nitrogen sources for the animal kingdom. During the digestion of plant materials by animals, these nitrogenous compounds are partially broken down, but the nitrogen remains mostly in the reduced form. Animals excrete some of this injested nitrogen (as ammonia, urea, or uric acid), but most of it is incorporated into their proteins and nucleic acids during growth or repair of tissue. Most of the fixed nitrogen remains in an animal's tissues until death when the process of microbial decomposition releases N_2 into the atmosphere, and the nitrogen cycle begins again.

As soon as a plant or animal dies, its tissues are immediately attacked by microorganisms, and the nitrogenous compounds are decomposed. The first

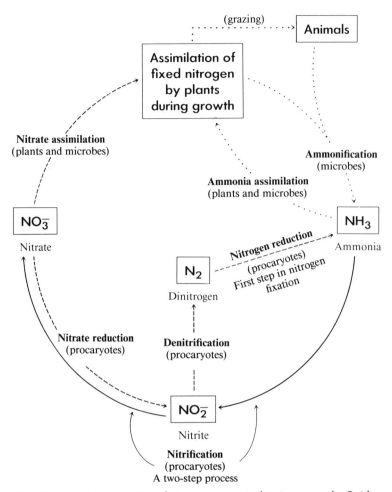

FIGURE 54-1 Participation of microorganisms in the nitrogen cycle. Oxidative reactions (removing high-energy electrons from hydrogen atoms during catabolism) are shown with solid arrows (→). Reduction reactions (using inorganic compounds as terminal electron acceptors during respiratory catabolism) are shown with dashed arrows (-→). Reactions having no exchange of hydrogens (electrons)—in other words, those that are neither oxidation nor reduction reactions—are shown with dotted arrows (···→).

steps in the decomposition process involve hydrolysis of proteins and nucleic acids into their building blocks (amino acids, purines, and pyrimidines). Eventually, the amino groups of these building blocks are removed, and ammonia is released (see Figure 54-1). This process is called **ammonification** (see Exercise 56). The ammonia released by this process can be used directly as a nitrogen source for plants (**ammonia assimilation**) or to support growth of microorganisms in the soil. Plant roots probably use very little of the ammonia liberated from dead animal and plant tissue because most of it is taken up by microorganisms.

Two highly specialized groups of soil bacteria are able to oxidize ammonia (NH_3) to nitrite (NO_2^-) and then oxidize nitrite to nitrate (NO_3^-). The whole oxidative process is called **nitrification** and is accomplished by strictly aerobic bacteria, using high-energy electrons removed from ammonia or nitrite as sources of energy for their growth (see Figure 54-1). Since this energy source is inorganic, these bacteria are said to be **chemoautotrophic** (chemolithotrophic). In well-aerated, moist, warm soils, the microbial oxidation of ammonia to nitrate occurs very rapidly, making nitrate the dominant nitrogen source in such soils. The process by which plants take up and use nitrate as their principal source of nitrogen is called **nitrate assimilation** (see Figure 54-1).

The practice of plowing residual plant materials and animal wastes into the soil accomplishes several things for the farmer and home gardener. It provides dead, partially digested tissue for soil microbes to break down to ammonia (ammonification); it mixes oxygen into the soil so that obligate aerobes can oxidize ammonia to nitrate for plant use (nitrification);

and since part of such tissue degrades slowly, its presence improves the texture and moisture-holding properties of the soil.

Microbial production of nitrate has also served the destructive aspects of human nature: Gun powder is a mixture of sulfur, carbon, and saltpeter (KNO_3). The British blockade of France during the Napoleonic wars caused France to run short of nitrate. So they developed "nitrate gardens" in which manure and soil were mixed, spread on the ground, and frequently turned for aeration. The organic nitrogen mineralized to nitrates, which were then extracted from the residue for gun-powder manufacture. Large *natural* accumulations of saltpeter (potassium nitrate) also occur. For example, a deposit of saltpeter existed just inside the natural entrance to Mammoth cave (Kentucky). This deposit was mined during the U.S. Civil War for gun-powder manufacture.

Whenever organic matter is decomposed in soil and oxygen is depleted by aerobic microorganisms, nitrite or nitrate can substitute for oxygen as an electron acceptor at the end of the electron-transport chain of some soil microorganisms. These microbes continue to respire in a process known as **anaerobic respiration** (see Exercise 18). Some microbes capable of anaerobic respiration can reduce nitrate only to nitrite **(nitrate reduction)**, but others can reduce nitrate completely to dinitrogen **(denitrification)**, which is released into the atmosphere.

Dentrification depletes soil of nitrates (see Exercise 57). Some researchers have estimated that up to 50 percent of all the artificial nitrogen fertilizer applied to soils is lost as dinitrogen because of anaerobic microbial respiration. However, denitrification is not all bad. Nitrate ions are very soluble and, thus, are easily leached from the soil after a rain, eventually ending up in the oceans. The earth's entire supply of nitrogen would end up in the oceans as nitrate except that microbial denitrification in the oceans again returns dinitrogen to the atmosphere, where it can once again be fixed by terrestrial bacteria . . . and the cycle continues.

54 INTRODUCTION TO THE NITROGEN CYCLE

NAME _____

DATE _____

SECTION _____

QUESTIONS

Completion

1. The phenomenon of complete breakdown of living tissue to its chemical elements, such as CO_2, P, N_2, and S, is called _____.

2. The physical state of molecular nitrogen (dinitrogen) is _____, and, in this state, it makes up about 80 percent of the earth's _____.

3. From a chemical standpoint, nitrogen fixation refers to the chemical _____ of dinitrogen to a chemical compound called _____, which can then be incorporated into cell material during bacterial growth.

4. Certain chemoautotropic bacteria are able to oxidize ammonia to nitrite or nitrite to nitrate. This process of biologically converting ammonia to nitrate is called _____.

True – False (correct all false statements)

1. Bacteria that oxidize ammonia to nitrite or nitrite to nitrate are using the electrons from either ammonia or nitrite as a source of energy for growth. _____

2. Bacteria that reduce nitrate to nitrite or nitrite to N_2 are using either nitrate or nitrite as a terminal electron acceptor for anaerobic respiration. _____

3. *Ammonification* is a term that refers to the reduction of amonia to nitrite. _____

4. Nitrogen fixation by bacteria accounts for about 96 percent of all of the nitrogen reduced to ammonia on our planet. _____

Multiple Choice (circle all that apply)

1. Which of the following polymers from plant or animal tissue yields ammonia during microbial breakdown (hydrolysis)?
   a. Proteins
   b. Polysaccharides
   c. Lipids
   d. Nucleic acids

2. Consider the parts of the nitrogen cycle listed below. Circle the letters before those processes that take place only when oxygen is present.
   a. Ammonification
   b. Nitrification
   c. Nitrate assimilation
   d. Nitrate reduction
   e. Denitrification
   f. Nitrogen fixation

3. Place an arrow after those parts of the nitrogen cycle (listed in 2) that are accomplished by procaryotic cells.

55

Nitrogen Fixation — Reduction of Dinitrogen to Ammonia by Procaryotes

Overview of nitrogen fixation

Nitrogen fixation refers to the reduction of atmospheric dinitrogen (N_2) and subsequent incorporation of that reduced nitrogen (NH_3) into cellular material, such as proteins and nucleic acids. The overall reaction is shown in Figure 55-1. Note that this nitrogen reduction is catalyzed by an enzyme complex called **nitrogenase,** which catalyzes a series of three reductions, as detailed in Figure 55-2. Also, note that the complete series of reductions requires three pairs of high-energy electrons and three pairs of protons to complete the reduction of N_2 to NH_3. Note also that a large amount of energy (in the form of ATP) is required to carry out the reduction of nitrogen.

As soon as ammonia is formed by the bacterial cell, it is incorporated into organic compounds, such as amino acids, that are continuously being made during microbial growth.

The exact way in which microbially fixed nitrogen is given to the growing plant is not well understood. A small amount of the nitrogen fixed by bacterial cells may be released as cellular excretions during growth, but most of it is liberated only after the bacterial cells die and decompose.

Much of the nitrogen fixation in nature is accomplished by procaryotes that exist in a *mutualistic* partnership with plants. This type of nitrogen fixation is called **symbiotic,** whereas nitrogen fixation by free-living procaryotes is called **nonsymbiotic.**

Symbiotic nitrogen fixation is thought to be the more important because the symbiotic process accounts for most of the fixed nitrogen. For example, bacteria called **rhizobia** (from the genus *Rhizobium*) live and grow in the root tissue of clover and, in so doing, form growths on the roots called **nodules.** It is estimated that these rhizobia, living symbiotically with clover-root cells, fix about 220 pounds of nitrogen per acre per year, whereas the most efficient free-living procaryotes are capable of fixing only about 22 pounds per acre per year. Although the rhizobia associated with clover and other leguminous crops are the best-known symbiotic nitrogen-fixing bacteria, other types of bacteria appear to live symbiotically with tree roots and fix enough nitrogen to significantly affect forest productivity.

Research on nitrogen fixation has greatly increased in the last few years due to circumstances such as the mushrooming human population and the world food shortage, but, most of all, due to the rising costs of synthetic nitrogen fertilizer. Stimulated by technological advances in genetic manipulation (genetic engineering) of plants and bacteria, great strides are now being made in our understanding of the metabolism and genetics of nitrogen fixation. Who knows? In the future, humans may be able to cultivate nonleguminous plants in nitrogen-poor soils by inoculating the soils or plants with bacteria that have been engineered to form symbiotic relationships with the plants.

Investigators use the **acetylene reduction assay** to detect the enzyme nitrogenase in microbial cells and to measure the cells' ability to fix dinitrogen. This assay works because the nitrogenase is not substrate specific; it reduces compounds such as acetylene (HC≡CH) as well as dinitrogen, cyanide, and several others. When acetylene is reduced to ethylene ($H_2C=CH_2$) by nitrogenase, the quantity of ethylene is then measured with a *gas chromatograph*. There-

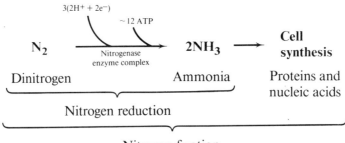

FIGURE 55-1 Overall representation of nitrogen fixation by bacteria. Nitrogen is reduced by an enzyme complex called nitrogenase, which catalyzes a series of reactions that convert dinitrogen to ammonia. Note that ammonia is never released but is immediately used by the bacterial cell to make nitrogen-containing building blocks (such as amino acids, purines, and pyrimidines) for making the proteins and nucleic acids that the bacterial cell needs. For an overview of how nitrogen fixation fits into the nitrogen cycle, see Figure 54-1.

fore, the acetylene reduction assay measures the activity of nitrogenase in microbial cells and, indirectly, the cells' ability to fix nitrogen.

In this exercise, you will not measure nitrogen fixation in this way because few introductory microbiology classes are equipped with gas chromatographs. Instead, you will examine root-nodule tissue for the presence of *Rhizobium,* which are nitrogen-fixing bacteria that symbiotically associate with the root cells of leguminous plants.

Symbiotic nitrogen fixation

Symbiotic nitrogen fixation is the phenomenon whereby bacteria of the genus *Rhizobium* grow together with the root tissue of a type of plant called a legume. **Legumes** are herbaceous and woody plants, such as peas, beans, clover, lupine, locust, and mimosa, that produce seeds in pods. The relationship between these bacteria and a plant's root tissue is called a *symbiosis,* which means merely that the bac-

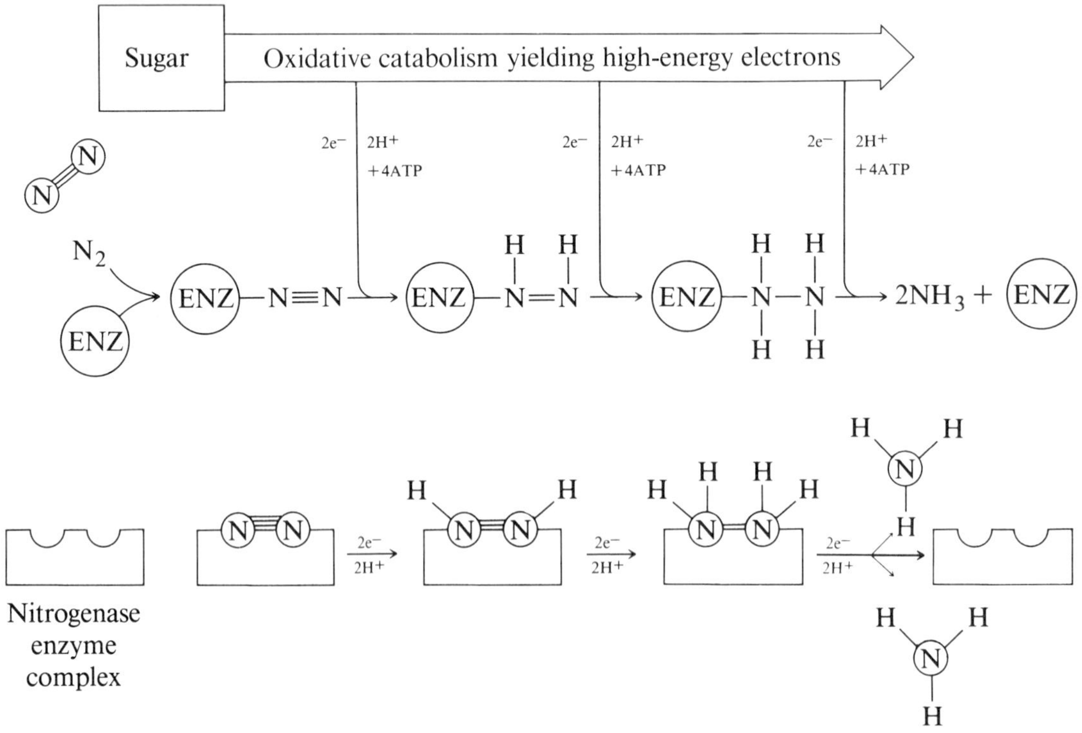

FIGURE 55-2 Proposed mechanism of nitrogen reduction to ammonia in *Clostridium*. Oxidative catabolism (such as the metabolic breakdown of sugars) provides the high-energy electrons (and the protons) used by the enzyme nitrogenase (ENZ) to reduce dinitrogen (N≡N). It presently appears that the reduction occurs in three sequential steps. The ammonia formed seems to be immediately incorporated into the microbial cell during synthesis of organic compounds.

teria and the plant cells live together. But since it is believed that both organisms benefit from this association, their form of symbiosis is called *mutualism*. The plant's root cells provide nutrients for the bacteria, and the bacteria fix atmospheric N_2, thus providing fixed nitrogen to the plant.

The type of bacteria that live symbiotically with the roots of legumes are collectively called rhizobia, and all species are placed in the genus *Rhizobium*. All *Rhizobium* species are gram-negative, non-spore-forming, aerobic rods that contain the enzyme complex called nitrogenase and that are typically motile. Their cellular morphology and biochemical characteristics are very similar to those of the nonsymbiotic nitrogen-fixing bacteria called *Azotobacter*.

The primary distinguishing characteristic of *Rhizobium* species seems to be their ability to nodulate leguminous plants. The rhizobia appear to be host specific; in other words, one species of *Rhizobium* seems able to nodulate only one species of legume.

Although *Rhizobium* species are able to grow on artificial media, only a few of them can fix very much nitrogen outside the root nodules. It appears that the nodule provides an environment that favors nitrogen fixation.

When grown on artificial media, cells from *Rhizobium* cultures are rod-shaped; but in a root nodule, they often have shapes like clubs or pears, or even branched like the letter *Y*. Pure cultures of *other* types of bacteria that exhibit various shapes are said to be **pleomorphic**. Such *Rhizobium* cells, however, are called **bacteroids** (bacterium-like).

It is common practice before planting the seeds of leguminous plants to inoculate with rhizobia because not all soils contain the right species of *Rhizobium* for optimum symbiosis with a given legume.

The first step in **nodulation** appears to be the excretion by the root of products that *attract* rhizobia, which then *aggregate* on root hairs at distinct sites where there is specific recognition between the plant and bacterial cells. It then appears that there is *binding* between the polysaccharide capsule around the bacterium and the surface of the root hair. Next, the rhizobia *invade* the tissue of the root hair, some of which seem to turn inside out (forming what is called an *infection thread*). Thus, instead of protruding outward, the root hair (infection thread) grows back into the root. At this point, the rhizobia in the infection thread begin to produce *auxins* (plant growth-regulating hormones) that stimulate cell growth around the infection thread, and these rapidly growing cells form the *nodule*. Finally, the infection thread branches into the developing nodule, the bacteria continue to multiply, and, eventually, the bacteria are found in the surrounding root cells.

Sections cut from a mature nodule show rhizobia surrounded by a membrane (presumably of plant origin). Often, four to six rhizobia (bacteroids) are found within a single membrane.

Remember that most rhizobia are *strict aerobes* and contain the enzyme complex called *nitrogenase;* therefore, they are best able to fix atmospheric nitrogen in well-aerated soils. The exact mechanism by which the rhizobia in the root nodules transmit fixed nitrogen to the plant is not fully understood; however, it is easy to show that **plant productivity** increases when the rhizobia are present. For example, when legume seeds are planted in sterile sand and treated with all nutrients essential for plant growth except nitrogen, they show very little growth and no nodule formation. But if the sand is inoculated with the correct species of *Rhizobium,* the plants show luxuriant growth and well-nodulated root systems.

LEARNING OBJECTIVES

• Become familiar with the concept of mutualistic symbiosis and the nature of rhizobia.

• Observe the presence of root nodules on leguminous plants.

• Observe the presence of *Rhizobium* bacteroids in root-nodule tissue.

MATERIALS

Cultures	*Rhizobium japonicum* (or a demonstration slide of a pure culture of this bacteium)
Media	None
Supplies	Legume roots with nodules Razor blades Ethanol (95 percent) and forceps Microscope slides Gram-staining reagents
Demonstrations	Phase-contrast microscope showing bacteroids in a wet-mount preparation of a squashed-root nodule How to ethanol incinerate forceps, microscope slides, and razor blades

PROCEDURE

1. **Cut off a nodule** from the root system with a clean ethanol-incinerated razor blade and a forceps.
2. **Place the nodule and one drop of water on a clean ethanol-incinerated microscope slide,** and put the slide on the laboratory bench.
3. **Clean and ethanol incinerate a second microscope slide, and crush the nodule between the two slides. Stir** the crushed mixture with a flamed inoculating loop.
4. **Pick out the chunks of uncrushed tissue** with an ethanol-incinerated forceps, and **smear the remaining cloudy fluid** on the slide with a flamed inoculating loop. Allow the smear to air dry.
5. **Heat-fix and Gram stain this smear; then label the slide.**
6. **Gram stain a smear of** *Rhizobium japonicum,* or examine a demonstration slide of a pure *R. japonicum* culture.
7. **Use the phase-contrast microscope to examine a demonstration slide** of a crushed root nodule. Make sure that you can differentiate the bacteria and the plant cells before you examine your own Gram-stained preparations.
8. **Examine your Gram-stained slide(s) with the oil-immersion objective.** Look especially for bacteroids in your crushed nodule preparation, and compare the shapes of these with the cell shapes found in the pure *Rhizobium* culture.
9. **Record your results and properly dispose of your slides and cultures.**

Early chemists describe the first dirt molecule

55 NITROGEN FIXATION — REDUCTION OF DINITROGEN

NAME _____

DATE _____

SECTION _____

RESULTS

Illustrate the observed cell shapes, give Gram-staining reactions, and describe what was typical of each preparation.

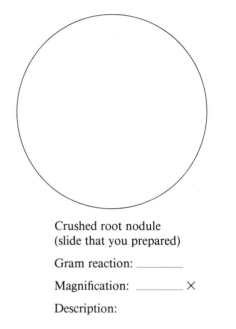

Crushed root nodule
(slide that you prepared)

Gram reaction: _____

Magnification: _____ ×

Description:

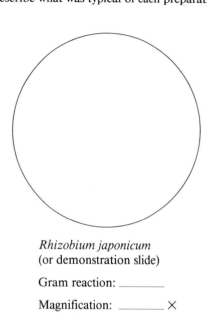

Rhizobium japonicum
(or demonstration slide)

Gram reaction: _____

Magnification: _____ ×

Description:

CONCLUSIONS

QUESTIONS

Completion

1. Nitrogen reduction refers to the chemical reduction of _____ to form _____, which is immediately used to form essential building blocks for cell synthesis. Nitrogen fixation refers to _____.

2. Two broad categories of bacteria that fix nitrogen are those called *mutualistic,* which form a mutually beneficial arrangement with the root cells of certain plants, and those called _____, which always seem

to grow without associating themselves closely with other living things. Of these two categories, the type that seems to fix the most nitrogen is _____.

3. Symbiotic nitrogen fixation is accomplished by a genus of bacteria called _____.

4. Bacteria growing inside leguminous root nodules often exhibit pleomorphic shapes unlike those found when the same cells are grown on artificial media. These forms are called _____.

5. Rhizobia grown in pure culture are gram-_____, their cellular shape is _____, and they are _____ with respect to their need or tolerance for oxygen.

True – False (correct all false statements)

1. The assay used to measure nitrogenase activity in cells is called a sillyene reduction assay, and the reduction product is measured with an instrument called the paper chromatograph. _____

2. As a genus, symbiotic nitrogen-fixing bacteria are strictly parasitic. _____

3. The attraction of nitrogen-fixing bacteria to leguminous roots is probably chemotactic, with the motile bacteria responding to plant excretion products. _____

4. Rhizobia excrete toxins that stimulate plant-cell growth forming the nodule. _____

5. Legumes are defined as woody plants that produce seeds within pods. _____

6. Mutualism is defined as a condition in which two dissimilar organisms live together in close association, both deriving benefit from this association. _____

7. Nitrogen fixation is an energy-requiring reduction of N_2 that occurs in bacteria that have an enzyme called nitrate reductase. _____

Multiple Choice (circle all that apply)

1. Species from each of the following procaryotic genera contain the enzyme nitrogenase. Which of them are capable of symbiotic nitrogen fixation?
   a. *Enterobacter*
   b. *Clostridium*
   c. *Azotobacter*
   d. *Spirillum*
   e. *Bacillus*
   f. *Rhodopseudomonas*
   g. *Chromatium*
   h. *Rhizobium*
   i. *Klebsiella*
   j. *Anabaena*

2. Which of the following plants would you expect to harbor *Rhizobium* species in nodules on their roots if soil and environmental conditions are suitable for nodule development?
   a. Soybeans
   b. Sweet peas
   c. Crown vetch
   d. Black locust
   e. Red Clover
   f. Potatoes

56

Ammonification — Microbial Deamination of Nitrogenous Organic Compounds

Plant growth requires more nitrogen than any other nutrient, but plants assimilate nitrogen only as inorganic nitrate or ammonium ions. However, most nitrogen in soil is bound to organic molecules from decayed plant or animal tissue or dead microorganisms. The release of this organically bound nitrogen as inorganic ammonium ions, called **ammonification** or **deamination,** is the consequence of microbial activity.

The chemistry of organically bound nitrogen in soils is not well understood. However, *amino acids* (resulting from plant and animal protein degradation) make up 20 to 50 percent of the bound nitrogen in soil, *amino sugars* such as glucosamine and galactosamine (resulting from bacterial cell-wall peptidoglycan degradation) account for 5 to 10 percent, and about 1 percent of the bound nitrogen is from *purine* and *pyrimidine bases* (resulting from nucleic acid breakdown). The chemical identity of the remaining bound nitrogen is unknown.

Nor is the microbiology of ammonification in soils well understood. Population estimates on many different types of soils reveal that about 10^5 to 10^7 microbes per gram of soil are capable of ammonification. The breakdown of proteins and other nitrogen-containing organic compounds in the soil probably represents the biochemical activity of many microbial types, each of which accomplishes only part of the degradation.

The initial steps of **protein** degradation are carried out by *extracellular proteolytic enzymes* excreted by microorganisms such as fungi, typical aerobic bacteria, actinomycetes, certain facultatives, and strict anaerobes. These excreted extracellular enzymes break down the proteins in dead animal and plant tissue into smaller molecular weight *polypeptides* and *amino acids,* which are then transported into the cells where they are broken down further and used as sources of energy and as intermediates for biosynthesis (Figures 56-1 and 56-2). Ammonification (the soil microbiologist's term for the overall process), or deamination (the biochemist's term for the enzymatically catalyzed reaction), is one of the initial steps in intracellular amino acid degradation. Although the mechanism varies (see Figure 56-2), deamination cleaves *amine groups* from amino acids, thereby releasing *ammonia.* Some ammonia is used by the bacterial cell for synthesis of its own proteins and nucleic acids, but an excessive quantity is almost always formed. This excess ammonia is excreted from the bacterial cell (see Figure 56-1), then most of the excreted ammonia is used as an energy source by the nitrifying bacteria in the soil (see Figure 54-1).

Another source of organically bound nitrogen in the soil are the **nucleic acids,** (DNA and RNA), from dead organisms. These compounds are also degraded to release nitrogen as ammonia (Figure 56-3). Many soil microorganisms produce *extracellular nucleases,* enzymes that break down nucleic acids. For example, extracellular *ribonucleases* (which attack only RNA) are formed by some bacteria (such as species of *Bacillus, Pseudomonas,* and *Mycobacterium*) and by some fungi (such as species of *Aspergillis, Cephalosporium, Fusarium, Mucor, Penicillium,* and *Rhizopus*). Extracellular *deoxyribonucleases* (which attack only DNA) are formed by certain bacteria (such as *Arthobacter, Bacillus, Clostridium,* and *Pseudomonas* species) and some fungi (such as *Cladosporium* and *Fusarium* species). The *mononucleotides* from RNA and DNA hydrolysis are then transported into the cells; inside, the phosphate and sugar are

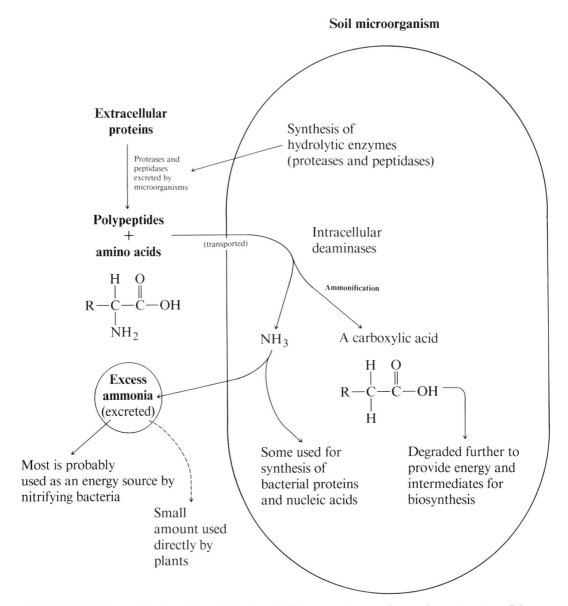

FIGURE 56-1 Ammonification of proteins in the soil. Microorganisms produce and excrete extracellular enzymes (proteases and peptidases) that hydrolyze proteins into peptides and amino acids; these smaller compounds are transported into the cell where they are used as sources of energy and carbon for synthesis. Intracellular enzymes deaminate the amino acids and break down (oxidize) the remainder of the molecule. Some ammonia is used intracellularly for synthesis of the cell's own proteins and nucleic acids; the remainder of the ammonia is excreted from the cells and most of that is absorbed by the soil moisture.

cleaved off the molecule, and then the purine or pyrimidine rings are broken and degraded to ammonia and other compounds (see Figure 56-3).

Regardless of the type of organically bound nitrogen degraded, the physical and chemical characteristics of soils (such as temperature, moisture, pH, aeration, and nutrient supply) control the rate of microbial ammonification and the types of microbes involved. For example, bacterial species of *Pseudomonas, Bacillus, Clostridium, Serratia,* and *Micrococcus* appear to contribute the most to ammonification in neutral to alkaline soils, whereas fungi seem predominately responsible for ammonification in acid soils. Actinomycetes also play a part, but their metabolism and growth in soil is so slow that their role in ammonification is presumed to be small. Since soil contains aerobic and anaerobic, acid-sensitive and non-acid-sensitive, spore-forming and non-spore-forming microorganisms that are capable of ammonification, it seems reasonable that at least some segment of the microbial population is active regardless of the changes that might occur.

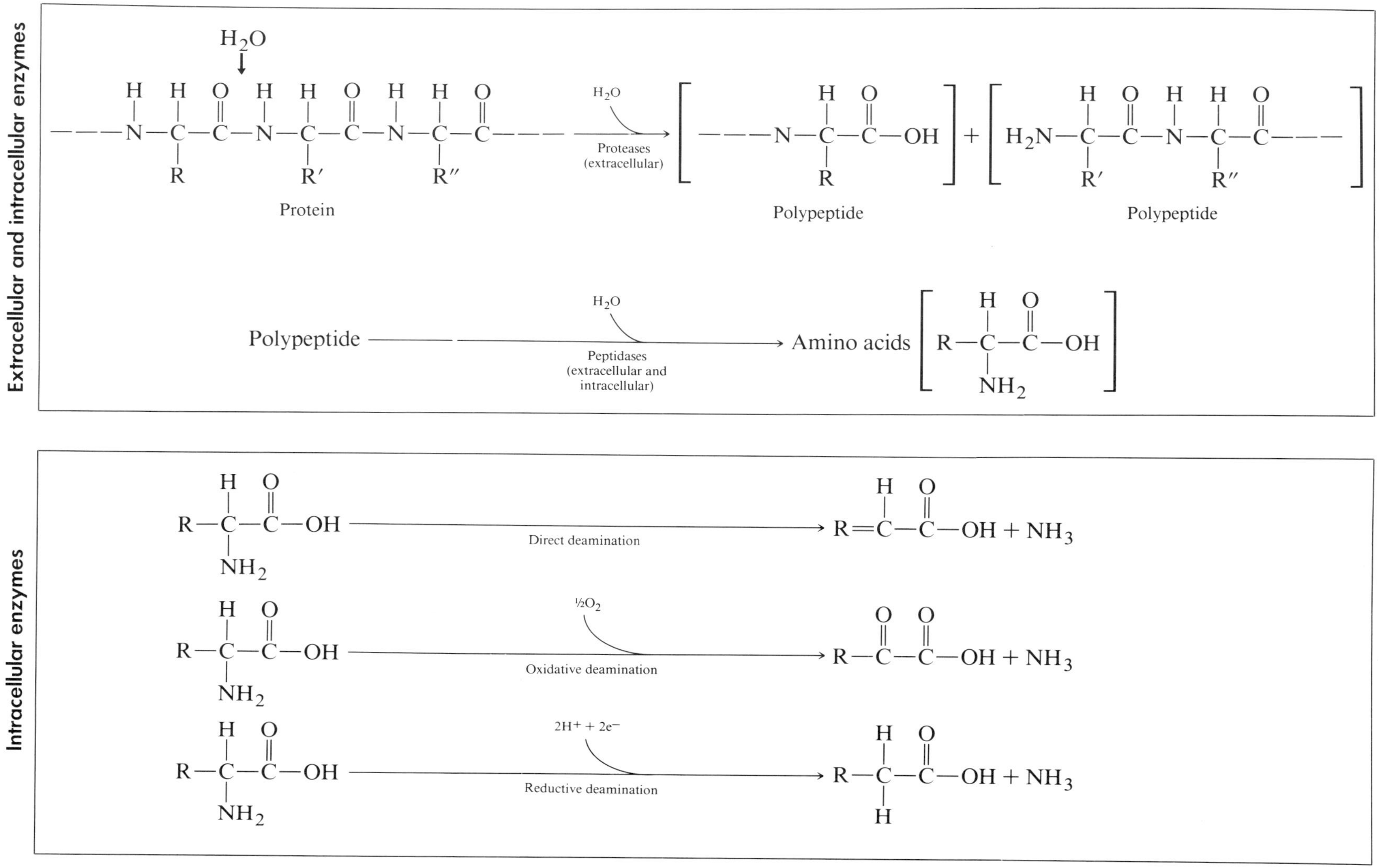

FIGURE 56-2 Enzyme-catalyzed reactions for protein breakdown by soil microorganisms. Proteins are hydrolyzed to smaller molecular weight polypeptides and amino acids by extracellular enzymes called proteases and peptidases. The resulting amino acids are then taken into the cell where they are deaminated and oxidized further by intracellular enzymes. Excess ammonia (that not needed for microbial growth) is excreted from the cell into the surrounding soil.

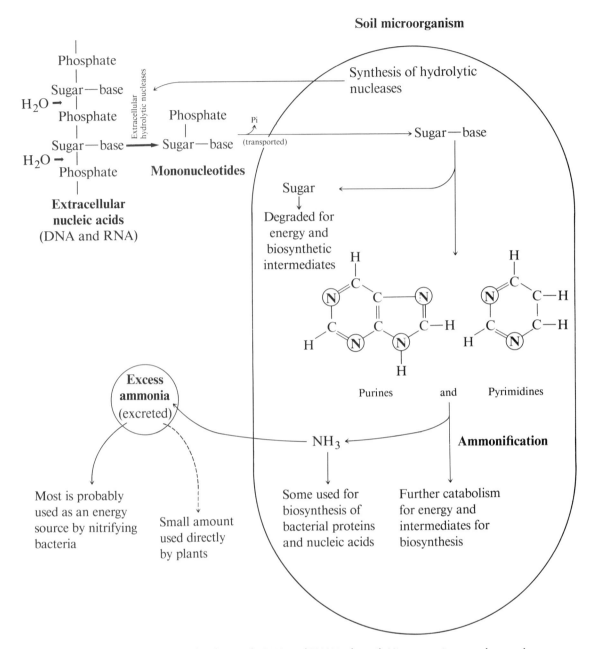

FIGURE 56-3 Ammonification of nucleic acids (DNA and RNA) in the soil. Microorganisms produce and excrete extracellular hydrolytic nucleases that break nucleic acids into mononucleotides that are transported into the cell where they are used as sources of energy and carbon for synthesis. Intracellular enzymes split the sugar from the purine or pyrimidine base and then hydrolytically cleave the ring structures. One of the products of this cleavage is ammonia, some of which is used for new cell synthesis; the excess is excreted from the cell.

In this exercise, you will evaluate the ability of microbes in the soil and also several pure cultures to deaminate amino acids and excrete ammonia. You will inoculate both soil and pure cultures into 4 percent *peptone,* which is the dried, water-soluble product of enzymatic (pancreatic enzymes) digestion of meat. Peptone contains everything from polypeptides of various sizes to amino acids, but no sugars; thus, microbes that grow on peptone *must* use amino acids as their only source of carbon and energy. Therefore, 4 percent peptone is a good medium to test for a cell's ability to transport and deaminate amino acids and excrete ammonia.

Because of the extremely solubility of ammonia in water (90 grams per 100 milliliters of cold water) the ammonia is normally trapped by the moisture in soil after it is excreted by soil microorganisms. Most of that excreted by microbes growing in liquid media

56 AMMONIFICATION—MICROBIAL DEAMINATION

remains within that liquid. Microbiologists test for ammonification by looking for dissolved ammonia in liquid culture media with a solution called **Nessler's reagent,** which contains potassium iodide (KI), mercuric iodide (HgI_2), and sodium hydroxide (NaOH). The presence of dissolved ammonia is indicated by a dark yellow, orange, or brown color formed when Nessler's reagent is added to the culture medium. The intensity of the color is directly proportional to the concentration of ammonia dissolved in the culture medium. The chemistry of these colored materials is not known, but they are believed to be various salts that are soluble in alkali, such as aquabasic mercuric iodide (OHg_2NH_2I), diamine mercuric iodide [$Hg_2(NH_3)_2I$], and ammonium triiodide (NH_4I_3).

In this exercise, you will test for the ability of certain pure cultures or unknown soil microorganisms to deaminate the amino acids in peptone broth.

LEARNING OBJECTIVES

- Know the four classes of organic compounds that give rise to ammonia during microbial ammonification in the soil and how these compounds get into the soil.
- Understand how microbial ammonification fits into the nitrogen cycle.
- Know how to test for microbial ammonification.

MATERIALS

Cultures *Bacillus cereus*
          Pseudomonas fluorescens
          Garden (greenhouse) soil, high in organic matter

Media   Sterile 4 percent peptone broth in tubes

Supplies Multiple-depression ceramic spot plate
          Small spatula
          Nessler's reagent
          Pasteur pipettes (nonsterile) and rubber pipetting bulbs

PROCEDURE: FIRST LABORATORY PERIOD

1. **Separately inoculate three peptone-broth tubes** with the following cultures or materials:
   a. *Bacillus cereus* (use aseptic technique)
   b. *Pseudomonas fluorescens* (use aseptic technique)
   c. Enough garden soil to barely cover the tip of a small spatula (aseptic technique not necessary)
2. **Appropriately label each tube.**
3. **Ask your laboratory instructor to set aside several uninoculated tubes of 4 percent peptone broth for use as controls** during the second laboratory period.
4. **Incubate at 30°C for 48 hours** or longer.

PROCEDURE: SECOND LABORATORY PERIOD

Test for ammonification (deamination) in the following way.

1. **Label three depressions** on the ceramic spot plate to correspond to each of the three peptone-broth tubes inoculated during the first laboratory period. See Results section for a suggested number code.
2. **Label a fourth depression** as the *uninoculated control.* Read the safety note before you proceed.

SAFTEY NOTE

Nessler's reagent contains mercury; therefore it is toxic to humans. **Never mouth-pipette this reagent!** Avoid spills! Discard extra reagent as directed by your instructor.

---

3. **Transfer two drops of Nessler's reagent** to each of the four labeled but empty depressions in the ceramic spot plate using a clean Pasteur pipette and rubber bulb.

> **TECHNIQUE NOTE**
>
> The Nessler's reagent should be the first thing transferred to the clean and empty depressions in the spot plate. If the cultures were added first and the pipette touched one positive sample in a depression, then the pipette would be contaminated with ammonia; and all other samples would be cross-contaminated by the ammonia carried over by the pipette. This would give you invalid results. The Nessler's reagent might also become contaminated with ammonia from one of the cultures, and if some of that contaminated reagent were placed back into the reagent bottle, the contaminated reagent would give invalid results for those who follow you.

4. **Transfer two drops from one culture to the appropriately labeled depression** in the ceramic spot plate using a clean Pasteur pipette and bulb. After you transfer two drops of culture to the spot plate, place the remaining culture fluid in the disinfectant jar. Your laboratory instructor should tell you how to dispose of the Pasteur pipette.
5. **Separately transfer two drops of each of the other two cultures to the appropriately labeled depressions** in the spot plate using separate Pasteur pipettes. Dispose of extra culture fluids and the used pipettes as instructed.
6. **Transfer two drops of sterile peptone broth to the fourth depression,** labeled *uninoculated control* in the same manner.

> **INTERPRETATION NOTES**
>
> The presence of ammonia in a solution reacting with Nessler's reagent is shown by the development of a yellow, orange, or brown color. The intensity of the color is greatest when the concentrations of dissolved ammonia are highest. The reaction should occur instantaneously.

7. **Record your results.**
8. **Clean the spot plate** as directed by your laboratory instructor.
9. **Clean and discard your tubes** in the proper manner.

56 AMMONIFICATION—MICROBIAL DEAMINATION

NAME _____

DATE _____

SECTION _____

RESULTS

Detection of dissolved ammonia in cultures using Nessler's reagent:

Depression number	Culture inoculum	Color and intensity
1	Uninoculated broth	
2	*Bacillus cereus*	
3	*Pseudomonas fluorescens*	
4	Soil sample	

CONCLUSIONS

Note especially which inoculum showed the greatest amount of deamination.

QUESTIONS

Completion

1. The microbial process of breaking down organic, nitrogen-containing polymers and excreting ammonia into the soil may be called either _____ or _____.

2. The three cellular polymers most commonly broken down to yield ammonia are _____, _____, and _____.

3. One should expect to find, per gram of an average soil sample, about _____ (how many) microorganisms that are capable of ammonification.

4. The one word that describes extracellular enzymes that hydrolyze proteins is _____, and the comparable term for those extracellular enzymes that hydrolyze nucleic acids is _____.

True – False (correct all false statements)

1. Ammonia excreted from microorganisms growing in broth either bubbles to the surface or collects in an inverted Durham tube. _____

2. Nessler's reagent is made from salts that contain mercury and iodine. _____

3. Peptone contains hydrolyzed nucleic acids whose amino acids are deaminated by microbial intracellular enzymes. _____

4. Ammonia released to the soil during ammonification is used by the chemolithotrophic nitrifiers, which use ammonia as a source of energy for growth. _____

Questions about Procedures

1. Why was it necessary to use aseptic technique to transfer the pure cultures into the peptone broth when another sterile broth tube was inoculated nonaseptically with soil?

2. Why might soil cause deamination of the tryptone broth?

57

Denitrification — Complete Reduction of Nitrate and Nitrite to Dinitrogen

To be used as a nutrient for plants, nitrogen must be available in the soil as the *nitrate* ion (NO_3^-). Other sources of nitrogen can be used indirectly by plants if they can first be converted to nitrates by the soil bacteria (see Figure 54-1).

Other types of bacteria and other microorganisms in the soil compete with plants for the available nitrates because those microbes can take up and *use* nitrates. Therefore, it is important to understand which bacteria deplete soil nitrates and what can be done to avoid that loss.

Microorganisms chemically reduce nitrates by one of two methods (Figure 57-1). In **assimilative** nitrate reduction, the microorganisms reduce nitrate first to nitrite and then eventually to ammonia (NH_3). However, the ammonia is immediately used by the cell for making amino acids and other nitrogen-containing organic building blocks; that is, the reduced nitrogen is immediately incorporated (assimilated) into new cell material as part of the biosynthetic metabolism of these microorganisms.

In **dissimilative** nitrate reduction (see Figure 57-1), some microorganisms reduce nitrate to nitrite, but that microbe or other types may reduce the nitrite to nitrous oxide and dinitrogen (N_2). This, you may now recognize, is the process we earlier called *denitrification:* a sequence of catabolic steps that convert nitrate to the gases nitrous oxide (N_2O) and dinitrogen (N_2). The final product of denitrification is excreted and escapes to the air, resulting in a net loss of nitrates from the soil. This microbial process is undesirable if you are dependent on cultivating nonleguminous plants as a source of food.

If the gaseous products of dissimilative nitrate reduction are excreted from the microbial cell, what is the function of this process for the microbe? Dissimilative nitrate reduction (denitrification) is part of a metabolic process called **anaerobic respiration** (see Exercise 33), in which facultatively anaerobic microbes use nitrate as the terminal electron acceptor at the end of their electron-transport chain when they run out of oxygen and when nitrate is present. Although the exact biochemical mechanisms are not well understood, it appears that the pathway used for dissimilative nitrate reduction by some microbes is shown in Figure 57-2. Current evidence shows that this mechanism supports oxidative phosphorylation in a manner similar to aerobic respiration, although less ATP is formed with nitrate than with oxygen. Note that cells using this pathway employ cytochromes and other electron-transport proteins. Therefore, dissimilative nitrate reduction (denitrification) should be considered one form of respiration.

Several environmental factors influence denitrification, such as temperature, quantity of moisture, pH, amount and type of organic matter, and the amount of oxygen. Oxygen is especially interesting because of the way it affects both the nitrifying and the denitrifying bacteria. If oxygen (and ammonia) are present in the soil, the nitrifying bacteria will oxidize ammonia and excrete nitrate into the soil (see Figure 54-1). The presence of oxygen will also allow those bacteria capable of denitrification to respire using oxygen instead of nitrate as a terminal electron acceptor. Thus, nitrates are formed and not depleted from the soil in the presence of oxygen. One major reason why farmers cultivate the soil around desirable plants is to encourage microbial nitrification and discourage microbial denitrification so that more nitrate is available for plant productivity.

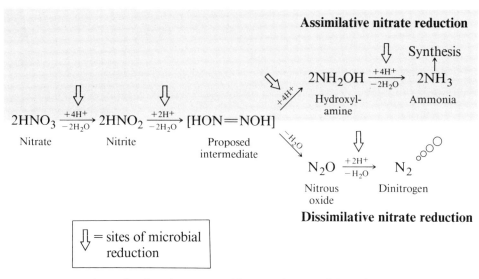

FIGURE 57-1 Generalized mechanisms used for assimilative and dissimilative nitrate reduction.

Denitrification is also related to the moisture content of the soil, being greatest in soils in which the water content is highest because water reduces the diffusion of oxygen to the soil microorganisms. Therefore, denitrification is minimized in well-drained soils having textures that encourage oxygen flow. In soils subjected to cyclic availabilities of oxygen, there are also cyclic fluctuations in nitrate concentrations. For example, in uncultivated soils that are periodically flooded (such as tidal areas), there are cyclic availabilities of oxygen and cyclic nitrification and denitrification.

Nitrates are often artificially added to agricultural soils to improve plant productivity, but, if the nitrates get out of place, they can be a damaging form of environmental pollution. For example, affluent farmers in technologically developed countries add nitrates to their soil to encourage growth of nonleguminous plants. Heavy rains can wash these nitrates from the fields into streams and rivers, where this nutrient encourages too much aquatic plant growth, upsetting the natural oxygen balance and killing other forms of aquatic life in the waterways. This result can be avoided by proper cultivation techniques and also

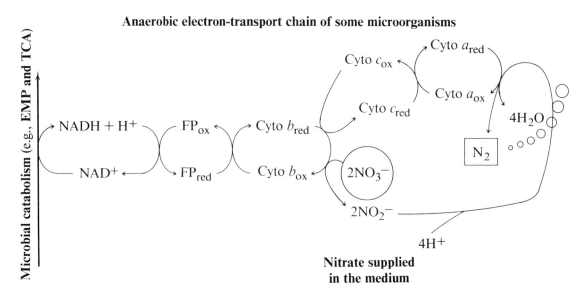

FIGURE 57-2 The mechanism by which dissimilative nitrate reduction is thought to enter into the electron-transport chain of denitrifying bacteria. Note that oxygen (not nitrate) is the *preferred* terminal electron acceptor for all of these bacteria.

57 DENITRIFICATION—PRODUCTION OF DINITROGEN

by creating holding ponds to capture runoff and allow the denitrifying bacteria in the pond to convert nitrates into dinitrogen.

Technology has created other potential forms of nitrate pollution. For example, many industrial processes require great quantities of nitric acid. The acid is easily neutralized by adding lime ($CaCO_3$), but the nitrate ions remain to pollute both ground and surface waters. To prevent this, industries pump nitrate-containing solutions into large tanks, add denitrifying bacteria, and let these microbes produce N_2 from the nitrates. The N_2 escapes into the atmosphere where it adds insignificant amounts to the 80 percent N_2 that already exists in the atmosphere. This illustrates that microbial denitrification can be put to good use once we understand how and why it occurs.

In this exercise, you will test the ability of pure cultures and microbes in soil to reduce nitrate completely to dinitrogen. You will not chemically test for the presence of N_2; instead, you will look for the accumulation of large quantities of gas when cultures are grown in the presence but not the absence of nitrate. All cultures will be grown under conditions where oxygen is prevented from entering the medium to encourage cultures to respire using nitrate, instead of oxygen, as a terminal electron acceptor.

LEARNING OBJECTIVES

- Understand the value of denitrification for our environment and the role that microbial denitrification plays in agriculture and industry.

- Know the metabolic role of nitrate and nitrite for denitrifying microorganisms.

- Know how to test for microbial denitrification.

MATERIALS

Cultures *Pseudomonas stutzeri*
Staphylococcus aureus
Moist garden (greenhouse) soil, rich in organic matter

Media Semisolid nitrate medium (0.1 percent agar)
Semisolid nonnitrate medium (same as above but without nitrate)
Agar overlay tubes
Sterile 1.5 percent agar (2 ml per tube for overlays)

Supplies Waterbath set at 45°C
Ice buckets containing ice

PROCEDURE: FIRST LABORATORY PERIOD

The semisolid media used in this exercise contain 0.1 percent agar rather than the 1.5 to 2.0 percent agar used to make agar plates, deeps, and slants. You will note that these semisolid media are more viscous than a broth but less solid than an agar deep at room temperature.

1. **Separately inoculate three tubes of semisolid nitrate medium** with the following cultures or materials:
   a. *Pseudomonas stutzeri* (use aseptic technique to inoculate; then mix well with the inoculating loop).
   b. *Staphylococcus aureus* (use aseptic technique to inoculate; then mix well with the inoculating loop).
   c. Enough garden soil to barely cover the tip of a small spatula. (Aseptic technique is not necessary; however, after inoculation the medium should be mixed with a sterile inoculating loop.) Ask your instructor to demonstrate how much soil should be used as an inoculum.

2. **Inoculate one tube of semisolid nonnitrate medium with** *Pseudomonas stutzeri*. You now have two tubes inoculated with this organism: One contains large quantities of nitrate and the other does not.

3. **Label each tube as soon as you inoculate it.**

4. **Place inoculated tubes in an ice bath** for 10 to 15 minutes or until the medium is completely chilled. Cold makes the semisolid medium firmer so that the overlay agar will stay on top.

5. **Aseptically pour the entire contents of one agar-overlay tube into each** inoculated tube of chilled medium. This should be poured so that it *gently* runs down the side of the tube and slowly layers over the top of the nitrate medium. Ask your instructor to demonstrate this technique.

6. **Set each tube in a rack until the overlay agar has solidified.** The purpose of the overlay is to form a seal over the medium to trap bubbles of nitrogen so that it can be observed after incubation.

7. **Ask your instructor to set aside several uninoculated tubes of overlayed nitrate medium** for you to observe as uninoculated controls during the second laboratory period of this exercise. **These should be prepared by adding agar overlays today.**

8. **Incubate all inoculated and uninoculated tubes at 30°C for 48 hours** or longer.

PROCEDURE: SECOND LABORATORY PERIOD

1. **Test for complete denitrification** by observing each tube for the presence of nitrogen bubbles trapped between the nitrate medium and the solid agar overlay. All four inoculated tubes and an uninoculated control should be examined. Don't forget to check for growth.

2. **Record your results.**

3. **Clean and discard your tubes** in the proper manner.

INTERPRETATION NOTES

The presence of gas in nitrate-containing media but not in media lacking nitrate is evidence for complete reduction of nitrate (NO_2^-) to N_2, and this is all we are testing for here. Many bacteria reduce nitrate or nitrite only to nitrous oxides but all the way to dinitrogen. Although the nitrous oxides are gases, they are also *very soluble* in aqueous solutions. Therefore, the presence of gas bubbles is indicative of N_2 formation and not nitrous oxide formation.

The agar plug has two main purposes in this procedure. First, it keeps atmospheric oxygen away from the growth medium (when strictly respiring microbes grow in the medium under the agar plug, they quickly use up the dissolved oxygen and no more oxygen can enter). The second function of the agar plug is to trap gases evolved from the microbes growing under the plug. Once the oxygen is depleted, respiring microbes (that have this ability) will be forced to substitute nitrate for oxygen as their electron acceptor at the end of their electron-transport chain. Those able to reduce nitrate completely to dinitrogen will release this gas, which will collect under the agar plug.

57 DENITRIFICATION—PRODUCTION OF DINITROGEN

NAME _____

DATE _____

SECTION _____

RESULTS

The following table contains test results for complete denitrification of nitrate in cultures incubated for _____ hours at 30°C.

Microorganism	Nitrate	Growth (+ or −)	Gas trapped in tubes (+ or −)	Approximate volume (ml)
Pseudomonas stutzeri	Present			
Pseudomonas stutzeri	Absent			
Staphylococcus aureus	Present			
Soil sample	Present			
Uninoculated broth	Present			

CONCLUSIONS

QUESTIONS

Completion

1. The nitrate ion is chemically written as _____. Denitrification is the stepwise _____ of nitrate to form either _____ or _____.

2. Reduction of nitrate to ammonia is called _____, whereas reduction of nitrate to N_2 is called _____.

3. In the metabolism of bacteria that reduce nitrate to N_2, the nitrate serves as a _____, and this entire process is known as anaerobic _____. Therefore, nitrate serves a similar function to _____ in aerobic respiration.

4. Denitrification occurs best when oxygen is _____.

5. Nitrates are used by plants as a nutritional source of _____.

True – False (correct all false statements)

1. Plowing and harrowing will mix oxygen into the soil; therefore, they favor nitrification (nitrate formation) over denitrification (nitrate removal). _____

2. Respiration only takes place in the presence of oxygen. _____

3. The cultivation of denitrifying microbes is not recommended. _____

4. Nitrates can be oxidized to nitrites, nitrous oxides, ammonia, and dinitrogen. _____

5. The gas produced during denitrification is an example of a fermentation product. _____

Questions about the Experiment

1. *Pseudomonas stutzeri* is capable only of respiration. Based upon this fact, how should you explain the results you obtained with this organism under anaerobic conditions?

2. Based upon the design of this experiment, how can you tell that the gas accumulating under the agar plug is the result of nitrate respiration and not of fermentation?

XIV

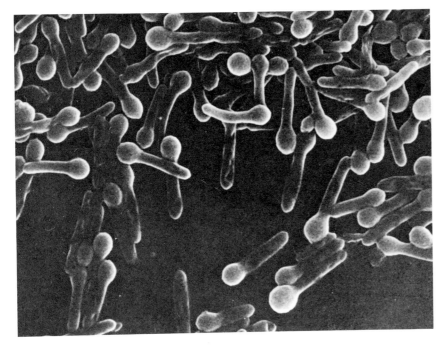

Clostridian tetani containing terminal endospores.

Les propriétés des bactéries sont variable.
— E. Brochu, et al.

How is that for an understatement?
— G. W. Claus

MICROBIAL IDENTIFICATION

The purpose of Part XIV is to allow you to test your skills and to pull together the information you have accumulated on the characteristics of microorganisms. You will need to isolate (purify) an organism; plan, run, and interpret diagnostic tests; and draw conclusions. If you have approached your studies seriously and kept good records. Exercise 58 will be fun.

58

Identification of Unknown Microorganisms

For your Final Unknown, you will receive a *mixture* of two or more microbes suspended in one broth tube. You will also be provided with a *table* that contains a list of possible microorganisms and their known (recently tested!) morphological and physiological characteristics. These characteristics are based on results obtained after incubation at 30°C for 48 hours unless otherwise noted. Use the list of microorganisms presented in this table to construct the *flow charts* mentioned in the following section.

THINGS TO DO BEFORE RECEIVING YOUR UNKNOWN

1. **Thoroughly study** this entire exercise before you begin work.
2. **Prepare flow charts for the following three types of microorganisms on separate sheets of paper:** yeast, gram-positive rod, and gram-positive coccus. Note that a sample identification flow chart for some gram-negative rods is given as Figure 58-1 for you to use as an example.

   The purpose of an identification flow chart is to show you what tests are needed to identify a microorganism *after* it is purified and you know the correct Gram reaction and cell morphology. It should be constructed like an inverted tree and contain all of the microbes of the type (such as yeasts) given in the table of organisms and characteristics supplied to you. Each step in the flow chart should show one (or more) test(s) that will subdivide this type of microbe into two smaller groups. This stepwise series of tests continues until all possible microbes are eliminated except one.

   If you know how to prepare an identification flow chart, you also understand the rationale in using microbial characteristics to identify unknown microorganisms. Use of a flow chart will help you minimize the number of tests necessary for identification; consequently, its use will save you time and will save us expense.

   You will use one or more of these flow charts to identify your unknowns. Your instructor will ask you to prepare and submit these flow charts prior to giving you unknowns to identify. You will be asked to return the instructor-initialed flow charts used to identify your unknown with your Report form. It is essential that you carefully plan your work to avoid wasting time and materials. **Run only those tests that will help you identify or confirm the identity of each unknown.** Points will be taken off for performing unnecessary tests.

   Each of your flow charts must be approved (and initialed) by your instructor before you are allowed to begin testing your two purified cultures.

SUGGESTIONS FOR ISOLATION

1. **Thoroughly mix your broth culture.** Before you proceed, make sure there is not a heavier concentration of cells at the bottom of the tube. Some of these microorganisms easily settle out of suspension. Failure to mix cul-

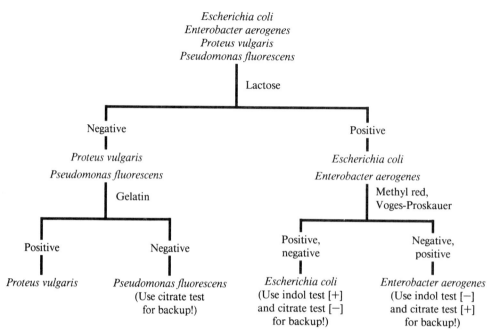

FIGURE 58-1 Typical planning flow chart for four gram-negative rods. Such a flow chart helps you easily plan your work and conserve media and reagents. The chart shown here gives the type and order of tests that may be used to separate only those bacteria given here. You should construct a separate but similar flow chart for each major group of microorganism from the list given to you, such as the gram-positive rods, the gram-positive cocci, the gram-negative cocci, and the yeasts. Your flow charts should contain only those tests that will allow you to clearly differentiate the microorganisms in each group. *Do not base an identification on all negative tests.*

tures prior to Gram staining or streaking plates is one of the most common causes of failure to see and isolate all unknowns in a microbial mixture.

2. **Prepare a Gram stain of your mixture. Fix with methanol;** do not heat fix this smear. You will be provided with a gram-positive and a gram-negative control culture to help you determine that you are doing your Gram stains correctly.

   Also prepare a wet mount for examination **with a phase-contrast microscope.** This, like the Gram-stained preparation, will help you determine whether yeasts are present in your mixture.
   a. If your Gram stain or wet mount shows that *yeasts are absent,* prepare only one streak plate using T-soy agar.
   b. If your Gram stain or wet mount shows that *yeasts are present,* separately streak one T-soy plate and one SAB plate.

CAUTIONARY NOTE

Colonies of most microbes do not look the same on T-soy and an Sabouraud's agar. Therefore, you must keep careful, written records about the appearance of each colony type and the Gram morphology of cells from that colony type. *Keep careful records of all that you do!*

3. **Refrigerate your broth culture** until you have successfully achieved separation of both microbes. If you need to use this culture again, mix the broth well to make sure that all types are well suspended.

CAUTIONARY NOTE

This mixture can be preserved for only a few days in the refrigerator, after which only one microbe may survive. Therefore, *make sure that you have separation within four days of receiving your culture.*

4. **Incubate streak plates at 30°C for 24 hours;** then carefully examine them. **If you see two well-isolated colony types** after 24 hours, refrigerate your plates until the next laboratory

period. **If you have achieved good colony separation but do not see two different colony types,** continue the 30° incubation for an additional 24 hours.

> **CAUTIONARY NOTE**
>
> Some cell types listed in your table of possible microbes (and correct diagnostic tests) may not yield visible colonies after 24 hours incubation. These microbes form very small colonies whose maximum size is achieved after 48 hours incubation.

5. **Prepare a written description of each colony type and then gram stain each type.** If you find different colony types and different cell shapes or different Gram-staining characteristics, you should be able to tell which cell characteristic belongs with which colony type. Your stains should also help you to determine whether these colonies are pure.

6. **Restreak each colony type** on an appropriate agar plate; use either T-soy agar for bacterial isolates or Sabouraud's (SAB) agar for yeast. *Do this to make sure that you have a pure culture.* If this second plate exhibits only one colony type (the same as previously described), and these colonies show only one type of cell morphology and/or Gram reaction (as previously described), then you can proceed with identifying your pure cultures.

> **CAUTIONARY NOTE**
>
> It is *imperative* that you correctly gram stain each culture and that you *work only with pure cultures.* Most of those who fail to correctly identify their unknown cultures either perform their Gram stain incorrectly or perform their diagnostic tests on a mixed culture.

7. **After you have isolated pure cultures** (one streak plate beyond initial separation), **transfer each culture to a separate agar slant.** Use Sabouraud's agar only if you isolated a yeast; otherwise use T-soy agar slants. **Incubate each slant at 30°C only until you begin to see growth** on the slant (usually no more than 24 hours); then **store each slant in the refrigerator.**

   Label each slant clearly so that you can easily retrieve it at a later time. Small pieces of tape containing your initials placed on the top of the cap will help you find *your* tubes when the rack is filled. You should use these slants as *stock cultures* and inoculate all subsequent test media from them.

> **HELPFUL HINT**
>
> After preparing and refrigerating your slant (stock) cultures, tape your final *streak plates* together and ask your instructor to place them in a plastic bag (to slow evaporation) and put them in cold storage for safekeeping. If you do this, you will be able to go back to these plates if your slant cultures are misplaced or accidently discarded.

INOCULATING PURE CULTURES INTO TEST MEDIA

1. **Use a very small inoculum** for each test medium unless otherwise noted below. Just *touch* the inoculating loop or needle to the growth on the slant, and transfer this to the sterile test medium. You should not be able to see the inoculum on your loop or needle. If you merely touch the needle (without picking up a visible quantity of cells), you will still transfer more that a million cells to the new medium. If you transfer too many cells, the test medium will look turbid immediately after inoculation, and you will not be able to tell if the cells grew and reacted properly in the test medium.

   TWO EXCEPTIONS: Use a large number of cells when you inoculate *coagulase plasma* and tubes containing *gelatin*.

2. **Perform all test procedures** exactly as described in the preceding exercises. Any exceptions on *incubation times* will be shown in the supplemental table of possible microbes and correct diagnostic test results. Be sure that you read the interpretation notes provided with your table.

SPECIAL INFORMATION

Laboratory hours: You are given _____ weeks (of class time) to identify your unknowns. The microbes were selected so that a well-prepared person can easily accomplish this in the allotted laboratory periods.

You may want to remove your cultures from the incubator and refrigerate them at times other than when your laboratory section is meeting. If so, do not disturb the other laboratory section. You may, however, place cultures in the refrigerator located in _____.

The laboratory and incubator will be open at special hours during the time you will be working on your unknowns (weekend and evenings). These special hours will be posted on the door to the teaching laboratory as soon as they are available.

Assistance: You are to receive *no assistance* from your fellow students, your laboratory instructor, the prep room staff, or your lecturer on *interpretation* of Gram stains, streak plates, or diagnostic tests. The notes you have made for each test during the academic term and the information provided in test summaries should be adequate for correct interpretation of your results. *Work independently; do not consult with others.*

If you have nontechnical questions, please ask only your own laboratory instructor. Do not ask other instructors unless they are there to monitor the teaching laboratory at times other than when their own class is in session. Do not consult with the prep room staff unless you have an emergency.

Reports due: Your completed Report form (with attached identification flow charts) is due on _____. So that there is no misunderstanding, check with your instructor and write the date on the preceding blank line.

58 IDENTIFICATION OF UNKNOWN MICROORGANISMS

NAME _____

DATE _____

SECTION _____

RESULTS — REPORT FORM

Your unknown number _____

Note that many (but not all) tests are suggested here. You should perform only those tests given on your flow chart (prepared after determining the size, shape, and Gram-staining characteristics for each of your unknowns). Use blank spaces to add other characteristics necessary for identification. Perform only those tests necessary to identify your unknowns!

Characteristics	Unknown #1	Unknown #2
Microorganism's complete name		
Gram stain (+ or −)		
Cell shape (rod, sphere, etc.)		
Colony morphology (size, color, edge, etc.)		
Glucose fermentation (A, G, or AG)		
Lactose fermentation (A, G, or AG)		
Indole formation (+ or −)		
Methyl red test (+ or −)		
Voges-Proskauer (+ or −)		
Gelatin hydrolysis (+ or −)		
Citrate utilization (+ or −)		
Catalate (+ or −)		
Coagulase test (+ or −)		

Staple your *two instructor-initialled flow charts* to the *back* of this form.

Please check here _____ *if* you have used the reverse side of this page for comments or close observations.

Tape your Gram-stained slide(s), below. Label the slide(s) with your name and clearly identify each smear.

XV

APPENDIXES

Media

NOTES FOR STUDENTS AND FACULTY

• All media used by students (and by preparation-room staff in preparing cultures) are listed alphabetically on the following pages.

• The *composition* of each medium is given here to satisfy the curiosity of the better student. When *pH* is given, it is the final pH of the medium (±0.2 pH units at 25°C) after sterilization without adjustment. Composition is usually given as grams of componant per liter of distilled and/or deionized water.

• Students should note that tap water is *not* used for preparing culture media. Use water that is deionized and/or distilled so that it is free of chlorine, copper, lead, and detergents, and so that it has a suitable conductivity and ion content as specified for *purified water*, by the *United States Pharmacopoeia* (U.S.P.), Vol. 20, 1980.

• Some media used to culture microorganisms can be purchased in a dehydrated, premixed form. The two largest manufacturers of premixed culture media and componants for preparing microbiological culture media are BBL Microbiology Systems (previously called Baltimore Biological Laboratories) and Difco Laboratories. If the entire medium (or part of it) is available in premixed form, that is indicated under *"Commercial availability."* If the name of the medium used in this text is listed by the manufacturer under a different name, then the manufacturer's name for the medium is also listed in *italics* in the "Commercial availability" section.

• *Instructions for preparing dehydrated, premixed media* are always provided by the manufacturer on the container; therefore these instructions are not repeated in this appendix.

• *Instructions for preparing all other media* used by the students (and by the preparation-room staff for preparing cultures) is included on the following pages.

• An antifoaming agent (*Pourite* obtained from American Scientific Products) is routinely added to all agar media used to pour plates. This reduces foaming and the need to flame bubbles on the surface of poured plates. We use one drop for volumes up to 800 ml, and one more drop for each additional 500 to 800 ml. Pourrite is also added to litmus milk medium (one drop per liter) to reduce foaming during preparation and dispensing.

• Unless otherwise specified, all liquid media are dispensed as 5 ml volumes into plain-lipped 15 × 125 mm (outer diameter) test tubes and covered with plastic caps (tradename, *Kaputs*). The use of different colored caps for different types of liquid media greatly facilitates visual identification of media in the teaching laboratory.

• *Additional assistance* for preparing materials for laboratories will be found in the *Instructor's Manual* for this text. Instructors who adopt this laboratory text may obtain a complementary copy of the *Instructors Manual* by contacting the W. H. Freeman and Company, 41 Madison Avenue, New York, NY 10010 (phone 1-212-576-9400).

AGAR (1.5%) OVERLAY TUBES

Used for: *Ex. 57* (forming a plug over semisolid nitrate medium [or control tubes] to trap dinitrogen in the case of the complete microbial reduction of nitrate [N_2 evolution])
Composition: (ingredients per liter)
Agar   15 g
Commercial availability: BBL and Difco
Preparation: Suspend 15 g of dehydrated agar in each liter of distilled water. Heat to liquify the agar, and mix well. Dispense 3 ml per tube. Sterilize in the autoclave by heating at 121°C for 15 minutes. Temper to 45°C for class use.

AMES MINIMAL AGAR

Used for: *Ex. 38* (base-layer medium for Ames-test plates to test chemicals as mutagens)
Composition: (ingredients per liter)
40% glucose solution   50 ml
Vogel-Bonner salts solution 20 ml
Agar   15 g

Vogel-Bonner salts solution (ingredients per liter)

$MgSO_4 \cdot 7H_2O$	10 g
Citric acid (monohydrate)	100 g
K_2HPO_4 (anhydrous)	500 g
$NaHNH_4PO_4 \cdot 4H_2O$	175 g

Commercial availability: None known
Preparation:

    Vogel-Bonner solution: Warm about 1.5 liters of distilled water to 45°C. In the order shown above, separately dissolve each of the dry ingredients in about 600 ml of the warm water. Adjust the final volume to 1000 ml with warm distilled water. Dispense into screw-capped containers. Sterilize in the autoclave. Tighten caps when cool.

    Minimal agar: Autoclave 100 ml of a solution that contains 40 grams of glucose, then temper at 50°C. Place 15 g of dehydrated agar, a large magnetic stirring bar, and 930 ml of distilled water in a 2-liter flask; loosely cap; autoclave at 121°C for 20 minutes; temper at 50°C. Temper 20 ml of the sterile Vogel-Bonner solution to 50°C. Add 20 ml of the warm V-B solution and 50 ml of the warm 40% glucose solution to the flask of warm agar while stirring thoroughly to mix. Pour about 30 ml per plate.

AMES SOFT OVERLAY AGAR (WITH TRACE HISTIDINE)

Used for: *Ex. 38* (overlay agar medium for Ames test to determine the mutagenic capacity of chemicals)
Composition: (mix each volume together after tempering)

Agar mixture	100 ml
Biotin-histidine solution	10 ml

  Agar mixture (ingredients per liter)

Agar	6 g
NaCl	5 g

  Trace biotin-histidine solution (ingredients per liter)

D-Biotin	124 g
L-Histidine	96 g

Commercial availability: None known
Preparation:

    Agar mixture: Dissolve sodium chloride in 500 ml of distilled water. Bring up to 1 liter with distilled water. Suspend and liquify agar. Mix well, then dispense 100 ml into 250-ml dilution bottles. Sterilize in the autoclave for 20 minutes. Cool and tighten caps.

    Trace solution: Dissolve biotin and histidine in distilled water. Sterilize in the autoclave for 20 minutes. Store in glass bottles at 4°C.

Melt and temper agar mixture, and warm trace solution in the tempering water bath. Mix the trace solution into the agar mixture, and keep tempered for student use.

BLOOD (SHEEP) AGAR

Used for: *Exs. 46, 48* (differential medium showing red blood cell hemolysis)
Composition:

Trypticase soy agar (tempered)	100 ml
Blood, defibrinated, sheep	4 ml

Commercial availability: BBL (as prepared plates only); BBL and Difco make trypticase soy agar; Other companies supply dedribinated sheep blood or prepared plates
Preparation: Prepare trypticase-soy agar and sterilize in the autoclave as usual. Temper to 45°C. Add 4.0 ml of warm, defibrinated sheep blood for each 100 ml of tempered T-soy agar. To accomplish this, place the tempered agar on a magnetic stirrer; let blood run down the side of the flask; stir until well mixed; then pour plates.

BRISTOL'S BROTH

Used for: *Ex. 22* (enrichment of algae and cyanobacteria [photolithotrophs] from other microbes in soil; this medium lacks a reduced organic carbon source)
Composition: (ingredients per liter)

$NaNO_3$	0.250 g
$MgSO_4 \cdot 7H_2O$	0.075 g
K_2HPO_4	0.075 g
$FeCl_3$	0.050 g
$CaCl_2$	0.025 g
NaCl	0.025 g
KH_2PO_4	0.018 g

Commercial availability: None known
Preparation: Dissolve each componant in a small amount of distilled water. Add all solutions together, and bring up to 1 liter with distilled water. Dispense into tubes, and sterilize in the autoclave.

CHOPPED MEAT MEDIUM

Used for: *Ex. 25* (preparing class cultures of *Clostridium* species for students to use as inoculum)
Composition: (ingredients per liter)

Nutrient broth	8 g
Yeast extract	5 g
Cooked meat medium	(see below)

Commercial availability: None known
Preparation: Make up the medium in 15 × 125 mm (outside diameter) screw-capped test tubes. Add enough dehydrated cooked meat medium (BBL or Difco) to just cover the bottom of the tube (this expands about three-fold upon rehydration). Dissolve the nutrient broth and yeast extract in 1 liter of distilled water, and add 8 ml of this solution to each tube (this nearly fills the tube). Sterilize in the autoclave for 20 to 25 minutes. Store in cold room. Steam for 5 to 10 minutes to drive off oxygen before using.

CHU'S BROTH

Used for: *Ex. 22* (enrichment of only cyanobacteria from all other microbes in soil; this medium lacks a reduced form of nitrogen, so microbes that grow must be capable of nitrogen fixation [N_2 reduction])
Composition: (ingredients per liter)

$MgSO_4 \cdot 7H_2O$	0.025 g
Na_2SiO_3	0.025 g
$NaCO_3$	0.020 g
K_2HPO_4	0.010 g
Ferric citrate	0.003 g
Citric acid	0.003 g

Commercial availability: None known
Preparation: Dissolve each component in a small amount of distilled water. Add all solutions together, and bring up to 1 liter with distilled water. Dispense into tubes, and sterilize in the autoclave.

EGG YOLK AGAR

Used for: *Ex. 29* (differentiating *Clostridium* species on the basis of their lecithinase production)
Composition: (ingredients per liter)
  Egg yolk, sterile   1 yolk
  McClung-Toabe agar base   500 ml
    McClung-Toabe agar base
      Proteose peptone   40.0 g
      Glucose (dextrose)   2.0 g
      Sodium phosphate, dibasic   5.0 g
      Potassium phosphate, monobasic   1.0 g
      Sodium chloride   2.0 g
      Magnesium sulfate   0.1 g
      Agar   25.0 g

Commercial availability: Difco (McClung-Toabe agar base)
Preparation: Rehydrate enough dehydrated McClung-Toabe egg-yolk agar base (directed on the bottle) to make up 500 ml (or multiples of 500 ml). Stir on a magnetic stirrer. Adjust pH of rehydrated agar base to 7.3 to 7.4 prior to autoclaving (as directed on bottle). Leave the stirring bar in the flask during autoclaving. Sterilize in the autoclave by heating at 121°C for 15 minutes. Temper the sterile agar to 50°C.

Soak chicken eggs for about 30 minutes in a beaker containing 95% ethyl alcohol (this disinfects them and also brings them to room temperature. Remove eggs and allow them to dry briefly. Thoroughly scrub hands. As aseptically as possible, break eggs and separate yolks from whites using the egg's shell. Place tempered agar base on a magnetic stirrer and turn stirring bar at moderate speed. Transfer one yolk to each 500 ml of tempered medium, and thoroughly mix. Pour plates. *Store plates in an anaerobe jar until used.*

EOSINE METHYLENE BLUE (EMB) AGAR

Used for: *Exs. 51, 53* (detection and isolation of gram-negative enteric bacteria)
Composition: (ingredients per liter)
  Peptone   10.000 g
  Lactose   5.000 g
  Sucrose   5.000 g
  Dipotassium phosphate   2.000 g
  Eosine Y   0.400 g
  Methylene blue   0.065 g
  Agar   13.500 g
Final pH: 7.2
Commercial availability: BBL and Difco
Preparation: According to instructions on bottle.

GELATIN (4%) IN TRYPTICASE SOY BROTH

Used for: *Exs. 28, 58* (detecting the microbial hydrolysis of an animal protein [gelatin])
Composition: (ingredients per liter of trypticase soy broth)
  Gelatin   40 g
  Trypticase (Tryptic) soy broth (dehydrated)   40 g
Commercial availability: None known
Preparation: Rehydrate trypticase-soy broth according to instructions on bottle (40 g per liter). Once these components are completely dissolved, *slowly* add 40 g of granular gelatin to *cold* distilled water. Bring this to a boil. Dispense 5 ml per tube, and sterilize in autoclave by heating at 121°C for 15 minutes.

GLUCONOBACTER AGAR

Used for: *Ex. 11* (preparing *Gluconobacter oxydans* cultures for student use)
Composition: (ingredients per liter)
  Sorbitol   50 g
  Yeast extract   10 g
  Peptone   10 g
  Agar   15 g
Adjust pH: to 6.0
Commercial availability: None known
Preparation: Dissolve the sorbitol, yeast extract, and peptone in about 750 ml of distilled water. Bring total volume to 1 liter with distilled water. Adjust pH to 6.0. Suspend 15 g of agar in this volume. Liquify agar, mix well, and dispense into screw-caped tubes. Cap loosely, and sterilize by heating at 121°C for 15 minutes in the autoclave. Slant tubes, and cool.

GLUCOSE BROTH WITH PHENOL RED (FERMENTATION TUBES)

Used for: *Exs. 31, 58* (detecting acid and gas excreted during the microbial fermentation of glucose)
Composition: (ingredients per liter)
  Peptone (carbohydrate free)   10.000 g
  Beef extract   1.000 g
  Glucose (dextrose)   5.000 g
  Sodium chloride   5.000 g
  Phenol red   0.018 g
Final pH: 7.4
Commercial availability: BBL—*phenol red carbohydrate media;* Difco—*phenyl red broths with carbohydrates*
Preparation: According to instructions on bottle. Add an inverted Durham (fermentation) tube to each tube of medium before autoclaving.

HIGH SALT (7.5%) AGAR

Used for: *Ex. 21* (a selective medium that inhibits most bacteria, but allows growth of those that can tolerate higher than normal concentrations of sodium chloride [such as *Staphylococcus* species])
Composition: (ingredients per liter)
  Trypticase soy agar (Difco dehydrated)   40 g
  Sodium chloride   70 g
(Note that rehydrated T-soy agar already contains 5 g of NaCl per liter.)
Commercial availability: None known

Preparation: Rehydrate trypticase-soy agar according to instructions on bottle (40 g per liter). Add sodium chloride. Sterilize in autoclave for 15 minutes at 121°C, and temper to 50°C before pouring plates.

LACTOSE BROTH WITH PHENOL RED (FERMENTATION TUBES)

Used for: *Exs. 31, 58* (detecting acid and gas excreted during the microbial fermentation of lactose)
Composition: (ingredients per liter)

Peptone (carbohydrate free)	10.000 g
Beef extract	1.000 g
Lactose	5.000 g
Sodium chloride	5.000 g
Phenol red	0.018 g

Final pH: 7.4
Commercial availability: BBL—*phenol red carbohydrate media;* Difco—*phenol red broths with carbohydrates*
Preparation: According to instructions on bottle. Add an inverted Durham (fermentation) tube to each tube of medium before autoclaving.

LAURYL TRYPTONE BROTH

Used for: *Ex. 53* (detecting coliform bacteria in water, waste water, and food)
Composition: (ingredients per liter)

Tryptose (trypticase peptone)	20.00 g
Lactose	5.00 g
Potassium phosphate, dibasic	2.75 g
Potassium phosphate, monobasic	2.75 g
Sodium chloride	5.00 g
Sodium lauryl sulfate	0.10 g

Final pH: 6.8
Commercial availability: BBL—*lauryl sulfate broth;* Difco—*lauryl tryptose broth*
Preparation: According to instructions on bottle. Add an inverted Durham (fermentation) tube to each tube of medium before autoclaving.

LITMUS MILK MEDIUM

Used for: *Ex. 35* (determining the action of bacteria on milk)
Composition:
Dehydrated skim milk and sufficient litmus to produce a good lavender color
Commercial availability: BBL and Difco
Preparation: According to instructions on bottle. Add 1 drop of *Pourite* per liter of medium during preparation to reduce foaming. Make up about one week prior to class to allow for reoxidation of litmus (color redevelopment).

LOWENSTEIN-JENSEN MEDIUM

Used for: *Ex. 8* (preparing *Mycobacterium smegmatis* cultures for student use)
Composition: (total ingredients)

KH_2PO_4	2.40 g
$MgSO_4 \cdot H_2O$	0.24 g
Magnesium citrate	0.60 g
Asparagine	3.60 g
Potato flour	30.00 g
Glycerol	12.00 g
Distilled water	600 ml
Homogenized whole eggs	1,000 ml
Malachite green (2% aq. sol.)	200 ml

Commercial availability: May be purchased from BBL and Difco as preprepared slants *(purchase recommended!)*
Preparation: Dissolve the salts and asparagine in the water. Add the glycerol and potato flour, and autoclave this potato-salt mixture at 121°C for 30 minutes. Cleanse whole eggs, not more than one-week old, by scrubbing with 5% soap solution. Allow to stand for 30 minutes in a clean soap solution, then rinse thoroughly in cold running water. Immerse eggs in 70% ethyl alcohol for 15 minutes, remove, and break into a sterile flask. Homogenize the eggs by shaking with sterile glass beads. Filter through four layers of sterile gauze. Add 1 liter of the homogenized eggs to the flask of cooled potato-salt mixture. Add the aqueous solution of malachite green. Mix well and dispense 6 to 8 ml into sterile, screw-capped, 20 × 150 ml tubes. Slant these tubes, and heat at 85°C for 50 minutes to thicken the mixture. *(Note that this medium contains egg as the only thickening agent.)* Incubate for 48 hours at 37°C to check sterility. Tightly cap the tubes and store in a refrigerator.

MANNITOL SALT AGAR

Used for: *Ex. 21* (selective isolation of pathogenic staphylococci)
Composition: (ingredients per liter)

Peptone	10.000 g
Beef extract	1.000 g
D-Mannitol	10.000 g
Sodium chloride	75.000 g
Phenol red	0.025 g
Agar	15.000 g

Final pH: 7.4
Commercial availability: BBL and Difco
Preparation: According to instructions on bottle.

METHYL-RED/VOGES-PROSKAUER (MR/VP) BROTH

Used for: *Exs. 32, 58* (differentiating strains of coliform bacteria)
Composition: (ingredients per liter)

Buffered peptone	7 g
Dipotassium phosphate	5 g
Glucose (dextrose)	5 g

Final pH: 6.9
Commercial availability: BBL and Difco
Preparation: According to instructions on bottle.

MUELLER-HINTON AGAR

Used for: *Ex. 44* (testing the susceptibility of microbes to antimicrobial agents)

Composition: (ingredients per liter)
  Beef infusion   300.0 g
  Case amino acids   17.5 g
  Starch   1.5 g
  Agar   17.0 g
Final pH: 7.4
Commercial availability: BBL and Difco
Preparation: According to instructions on bottle.

NUTRIENT AGAR

Used for: *Ex. 21* (general purpose growth medium)
Composition: (ingredients per liter)
  Beef extract   3 g
  Peptone   5 g
  Agar   15 g
Final pH: 6.8
Commercial availability: BBL and Difco
Preparation: According to instructions on bottle.

NUTRIENT BROTH

Used for: *Ex. 22* (general purpose growth medium)
Composition: (ingredients per liter)
  Beef extract   3 g
  Peptone   5 g
Final pH: 6.8
Commercial availability: BBL and Difco
Preparation: According to instructions on bottle.

NUTRIENT SPORULATION MEDIUM (NSM)

Used for: *Ex. 9* (growth and sporulation of *Bacillus* species)
Composition: (ingredients per liter)
  Nutrient Agar (Difco dehydrated)   23.0 g
  Salt solution   5.0 ml
  Yeast extract   0.5 g
    Salt solution: (ingredients per liter)
      Calcium chloride   15.54 g
      $MnCl_2 \cdot 4H_2O$ 1.98 g
      $MgCl_2 \cdot 6H_2O$ 40.66 g
Commercial availability: None known
Preparation: Prepare an adequate amount of salt solution, and set this aside. Separately rehydrate nutrient agar (23 g per liter), and add yeast extract (0.5 g per liter). Add salt solution (5 ml per liter). Sterilize in the autoclave for 15 minutes at 121°C.

PEPTONE (4%) BROTH

Used for: *Ex. 56* (detecting microbial deamination of nitrogenous organic compounds [ammonification])
Composition: (ingredients per liter)
  Peptone (pancreatic digest of casein)   40 g
Commercial availability: BBL—*polypeptone;* Difco—*peptone*
Preparation: Dissolve 40 g of peptone per liter of distilled water. Dispense into tubes. Sterilize in the autoclave for 15 minutes at 121°C.

PEPTONE IRON AGAR

Used for: *Ex. 33* (indicates hydrogen sulfide production by microbes)
Composition: (ingredients per liter)
  Peptone   15.00 g
  Proteose peptone   5.00 g
  Ferric ammonium citrate   0.50 g
  Sodium glycerophosphate   1.00 g
  Sodium thiosulfate   0.08 g
  Agar   15 g
Final pH: 6.7
Commercial availability: Difco
Preparation: According to instructions on bottle; but, after liquifying agar, dispense 7 ml per 15 × 125 mm plain test tube. After capping tubes and sterilizing in the autoclave, allow to cool in the upright position and use as agar deeps.

SABOURAUD'S AGAR (WITH GLUCOSE)

Used for: *Exs. 16, 41, 42* (culture prep), *58* (culturing yeasts, molds, and aciduric microbes)
Composition: (ingredients per liter)
  Neopeptone (polypeptone)   10 g
  Glucose (dextrose)   40 g
  Agar   15 g
Final pH: 5.6
Commercial availability: BBL—*Sabouraud dextrose agar;* Difco—*Sabouraud dextrose agar*
Preparation: According to instructions on bottle.

SEMISOLID NITRATE MEDIUM

Used for: *Ex. 57* (detecting denitrification—complete reduction of nitrate to dinitrogen)
Composition: (ingredients per liter)
  Beef extract   3 g
  Peptone   5 g
  Potassium nitrate   1 g
  Agar   1 g
Commercial availability: Difco—*nitrate broth*
Preparation: Dissolve nitrate broth components according to instructions on the bottle. Suspend 1.0 gram of dehydrated agar in each liter of dissolved nitrate broth. Heat to liquify the agar, and mix well. Dispense into tubes and cap. Sterilize in the autoclave by heating at 121°C for 15 minutes. Cool in upright position.

SEMISOLID NON-NITRATE MEDIUM

Used for: *Ex. 57* (negative control for detecting microbial denitrification—complete reduction of nitrate to dinitrogen)
Composition: (ingredients per liter)
  Beef extract   3 g
  Peptone   5 g
  Agar   1 g
Commercial availability: BBL—*nutrient broth;* Difco—*nutrient broth*
Preparation: Dissolve nutrient broth components according to instructions on the bottle. Suspend 1.0 gram of dehy-

drated agar in each liter of dissolved nutrient broth. Heat to liquify the agar, and mix well. Dispense into tubes and cap. Sterilize in autoclave for 15 minutes at 121°C. Cool in upright position.

SIMMON'S CITRATE AGAR (WITH BROMTHYMOL BLUE)

Used for: *Exs. 30, 58* (differentiation of gram-negative enteric bacteria on the basis of citrate utilization)
Composition: (ingredients per liter)

Sodium citrate	2.00 g
Magnesium sulfate	0.20 g
Ammonium dihydrogen phosphate	1.00 g
Dipotassium phosphate	1.00 g
Sodium chloride	5.00 g
Brom thymol blue	0.08 g
Agar	15.00 g

Final pH: 6.8
Commercial availability: BBL—*Simmons citrate agar;* Difco—*Simmons citrate agar*
Preparation: According to instructions on bottle.

SKIM MILK AGAR

Used for: *Ex. 28* (detecting microbial hydrolysis of an animal protein [casein])
Composition: (ingredients per liter)

Tryptone	17 g
Soytone	3 g
Sodium chloride	5 g
Skim milk	100 g
Agar	15 g

Commercial availability: None known
Preparation:
  Solution A: Suspend 40 g of trypticase (tryptic) soy agar per 500 ml (to make a 2× solution). Use a magnetic stirring bar, and leave bar in flask while autoclaving. Sterilize in the autoclave by heating at 121°C for 15 minutes. Temper agar to 55°C (sets up rapidly).
  Solution B: Stir 500 ml of room-temperature distilled water on a magnetic stir plate. Slowly add 100 g of BBL or Difco skim milk. Stir until completely dissolved. This makes a 2× solution. Transfer to a clean flask for sterilization. Sterilize in autoclave for *12 minutes only* at 121°C. Aseptically remove any scum that forms on top of skim milk. Temper agar to 55°C.

Aseptically pour solution B down the side of flask containing solution A. Stir on magnetic stir plate at moderate speed to thoroughly mix without foaming. Immediately pour plates. If necessary, flame bubbles that appear on the surface of the poured agar surface before agar solidifies.

SOFT-AGAR (0.7%) OVERLAY MEDIUM

Used for: *Ex. 37* (overlaying base-layer plates after mixing with host cells and phage in the plaque assay)
Composition: (ingredients per liter)

Agar	7 g

Commercial availability: BBL and Difco
Preparation: Suspend 7 g of agar per liter of distilled water, and heat to liquify. Dispense 5 ml per tube. Sterilize in the autoclave by heating at 121°C for 15 minutes. Temper to 45°C for class use.

SPIRIT BLUE AGAR

Used for: *Ex. 29* (the detection, enumeration, and study of lipolytic microbes)
Composition: (ingredients per liter)

Tryptone (trypticase peptone)	10.00 g
Yeast extract	5.00 g
Spirit blue	0.15 g
Lipase reagent (or suitable lipid)	30 ml
Agar	20.00 g

Final pH: 6.8
Commercial availability: BBL—*spirit blue agar + suitable lipid emulsion;* Difco—*spirit blue agar + lipase reagent*
Preparation: According to instructions on bottle.

STANDARD PLATE COUNT AGAR

Used for: *Ex. 52* (enumeration of bacteria in water, waste water, dairy products, and food)
Composition: (ingredients per liter)

Tryptone (trypticase—BBL)	5.0 g
Yeast extract	2.5 g
Glucose (dextrose)	1.0 g
Agar	15.0 g

Final pH: 7.0
Commercial availability: BBL—*standard methods agar;* Difco—*plate count agar*
Preparation: According to instructions on bottle.

STARCH AGAR

Used for: *Ex. 27* (cultivating organisms being tested for starch hydrolysis)
Composition: (ingredients per liter)

Beef extract	3 g
Soluble starch	10 g
Agar	12 g

Final pH: 7.5
Commercial availability: Difco
Preparation: According to instructions on bottle.

SUCROSE BROTH WITH PHENOL RED (FERMENTATION TUBES)

Used for: *Exs. 31, 58* (detecting acid and gas excreted during the microbial fermentation of sucrose)
Composition: (ingredients per liter)

Peptone (carbohydrate free)	10.000 g
Beef extract	1.000 g
Sucrose (saccharose)	5.000 g
Sodium chloride	5.000 g
Phenol red	0.018 g

Final pH: 7.4
Commercial availability: BBL—*phenol red carbohydrate media;* Difco—*phenol red broths with carbohydrates*
Preparation: According to instructions on bottle. Add an

inverted Durham (fermentation) tube to each tube of medium before autoclaving.

TRYPTICASE SOY AGAR

Used for: *Exs. 13, 14, 15, 16, 18, 20, 25, 37, 39, 43, 47, 51* (a general purpose medium especially for isolation, sensitivity testing, and hemolysis determinations with fastidious microbes)

Composition: (ingredients per liter)

Tryptone (trypticase peptone; a pancreatic digest of casein)	15 g
Soytone (phytone peptone; a papaic digest of soybean meal)	5 g
Sodium chloride	5 g
Agar	15 g

Final pH: 7.3

Commercial availability: BBL—*trypticase soy agar;* Difco—*tryptic soy agar*

Preparation: According to instructions on bottle.

TRYPTICASE SOY AGAR WITH SODIUM CHLORIDE (0.5, 5, AND 20%)

Used for: *Ex. 24* (testing a microbe's ability to grow in the presence of high concentrations of sodium chloride)

Composition: (ingredients per liter)

Trypticase soy agar (Difco dehydrated)	40 g
Plus *one* of the following concentrations of NaCl:	
0.5% sodium chloride	(none)
5.0% sodium chloride	45 g
20.0% sodium chloride	195 g

(Note that rehydrated T-soy agar already contains 5 g of NaCl per liter.)

Commercial availability: None known

Preparation: *For 0.5% sodium chloride,* make one liter of trypticase soy agar only, since this medium already contains 0.5% sodium chloride. *For 5% sodium chloride,* add 40 g of dehydrated trypticase soy agar to abut 750 ml of distilled water, and heat while stirring to liquify the agar. Dissolve 45 g of sodium chloride in that liquified agar. Add more distilled water to bring the volume up to one liter. *For 20% sodium chloride,* add 40 g of dehydrated trypticase soy agar to about 600 ml of distilled water, and heat while stirring to liquify the agar. Dissolve 195 g of sodium chloride in that liquified agar. Add more distilled water to bring the volume up to one liter.

Sterilize in the autoclave by heating at 121°C for 15 minutes. Carefully swirl flask when removing from the autoclave to suspend sludge (liquid foams vigorously). Temper to 55°C (high concentrations solidify within 5 to 10 minutes at room temperature). Swirl frequently while pouring plates.

TRYPTICASE SOY AGAR WITH SUCROSE (0.5, 15, AND 60%)

Used for: *Ex. 24* (testing a microbe's ability to grow in the presence of high concentrations of sucrose [table sugar])

Composition: (ingredients per liter)

Trypticase soy agar (Difco dehydrated)	40 g
Plus *one* of the following concentrations of sucrose:	
0.5% table sugar	5 g
15.0% table sugar	150 g
60.0% table sugar	600 g

Commercial availability: None known

Preparation: *For 5% sucrose,* add 40 g of dehydrated trypticase soy agar to about 750 ml of distilled water, and heat while stirring to liquify the agar. Dissolve 5 g of sucrose in that liquified agar. Add more distilled water to bring the volume up to one liter. *For 15% sucrose,* add 40 g of dehydrated trypticase soy agar to about 750 ml of distilled water, and heat while stirring to liquify the agar. Dissolve 150 g of sucrose in that liquified agar. Add more distilled water to bring the volume up to one liter. *For 60% sodium chloride,* add 40 g of dehydrated trypticase soy agar to about 400 ml of distilled water, and heat while stirring to liquify the agar. Dissolve 600 g of sodium chloride in that liquified agar. Add more distilled water to bring the volume up to one liter.

Sterilize in the autoclave by heating at 121°C for 15 minutes. Carefully swirl flask when removing from the autoclave to suspend sludge (liquid foams vigorously). Temper to 55°C (high concentrations solidify within 5 to 10 minutes at room temperature). Swirl frequently while pouring plates.

TRYPTICASE SOY BROTH

Used for: *Exs. 13, 17, 19, 23* (general purpose medium for growing fastidious and nonfastidious microbes)

Composition: (ingredients per liter)

Tryptone (trypticase peptone; pancreatic digest of casein)	17.0 g
Soytone (phytone peptone; a papaic digest of soybean meal)	3.0 g
Sodium chloride	5.0 g
Dipotassium phosphate	2.5 g

Final pH: 7.3

Commercial availability: BBL—*trypticase soy broth;* Difco—*tryptic soy broth (without dextrose)*

Preparation: According to instructions on bottle.

TRYPTONE (1%) BROTH

Used for: *Exs. 34, 58* (growing cells for the indol test)

Composition: (ingredients per liter)

Tryptone	10 g

Commercial availability: Difco

Preparation: Dissolve 10 g of peptone per liter of distilled water. Sterilize in the autoclave by heating at 121°C for 15 minutes.

TRYPTONE (1%) BROTH WITH GLUCOSE (5%)

Used for: *Exs. 34, 58* (growing cells to show how catabolite inhibition interferes with the indol test)

Composition: (ingredients per liter)

Tryptone	10 g
Glucose (dextrose)	50 g

Commercial availability: None known
Preparation: Dissolve 10 g of peptone and 50 g of glucose per liter of distilled water. Sterilize in the autoclave by heating at 121°C for 15 minutes.

VAN DELDEN'S BROTH

Used for: *Ex. 22* (enrichment of sulfate-reducing bacteria from other microbes in the soil)
Composition: (ingredients per liter)

Sodium lactate	5.0 g
$MgSO_4 \cdot 7H_2O$	2.0 g
Asparagine	1.0 g
$CaSO_4$	1.0 g
K_2HPO_4	0.5 g
$FeSO_4 \cdot 7H_2O$	0.1 g

Commercial availability: None known
Preparation: Dissolve each componant in a small amount of distilled water. Add all solutions together, and bring up to volume with distilled water. Dispense into *rubber-stoppered* dilution bottles so that there is very little head space above the liquid. Make sure all stoppers are just resting on the tube opening before placing in the autoclave. Sterilize in the autoclave by heating at 121°C for 15 minutes. If the medium is not to be used immediately upon cooling, the medium should be steamed to drive off dissolved oxygen before giving to students.

YEAST MANNITOL AGAR

Used for: *Ex. 55* (preparing *Rhizobium japonicum* cultures for student use)
Composition: (ingredients per liter)

Yeast extract	0.40 g
Mannitol	10.00 g
Salts solution	25 ml g
Agar	3.75 g

Salts solution (ingredients per 250 ml)

K_2HPO_4	5 g
$MgSO_4 \cdot 7H_2O$	2 g
NaCl	1 g

(Adjust pH to 7.0 with HCL)
Commercial availability: None known
Preparation: Dissolve the yeast extract and mannitol in about 750 ml of distilled water. Add the salts solution. Add distilled water to bring volume up to 1.0 liter. Suspend the agar in this 1-liter volume. Liquify the agar, mix well, and dispense into tubes. Cap loosely, and sterilize by heating at 121°C for 15 minutes in the autoclave. Slant tubes, and cool.

B

Stains and Staining Reagents

NOTES FOR STUDENTS AND FACULTY

- This appendix alphabetically lists all dye solutions (stains) and reagents used for staining procedures described in this text.

- Each entry lists the exercise where the stain or staining reagent is used and the purpose of that stain. This is followed by instructions for preparing the stain or staining reagent with the components of each solution given in *italics*.

- All dyes listed here may be purchased from either Fisher Scientific Company (Pittsburgh) or Sigma Chemical Company (St. Louis).

- Instructors who use this text are also encouraged to obtain a complementary copy of the Instructors Manual by contacting the W. H. Freeman and Company, 41 Madison Avenue, New York, NY 10010 (phone 1-212-576-9400).

- For further information on dye chemistry, consult *H. J. Conn's Biological Stains,* 8th ed., by R. D. Lillie et al. (Williams & Wilkins Company, Baltimore, 1969).

ACID ALCOHOL

Used for: *Ex. 8* (acid fast stain for mycobacteria and microorganisms)
Preparation: Place 30 ml of concentrated *hydrochloric acid* in a 1-liter graduated cylinder, then add 970 ml of 95% *ethyl alcohol.*

ACID FAST STAIN
(see ACID ALCOHOL; CARBOL FUCHSIN; METHYLENE BLUE)

Used for: *Ex. 8* (differential stain for mycobacteria and other acid-fast microorganisms)

BASIC FUCHSIN

Used for: *Exs. 7* (Gram stain), *9* (simple stain for detection of endospores)
Preparation: Dissolve 1.0 g of *basic fuchsin* (90% dye content) in 1 liter of *distilled water.*

CARBOL FUCHSIN

Used for: *Ex. 8* (primary stain in the acid-fast staining procedure)
Preparation:
      Solution A: Dissolve 3.0 g of *basic fuchsin* (90% dye content) in 90 ml of 95% *ethyl alcohol.*
      Solution B: Mix 50 ml of *liquid phenol* (carbolic acid) in 850 ml of *distilled water.*
Mix solutions A and B. Let stand for two days before use.

CRYSTAL VIOLET (HUCKER MODIFICATION)

Used for: *Exs. 7* (primary stain in the Gram-staining procedure), *5, 8, 11* (simple stain)
Preparation:
      Solution A: Dissolve 2.0 g of *crystal violet* (85% dye content) in 20 ml of 95% *ethyl alcohol.*
      Solution B: Dissolve 0.8 g of *ammonium oxalate* in 80 ml of *distilled water.*
Mix solutions A and B.

FLAGELLA STAIN (PAULEY'S MODIFICATION OF LEIFSON'S STAIN)

Used for: *Ex. 10* (precipitates around procaryotic flagella allowing them to be seen with the brightfield light microscope)
Precautions: It is best to make up this stain fresh just prior to the laboratory period. Alternatively, solutions A, B, and E may be made fresh the night before the lab then mixed with the other reagents the next day, immediately prior to the laboratory period.
Preparation:
      Solution A: Add 33.0 g of *ammonium aluminum sulfate* $(NH_4Al(SO_4)_2 \cdot 24H_2O)$ to 180 ml of *distilled water.* [Alternatively, add 25.0 g of $NH_4Al(SO_4)_2 \cdot 12H_2O$ to 180 ml of distilled water.] Heat to dissolve. This creates a saturated, aqueous solution.

Solution B: Slowly add 20.0 g of *tannic acid* to 100 ml of *distilled water* while continuously stirring.

Solution C: Add 2.0 g of *basic fuchsin* (90% dye content) to 30 ml of *95% ethyl alcohol*.

For 535 ml of stain, add while stirring the following (in order) *to the flask containing solution C:* Add solution A *while it is still warm*. Add solution B. Add 135 ml of 95% ethyl alcohol. Add 90 ml of distilled water.

GRAM STAIN
(see CRYSTAL VIOLET; IODINE SOLUTION; 95% ETHYL ALCOHOL; BASIC FUCHSIN)

Used for: *Exs. 7* (Gram stain), *9, 15, 21, 46, 47, 48, 50, 51, 55, 58*.

INDIA INK

Used for: *Ex. 11* (capsule detection)
Precautions: It is essential that you choose an ink that is composed of finely divided, well-dispersed carbon particles. Otherwise, larger clumps of carbon will prevent the cover-glass from coming in close contact with the cells and microscope slide, and the pressed suspension of India-ink particles and cells will be too thick to see the capsule surrounding the cells. We have had good success with Higgins Black Magic Waterproof India Ink #4466 (Faber Castell, Corp.).

IODINE SOLUTION (GRAM'S IODINE; LUGOL'S SOLUTION)

Used for: *Exs. 7* (a mordant in the Gram Stain), *27* (detecting microbial hydrolysis of starch using starch-agar plates), *42* (staining of intracellular starch vacuoles in yeast)

Preparation: Dissolve 2.0 g of *potassium iodide* (KI) in 300 ml of *distilled water*, then add 1.0 g of finely ground (with mortar and pestle) *iodine crystals* (I). Stir at room temperature until completely dissolved. For use as a mordant or to test for *starch hydrolysis*, use in the undiluted form. *For staining starch granules in yeast*, dilute equal parts of this iodine solution and distilled water.

MALACHITE GREEN

Used for: *Exs. 9* (staining bacterial endospores), *42* (staining ascospores in yeast)
Preparation: Dissolve 5.0 g of *malachite green (oxalate salt)* in 100 ml distilled water.

METHYLENE BLUE

Used for: *Exs. 8* (a counter stain in the acid-fast staining procedure), *22* (a simple stain for examining enrichment cultures)
Preparation: Dissolve 1.0 g of *methylene blue* (90% dye content) in 100 ml of *distilled water*.

SAFRANIN

Used for: *Ex. 9* (a counter stain in the hot malachite green spore staining procedure)
Preparation: Dissolve 1.0 g of *safranin-O* in 100 ml of *distilled water*.

SPORE STAIN
(see MALACHITE GREEN; SAFRANIN)

Used for: *Ex. 9* (hot malachite green method for staining bacterial endospores)

Other Reagents

NOTES FOR STUDENTS AND FACULTY

- This appendix alphabetically lists all reagents (other than those used for staining procedures) that are used in this text.

- Each entry lists the exercise in which that reagent is used, the purpose of that reagent, its components, and the method of its preparation. Some reagents are purchased ready to use from BBL Microbiology Systems (previously called Baltimore Biological Laboratories), Difco Laboratories (Detroit, MI), or Marion Scientific (Kansas City, MO).

- For more information and assistance on preparing for laboratory classes, instructors are encouraged to obtain a complementary copy of the Instructors Manual by contacting the W. H. Freeman and Company, 41 Madison Avenue, New York, NY 10010 (phone 212-576-9400).

ALPHA-NAPHTHOL SOLUTION (one of two reagents used for the Voges-Proskauer test)

Used for: *Ex. 32* (testing for the presence of acetoin [acetyl methyl carbinol] in the butylene-glycol type of fermentation; used to differentiate the enteric bacteria)
Composition:
  alpha-Napthol   5.0 g
  Ethyl alcohol   100 ml
Preparation: Dissolve the *alpha*-napthol in the ethyl alcohol. Note that this reagent is relatively *unstable*, so it should be made fresh just prior to class. It is also *light and temperature sensitive*, so it should be placed in a brown (or covered) bottle, and stored in the refrigerator for no more than one week. (To complete the VP test, this reagent is used along with the potassium-hydroxide/creatine solution.) *THIS IS A POTENTIAL CARCINOGEN, so handle with extreme care!*

COAGULASE PLASMA (EDTA)

Used for: *Ex. 45* (testing for microbial production of coagulase)
Availability: Coagulase plasma (EDTA) is purchased directly from Difco or BBL. The EDTA (ethylenediaminetetra-acetic acid) is added to the plasma to prevent clotting (an anticoagulant); unlike some other anticoagulants, EDTA is not degraded by microorganisms.

95% ETHYL ALCOHOL

Used for: *Exs. 22, 39, 43, 44, 52, 55, 58* (incineration of spatulas, forceps, and spreading rods)
Availability: Certified A.S.C. ethyl alcohol may be purchased as a 95% solution from many different sources.

FACTOR 9 (GROUP D) ANTISERUM (for *Salmonella enteritidis*)

Used for: *Ex. 49* (slide-agglutination test)
Preparation: Purchased in a dehydrated form from Difco as *Salmonella O.* antiserium, Factor 9.

HYDROGEN PEROXIDE (3%)

Used for: *Ex. 36* (catalase test)
Availability: A 3% solution of hydrogen peroxide can be purchased from many sources. Most drug stores carry this solution because it is commonly used as a topical disinfectant.

IODINE SOLUTION

Used for: *Ex. 27* (detecting microbial hydrolysis of starch using starch-agar plates) *Note:* This is similar to the solution used in the Gram stain, except that it is only one-half as concentrated.
Composition:
  Potassium iodide (KI)   2.0 g
  Distilled water   300 ml
  Iodine (I), crystaline 1.0 g
Preparation: Dissolve the potassium iodide in the distilled water; use a mortar and pestle to grind the iodine crystals to a very fine powder; add the powdered iodine to the potassium iodide solution; and stir at room temperature until completely dissolved.

KOVAC'S REAGENT (DIMEHTHYLAMINOBENZALDEHYDE)

Used for: Ex. 34 (detecting microbial production of indol from hydrolysis of the amino acid trypthphan)
Composition:

Amyl (or butyl) alcohol	150 ml
para-dimethylaminobenzaldehyde	10 g
Hydrochloric acid, concentrated	50 ml

Preparation: Slowly add the aldehyde to room-temperature alcohol using a magnetic stirrer, then add the acid. Protect from light, and store in the refrigerator.

METHYL ALCOHOL (METHANOL)

Used for: Exs. 5, 7, 8, 9, 15, 21, 22, 46, 47, 48, 50, 51, 55, 58 (for fixing cells from broth cultures to glass microscope slides)
Availability: Certified A.S.C. methyl alcohol may be purchased from a number of sources, such as Fisher Scientific Company (Pittsburg, PA). This is stored in the refrigerator until needed.

METHYL RED REAGENT

Used for: Ex. 32 (the methyl red test to detect significant acid production from the mixed-acid type of fermentation) Used to help differentiate the enteric bacteria; often used in conjunction with the Voges-Proskauer test.
Composition:

Methyl red	0.2 g
Ethyl alcohol, 95%	600 ml
Distilled Water	400 ml

Preparation: Slowly add the methyl red to room-temperature ethyl alcohol while mixing on a magnetic stirrer, then add the distilled water.

MUTAGEN SOLUTIONS

Used for: Ex. 38 (the Ames test; using the Ames mutant of Salmonella typhimurium called TA-98).
Suggestions: Use 4-Nitro-ortho-phenylenediamine (NPD) and/or 4-Nitroquinoline-N-oxide (NQNO).
Notes: NPD is a common hair dye component; it should show about 200 to 500 revertants of TA-98 per plate (spontaneous subtracted). HQNO should show about 100 to 200 revertants of TA-98 per plate (spontaneous subtracted). Both mutagens may be purchased from the Sigma Chemical Company, St. Louis, MO.
Preparation: Dissolve enough of the NPD or the NQNO in distilled water to achieve 10 micrograms per ml. (Students will use 0.1 ml per overlay.)
Handling mutagens: Every effort should be made to isolate areas where mutagens are handled in order to avoid contamination of the laboratory. Mutagenicity assays should be performed in a well-ventilated fume hood. Open-face hoods should have an average linear face velocity of 100 feet per minute. Solids and volatile liquids used to prepare the mutagen solutions should be handled in the hood. All weighings should be done by weight differences to avoid opening containers of mutagens in the laboratory outside of the hood. Persons handling mutagens should wear disposable plastic or latex gloves. Petri plates containing mutagens should be isolated in a separate incubator located adjacent to or (preferably) inside the hood. Place warning signs on the door of this incubator to prevent unauthorized persons from opening it. Open the incubator only while the hood is operating.
Disposal of mutagens: Wherever possible, use disposable test tubes, petri plates, and micropipettes with disposable tips. All contaminated material should be placed in a cardboard box lined with two heavy-duty plastic bags. The bags and boxes should be sealed with strapping tape and labeled with the contents. Disposal of mutagen waste is done by contract with a firm that disposes of radioactive waste from the laboratory. *Contact your college or university health and safety officer for full details and policy.*
Additional mutagen information: The Environmental Mutagen Information Center (EMIC) provides mutagenicity information, without charge, for any chemical tested with the Ames assay or any other mutagenicity tests that have been reported in the literature. For information, contact Mr. John Wassom, EMIC, Oak Ridge National Laboratories, Oak Ridge, TN 37830. Phone (615)574-7871.

NESSLER'S REAGENT

Used for: Ex. 56 (detecting microbial deamination of amino acids) This reagent tests for dissolved ammonia.
Preparation: Dissolve 5.0 g of *potassium iodide* in 5.0 ml of ammonia-free *distilled water*. To this solution, slowly add a saturated solution of *mercuric chloride* (about 2.0 g of HgI_2 in 35 ml of distilled water) until a precipitate begins to form. Then add 20 ml of 5 *N sodium hydroxide,* and dilute this entire solution to 100 ml with distilled water. Let any precipitate settle out, and draw off (or carefully decant) the clear supernatant fluid; it should have no yellow color. Store in a dark bottle; stable for up to 1 year in the refrigerator. *Note: this solution contains mercury; dispose according to EPA guidelines!*
Alternative: Nessler's solution may be purchased ready to use from Fisher Scientific Company (Pittsburgh, PA) and other sources.

POTASSIUM HYDROXIDE/CREATINE SOLUTION (one of two reagents used for the Voges-Proskauer test)

Used for: Ex. 32 (testing for the presence of acetoin [acetylmethylcarbinol] in the butylene-glycol type of fermentation used to differentiate the enteric bacteria)
Composition:

Potassium hydroxide	40.0 g
Creatine	0.5 g
Distilled Water	100 ml

Preparation: Dissolve the potassium hydroxide and creatine in distilled water at room temperature. (To complete the VP test, this reagent is used along with the alpha-naphthol solution.)

C OTHER REAGENTS

SALINE (sterile, physiological)

Used for: *Exs. 46, 47, 48* (moistening swabs before microbial sampling)
Composition:
Sodium chloride   8.5 g
Distilled Water   1000 ml
Preparation: Dissolve the NaCl in the distilled water. Dispense 2 ml per tube and cap. Sterilize in the autoclave by heating at 121°C for 15 minutes.

TETRA-METHYL-PARA-PHENYLENEDIAMINE

Used for: *Ex. 36* (cytochrome-*c* [oxidase] test)
Composition:
tetra-methyl-*para*-phenylenediamine
dihydrochloride   1.0 g
Distilled Water   100 ml

Note: Dimethyl-*para*-phenylenediamine hydrochloride may be used. It is less expensive, but also less sensitive.
Preparation: Dissolve the phenylenediamine in distilled water. This solution is unstable and should not be stored in the refrigerator for more than 1 week. This solution should be made up fresh daily.
Alternative: The premixed form of this reagent (*tetra*-methyl form) is available from Marion Scientific (Kansas City, MO) and from other commercial sources.

VASPAR

Used for: *Ex. 41* (sealing three sides of the cover glass covering the inoculated agar slab for moist-chamber growth of filimentous fungi)
Preparation: Melt together one part vaseline and one part paraffin. Store in a small bottle for culture curator use (demonstration) or class use.

Cultures

NOTES FOR STUDENTS AND FACULTY

- This appendix alphabetically lists all cultures that are used for successfully completing the exercises in this text. Abbreviations for various culture collections: ATCC = American Type Culture Collection (12301 Parklawn Drive, Rockville, MD 20852); VPI = Virginia Polytechnic Institute (% Culture Curator, Department of Anaerobic Microbiology, Virginia Tech [Va. Tech], Blacksburg, VA 24061); Va. Tech. = Biology Department, Va. Tech. (% Culture Curator, Microbiology Section, Biology Department, Virginia Tech, Blacksburg, VA 24061); RC = Dr. Roy Curtiss III (% Biology Department, Campus Box 1137, University of Washington, St. Louis, MO 63130); USDA = U.S. Department of Agriculture (Culture Collection, USDA, Beltsville, MD 20705).

- *Cell type* refers to the shape and Gram-staining characteristics (where applicable) of that culture and offers students a review of these known characteristics.

- *Agar growth* gives the type of solid medium routinely used at Virginia Tech to cultivate that strain, the colony size, and other colony characteristics when incubated for the stated time and at the stated temperatures. Abbreviations: TSA = trypticase soy agar.

- *Broth growth* gives the liquid medium routinely used at Virginia Tech and the characteristics of growth in that broth when incubated at the stated time and at the stated temperatures. Abbreviations: TSB = trypticase soy broth.

- *Comments* include other information about the culture that may be helpful for the instructor.

- *Uses* lists the exercises in this text where that culture is used.

- For further information and assistance regarding class cultures and class preparation, instructors are encouraged to obtain a complementary copy of the Instructor's Manual by contacting W. H. Freeman and Company, 41 Madison Avenue, New York, NY 10010 (phone: 212-576-9400).

ALCALIGENES VISCOLACTIS (Va. Tech)

Cell type: Bacterium; gram-negative rods with capsule
Agar growth: TSA for 24 hours at 30°C; colonies 0.5 to 1.0 mm diameter, entire edge, cream-colored, very slimy when probed with loop
Broth culture: TSB for 24 hours at 30°C; uniform turbidity
Comments: Growth difficult to remove from TSA slants; culture originally obtained from R. E. Benoit, Va. Tech
Uses: *Ex. 11*

BACILLUS CEREUS (ATCC 14579)

Cell type: Bacterium; long gram-positive rods
Agar growth: TSA for 24 hours at 30°C; colonies large, white, spreading
Broth culture: TSB for 24 hours at 30°C; uniformly turbid during exponential growth; clear with sediment in stationary phase (24 h as above)
Comments: NSM at 24 hours; spores central, elliptical, cell not swollen
Uses: *Exs. 7, 27, 28, 35, 56*

BACILLUS SPHAERICUS (ATCC 23857)

Cell type: Bacterium; gram-variable rods (majority gram-negative)
Agar growth: TSA for 24 hours at 30°C; colonies flat, entire, not shiny
Broth culture: TSB for 24 hours at 30°C; uniformly turbid
Comments: For producing spores, grow cells on Nutrient Sporulation Medium (see Appendix A) for 24 hours at 30°C; spores are spherical and subterminal and swell the cell; at 48 hours incubation, this culture will be almost 100% free of spores; if a 24-hour NSM slant is left in the refrigeration over the weekend, most of the endospores are released; (a *fresh* 24-hour NSM culture should be used); sporulation is inconsistent on TSA; strain deposited with ATCC by A. A. Yousten, Va. Tech, (his strain SS II-1)
Uses: *Ex. 9*

BACILLUS STEAROTHERMOPHILUS (ATCC 12980)

Cell type: Bacterium; long gram-positive rods
Agar growth: TSA for 24 hours at 55°C; colonies smooth, entire edge, cream-colored
Broth culture: TSB for 24 hours at 55°C; clear with fluffy sediment
Comments: NSM: spores oval, subterminal, with swollen cell; growth at 24 hours: − at 30° to 35°C; ± at 45°C; + at 55°C
Uses: *Ex. 23*

BACILLUS SUBTILIS (ATCC 6051)

Cell type: Bacterium; long gram-positive rods
Agar growth: TSA for 24 hours at 30°C; colonies large, rough (ground-glass) surface
Broth culture: TSB for 24 hours at 30°C; good pelicle with light sediment
Comments: NSM at 24 hours: spores oval, subterminal; seems to have a tendency to give two colony types; therefore, don't use when looking for purity
Uses: *Exs. 7, 8, 17*

BACILLUS SUBTILIS (ATCC 6633)

Cell type: Bacterium; gram-positive rods
Agar growth: TSA for 24 hours at 30°C
Broth culture: TSB for 24 hours at 30°C
Comments: Gives good positive lecithinase (phospholipase) reaction on egg-yolk agar
Uses: *Ex. 29*

BACILLUS SUBTILIS (Va. Tech)

Cell type: Bacterium; gram-positive rods with squared ends
Agar growth: TSA for 24 hours at 30°C; colonies dry, spreading, wavy edges
Broth culture: TSB for 24 hours at 30°C; fragile pelicle with light sediment
Comments: NSM at 24 hours shows oval, subterminal spores; culture obtained from A. A. Yousten (Va. Tech) as strain 168; ATCC strain 23857 is reported also to be strain 168 and could possibly be substituted for the Va. Tech strain
Uses: *Exs. 8, 16*

CLOSTRIDIUM SPOROGENES (VPI 8654E)

Cell type: Bacterium; gram-positive rods
Agar growth: TSA plates used for purity check only; incubate in an anaerobe jar at 30°C at 48 hours
Broth culture: Chopped meat medium (see Appendix A) for 24 hours at 37°C
Comments: This strain is positive for lipase and negative for lecithinase (phospholipase), so it may be substituted for *Staphylococcus epidermides* in Exercise 29
Uses: *Exs. 18, 25*

ENTEROBACTER AEROGENES (ATCC 13048)

Cell type: Bacterium; short, biolar-staining, gram-negative rods
Agar growth: TSA for 24 hours at 30° and 37°C; colonies round, smooth, off-white
Broth culture: TSB for 24 hours at 30° and 37°C; heavy turbidity
Comments: Phenol red fermentation broths show an acid bottom and red band on top after 48 hours at 30°C
Uses: *Exs. 30, 32, 34, 53*

ESCHERICHIA COLI (ATCC 9637)

Cell type: Bacterium; short gram-negative rods
Agar growth: TSA for 24 hours at 30° and 37°C; colonies round, cream-colored, pearly
Broth Culture: TSB for 24 hours at 30° and 37°C; turbid
Comments: This strain does not give a good metalic sheen on EMB plates
Uses: *Exs. 14, 16, 17, 18, 19, 20, 21, 23, 24, 25, 27, 28, 30, 31, 32, 33, 34, 35, 36, 39, 42, 43, 44, 49, 53*

ESCHERICHIA COLI STRAIN B (Va. Tech)

Cell type: Bacterium; short, bipolar staining, gram-negative rods
Agar growth: TSA for 24 hours at 30° and 37°C; colonies round, smooth, off-white
Broth culture: TSB for 24 hours at 30° and 37°C; slightly turbid with sediment
Comments: No distinct metalic sheen on EMB plates; this strain is an extrasensitive host for coliphage T-1
Uses: *Ex. 37*

ESCHERICHIA COLI STRAIN K12 (RC strain Chi-15)

Cell type: Bacterium; gram-negative rods
Agar growth: TSA for 24 hours at 30° and 37°C; colonies round, cream-colored; pearly
Broth culture: TSB for 24 hours at 30° and 37°C; uniformly turbid
Comments: Used for its excellent sheen on EMB; isolated by Roy Curtiss III (while a graduate student at the University of Chicago; hence *Chi*) from "original cultures" of K12; alternatively, one could use a natural isolate from a fresh fecal sample for Exercise 53
Uses: *Ex. 53*

ESCHERICHIA COLI PHAGE T-1 (Va. Tech)

Cell type: Not cells; bacterial viruses
Comments: Very clear plaques formed on *E. coli* strain B; this phage is extremely lytic for *E. coli* strain B
Uses: *Ex. 37*

GLUCONOBACTER OXYDANS (ATCC 621)

Cell type: Bacterium; short bipolarly staining gram-negative rods
Agar growth: Gluconobacter agar for 48 hours at 30°C; colonies very small (1 to 2 mm), translucent
Broth culture: Gluconobacter broth for 48 hours at 30°C; turbid with fragile pelicle
Comments: Does not grow (or grows very poorly) at 37°C;

no evidence of capsule formation under any circumstances
Uses: *Ex. 11*

LACTOBACILLUS PLANATARUM (ATCC 14917)

Cell type: Bacterium; single or paired gram-negative rods
Agar growth: TSA for 48 hours at 30° or 37°C; colonies very small, round, pearllike
Broth culture: TSB for 24 hours at 30°C; uniform turbidity
Comments: Colonies on TSA are only about 0.25 mm, white, and round after 48 hours at 30°C
Uses: *Ex. 9* (used as a nonsporulating control culture)

MICROCOCCUS CRYOPHILUS (ATCC 15174)

Cell type: Bacterium; single or paired gram-positive cocci
Agar growth: TSA for 8 to 10 days at 4° to 10°C; colonies small, round, raised, beige
Broth culture: TSB for 3 to 4 days at 4° to 10°C; lightly turbid
Comments: A heavy inoculation or a long incubation of class cultures is required; this culture can grow in TSB at a *cool* room temperature within 24 hours; will not grow at 30°C
Uses: *Ex. 23*

MICROCOCCUS LUTEUS (ATCC 10240)

Cell type: Bacterium; numerous tetrads, gram-positive cocci
Agar growth: TSA for 48 hours at 30°C; colonies bright yellow, smooth, round
Broth culture: TSB for 48 hours at 30°C; turbid with mucoid sediment
Comments: Colonies are very small and lack significant pigment by 24 hours at 30°C but they are well pigmented by 48 hours
Uses: *Exs. 18, 25, 39*

MYCOBACTERIUM SMEGMATIS (ATCC 14468)

Cell type: Bacterium; acid fast, poorly Gram-stained
Agar growth: TSA for 48 hours at 37°C; colonies rough, crinkled, may appear buff, cream-colored, or yellow
Broth culture: TSB for 48 hours at 37°C; particulate growth
Comments: Stock cultures kept on Lowenstein-Jensen medium (see Appendix A)
Uses: *Ex. 8*

NEISSERIA SUBFLAVA (ATCC 14221)

Cell type: Bacterium; gram-negative coccus; usually paired
Agar growth: TSA for 48 hours at 37°C; colonies small, dull, entire edge, yellow
Broth culture: TSB for 24 to 48 hours at 30°C; slightly turbid (more turbidity at 37°C)
Comments: Does not remain viable for more than 1 month on TSA slants or as glycerol stocks (see *Instructor's Manual*) stored at $-10°C$; routinely kept in liquid nitrogen until needed, often used as a gram-negative coccus for Gram-stain unknowns (Exercise 7)
Uses: *Ex. 7*

PENICILLIUM NOTATUM (Va. Tech)

Cell type: Filamentous fungus; deutromycete, but has ascomycetelike vegetative hyphae and asexually formed fruiting bodies
Agar growth: SAB for 3 to 4 days at 25° to 30°C; colonies fuzzy and white until sporulating, then have green centers
Broth culture: TSB for 4 to 6 days at 25° to 30°C; tough pelicle when undisturbed
Comments: Septate mycelium, asexually formed conidiospores; slide culture in moist chamber shows conidiospores after 24 to 48 hours at 30°C, slower at 25°C, unnumbered strain originally received from the collection at the University of Maryland
Uses: *Exs. 24, 41*

PROTEUS MIRABILIS (ATCC 14153)

Cell type: Bacterium: bipolarly staining gram-negative rod
Agar growth: TSA for 24 hours at 30°C; colonies appear clear, swarming
Broth Culture: TSB for 24 hours at 30°C; turbid
Comments: Seems to be less swarming on nutrient agar; flagella stain works best from young cells grown on TSA slants and removed with water run down the side of the tube opposite the slant surface
Uses: *Exs. 10, 31, 33, 35*

PROTEUS MIRABILIS (ATCC 25933)

Cell type: Bacterium; gram-negative rod
Agar growth: TSA for 24 hours at 30°C
Broth culture: TSB for 24 hours at 30°C
Uses: *Ex. 29*

PSEUDOMONAS AERUGINOSA (ATCC 15692)

Cell type: Bacterium; very small gram-negative rod
Agar growth: TSA for 24 hours at 30°C; colonies large, opaque, shiny, serrated edges
Broth culture: TSB for 24 hours at 30°C; uniformly turbid
Comments: Typical blue-green pigment on TSA
Uses: *Exs. 21, 36, 44*

PSEUDOMONAS FLUORESCENS (ATCC 13525)

Cell type: Bacterium; gram-negative rod
Agar growth: TSA for 24 hours at 30°C; colonies are smooth, entire edge
Broth culture: TSB for 24 hours at 30°C; turbid broth with slimy pellicle
Comments: Excellent for demonstrating motility; glucose is oxidized, but phenol-red/glucose reaction is not clear; green soluble pigment formed on TSA
Uses: *Exs. 6, 7, 56*

PSEUDOMONAS STUTZERI (ATCC 11607)

Cell type: Bacterium; slender gram-negative rod
Agar growth: TSA for 24 hours at 30°C; colonies are smooth, entire edge, yellow tint

Broth culture: TSB for 24 hours at 30°C; turbid broth with pellicle
Uses: *Ex. 57*

RHIZOBIUM JAPONICUM (USDA strain 6)

Cell type: Bacterium; gram-negative rods
Agar growth: Yeast-mannitol medium for 1 to 2 weeks at 25°C; colonies white, moist, gummy
Broth culture: No growth on TSB at 25°C
Comments: Growth from yeast-mannitol medium forms strings and clumps when suspended in broth
Uses: *Ex. 55*

RHIZOPUS NIGRICANS (Va. Tech)

Cell type: Filamentous fungus; most vegetative hypha are nonseptate; asexual reproduction forms sporangiospores within a sac; sexual reproduction forms thick-walled zygospores
Agar growth: SAB agar gives fuzzy growth and sporangiospores after 3 to 4 days at 25°C; about the same amount of growth will occur after 24 hours at 30°C; moist, gummy
Broth culture: TSB growth (submerged fuzz balls) after 3 to 4 days at 25°C (or earlier at 30°C)
Comments: Good example of morphology of the class *Zygomycetes;* cross-walls do form during sexual (zygote) formation
Uses: *Ex. 41*

SACCHAROMYCES CEREVISIAE (Va. Tech)

Cell type: Yeast (single-cell fungi); oval shape with budding
Agar growth: SAB agar after 24 hours at 30°C exhibits only tiny colonies; most colonies large, cream-colored after 48 hours at this temperature
Broth culture: TSB after 24 hours at 30°C; clear broth with heavy sediment
Comments: Use 24-hour culture to demonstrate budding; little if any ascospores formed under these conditions; optimum growth temperature is 26° to 30°C; to visualize internal starch vaccuoles, use 48-hour culture and Gram's iodine diluted 1:2 with distilled water
Uses: *Exs. 5, 6, 16, 17, 24, 42*

SALMONELLA ENTERITIDIS (ATCC 13076)

Cell type: Bacterium; gram-negative rods
Agar growth: TSA after 24 hours at 37°C should exhibit smooth colony types (see Comments)
Broth culture: TSB after 24 hours at 37°C gives turbidity with heavy sediment
Comments: Tendency to form two colony types (smooth and clear) on TSA plates; select the smooth colony type for serotyping (Exercise 49)
Uses: *Ex. 49*

SALMONELLA TYPHIMURIUM (Ames strain TA-98)

Cell type: Bacterium; gram-negative rods
Agar growth: TSA after 24 hours at 30°C exhibits light growth; use 48 hours incubation on plates for contamination check
Broth culture: TSB after 15 hours at 30°C gives barely visible turbidity; 24 hours incubation results in readily visable turbidity
Comments: The advantage of this strain is that it gives a low number of spontaneous revertants in absence of mutagens (for further information on this strain, see *Mutation Research* 113:173–215, 1983); this mutant is resistant to ampicillin; some investigators routinely keep stocks of this strain on agar containing ampicillin.
Uses: *Ex. 38*

SALMONELLA TYPHIMURIUM (Ames strain TA-102)

Cell type: Bacterium; gram-negative rods
Agar growth: TSA for 24 hours at 30°C
Broth culture: TSB for 24 hours at 30°C
Comments: The advantage of this strain is that it readily mutates in (is very sensitive to) the presence of UV light (for further information on this strain, see *Mutation Research* 113:173–215, 1983)
Uses: *Ex. 39*

SCHIZOSACCHAROMYCES OCTOSPORUS (Va. Tech)

Cell type: Yeast (single-cell fungi); oval shape
Agar growth: Fair growth on SAB agar after 24 hours at 25°C; good growth on SAB agar for 3 or 4 days at 30°C
Comments: Ascospore formation in 2 to 3 days at 25°C on SAB agar
Uses: *Ex. 42*

STAPHYLOCOCCUS AUREUS (ATCC 12600)

Cell type: Bacterium; gram-positive cocci in clusters
Agar growth: TSA for 24 to 48 hours at 30°C; colonies smooth, round, yellow
Broth culture: TSB for 24 hours at 30° or 37°C; uniformly turbid
Comments: Facultatively anaerobic growth; has a heat-stable DNAse; the ATCC strain 25923 has been substituted for this strain in Exercise 21
Uses: *Exs. 21, 24, 43, 44, 45, 57*

STAPHYLOCOCCUS EPIDERMIDIS (ATCC 12228)

Cell type: Bacterium; gram-positive cocci
Agar growth: TSA for 36 hours at 30°C
Broth culture: TSB for 24 hours at 30°C
Comments: Gives a strong positive reaction for lipases on spirit-blue agar and a negative reaction for lecithinase (pholpholipase) on egg-yolk agar
Uses: *Ex. 29*

STAPHYLOCOCCUS EPIDERMIDIS (ATCC 14990)

Cell type: Bacterium; gram-positive cocci
Agar growth: TSA for 36 hours at 30°C; colonies small, white

Broth culture: TSB for 24 hours at 30°C; uniformly turbid
Comments: Facultatively anaerobic growth; does not have a heat-stable DNAse; Florida State University used successfully for high-salt positive and mannitol-salt negative during fall 1987
Uses: *Exs. 5, 10, 21, 36, 45*

STREPTOCOCCUS FAECALIS (ATCC 19433)

Cell type: Bacterium; gram-positive coccus
Agar growth: TSA for 24 hours at 30 or 37°C; colonies small, smooth, raised, white
Broth culture: TSB for 24 hours at 30 or 37°C; uniformly turbid
Comments: Facultatively anaerobic growth
Uses: *Ex. 31*

STREPTOCOCCUS LACTIS (ATCC 11454)

Cell type: Bacterium; slightly elongated gram-positive cocci, in pairs
Agar growth: TSA for 24 hours at 30°C; colonies small, entire edge, smooth surface, white.
Broth culture: TSB for 24 hours at 30° or 37°C; uniformly turbid
Comments: Culture rapidly dies off in broth at about 24 to 32 hours because of strong acid production
Uses: *Ex. 36*

STREPTOCOCCUS PYOGENES (ATCC 12344)

Cell type: Bacterium; gram-positive cocci
Agar growth: Does not grow well on TSA; grows well on blood agar at both 30° and 37°C
Broth culture: TSB for 24 hours at 37°C shows light turbidity (probably grows in TSB because of the presence of sugar that is lacking in TSA; see Appendix A)
Uses: *Ex. 46*

STREPTOCOCCUS SALIVARIUS (ATCC 13419)

Cell type: Bacterium; gram-positive coccus
Agar growth: Grows poorly on TSA for 48 hours at 30°C
Broth culture: TSB for 24 hours at 30°C gives light turbidity
Comments: Does not grow well on TSA or on hydrolysis plates used for exercises 27 and 28; grows better on blood-agar plates (VT) incubated in a candle jar (FSU); Lancefield Group K; catalase negative.
Uses: *Exs. 27, 28, 43, 44, 46*

STREPTOCOCCUS SANGUIS (ATCC 10556)

Cell type: Bacterium; gram-positive cocci
Agar growth: TSA for 24 to 48 hours at 30°C
Broth culture: TSB for 24 to 48 hours at 30° or 37°C
Uses: *Ex. 46*

Index

Acid-fast test, 59-62
  summary, 62
Aerobic microorganisms, defined, 225
Agar, deeps, 110-112, 225-228
  defined, 108, 110-114
  plates, 110-114, 119-125
  preparing, 110-114
  slants, 110-112
  sterilizing, 114
Agglutination reaction, 447-448
Alcanigenes viscolactis, capsules in, 91-94
Alexopoulos, C. J., 352
Ames, Bruce, 325
Ames test, 323-329
  and ultraviolet radiation, 339-341
Ammonification, 499-504
  schematic diagrams, 500-502
  See also Nitrogen fixation
Anaerobic microorganisms, culturing, 159-162
  defined, 225
Anaerobic respiration, 285
  and denitrification, 507
  and the nitrogen cycle, 489
Animal protein (*see* Gelatin)
Antibiotic, defined, 405
Antibiotic evaluation, 397-399
  by Kirby-Bauer method, 405-410
Antiseptics, defined, 397
  effectiveness of, 397-399
Ascomycetes, 366-368
  reproduction in, 366-368
  structure and growth, 366-367
Aseptic techniques, 1-7
  See also, Sterilization methods
Aspergillus niger, 347
Auxotroph, 325-326

Bacillus anthracis, 395
Bacillus brevis, 11
Bacillus cereus, and ammonification, 503-504
  casein hydrolysis by, 249-250

gelatin hydrolysis by, 252
Gram staining, 52-54
starch hydrolysis by, 242-244
Bacillus sphaericus, 69
  structure of, 71-72
Bacillus stearothermophilus,
  temperature effects on, 214-215
Bacillus subtilis, 97
  growth of, 143-146, 152-154
  lipid hydrolysis by, 259-260
Bacteria, beneficial, 68
  growth, 143-228
  motility, 79-83
  morphology, 51-91
  physiology, 235-308
  reproduction, 67-68
  taxonomy, 67-68
Bacterial capsules, 89-93
Bartholomew-Mittwer procedure, 71
Basic fuchsin, as a counterstain, 51
BGLB (brilliant green lactose bile) broth, for detecting coliform bacteria, 480
Brewer-Allgeier technique, 160
Brightfield microscope, parts of, 13-15
  using, 15-16
Brilliant green, use in detecting coliform bacteria, 480
Bristol's medium, use in culturing algae and cyanobacteria, 205-207
Broths, 108-109
  culturing with, 151-154
  preparing, 108-110
  turbidity during growth in, 167-170
Brownian movement, defined, 38

Carbol fuchsin, in acid-fast staining, 59-62
Carcinogens, detecting using Ames test, 323-329
Casein, 297-455
  hydrolysis of, 248-250
  hydrolysis test summary, 250

Catalase test, 307-308
  summary, 308
Chu's medium, use in culturing cyanobacteria, 205-207
Citrate utilization, by enteric bacteria, 263-265
  test summary, 265
Clostridium, and human pathology, 421-422
Clostridium sporogenes, incubation, 161-162
  oxygen concentration and, 226-228
Clostridium tetani, 513
Coagulase test, 413, 415-417
  summary, 417
Cocci, 27
  See also specific bacteria
Coliform bacteria, 27
  confirmed test, 480-481
  completed test, 480-482
  detecting, standard test for, 479-484
  differential media in culturing, 196
  identifying, 264-265
  IMViC tests, 479
  persumptive test, 479-480
  See also Escherichia coli
Contamination, defined, 3
Crystal violet, using, 30-31
Cutaneous infections (*see* Skin, bacterial infection; Mycoses)
Culture media, BGLB (brilliant green lactose bile), 480
  broths, 108-109
  Chu's, 205-207
  complex, 107, 109
  differential, 196
  egg yolk agar, use in detecting phospholipases, 258, 260
  EMB (eosin methylene blue) agar, 480-482
  enrichment technique, 203-207
  Hay infusion, 40, 43
  Kligler's iron agar, 286-288
  liquid (*see* Broths)
  Mueller-Hinton agar, 406, 408-410

Culture media *(continued)*
  preparation of, 107–114
  Sabouraud's, 388–390
  Sabouraud's modified, 355
  seeded-agar overlay, 316–320
  selective, 195–196
  soft-agar deeps, 82–83
  solid, 110–114
  Simmons' citrate agar, 263, 265
  tributyrate agar, 257
  triple-sugar-iron (TSI) agar, 286–288
  Van Delden's, 207
Culture transfer, 5–7
Cultures, pure, 122–125, 131–134
Culturing microorganisms, 122–125, 143–146
  controls in, 124
  See also specific organisms
Cytochrome c (oxidase) test, 304–306
  test summary, 306

Denitrification, 507–510
Deuteromycetes, 368–370
  beneficial uses, 368
  mycoses caused by, 369
  reproduction, 369–370
  structure, 369
Differential culture media, 196
Differential stains,
  acid-fast stain, 59–62
  defined, 21
  Gram stain, 51–54
Disinfectants, defined, 397
  effectiveness of, 397–399
Dyes (*see* Stains and dyes)

Ehrlich, P., 304
EMB (eosin methylene blue) agar, in detecting coliform bacteria, 480–482
  molecular structure, 481
EMP (Embden-Meyerhoff-Parnas) pathway, in enteric bacteria fermentation, 278–279
  in fermentation, 270
  schematic diagram, 456–457
Endospore, defined, 67
  Gram stain, 69–70
  physical properties, 67–68
  reproduction, 67–68, 89–90
  stain, test summary, 73
  See also Clostridium; Bacillus
Enrichment techniques, for algae and cyanobacteria, 205–207
  for sulfate-reducing bacteria, 204–207
Enteric bacteria, citrate utilization of, 263–265

  characteristics of, 278
  differentiating, 263–265, 277–281, 293–294, 482–484
  hydrogen sulfide production in, 285–288
  indole production in, 291–294
  MR/VS (methyl-red/Voges-Proskauer) test, 277–281
Enteric diseases, 479
  See also specific coliform bacteria
Enterobacter aerogenes,
  differentiating from *E. coli,* 264–465, 279–281, 482–484
  indole production in, 293–294
Enzymes, amylase, 241–243
  catalase, 307–308
  cellulase, 241–242
  coagulase, 414
  cytochrome c, 304–306
  hydrolitic, 235–237
  lipase, 257–258
  oxidase (*see* Cytochrome c)
  peptidase, 247, 297
  proteinase, 247
  tryptophanase, 291
Escherichia coli, (*E. coli*), 27
  antibiotic sensitivity, 408–410
  antiseptic and disinfectant sensitivity, 398–399
  casein hydrolysis by, 249–250
  colonies, 122–125
  culturing, 161–162
  cytochrome c test, 303–308
  detecting, 479–484
  differential and selective media for, 197–198
  differentiating from *Enterobacter aerogenes,* 264–265, 279–281
  in fermentation, 272–274
  gelatin hydrolysis by, 252
  growth of, 143–146, 152–154, 169–170
  hydrogen sulfide production by, 288
  indole production by, 293–294
  and lactic-acid bacteria, 299–300
  morphology, 28
  nutrient and saline concentration and, 221–222
  oxygen concentration and, 226–228
  ruler, 389–390
  serotyping, 448–449
  starch hydrolysis by, 242–244
  temperature effects on growth, 214–215
  ultraviolet radiation and cell viability, 337–339
Escherichia coli phage T-1, in bacterial lysis, 318–320
Escherichia coli strain B, in viral parasitism, 318–320
Environmental microbiology, 477–510
  of soil, 487–510
  of water, 479–484

Facutatively anaerobic bacteria,
  differentiating, 277–281
  defined, 159, 225
  phenol red to identify, 198
  See also Enteric bacteria
Fermentation
  acid and gas production in, 269–274
  by enteric bacteria, 277–283
  features of, 269
  of milk, 455–460
  schematic diagram, 270
  of solid food, 463–466
  test summary, 274
  yeasts in, 380, 385
Filamentous fungi (see Molds)
Flagella staining, 80–84
Food microbiology, 453–471
  milk preservation, 455–460
  pathogens in, 469–471
  solid food preservation, 463–466
  See also Fermentation
Food preservation 455–470
  of milk, 455–460
  of solid food, 463–466
Fungi, 349–390
  beneficial, 359
  characteristics of, 349–350
  habitat, 358–359
  human pathogens (*see* Mycoses)
  laboratory cultivation, 355–356
  morphology, 350–351, 363–370, 379–382
  parasitism, 355
  physiology and growth, 353–355
  plant pathogens, 359
  reproduction, 356–358, 363–370
  taxonomy, 351–354, 384–386
Fungi imperfecti (see *Deuteromycetes*)

Gelatin, hydrolysis of, 250–252
  hydrolysis test summary, 252
Gluconobacter oxydans, 313
  lack of capsules in, 91–94
Gram, Christian, 49
Gram stain, 51–54
  and colony purity, 131–134
  for endospores, 69–70
  test summary, 54
  timing of, 53
Growth of bacteria, 143–228
  characteristics of, 154
  exponential phase, 213
  free oxygen, effects on, 225–228
  fungi, 353–355
  growth curves, 213–214
  nutrients and saline concentration effects on, 219–222
  optimum temperature, 213
  osmotic pressure and, 219–222
  temperature effects on, 213–215
  turbidity changes during, 167–170

Haeckel, Ernst, 352
Hartree, E. F., 304
Hemolysis, 421–425
 bacteria causing, 422–423
Hydrogen sulfide test summary, 288
Hydrolysis by microorganisms, 235–260
 of casein, 248–250
 of gelatin, 250–252
 of lipids, 235, 257–260
 of proteins, 235, 247–252
 starch, test summary, 244
 of starches, 235, 241–244

Immunology, 395–450
 of skin, 431–434
 of throat, 441–444
IMViC tests, 291
 alternatives to, 479–484
 for detecting coliform bacteria, 479
India ink, in structural staining, 89
Indole production, to differentiate enteric bacteria, 291–294
 Kovac's reagent in detecting, 292–294
 schematic diagram, 292
 test summary, 294
Ionizing radiation, 104
 as a method of sterilization, 104

Keilin, D., 304
Kirby-Bauer method, 405–410
Kligler's iron agar, for detecting hydrogen sulfide, 286–288
Koch, Robert, 117, 119, 250
Koser, S. A., 263
Kovac's method, in cytochrome c test, 306
Kovac's reagent, 292–294

Lactic-acid bacteria, in fermentation, 269
 in food preservation, 455–460, 463–466
 litmus-milk test, use in differentiating, 297
Lactobacillus plantarum, 69, 453
 structure of, 71–72
Leifson, Einar, 79
Leifson's stain, 81
Lipids, hydrolysis of, 235, 257–260
 hydrolysis test summary, 259
 phospholipid hydrolysis test summary, 260
 structure of, 258
Litmus, as an oxidation/reduction (O/R) indicator, 298–300
 as a pH indicator, 298
Litmus-milk test, 297–300
 in differentiating lactic-acid bacteria, 297–300

 summary, 300
Loeffler's methylene blue, 22
Lysis, bacterial, 315–320
 hemolysis, 421–425
 schematic diagram, 316
 viral agents in, 315–320

Malachite green, as a structural stain, 70–71
Media (*see* Culture media)
Medical microbiology, 395–444
 antibiotic effectiveness, 405–407
 antiseptic and disinfectant effectiveness, 397–398
 coagulase test, 413–417
 hemolysis, 421–424
 serotyping, 447–448
Methyl red, pH indicator in detecting fermentation, 277, 279–281
 test summary, 280
Methyl red/Voges-Proskauer (MR/VP) test, 279–281
Methylene blue, 21
 in acid-fast tests, 59–62
 in detecting coliform bacteria, 480–481
 oxidation of, 162
 timing stain, 31
Micrococcus cryophilus, temperature effects on growth, 214–215
Micrococcus luteus, anaerobic and aerobic incubation, 161–162
 oxygen concentration and growth of, 226–228
 ultraviolet radiation and cell viability, 337–339
Microorganisms, beneficial, 68
 identifying, 515–518
 physiology, 235–308
Microscope, brightfield, parts of, 13–15
 care of, 13–16
 phase-contrast, 35–42
 using, 13–16
Milk protein (*see* Casein)
Mims, C. W., 352
Molds, 27, 350–370
 asexual reproduction, 356
 defined, 350
 morphology, 363–370
 sexual reproduction, 363–370
 structure, 351
 See also specific molds
Morphology, of bacteria, 51–91
 defined, 27
 eucaryotic, 40
 of fungi, 350–351
 of molds, 363–370
 procaryotic, 40
 of yeasts, 379–382
Motility, 38, 79–84
Mueller-Hinton agar, use in Kirby-Bauer method, 406, 408–410

Mutagens, 325
 relation to carcinogens (*see* Ames test)
Mutation, 323–341
 and carcinogens, 323–329
 ultraviolet radiation, effect on, 333–337, 339–341
Mycobacterium smegmatis, acid-fast bacterium, 60
Mycoses, 358–359

Negative staining, 89
Nessler's reagent, in testing for ammonification, 503–504
Nigrosin, use in structural staining, 90
Nitrogen cycle, 487–489
Nitrogen fixation, 205, 493–496
 acetylene-reduction assay in detecting, 493
 and symbiosis, 493–495
 See also Ammonification

Osmotic pressure, and growth, 219–222
 solute concentration and, 220
Oxidase test (*see* Cytochrome c test)
Oxidation/reduction (O/R) indicator, 298–300
 methylene blue, 162
Oxygen, growth and, 225–228

Parasitism, bacteriophages, 315–320
 fungal, 355
 symbiotic, 355
Pathogenic bacteria, 421–425
 food and, 469–471
 hemolysis caused by, 421–425
 opportunistic, 59, 413
 virulence factors, 415
 See also specific bacterium
Penicillium notatum, differentiating from *Zygomycetes,* 371–74
 growth, 152–154
 nutrient and saline concentration and, 221–222
Petri, R. J., 119
pH indicators, bromothymol blue, 263, 265
 common, 271
 in detecting fermentation, 271, 277, 279–281
 litmus, 298–300
 methyl red, 277, 279–281
 phenol red, 198, 274
 spirit blue, 257
Phase-contrast microscope, 35–42
Phenol red as a pH indicator, 198
 in fermentation, 274
Physiology of microorganisms, 235–308
 acid production, 269–281

Physiology of microorganisms *(continued)*
 citrate utilization by enteric bacteria, 263–265
 enzymes, role of in, 235, 241–243, 247–249, 257–258, 286, 292, 297, 303, 307
 during fermentation, 269–281
 fungi, 353–355
 hydrogen sulfide production, 285–288
 hydrolysis of lipids, 235, 247–252
 hydrolysis of proteins, 247–252
 hydrolysis of starches, 241–244
 indole production, 291–294
 metabolic oxidation, 287
 respiratory catabolism, 303–308
 of yeasts, 385–387

Plaque assay, 315–322
Plating methods, pour plates, 179–180, 182–185
 spread plates, 179–182, 185
Polysaccharides *(see* Starches)
PPK (pentose phosphoketolase) pathway, 464
 schematic, 465
Proskauer, B., 277
Proteins, casein, hydrolysis of, 248–250
 gelatin, hydrolysis of, 250–252
 hydrolysis of, 235, 247–252
 schematic diagram of hydrolysis of, 249
 translocation, 236–237
Proteus mirabilis
 swarming of, 141
 in fermentation, 272–274
 hydrogen sulfide production in, 288–290
 and lactic-acid bacteria, 299–300
 lipid hydrolysis by, 259–260
 motility of, 80–83
Proteus vulgaris, gelatin hydrolysis by, 252
Pseudomonas species, antibiotic sensitivity, 408–410
 motility, 40, 43
Pseudomonas aeruginosa, 211
 in cytochrome *c* test, 303–308
 motility of, 80–82
 in selective and differential media, 197–198
Pseudomonas fluorescens, and ammonification, 503–504
 Gram staining, 52–54
Pseudomonas stutzeri, denitrification by, 509–510
Pure cultures, determining, 131–134
 producing, 122–125

Radiation, infrared, 333
 ionizing, 104, 333–334
 ultraviolet, 333–341

Reproduction, bacterial, 67–68, 89–90
 fungal, 356–358
 of molds, 364–371
 of yeasts, 379–384
Respiratory catabolism, 303–308
 schematic overview, 304
 See also EMP pathway
Rhizobium japonicum, nitrogen fixation and, 495–496
Rhizopus nigricans, differentiating from *Deuteromycetes,* 371–374
Rhodospirillium rubrum, motility of, 82
Rohrmann, F., 304

Sabouraud's medium, 388–390
 modified, 355
Saccharomyces species, motility of, 40, 43
Saccharomyces cerevisiae, growth of, 143–146, 152–154
 morphology of, 28
 morphology and reproduction of yeasts, 389–390
 nutrient and saline concentration effects, 221–222
Salmonella enteritidis, serotyping, 448
Salmonella typhimurium, and ultraviolet radiation, 339–341
Salmonella typhimurium, Ames strain TA98, in detecting carcinogens, 327–329
Schaeffer-Fulton procedure, 71
Schizosaccharomyces octosporus, morphology of yeasts and, 389–390
Selective culture media, 196
 See also specific media
Seliberia species, 477
Serial dilution, 175–185
 for phage enumeration, 318
 schematic diagram, 176
Serotyping, 447–448
Simmons, J. S., 263
Simmons' citrate agar, use in identifying *E. coli,* 263, 265
Simple stains, 21–23, 27–31
 acidic dye, 22–23
 basic dye, 22, 27–31
 defined, 21
 See also specific stains
Skin, anatomy of, 431–433
 bacterial infection of, 431–434. *See also* Mycoses
 normal microflora of, 431, 433
Soil microorganisms, ammonification by, 499–504
 nitrogen fixation by, 493–496
Spectrophotometer, 167
 measuring microbial growth with, 169–170
 schematic diagram, 168
Spirillium volutans, motility of, 82

Spirit blue agar, as pH indicator in lipid degradation, 257, 260
Spitzer, W., 304
Sporangium, defined, 67
Spores, fungal
 resistance of, 357–358
Stainer, Roger, 352
Staining, 21–31
Stains and dyes, acid-fast, 59–62
 acidic, 22–23
 basic, 22, 27–31
 basic fuchsin, 51
 brilliant green, 480
 bromothymol blue, 263–265
 capsule, 91–93
 carbol fuchsin, 31, 59–62
 crystal violet, 30–31
 differential, 51–54, 59–62
 endospore, 69–70
 flagella, 80–84
 Gram, 51–54, 69–70, 131–134
 India ink, 89
 Leifson's, 81
 Loeffler's methylene blue, 22
 malachite green, 70–71
 methyl red, 277–281
 methylene blue, 21, 31, 59–62, 162, 480–481
 nigrosin, 90
 phenol red, 198, 274
 simple, 21–23, 27–31
 spirit blue, 257, 260
 structural, 67–73, 80–84, 91–93
 Wuster's blue, 305–306
Staphylococci, 413–417, 421–422
 coagulase production by, 413–417
 and human pathology, 421–422
Staphylococcus aureus, 1, 17
 antibiotic sensitivity, 408–410
 antiseptic and disinfectant sensitivity, 598–599
 denitrification by, 509–510
 identifying pathogenic species, 415–417
 nutrient and saline concentration and growth of, 221–222
 in selective and differential media, 197–198
Staphylococcus epidermis, in catalase test, 303–308
 colonies, 122–125
 identifying pathogenic species, 415–417
 lipid hydrolysis by, 259–260
 morphology, 28
 motility of, 83
 in selective and differential media, 197–198
Starches, hydrolysis of, 235
 nutrient, 241
 structural, 241
Sterilization methods, 3, 100–104
 autoclaving, 101
 chemical, 103–104

filtration, 102–103
hot-air oven, 100–101
incineration, 1–7, 100
ionizing radiation, 104
See also Aseptic technique
Streak plates, use in culturing, 119–125
Streptococci, and human pathology, 421–425
Streptococcus faecalis, in fermentation, 272–274
Streptococcus lactis, in catalase test, 303–308
and lactic-acid bacteria, 299–300
Streptococcus mutans, 91
Streptococcus pneumoniae, 91
Streptococcus pyogenes, and hemolysis, 424–425
Streptococcus salivarius, 242
antibiotic sensitivity, 408–410
antiseptic and disinfectant sensitivity, 398–399
and hemolysis, 424–425
casein hydrolysis by, 249–250
starch hydrolysis by, 242–244
Streptococcus sanguis, and hemolysis, 424–425
Structural stains, capsule, 91–93
defined, 21
endospore, 67–73
flagella, 80–84
Sulfate-reducing bacteria, 285
Symbiosis, nitrogen fixation and, 493–495

Taxonomy, of fungi, 351–354
of yeasts, 384–386

Test summaries, acid-fast, 62
casein hydrolysis, 250
catalase, 308
citrate utilization, 205
coagulase, 417
cytochrome c, 306
endospore stain, 73
fermentation, 274
gelatin hydrolysis, 252
Gram stain, 54
hydrogen sulfide, 288
indole, 294
lipid hydrolysis, 259
litmus milk, 300
methyl red, 280
phospholipid hydrolysis, 260
starch hydrolysis, 244
Voges-Proskauer, 281
Thioglycolate, use in culturing anaerobes, 159–162
Throat, cultures, 443–444
microbiology of, 441–444
Transduction, 317
Triglyceride (*see* Lipid)
Triple-sugar-iron (TSI) agar, for detecting hydrogen sulfide, 286–288
Turbidity, defined, 167

Ultraviolet radiation, 333–341
and cell viability, 337–339
and DNA, 334–337
mutation caused by, 336–341

Van Delden's medium, use in culturing sulfate-reducing bacteria, 207

Virulence, 414–415
factors, 415
See also specific microorganisms
Viruses, bacteriophages, 315–320
plaque assay in quantifying, 315–320
Voges, O., 277
Voges-Proskauer test, 277, 280–281
See also Methyl red test
test summary, 281

Warburg, Otto, 304
Wet-mount method, 35–43
Woese, Carl, 352
Wuster's blue, in detecting cytochrome c, 305–306

Yeasts, 27, 350–351, 379–390
defined, 350
in fermentation, 380, 385
identification of, 382–383
laboratory cultivation of, 387–390
morphology, 379–382
physiology, 385–387
reproduction, 379–384
structure, 351

Zernicke, Frederick, 35
Zoogloea ramigera, 90
Zygomycetes, 363–366
reproduction, 364–366
structure and growth, 363–364
Zymomonas mobilis, 233